COHOMOLOGY OF VECTOR BUNDLES AND SYZYGIES

The central theme of this book is an exposition of the geometric technique of calculating syzygies. It is written from the point of view of commutative algebra; without assuming any knowledge of representation theory, the calculation of syzygies of determinantal varieties is explained. The starting point is a definition of Schur functors, and these are discussed from both an algebraic and a geometric point of view. Then a chapter on various versions of Bott's theorem leads to a careful explanation of the technique itself, based on a description of the direct image of a Koszul complex. Applications to determinantal varieties follow. There are also chapters on applications of the technique to rank varieties for symmetric and skew symmetric tensors of arbitrary degree, closures of conjugacy classes of nilpotent matrices, discriminants, and resultants. Numerous exercises are included to give the reader insight into how to apply this important method.

CAMBRIDGE TRACTS IN MATHEMATICS

General Editors

B. BOLLOBAS, W. FULTON, A. KATOK, F. KIRWAN,
P. SARNAK

149 Cohomology of Vector Bundles and Syzygies

Jerzy Weyman

Northeastern University

Cohomology of Vector Bundles and Syzygies

CAMBRIDGE
UNIVERSITY PRESS

PUBLISHED BY THE PRESS SYNDICATE OF THE UNIVERSITY OF CAMBRIDGE
The Pitt Building, Trumpington Street, Cambridge, United Kingdom

CAMBRIDGE UNIVERSITY PRESS
The Edinburgh Building, Cambridge CB2 2RU, UK
40 West 20th Street, New York, NY 10011-4211, USA
477 Williamstown Road, Port Melbourne, VIC 3207, Australia
Ruiz de Alarcón 13, 28014 Madrid, Spain
Dock House, The Waterfront, Cape Town 8001, South Africa

http://www.cambridge.org

First published 2003

Printed in the United States of America

Typeface Times 10/13 pt.   *System* LaTeX 2_ε [TB]

A catalog record for this book is available from the British Library.

Library of Congress Cataloging in Publication Data
Weyman, Jerzy, 1955–
Cohomology of vector bundles and syzgies / Jerzy Weyman.
    p. cm. – (Cambridge tracts in mathematics; 149)
Includes bibliographical references and index.
ISBN 0-521-62197-6
1. Syzygies (Mathematics)   2. Vector bundles.   3. Homology theory.   I. Title.   II. Series.

QA247 .W49   2003
512′.5 – dc21   2002074071

ISBN 0 521 62197 6 hardback

To Katarzyna

Contents

Preface

This book is devoted to the geometric technique of calculating syzygies. This technique originated with George Kempf and was first used successfully by Alain Lascoux for calculating syzygies of determinantal varieties. Since then it has been applied in studying the defining ideals of varieties with symmetries that play a central role in geometry: determinantal varieties, closures of nilpotent orbits, and discriminant and resultant varieties.

The character of the method makes it comparable to the symbolic method in classical invariant theory. It works in only a limited number of cases, but when it does, it gives a complete answer to the problem of calculating syzygies, and this answer is hard to get by other means.

Even though the basic idea is more than 20 years old, this is the first book treating the geometric technique in detail. This happens because authors using the geometric method have usually been interested more in the special cases they studied than in the method itself, and they have used only the aspects of the method they needed. Therefore the basic theorems from chapter 5 stem from the efforts of several mathematicians.

The possibilities offered by the geometric method are not exhausted by the examples treated in the book. The method can be fruitfully applied to any representation of a linearly reductive group with finitely many orbits and with actions such that the orbits can be described explicitly. The varieties treated in chapters 7 and 9 show that the scope of the method is not limited to such actions.

The book is written from the point of view of a commutative algebraist. We develop the rudiments of representation theory of general linear group in some detail in chapters 2–4. At the same time we assume some knowledge of commutative algebra and algebraic geometry, including sheaves and their cohomology—for example, the notions covered in chapters II and III of the book [H1] of Hartshorne. The notions of Cohen–Macaulay and Gorenstein rings and rational singularities are also used, and their definitions are briefly recalled in chapter 1.

Some parts of the book demand more advanced knowledge. One statement in chapter 5 is an application of Grothendieck duality, whose statement is briefly recalled in chapter 1. We advise less experienced readers to just accept Theorem (5.1.4) and look first at its applications.

In some sections of chapters 5 and 8 we assume familiarity with highest weight theory and some facts on linear algebraic groups. At the same time, in exercises we use the geometric technique to develop some of that theory for the classical groups. Still, the reader not familiar with these notions should be able to understand all the remaining chapters.

Let us describe briefly the contents of the book. The first chapter discusses preliminaries. We recall elementary notions from multilinear algebra and combinatorics. The section on commutative and homological algebra covers briefly the definitions of depth, Koszul complexes, the Auslander–Buchsbaum–Serre theorem, and Cohen–Macaulay and Gorenstein rings. There is also a section on de Jong's algorithm for explicit calculation of normalization, on the exactness criterion of Buchsbaum and Eisenbud, and on Grothendieck duality. The final section is devoted to a brief review of the notion of the determinant of a complex.

Chapters 2, 3, and 4 are devoted to developing the representation theory of general linear groups and to the proof of Bott's theorem on the cohomology of line bundles on homogeneous spaces. This provides the basic tools for the calculations to be performed in later chapters.

Our approach is based on Schur and Weyl functors, introduced in the first section of chapter 2. They are defined by generators and relations. The relation to highest weight theory and Schur–Young theory is discussed. This is followed by a discussion of Cauchy formulas, the Littlewood–Richardson rule, and plethysm. The final section of chapter 2 discusses Schur complexes.

In chapter 3 we relate Schur functors to geometry by realizing them as multihomogeneous components of homogeneous coordinate rings of flag varieties. This is followed by a proof of the Cauchy formula based on restriction of the straightening from the Grassmannian to its affine open subset, and by a section on tangent bundles of Grassmannians and flag varieties.

In chapter 4 we prove Bott's theorem on cohomology of line bundles on flag varieties. We follow the approach of Demazure. In the last section we formulate the theorem for arbitrary reductive groups and give explicit interpretations for classical groups.

Chapter 5 is devoted to the description of terms and properties of the direct images of Koszul complexes. We study the basic setup, i.e., the diagram

$$
\begin{array}{ccc}
Z & \subset & X \times V \\
\downarrow q' & & \downarrow q \\
Y & \subset & X
\end{array}
$$

where X is an affine space, V is a nonsingular projective variety, Z is a total space of a vector subbundle S of the trivial bundle $X \times V$, and $Y = q(Z)$. The variety Z can be described as the vanishing set of a cosection $p^*(\xi) \to \mathcal{O}_{X \times V}$, and it is a locally complete intersection. Here ξ is the dual of the factor $(X \times V)/S$. The original idea of Kempf was that the study of a direct image of the resulting Koszul complex can be used to prove results about the defining equations and syzygies of the subvariety Y. We give the basic properties of these direct images in the general case and in important special cases, for example when Z is a desingularization of Y. We also treat the more general case of twisted Koszul complexes.

The remaining chapters are devoted to examples and applications. In each of these chapters a different aspect of the method is illustrated.

In chapter 6 we study the case of determinantal varieties. Here we show how to handle basic calculations in simple cases when the bundle ξ is a tensor product of tautological bundles. Apart from the proof of Lascoux's theorem and the calculation of syzygies of determinantal ideals for symmetric and skew symmetric matrices, we also give results on the equivariant modules supported in determinantal varieties.

Chapter 7 is devoted to the rank varieties for tensors of degree higher than two. This illustrates that the method can be applied in cases when the variety X or even Y does not have finitely many orbits with respect to some action of the reductive group.

In chapter 8 the study of nilpotent orbit closures allows us to understand how to handle the situation when the cohomology groups needed to get the syzygies cannot be calculated directly, but still partial results can be recovered by estimating the terms in a spectral sequence associated to filtrations on a basic bundle ξ. We also prove the Hinich–Panyushev theorem on rational singularities of normalizations of nilpotent orbits for a general simple group.

Chapter 9 illustrates the use of twisted modules supported in resultant and discriminant varieties, which allow one to get natural determinantal expressions for the defining equation.

Each chapter is followed by exercises. They should allow readers to learn how to apply the geometric method on their own. At the same time they illustrate further applications of the method. In particular the exercises to chapter 6 deal with the analogues of the determinantal varieties for the symplectic and orthogonal groups. In exercises to chapter 7 we give some calculations of minimal resolutions of Plücker ideals.

The book can be read on several levels. For the reader who is not familiar with representation theory and/or derived categories and Grothendieck duality, we suggest first reading chapters 2 through 4. Then one can proceed with the proof of the statement of Theorem 5.1.2 given in section 5.2. At that point most of the following chapters (with the exception of section 8.3) can be understood using only that statement.

In such a way the book could be used as the basis of an advanced course in commutative algebra or algebraic geometry. It could also serve as the basis of a seminar. A lot of general notions and theories can be nicely illustrated in the special cases treated in later chapters by the methods given in the book.

A reader familiar with representation theory can just skim through chapters 2 through 4 to get familiar with the notation, and then proceed straight to chapter 5 and study the applications.

I am indebted to many people who introduced me to the subject, especially to David Buchsbaum, Corrado De Concini, Jack Eagon, David Eisenbud, Tadeusz Józefiak, Piotr Pragacz, and Joel Roberts. I also benefited from conversations on some aspects of the material with Kaan Akin, Giandomenico Boffi, Michel Brion, Bram Broer, Andrzej Daszkiewicz, Steve Donkin, Toshizumi Fukui, Laura Galindo, Wilberd van der Kallen, Jacek Klimek, Hanspeter Kraft, Witold Kraśkiewicz, Alain Lascoux, Steve Lovett, Olga Porras, Claudio Procesi, Rafael Sanchez, Mark Shimozono, Alex Tchernev, and Andrei Zelevinsky. Throughout my work on the book I was partially supported by grants from the National Science Foundation.

1

Introductory Material

1.1. Multilinear Algebra and Combinatorics

1.1.1. Exterior, Divided, and Symmetric Powers; Multiplication and Diagonal Maps

Let $\mathbf{K}$ be a commutative ring, and let E be a free $\mathbf{K}$-module with a basis $\{e_1, \ldots, e_n\}$.

We define *the r-th exterior power $\bigwedge^r E$ of E* to be the r-th tensor power $E^{\otimes r}$ of E divided by the submodule generated by the elements:

$$u_1 \otimes \ldots \otimes u_r - (-1)^{\operatorname{sgn} \sigma} u_{\sigma(1)} \otimes \ldots \otimes u_{\sigma(r)}$$

for all $\sigma \in \Sigma_r$, $u_1, \ldots, u_r \in E$. We denote the coset of $u_1 \otimes \ldots \otimes u_r$ by $u_1 \wedge \ldots \wedge u_r$.

The following basic properties of exterior powers are proved in [L, chapter XIX, section 1].

(1.1.1) Proposition.

(a) *Let $\{e_1, \ldots, e_n\}$ be an ordered basis of E. Then the elements $e_{i_1} \wedge \ldots \wedge e_{i_r}$ for $1 \leq i_1 < \ldots < i_r \leq n$ form a basis of $\bigwedge^r E$. In particular, $\bigwedge^r E$ is a free $\mathbf{K}$-module of dimension $\binom{n}{r}$.*

(b) *(Universality property of exterior powers) We have a functorial isomorphism*

$$\theta_M : \operatorname{Alt}^r(E^r, M) \to \operatorname{Hom}_{\mathbf{K}}\left(\bigwedge^r E, M\right)$$

where $\operatorname{Alt}^r(E^r, M)$ denotes the set of multilinear alternating maps from $E^{\times r}$ to M, given by the formula $\theta_M^r(f)(u_1 \wedge \ldots \wedge u_r) = f(u_1, \ldots, u_r)$.

(c) We have natural isomorphisms

$$\alpha^r : \overset{r}{\bigwedge}(E^*) \to \left(\overset{r}{\bigwedge} E\right)^*$$

sending the exterior product $l_1 \wedge \ldots \wedge l_r$ to the linear function l on $\bigwedge^e E$ defined by the formula

$$l(u_1 \wedge \ldots \wedge u_r) = \sum_{\sigma \in \Sigma^r}(-1)^{\mathrm{sgn}\,\sigma} l_{\sigma(1)}(u_1)\ldots l_{\sigma(r)}(u_r).$$

The r-th exterior power is an endofunctor on the category of free **K**-modules and linear maps. More precisely, for two free **K**-modules E, F and a linear map $\phi : E \to F$ we have a well-defined linear map

$$\overset{r}{\bigwedge}\phi : \overset{r}{\bigwedge} E \to \overset{r}{\bigwedge} F$$

defined by the formula $\bigwedge^r \phi(u_1 \wedge \ldots \wedge u_r) = \phi(u_1) \wedge \ldots \wedge \phi(u_r)$. Let us denote $m = \dim F$. Let $\{e_1, \ldots, e_n\}$ be a basis of E and let $\{f_1, \ldots, f_m\}$ be a basis of F. In these bases ϕ correspond to the $m \times n$ matrix $(\phi_{j,i})$ where

$$\phi(e_i) = \sum_{j=1}^{m}\phi_{j,i} f_j.$$

The map $\bigwedge^r \phi$ can be written in the corresponding bases of $\bigwedge^r E$, $\bigwedge^r F$ as follows:

$$\overset{r}{\bigwedge}\phi(e_{i_1} \wedge \ldots \wedge e_{i_r})$$
$$= \sum_{1 \le j_1 < \ldots < j_r \le m} M(j_1, \ldots, j_r | i_1, \ldots, i_r; \phi) f_{j_1} \wedge \ldots \wedge f_{j_r},$$

where $M(j_1, \ldots, j_r | i_1, \ldots, i_r; \phi)$ denotes the $r \times r$ minor of the matrix $(\phi_{j,i})$ corresponding to the rows $j_1, \ldots, j_r$ and columns $i_1, \ldots, i_r$.

The vector space

$$\overset{\bullet}{\bigwedge}(E) := \bigoplus_{r \ge 0} \overset{r}{\bigwedge} E$$

has a natural multiplication

$$m : \overset{\bullet}{\bigwedge}(E) \otimes \overset{\bullet}{\bigwedge}(E) \to \overset{\bullet}{\bigwedge}(E)$$

given by the formula

$$m(u_1 \wedge \ldots u_r \otimes v_1 \wedge \ldots \wedge v_s) = u_1 \wedge \ldots \wedge u_r \wedge v_1 \wedge \ldots \wedge v_s.$$

This gives $\bigwedge^\bullet(E)$ the structure of associative, graded commutative algebra (meaning that the commutative law reads $fg = (-1)^{\deg(f)\deg(g)}gf$). We call this algebra *the exterior algebra on* E. The algebra $\bigwedge^\bullet(E)$ has a unit $\eta : \mathbf{K} \to \bigwedge^\bullet(E)$.

The components of the multiplication map will be denoted by $m : \bigwedge^r E \otimes \bigwedge^s E \to \bigwedge^{r+s} E$.

The diagonal map $\Delta : E \to E \oplus E$ induces an algebra map

$$\Delta : \bigwedge^\bullet(E) \to \bigwedge^\bullet(E \oplus E) \cong \bigwedge^\bullet(E) \otimes \bigwedge^\bullet(E)$$

which we will call *the diagonal* (or *comultiplication*) *map*.

The components of Δ will be denoted by $\Delta : \bigwedge^{r+s} E \to \bigwedge^r E \otimes \bigwedge^s E$. In terms of elements we have

$$\Delta (u_1 \wedge \ldots \wedge u_{r+s})$$
$$= \sum_{\sigma \in \Sigma_{r+s}^{r,s}} (-1)^{\text{sgn } \sigma} u_{\sigma(1)} \wedge \ldots \wedge u_{\sigma(r)} \otimes u_{\sigma(r+1)} \wedge \ldots \wedge u_{\sigma(r+s)}$$

where $\Sigma_{r+s}^{r,s} = \{\sigma \in \Sigma_{r+s} \mid \sigma(1) < \ldots < \sigma(r); \; \sigma(r+1) < \ldots < \sigma(r+s)\}$.

Finally we have the counit map

$$\epsilon : \bigwedge^\bullet(E) \to \mathbf{K},$$

defined to be zero on all spaces $\bigwedge^r E$ for $r > 0$, and satisfying $\epsilon\eta(1) = 1$.

The following proposition is an elementary calculation.

(1.1.2) Proposition.
   (a) The maps $m, \Delta, \epsilon, \eta$ define on $\bigwedge^\bullet(E)$ the structure of commutative, cocommutative bialgebra.

   (b) The map $\alpha : \bigwedge^\bullet(E^*) \to (\bigwedge^\bullet E)^*$ defined in (1.1.1) (c) is an isomorphism of bialgebras.

Part (b) of the proposition means that the dual map to the multiplication map m on $\bigwedge^\bullet(E)$ is the diagonal map Δ on $\bigwedge^\bullet(E^*)$ and vice versa.

We define *the r-th symmetric power $S_r E$ of E* to be the r-th tensor power $E^{\otimes r}$ of E divided by the submodule generated by the elements

$$u_1 \otimes \ldots \otimes u_r - u_{\sigma(1)} \otimes \ldots \otimes u_{\sigma(r)}$$

for all $\sigma \in \Sigma_r$, $u_1, \ldots, u_r \in E$. We denote the coset of $u_1 \otimes \ldots \otimes u_r$ by $u_1 \ldots u_r$.

The following basic properties of symmetric powers are proved in [L, chapter XVI, section 8].

(1.1.3) Proposition.

 (a) *Let* $\{e_1, \ldots, e_n\}$ *be an ordered basis of E. Then the elements* $e_1^{i_1} \ldots e_n^{i_n}$
  for $i_1 + \ldots + i_n = r$ *form a basis of* $S_r E$. *In particular* $S_r E$ *is a free*
  K-*module of dimension* $\binom{n+r-1}{r}$.

 (b) *(Universality property of symmetric powers) We have a functorial*
  isomorphism

$$\theta_M : \mathrm{Sym}^r(E^r, M) \to \mathrm{Hom}_{\mathbf{K}}(S_r E, M)$$

 where $\mathrm{Sym}^r(E^r, M)$ *denotes the set of multilinear symmetric maps*
 from $E^{\times r}$ *to M, given by the formula* $\theta_M^r(f)(u_1 \ldots u_r) = f(u_1, \ldots, u_r)$.

The r-th symmetric power is an endofunctor on the category of free **K**-modules and linear maps. More precisely, for two free **K**-modules E, F and a linear map $\phi : E \to F$ we have a well-defined linear map

$$S_r \phi : S_r E \to S_r F$$

defined by the formula $S_r \phi(u_1 \ldots u_r) = \phi(u_1) \ldots \phi(u_r)$. Let us denote $m = \dim F$. Let $\{e_1, \ldots, e_n\}$ be a basis of E, and let $\{f_1, \ldots, f_m\}$ be a basis of F. In these bases ϕ correspond to the $m \times n$ matrix $(\phi_{j,i})$ where

$$\phi(e_i) = \sum_{j=1}^{m} \phi_{j,i} f_j.$$

The map $S_r \phi$ can be written in the corresponding bases of $S_r E$, $S_r F$ as follows:

$$S_r \phi(e_{i_1} \ldots e_{i_r}) = \sum_{1 \le j_1 < \ldots < j_r \le m} P(j_1, \ldots, j_r \mid i_1, \ldots, i_r; \phi) f_{j_1} \ldots f_{j_r},$$

where $P(j_1, \ldots, j_r \mid i_1, \ldots, i_r; \phi)$ denotes the permanent of the $r \times r$ submatrix of the matrix $(\phi_{j,i})$ corresponding to the (possibly repeated) rows $j_1, \ldots, j_r$ and (possibly repeated) columns $i_1, \ldots, i_r$. More precisely, if the columns $(i_1, \ldots, i_r)$ with repetitions are written as $i_1^{b_1}, \ldots, i_s^{b_s}$ with $b_1 + \ldots + b_s = r$, we have

$$P(j_1, \ldots, j_r \mid i_1^{b_1}, \ldots, i_s^{b_s}) = \sum_{\sigma \in \Sigma_r / (\Sigma_{b_1} \times \Sigma_{b_s})} \phi(j_1, i_{\sigma(1)}) \ldots \phi(j_r, i_{\sigma(r)}).$$

where $\Sigma_{b_1} \times \ldots \times \Sigma_{b_s}$ is the subgroup of permutations from Σ_r preserving the groups of repeating symbols among $j_1, \ldots, j_r$.

 The vector space

$$\mathrm{Sym}(E) := \bigoplus_{r \ge 0} S_r E$$

has a natural multiplication

$$m : \mathrm{Sym}(E) \otimes \mathrm{Sym}(E) \to \mathrm{Sym}(E)$$

given by the formula

$$m(u_1 \ldots u_r \otimes v_1 \ldots v_s) = u_1 \ldots u_r v_1 \ldots v_s.$$

This gives $\mathrm{Sym}(E)$ the structure of associative, commutative algebra. We call this algebra *the symmetric algebra on* E. It can be identified with the polynomial ring over $\mathbf{K}$ in n variables $e_1, \ldots, e_n$. In order to keep the notion of commutativity the same as for the exterior algebras, we assume that $\mathrm{Sym}(E)$ is generated by elements of degree 2.

The components of the multiplication map will be denoted by $m : S_r E \otimes S_s E \to S_{r+s} E$.

We also have an obvious unit map $\eta : \mathbf{K} \to \mathrm{Sym}(E)$ sending $\mathbf{K}$ to the degree zero component of $\mathrm{Sym}(E)$.

The diagonal map $\Delta : E \to E \oplus E$ induces an algebra map

$$\Delta : \mathrm{Sym}(E) \to \mathrm{Sym}(E \oplus E) \cong \mathrm{Sym}(E) \otimes \mathrm{Sym}(E),$$

which we will call *the diagonal (or comultiplication) map*.

The components of Δ will be denoted by $\Delta : S_{r+s} E \to S_r E \otimes S_s E$. In terms of elements we have

$$\Delta(u_1 \ldots u_{r+s}) = \sum_{\sigma \in \Sigma_{r+s}^{r,s}} u_{\sigma(1)} \ldots u_{\sigma(r)} \otimes u_{\sigma(r+1)} \ldots u_{\sigma(r+s)}$$

where $\Sigma_{r+s}^{r,s} = \{\sigma \in \Sigma_{r+s} \,|\, \sigma(1) < \ldots < \sigma(r); \sigma(r+1) < \ldots < \sigma(r+s)\}$.

Finally we have the counit map

$$\epsilon : \mathrm{Sym}(E) \to \mathbf{K}$$

defined to be zero on all spaces $S_r E$ for $r > 0$, and satisfying $\epsilon \eta(1) = 1$.

We have the following analogue of (1.1.2) (a).

(1.1.4) Proposition. *The maps* m, Δ, ϵ, η *define on* $\mathrm{Sym}(E)$ *the structure of a commutative, cocommutative bialgebra.*

Let us investigate the duality. The algebra $\mathrm{Sym}(E) = \bigoplus_{r \geq 0} S_r E$ is not finite dimensional, so instead of the dual we have to work with the graded dual

$$\mathrm{Sym}(E)_{gr}^* := \bigoplus_{r \geq 0} (S_r E)^*.$$

The module map

$$E^* = (S_1 E)^* \rightarrow \mathrm{Sym}(E)^*_{gr}$$

induces by universality an algebra map

$$\beta : \mathrm{Sym}(E^*) \rightarrow \mathrm{Sym}(E)^*_{gr}.$$

This map β is an isomorphism only when $\mathbf{K}$ contains a field of rational numbers. In fact it is given by the formula

$$\beta(l_1 \ldots l_r)(u_1 \ldots u_r) = \sum_{\sigma \in \Sigma_r} l_{\sigma(1)}(u_1) \ldots l_{\sigma(r)}(u_r).$$

In particular, when $l_1 = \ldots = l_r$, $u_1 = \ldots = u_r$ we see that $\beta(l_1^r) = r!(u_1^r)^*$.

In order to describe the graded dual of the symmetric algebra we introduce the divided powers.

We define *the r-th divided power* $D_r(E)$ as the dual of the symmetric power.

$$D_r(E) := (S_r(E^*))^*.$$

Its basis is the dual basis to the natural basis of the symmetric power. If $\{e_1, \ldots, e_n\}$ is a basis of E, we define $e_1^{(i_1)} \ldots e_n^{(i_n)}$ to be the element of the dual basis to the basis $\{(e_1^*)^{j_1} \ldots (e_n^*)^{j_n}\}$, dual to $(e_1^*)^{i_1} \ldots (e_n^*)^{i_n}$.

For every $u \in E$ we can define its *r-th divided power* $u^{(r)} \in D_r E$. It is given by the formula

$$\left(\sum_{i=1}^n u_i e_i \right)^{(r)} = \sum_{p_1 + \ldots + p_n = r} u_1^{p_1} \ldots u_n^{p_n} e_1^{(p_1)} \ldots e_n^{(p_n)}.$$

It is easy to check that this definition does not depend on the choice of basis $\{e_1, \ldots, e_n\}$.

(1.1.5) Proposition. *The divided powers have the following properties:*

(a) $u^{(0)} = 1$, $u^{(1)} = u$, $u^{(r)} \in D_r E$,
(b) $u^{(p)} u^{(q)} = \binom{p+q}{q} u^{(p+q)}$,
(c) $(u + v)^{(p)} = \sum_{k=0}^p u^{(k)} v^{(p-k)}$,
(d) $(uv)^{(p)} = u^{(p)} v^{(p)}$,
(e) $(u^{(p)})^{(q)} = [p, q] u^{(pq)}$ *for* $u \in E$; $[p, q] = [(pq)!]/(q! p^q !)$.

(1.1.6) Remark. *In the notation used above, $e_1^{(i_1)} \ldots e_n^{(i_n)}$ has a double meaning. It is the element of the dual basis to the basis in the symmetric power,* *and it is the product of divided powers. It is not difficult to see that the two elements coincide.*

The r-th divided power is an endofunctor on the category of free **K**-modules and linear maps. More precisely, for two free **K**-modules E, F and a linear map $\phi : E \to F$ we have a well-defined linear map

$$D_r\phi : D_r E \to D_r F$$

which is best described as the transpose of the map $S_r(\phi^*) : S_r(F^*) \to S_r(E^*)$. This also gives the description of the matrix coefficients for $D_r\phi$ as polynomials in the entries of ϕ, which we leave to the reader.

The divided power algebra $D(E) := \bigoplus D_r(E)$ on E is a commutative, cocommutative algebra because it is a graded dual of the symmetric algebra on E^*. Again we denote the components of the multiplication map by

$$m : D_r E \otimes D_s E \to D_{r+s} E,$$

and the components of the comultiplication by

$$\Delta : D_{r+s} E \to D_r E \otimes D_s E.$$

Let us record the duality statements.

(1.1.7) Proposition.

   (a) The multiplication map

$$m : D_r E \otimes D_s E \to D_{r+s} E$$

    is the dual of the diagonal map

$$\Delta : S_{r+s} E^* \to S_r E^* \otimes S_s E^*.$$

   (b) The diagonal map

$$\Delta : D_{r+s} E \to D_r E \otimes D_s E$$

    is the dual of the multiplication map

$$m : S_r E^* \otimes S_s E^* \to S_{r+s} E^*.$$

   (c) The diagonal map $\Delta : D_{r+s} E \to D_r E \otimes D_s E$ is given by the formula

$$\Delta(e_1^{(i_1)} \ldots e_n^{(i_n)})$$
$$= \sum_{j_1+\ldots+j_n=r, \ 0 \le j_s \le i_s \text{ for } s=1,\ldots,n} e_1^{(j_1)} \ldots e_n^{(j_n)} \otimes e_1^{(i_1-j_1)} \ldots e_n^{(i_n-j_n)}.$$

1.1.2. Partitions, Skew Partitions. Combinatorics
of Z_2-Graded Tableaux.

Let n be a natural number. *A partition* λ of n is a sequence $\lambda = (\lambda_1, \ldots, \lambda_s)$ of natural numbers such that $\lambda_1 \geq \lambda_2 \geq \ldots \geq \lambda_s \geq 0$ and $\lambda_1 + \lambda_2 + \ldots + \lambda_s = n$. We identify the partitions $(\lambda_1, \ldots, \lambda_s)$ and $(\lambda_1, \ldots, \lambda_s, 0)$. To each partition λ we associate its Young frame (or Ferrers diagram) $D(\lambda)$. It can be defined as

$$D(\lambda) = \{(i, j) \in \mathbf{Z} \times \mathbf{Z} \mid (1 \leq i \leq s, 1 \leq j \leq \lambda_i \}.$$

To represent the Young frames graphically we think of them as contained in the fourth quadrant. A Young frame is a set of boxes with λ_i boxes in the i-th row from the top. Formally it could be achieved by considering the point $(j, -i)$ instead of (i, j).

(1.1.8) Example. $\lambda = (4, 2, 1)$:

$$D(\lambda) = \quad \begin{array}{c}\text{[Young diagram]}\end{array} \quad .$$

Formally the boxes of $D((4, 2, 1))$ correspond to the set of points

$$\{(1, -1), (2, -1), (3, -1), (4, -1), (1, -2), (2, -2), (1, -3)\}$$

in the grid $\mathbf{Z} \times \mathbf{Z}$.

Let λ be a partition. We say that λ has *a Durfee square of size* r (or rank $\lambda = r$) if $\lambda_r \geq r$, $\lambda_{r+1} \leq r$, i.e., if the biggest square fitting inside of λ is an $r \times r$ square.

Let λ be a partition, and let X be a box in λ. The set of boxes to the right of X (including X) is called an *arm of X*. The set of boxes below X (including X) is called the *leg of X*. The arm length (leg length) of X are defined as the numbers of boxes in the arm (leg) of X.

The arm and leg of X form *a hook* of X. The number of boxes in the hook of X is called the hook length of X.

Let λ be a partition of rank r. Let a_i (b_i) be the arm length (leg length) of the i-th box on the diagonal of λ. The partition λ is uniquely determined by its rank r and the numbers a_i, b_i $(1 \leq i \leq r)$. These numbers satisfy the conditions $a_1 > \ldots > a_r > 0$, $b_1 > \ldots, b_r > 0$.

We will sometimes denote by $\lambda = (a_1, \ldots, a_r | b_1, \ldots, b_r)$ the partition with diagonal arm lengths a_i and diagonal leg lengths b_i. We refer to this as a *Frobenius* (or *hook*) *notation* for λ.

(1.1.9) Example. *The partition* $\lambda = (4, 3, 2)$ *in the hook notation is* $(4, 2|3, 2)$. *The boxes in the arm (leg) of the i-th diagonal box are filled with symbol i* $(\bar{i})$:

$$
\begin{array}{|c|c|c|c|}
\hline
\text{X} & 1 & 1 & 1 \\
\hline
\bar{1} & \text{X} & 2 \\
\cline{1-3}
\bar{1} & \bar{2} \\
\cline{1-2}
\end{array}
\quad .
$$

Let λ be a partition. The conjugate (or dual) partition λ' is defined by setting

$$\lambda'_i = \text{card}\{t \mid \lambda_t \geq i\}.$$

The Young frame of λ' is obtained from the Young frame of λ by reflecting in the line $y = -x$.

(1.1.10) Example. $\lambda = (4, 2, 1)$, $\lambda' = (3, 2, 1, 1)$:

$$D(\lambda') = \quad$$ 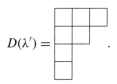 $$\quad .$$

Let λ and μ be two partitions. We say that μ is contained in λ (denoted $\mu \subset \lambda$) if for each i we have $\mu_i \leq \lambda_i$. Let λ and μ be two partitions with $\mu \subset \lambda$. We refer to such a pair as a *skew partition* λ/μ.

We associate to a skew partition λ/μ the skew Young frame

$$D(\lambda/\mu) := D(\lambda) \setminus D(\mu).$$

Graphically we can represent it as a Young frame of λ with the boxes corresponding to μ missing.

(1.1.11) Example. $\lambda = (4, 2, 2, 1, 1)$, $\mu = (3, 1)$:

$$D(\lambda/\mu) = \quad$$ 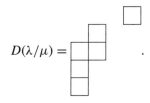 $$\quad .$$

Let $A = (A_0, A_1)$ be a $\mathbf{Z}_2$-graded set, i.e. the pair of sets indexed by $\{0, 1\}$. Assume that the set A is ordered by a total order $\lhd$. A *tableau of shape λ/μ with values in A* is a function $T : D(\lambda/\mu) \to A$.

(1.1.12) Definition.

(a) *A tableau T of shape λ/μ with values in A is row standard if for each (u, v) we have $T(u, v) \lhd T(u, v + 1)$ with equality possible if $T(u, v) \in A_1$.*

(b) *We say that a tableau T of shape λ/μ with values in A is column standard if $T(u, v) \lhd T(u + 1, v)$ with equality possible when $T(u, v) \in A_0$.*

(c) *A tableau T of shape λ/μ with values in A is standard if it is both column standard and row standard.*

(1.1.13) Notation. *We denote by* $\mathrm{RST}(\lambda/\mu, A)$ $(\mathrm{CST}(\lambda/\mu, A), \mathrm{ST}(\lambda/\mu, A))$ *the set of row standard (column standard, standard) tableaux of shape λ/μ with values in A. We denote by $[1, m] \cup [1, n]'$ the $\mathbf{Z}_2$-graded set A with $A_0 = [1, m]$, $A_1 = [1', n']$ and with the order $\lhd$ defined to be the natural order on A_0 and A_1 with A_0 preceding A_1. Similarly we define the $\mathbf{Z}_2$-graded set $[1, n]' \cup [1, m]$ (here A_1 preceeds A_0).*

(1.1.14) Examples. *Let $\lambda = (4, 2, 2, 1, 1)$, $\mu = (2, 1)$. Let $A = [1, 2] \cup [1, 3]'$.*

(a) *The tableau*

$$T_1 = \begin{array}{cc} & \boxed{1}\,\boxed{2} \\ \boxed{1'} & \\ \boxed{1'}\,\boxed{1'} & \\ \boxed{2'} & \\ \boxed{2'} & \end{array}$$

is row standard but not column standard.

(b) *The tableau*

$$T_2 = \begin{array}{cc} & \boxed{1}\,\boxed{1} \\ \boxed{1'} & \\ \boxed{1'}\,\boxed{2'} & \\ \boxed{2'} & \\ \boxed{3'} & \end{array}$$

is column standard but not row standard.

(c) The tableau

$$T_3 = \begin{array}{|c|c|} \hline 1 & 2 \\ \hline \end{array}$$

is standard.

Let λ/μ be a skew partition, and let $A = (A_0, A_1)$ be a $\mathbf{Z}_2$-graded set ordered by the total order $\lhd$. We define the orders $\preceq$ (relative to $\lhd$) on the sets of row standard (column standard, standard) tableaux as follows.

Consider the set $\mathrm{RST}(\lambda/\mu, A)$. Given two tableaux T, U, we have $T \preceq U$ if $T = U$. Assume that $T \neq U$. Let us write them as $T = (T_1, \ldots, T_s)$, $U = (U_1, \ldots, U_s)$ with T_i (U_i) being the part of T (U) from the i-th row of λ/μ. Let j be the minimal i for which $T_i \neq U_i$. We have $T_j = (T(j, 1), \ldots, T(j, \lambda_j - \mu_j))$, $U_j = (U(j, 1), \ldots, U(j, \lambda_j - \mu_j))$. Now let k be the smallest index for which $T(j, k) \neq U(j, k)$ (such a k exists by the choice of j). We say that $T \preceq U$ if and only if $T(j, k) \lhd U(j, k)$.

The order $\preceq$ on $\mathrm{ST}(\lambda/\mu, A)$ is defined to be the restriction of $\preceq$ from $\mathrm{RST}(\lambda/\mu, A)$.

Finally we define the order $\preceq$ on $\mathrm{CST}(\lambda/\mu, A)$. Given two tableaux T, U from $\mathrm{CST}(\lambda/\mu, A)$, then $T = U$ implies $T \leq U$. Assume $T \neq U$. We write $T = (T^1, \ldots, T^s)$, $U = (U^1, \ldots, U^s)$ with T^i (U^i) being the part of T (U) from the i-th column of λ/μ. Let j be the minimal i for which $T^i \neq U^i$. We have $T^j = (T(1, j), \ldots, T(\lambda'_j - \mu'_j, j))$, $U^j = (U(1, j), \ldots, U(\lambda'_j - \mu'_j, j))$. Now let k be the smallest index for which $T(k, j) \neq U(k, j)$ (such k exists by the choice of j). We say that $T \preceq U$ if and only if $T(k, j) \lhd U(k, j)$.

Note. *The order $\preceq$ on $\mathrm{ST}(\lambda/\mu, A)$ is the restriction of the order $\preceq$ on $\mathrm{RST}(\lambda/\mu, A)$. It is different from the restriction of $\preceq$ on $\mathrm{CST}(\lambda/\mu, A)$.*

(1.1.15) Examples. *Let $\lambda = (2, 1)$, $\mu = (0)$. Set $A = [1, 2] \cup [1, 2]'$. In RST $(\lambda/\mu, A)$ we have*

$$\begin{array}{|c|c|} \hline 1 & 2 \\ \hline 1' \\ \hline \end{array} \preceq \begin{array}{|c|c|} \hline 1 & 1' \\ \hline 2 \\ \hline \end{array} \preceq \begin{array}{|c|c|} \hline 2 & 1' \\ \hline 1 \\ \hline \end{array}.$$

In $CST(\lambda/\mu, A)$ *we have*

$$\begin{array}{|c|c|}\hline 1 & 1' \\\hline 2 \\\hline\end{array} \preceq \begin{array}{|c|c|}\hline 1 & 2 \\\hline 1' \\\hline\end{array} \preceq \begin{array}{|c|c|}\hline 2 & 1 \\\hline 1' \\\hline\end{array}.$$

In $ST(\lambda/\mu, A)$ *we have*

$$\begin{array}{|c|c|}\hline 1 & 2 \\\hline 1' \\\hline\end{array} \preceq \begin{array}{|c|c|}\hline 1 & 1' \\\hline 2 \\\hline\end{array}.$$

1.2. Homological and Commutative Algebra

1.2.1. Regular Sequences, Koszul Complexes, Depth

Let R be a commutative Noetherian ring. Let M be an R-module. *The dimension* $\dim M$ of M is defined to be the Krull dimension of $R/\mathrm{Ann}(M)$, where

$$\mathrm{Ann}(M) = \{x \in R \mid xM = 0\}$$

is the annihilator of M.

Let I be an ideal in R. If $IM \neq M$, we define the I-depth of M as

$$\mathrm{depth}_R(I, M) = \min\{i \mid \mathrm{Ext}_R^i(I, M) \neq 0\}.$$

In the case $IM = M$ we define $\mathrm{depth}_R(I, M) = \infty$. For a finitely generated R-module M we have $IM \neq M$ if and only if $\mathrm{depth}_R(I, M) < \infty$ if and only if $\mathrm{depth}_R(I, M) \leq \dim M$.

A sequence $\underline{a} = (a_1, \ldots, a_n)$ of elements from R is an M-sequence (or a regular sequence on M) if $M \neq (a_1, \ldots, a_n)M$ and the multiplication $a_i : M_{i-1} \to M_{i-1}$ is injective for $i = 0, 1, \ldots, n-1$, where $M_i := M/(a_1, \ldots, a_i)M$.

The connection between these notions is expressed in

(1.2.1) Theorem. *Let R be a Noetherian ring, and M a finitely generated R-module. Let I be an ideal in R. The following conditions are equivalent:*

(a) $\mathrm{depth}_R(I, M) \geq n$.
(b) $\mathrm{Ext}_R^i(R/I, M) = 0$ for $i < n$.
(c) There exists an M-sequence $\underline{a} = (a_1, \ldots, a_n)$ of length n with $a_i \in I$ for $i = 1, \ldots, n$.

A regular sequence $(a_1, \ldots, a_n)$ is *a maximal regular M-sequence* if there is no b such that $(a_1, \ldots, a_n, b)$ is an M-sequence. In particular the theorem implies that two maximal regular M-sequences with terms from I must have the same length, equal to $\mathrm{depth}(I, M)$.

Let M be an R-module, and let $\underline{a} = (a_1, \ldots, a_n)$ be a sequence of elements from R. We define the *Koszul complex* $K(\underline{a}, M)_\bullet$ as follows. For an n-dimensional free R-module $E = R^n$ with a basis $e_1, \ldots, e_n$ we set $K(\underline{a}, M)_i = \bigwedge^i E \otimes_R M$, and the differential

$$d : \bigwedge^i E \otimes_R M \to \bigwedge^{i-1} E \otimes_R M$$

is defined by the formula

$$d(e_{j_1} \wedge \ldots \wedge e_{j_i} \otimes m) = \sum_{u=1}^{i} (-1)^{u+1} e_{j_1} \wedge \ldots \wedge \hat{e}_{j_u} \wedge \ldots \wedge e_{j_i} \otimes a_{j_u} m.$$

Let M be a finitely generated R-module. We define the codimension of M,

$$\mathrm{codim}_R(M) := \mathrm{ht}\, \mathrm{Ann}(M),$$

where ht denotes the height of an ideal. We also define the grade of M,

$$\mathrm{grade}_R(M) = \mathrm{depth}_R(\mathrm{Ann}(M), R).$$

The homological properties of Koszul complex include the information about the depth.

(1.2.2) Theorem. *Let M be a finitely generated R-module, and let $\underline{a} = (a_1, \ldots, a_n)$ be a sequence of elements from R. Denote $I = (a_1, \ldots, a_n)$. Then*

$$\mathrm{depth}_R(I, M) = n - \max\{\, i \mid H_i(K(\underline{a}, M)) \neq 0 \,\}.$$

(1.2.3) Corollary. *Let R be a commutative ring. Assume that $I = (a_1, \ldots, a_n)$ is an ideal generated by a regular sequence. Then the Koszul complex $K(\underline{a}, R)_\bullet$ is a free resolution of the R-module R/I.*

The ideal I is a *complete intersection ideal of codimension n* if there exists a regular sequence $(a_1, \ldots, a_n)$ such that $I = (a_1, \ldots, a_n)$. Thus the finite free resolutions of complete intersection ideals are provided by Koszul complexes.

The projective dimension, codimension, and grade of an R-module are related.

(1.2.4) Theorem. *For an R-module $M \neq 0$ we have*

$$\mathrm{pd}_R(M) \geq \mathrm{codim}(M) \geq \mathrm{grade}_R(M).$$

A finitely generated R-module N is *perfect* if $\mathrm{pd}_R(N) = \mathrm{grade}(N)$. In that case the inequalities in (1.2.4) become equalities. We call $\mathrm{codim}(N)$ the *codimension of N*. Sometimes by abuse of notation we call the grade of R/I the *grade of the ideal I*.

Let us note the following consequence of Theorem (1.2.4) applied to $N = R$.

(1.2.5) Proposition. *Let I be an ideal of codimension n. The functor $N \mapsto \mathrm{Ext}_R^n(N, R)$ is an exact contravariant involution on the category of perfect modules N with $\mathrm{Ann}(N) = I$ up to radical.*

If R/I is a perfect module, we call I a *perfect ideal*. An ideal I is *Gorenstein* if I is perfect and $\mathrm{Ext}^n(R/I, R) \cong R/I$ for $n = \mathrm{codim}(R/I)$.

1.2.2. Cohen–Macaulay Rings and Modules, Gorenstein Rings

Let (R, m) be a local ring. The depth of a module M is defined as

$$\mathrm{depth}_R(M) := \mathrm{depth}_R(m, M).$$

We have the following inequalities:

(1.2.6) Proposition. *Let M be a finitely generated module over a local ring R. Then*

$$\mathrm{depth}(M) \leq \dim M \leq \dim R.$$

An R-module M is *Cohen–Macaulay* if $\mathrm{depth}(M) = \dim M$. If $\mathrm{depth}(M) = \dim R$, we say that M is *maximal Cohen–Macaulay*. The zero module is by definition maximal Cohen–Macaulay. The ring R is Cohen–Macaulay if it is Cohen–Macaulay as a module over itself.

The projective dimension and depth of a module over a local ring are complementary to each other.

(1.2.7) Theorem (Auslander–Buchsbaum Formula). *Let R be a Noetherian local ring. Assume that $\mathrm{pd}_R(M) < \infty$. Then we have*

$$\mathrm{pd}_R(M) + \mathrm{depth}(M) = \mathrm{depth}(R).$$

It follows that if R is Cohen–Macaulay and M is a maximal Cohen–Maculay module of finite projective dimension over R, then M is R-free.

If R is a Cohen–Macaulay local ring and M is a finitely generated R-module of finite projective dimension, then M is Cohen–Macaulay if and only if it is perfect.

If R is a Cohen–Macaulay local ring, then dim $R = $ dim R/P for every associated prime P of R. This means that R is equidimensional.

A local ring (R, m) is *Gorenstein* if an only if R has a finite injective dimension as an R-module.

(1.2.8) Theorem. *Let (R, m) be a local ring of dimension d. The following conditions are equivalent:*

    (a) R is Gorenstein,
    (b) for $i \neq d$ we have $\mathrm{Ext}^i_R(K, R) = 0$, $\mathrm{Ext}^d(K, R) = K$,
    (c) there exists $i > d$ such that $\mathrm{Ext}^i_R(K, R) = 0$,
    (d) $\mathrm{Ext}^i_R(K, R) = 0$ for $i < d$, $\mathrm{Ext}^d_R(K, R) = K$,
    (e) R is Cohen–Macaulay and $\mathrm{Ext}^d_R(K, R) = K$.

Recall that the *embedding dimension* of a local ring is emdim $(R) = \dim_K m/m^2$. A local ring R is *regular* if $\mathrm{emdim}(R) = \dim R$.

We denote by gl dim R *the global dimension* of R.

(1.2.9) Theorem (Auslander and Buchsbaum, Serre). *Let (R, m) be a local ring of dimension d. Then the following are equivalent:*

    (a) R is a regular local ring,
    (b) gl.dim $R < \infty$,
    (c) gl.dim $R = d$,
    (d) $\mathrm{pd}_R K = d$,
    (e) m is generated by a regular sequence of length d.

The connection between the notions of Cohen–Macaulay (Gorenstein) ring and perfect (Gorenstein) ideal is stated in the next proposition.

(1.2.10) Proposition.
    (a) Let R be a Cohen–Macaulay local ring. Then the ring R/I is Cohen–Macaulay if and only if I is perfect,
    (b) Let R be a Gorenstein local ring. Then R/I is Gorenstein if and only if I is a Gorenstein ideal.

The theory outlined above for local rings has an analogue for graded rings and graded modules. Let us state the corresponding statements.

Let R be a graded ring $R = \bigoplus_{i \geq 0} R_i$ where $R_0 = K$ is a field and R_i are finite dimensional vector spaces over K. We assume that R is generated as a K-algebra by elements of degree 1, which implies that R is Noetherian. We denote by m the maximal ideal $m = R_+ = \bigoplus_{i > 0} R_i$. For a graded R-module M we denote $\mathrm{depth}_R(M) := \mathrm{depth}_R(m, M)$.

Then the following statements hold.

(1.2.6)′ Proposition. *Let M be a finitely generated graded module over a graded ring R. Then*

$$\mathrm{depth}_R(M) \leq \dim M \leq \dim R.$$

(1.2.7)′ Theorem (Auslander–Buchsbaum formula). *Let R be a graded ring, and let M be a graded R-module. Assume that $\mathrm{pd}_R(M) < \infty$. Then we have*

$$\mathrm{pd}_R(M) + \mathrm{depth}_R(M) = \mathrm{depth}_R(R).$$

(1.2.8)′ Theorem. *Let R be a graded ring of dimension d with the maximal ideal $m = R^+$. The following conditions are equivalent:*

(a) R is Gorenstein,
(b) for $i \neq d$ we have $\mathrm{Ext}^i_R(K, R) = 0$, $\mathrm{Ext}^d(K, R) = K$,
(c) there exists $i > d$ such that $\mathrm{Ext}^i_R(K, R) = 0$,
(d) $\mathrm{Ext}^i_R(K, R) = 0$ for $i < d$, $\mathrm{Ext}^d_R(K, R) = K$,
(e) R is Cohen–Macaulay and $\mathrm{Ext}^d_R(K, R) = K$.

The theorem characterizing the regular rings differs because the only graded regular ring is a polynomial ring.

The embedding dimension of a graded ring R is $\mathrm{emdim}(R) = \dim_K m/m^2$. A graded ring R is *regular* if $\mathrm{emdim}(R) = \dim R$.

(1.2.9)′ Theorem. *Let R be a graded ring of dimension d with the maximal ideal $m = R^+$. Then the following are equivalent:*

(a) R is a regular graded ring,
(b) gl.dim $R < \infty$,
(c) gl.dim $R = d$,
(d) $\mathrm{pd}_R K = d$,
(e) m is generated by a regular sequence of length d,
(f) R is a polynomial ring over K in d variables.

The definitions of Cohen–Macaulay, Gorenstein, and regular rings generalize to the global case. We define a commutative ring R to be regular (Cohen–Macaulay, Gorenstein) if for every prime ideal $P \in \mathrm{P}$ in $\mathrm{R}(R)$ the localization R_P is regular (Cohen–Macaulay, Gorenstein).

1.2.3. Minimal Resolutions

We will be working throughout the book with complexes

$$F_\bullet : \ldots \to F_n \overset{d_n}{\to} \ldots \to F_1 \overset{d_1}{\to} F_0 \ldots$$

of free R-modules. We use the following notation. The rank of the free module F_i will be denoted by f_i. Sometimes we will choose bases $\{e_j^{(i)}\}_{1 \le j \le f_i}$ in free modules F_i. Then the homomorphism d_i can be identified with $f_{i-1} \times f_i$ matrix $(\phi_{k,j}^{(i)})_{1 \le k \le f_{i-1}, 1 \le j \le f_i}$ where

$$d_i(e_j^{(i)}) = \sum_{k=1}^{f_{i-1}} \phi_{k,j}^{(i)} e_k^{(i-1)}.$$

We define the rank r_i of d_i to be the biggest integer r such that there exists a nonzero $r \times r$ minor of $\phi^{(i)}$. We denote by $I_{r_i}(d_i)$ the ideal of $r_i \times r_i$ minors of the matrix $\phi_{k,j}^{(i)}$.

Assume that R is a local (resp. graded) ring, and let m denote the maximal ideal (resp. $m = R_+$).

A complex

$$F_\bullet : \ldots \to F_n \overset{d_n}{\to} \ldots \to F_1 \overset{d_1}{\to} F_0$$

of free R-modules is *minimal* if for each i, $1 \le i \le n$, we have $d_i(F_i) \subset m F_{i-1}$. Equivalently, after choosing bases in F_i we can identify the differential d_i with a matrix with entries in R. The minimality condition says that all entries of matrices of differentials d_i are in the maximal ideal m.

(1.2.11) Proposition. *Let (R, m) be a local ring. Denote $K = R/m$. Let M be an R-module.*

(a) *The module M has a minimal free resolution $F_\bullet$. The resolution $F_\bullet$ is unique up to isomorphism.*

(b) *$F_i \otimes_R R/m = \mathrm{Tor}_i^R(R/m, M)$, so rank $F_i = \dim_K \mathrm{Tor}_i^R(R/m, M)$.*

Let us notice that one can construct the minimal free resolution of a finitely generated module M from short exact sequences

$$0 \to \Sigma_{i+1}(M) \to F_i \overset{\pi_i}{\to} \Sigma_i(M) \to 0$$

where the modules $\Sigma_i(M)$ and maps π_1 are constructed inductively as follows. We take $\Sigma_0(M) = M$ and consider the vector space $V(M) = M/mM$. We choose a basis $\{v_1^{(0)}, \ldots v_{f_0}^{(0)}\}$ of the vector space $V(M)$ and take as F_0 a free R-module with a basis $\{e_1^{(0)}, \ldots e_{f_0}^{(0)}\}$. We define $\pi_0(e_j^{(0)}) := m_j^{(0)}$ where $m_j^{(0)} \in \Sigma_0(M)$ is an element whose representative modulo mM is $v_j^{(0)}$. The homomorphism π_0 is onto by Nakayama's lemma. Next we define $\Sigma_1(M) = \mathrm{Ker}\ \pi_0$ and continue the procedure with $\Sigma_1(M)$.

The module $\Sigma_i(M)$ is the *i-th syzygy module of M*. It is unique up to isomorphism.

The uniqueness of minimal free resolution occurs also in the case of graded rings and graded modules. Let $R = \bigoplus_{i \geq 0} R_i$ be a graded ring, as defined in the previous section. We denote by $m = R_+$ the maximal ideal of elements of positive degree. Let $M = \bigoplus_{i \geq 0} M_i$ be a graded R-module. Recall that we define the shifted module $M(n)$ by setting $M(n)_i := M_{n+i}$. The analogue of (1.2.11) is true in the graded case.

Then one can construct the minimal free resolution as sketched above, as the analogue of Nakayama's lemma holds, and the analogue of (1.2.11) is true. Let $F_\bullet$ be a minimal resolution of M. We will write

$$F_i = \bigoplus_{j \geq i} R(-j)^{f^{(i,j)}}.$$

The numbers $f^{(i,j)}$ are the dimensions of the graded pieces of graded K-vector spaces $\mathrm{Tor}_i^R(K, M)$. Sometimes they are called the graded Betti numbers of M.

We have a very useful exactness criterion for acyclicity of a finite complex of free R-modules.

(1.2.12) Theorem (Buchsbaum–Eisenbud acyclicity criterion, [BE1]). *Let R be a Noetherian ring, and let*

$$F_\bullet : 0 \to F_n \overset{d_n}{\to} \ldots \to F_1 \overset{d_1}{\to} F_0$$

be a complex of free finitely generated R-modules. Then $F_\bullet$ is acyclic (i.e. $H_i(F_\bullet) = 0$ for $i > 0$) if and only if the following two conditions hold:

(a) $r_i + r_{i-1} = f_i$ for $1 \leq i \leq n+1$ (with the convention $r_{n+1} = 0$),
(b) $\mathrm{depth}(I(d_i)) \geq i$ for $1 \leq i \leq n$.

(1.2.13) Proposition. *Let $F_\bullet$ be a complex from (1.2.12).*

(a) We have $\sqrt{I(d_i)} \subset \sqrt{I(d_{i+1})}$ for $1 \leq i \leq n-1$,

(b) Assume that $F_\bullet$ is a resolution of a perfect module M, and let I be the defining ideal of the support of M. Then for every i, $1 \le i \le n$, we have $\sqrt{I(d_i)} = I$.

Proof. Assume that $I(d_i) = R$. Then the map d_i splits and it follows that all maps d_j have to split for $j > i$. This means that $I(d_j) = R$ for $j > i$. Applying localization, we get (a). To prove (b) we notice that if $F_\bullet$ is a resolution of a perfect module M, then F^* is also a resolution of the perfect module with the same support by (1.2.5). Applying (a), we see that all radicals of ideals $I(d_i)$ are equal and therefore have to be equal to I. ∎

This statement implies the following result of Eagon and Northcott.

(1.2.14) Theorem (Generic Perfection Theorem, [EN2]). *Let* **K** *be a commutative ring, and let* $R = \mathbf{K}[X_1, \ldots, X_n]$ *be a polynomial ring over* **K**. *Let*

$$F_\bullet : 0 \to F_m \to F_{m-1} \to \ldots \to F_1 \to F_0$$

be a free resolution of an R-module M. Assume that the complex $F_\bullet$ is perfect, i.e., $m = $ depth $\mathrm{Ann}_R(M)$. Assume that M is free as a **K**-*module. Then for every ring homomorphism $\phi : R \to S$ such that $m = $ depth $\mathrm{Ann}_S(M \otimes_R S)$, the complex $F_\bullet \otimes_R S$ is a free resolution over S of an S-module $M \otimes_R S$.*

Proof. Applying Proposition (1.2.13) (b) to the complex $F_\bullet$, we see that all ideals $I(d_i)$ for this complex are equal up to a radical to $\mathrm{Ann}_R(M)$. We deduce that the same is true for the complex $F \otimes_R S$, and the depth assumption implies that the complex $F \otimes_R S$ is acyclic by the Buchsbaum–Eisenbud criterion. ∎

1.2.4. Effective Calculation of Normalization

We describe a very useful algorithm due to Grauert and Remmert and to de Jong ([GRe], [dJ]) for calculating the normalization of a reduced affine ring. The algorithm is based on the following criterion of normality.

Consider a radical ideal J containing a nonzero divisor, whose zero set contains the nonnormal locus of R. Then we have canonical inclusions

$$R \subset \mathrm{Hom}_R(J, J) \subset \bar{R}$$

given by the maps

$$r \mapsto \psi_r, \psi \mapsto \frac{\psi(x)}{x},$$

where ψ_r is a multiplication by r and $x \in J$ is a nonzero divisor.

(1.2.15) Proposition ([DGJP]). *Let R be a reduced Noetherian ring. Let J be a radical ideal in R containing a nonzero divisor, whose zero set $V(J)$ contains the nonnormality locus of R. Then R is normal if and only if $R = \mathrm{Hom}_R(J, J)$.*

The proposition implies that if R is not normal, then $\mathrm{Hom}_R(J, J)$ is an intermediate ring between R and $\bar{R}$, different from R. The algorithm consists of finding J and then replacing R by the bigger ring $\mathrm{Hom}_R(J, J)$. The authors prove that after finitely many steps we reach $\bar{R}$.

One also has an interesting presentation of $\mathrm{Hom}_R(J, J)$ as a ring. One starts with the R-module generators $u_0 = x, u_1, \ldots, u_s$ of $\mathrm{Hom}_R(J, J)$. Since $\mathrm{Hom}_R(J, J)$ is an algebra, we have the quadratic relations

$$\frac{u_i}{x} \frac{u_j}{x} = \sum_{k=0}^{s} a_k^{i,j} \frac{u_k}{x}.$$

We also have the linear relations between $u_0, \ldots, u_s$. Let us assume that the relations $\sum_{k=0}^{s} b_k^j u_k$ $(j = 1, \ldots, m)$ are the generators of the first syzygies between $u_0, \ldots, u_s$. We define an epimorphism of commutative R-algebras

$$\theta : R[T_1, \ldots, T_s] \to \mathrm{Hom}_R(J, J).$$

(1.2.16) Proposition ([DGJP]). *The kernel of the homomorphism θ is generated by the linear relations $\sum_{k=1}^{s} b_k^j T_j$ $(j = 1, \ldots, m)$ and the quadratic relations $T_i T_j - \sum_{k=0}^{s} a_k^{i,j} T_k$ $(1 \leq i, j \leq s)$.*

The above algorithm and presentation allow to write down the normalization quite explicitly.

1.2.5. Duality for Proper Morphisms and Rational Singularities

In this subsection we use the notions related to derived categories. Our principal references are [H2], [GM]. Apart from the notion of rational singularities, these results will be used only in the proof of the duality statement for complexes $F_\bullet(\mathcal{V})$ (Theorem (5.1.4)) and in the proof of the Hinich–Panyushev theorem in section 8.3. Thus this subsection can skipped in the first reading.

Let X be a locally Noetherian scheme. We denote by $D^*(X)$ (where $* = \emptyset, +, -, b$) the derived category $D^*(\mathcal{A})$, where $\mathcal{A}$ is the category of $\mathcal{O}_X$-modules. By $D^*_{\text{Qco } X}(X)$ we denote the thick subcategory $D^*_{\mathcal{A}'}(\mathcal{A})$ where $\mathcal{A}$ is the category of $\mathcal{O}_X$-modules and $\mathcal{A}' = \text{Qco } X$ is a category of quasicoherent $\mathcal{O}_X$-modules. Note that by [H2, Proposition I.4.8] the natural embedding $D^*(\text{Qco } X) \to D^*_{\text{Qco } X}(X)$ is an equivalence of categories, for a quasicompact scheme X and for $* = +, \emptyset$. We will use the abbreviation $D^*_{qc}(X)$.

We start with the discussion of dualizing complexes. Let X be a Noetherian scheme. A complex of quasicoherent $\mathcal{O}_X$-modules $\mathcal{I}^\bullet$ is a *dualizing complex of X* if $\mathcal{I}^\bullet$ is bounded, each term of $\mathcal{I}^\bullet$ is an injective module, each cohomology group is coherent, and the canonical map

$$\mathcal{O}_X \to \text{Hom}^\bullet_{\mathcal{O}_X}(\mathcal{I}^\bullet, \mathcal{I}^\bullet)$$

is a quasiisomorphism. A dualizing complex is treated as an object in the derived category, so any complex isomorphic to a dualizing complex in $D^+_{\text{Qco } X}(X)$ is also called a dualizing complex.

If $\mathcal{I}^\bullet$ is a dualizing complex of X, then for any complex $\mathcal{F}^\bullet$ of $\mathcal{O}_X$-modules with coherent cohomology groups, the canonical map

$$\mathcal{F}^\bullet \to \text{Hom}^\bullet_{\mathcal{O}_X}(\text{Hom}^\bullet_{\mathcal{O}_X}(\mathcal{F}^\bullet, \mathcal{I}^\bullet), \mathcal{I}^\bullet)$$

is a quasiisomorphism.

Dualizing complexes are unique in the following sense.

(1.2.17) Theorem ([H2, Theorem V.3.1]). *Let $\mathcal{I}^\bullet$ be a dualizing complex on X, and $\mathcal{I}'^\bullet$ a complex of $\mathcal{O}_X$-modules bounded above with coherent cohomology groups. Then $\mathcal{I}'^\bullet$ is dualizing if and only if there exists an invertible sheaf $\mathcal{L}$ and an integer n such that $\mathcal{I}'^\bullet$ is isomorphic to $\mathcal{I}^\bullet \otimes_{\mathcal{O}_X} \mathcal{L}[n]$ in $D^+_{\text{Qco } X}(X)$. In this case $\mathcal{L}$ and n are determined by*

$$\mathcal{L}[n] = \underline{R}\,\text{Hom}^\bullet_{\mathcal{O}_X}(\mathcal{I}^\bullet, \mathcal{I}'^\bullet).$$

Let X be a Noetherian scheme with a dualizing complex $\mathcal{I}^\bullet_X$. Let $r := \min\{i \in \mathbf{Z} \mid H^i(\mathcal{I}^\bullet_X) \neq 0\}$. We define *the canonical sheaf* $\omega_X := H^r(\mathcal{I}^\bullet_X)$. The coherent sheaf ω_X is defined up to tensoring with an invertible sheaf. If the scheme X is affine, this means that the canonical module ω_X is defined uniquely up to isomorphism as an $\mathcal{O}_X$-module.

(1.2.18) Proposition. *Let X be a Noetherian scheme.*

 (a) *If X is a Cohen–Macaulay scheme, then the dualizing complex $\mathcal{I}^\bullet_X$ has only one cohomology, so $\omega_X = \mathcal{I}^\bullet_X$.*
 (b) *If X is a Gorenstein scheme, then $\omega_X = \mathcal{O}_X$.*

It follows that in the case of affine Cohen–Macaulay schemes we recover the theory of canonical modules as described in [HKu].

Let X, Y be Noetherian schemes. The following theorem was proved by Nagata [N].

(1.2.19) Theorem. *Let $f : X \to Y$ be a morphism of finite type between Noetherian schemes. Then f is compactifiable, i.e., there exists a scheme $\tilde{X}$, a proper morphism $p : \tilde{X} \to Y$, and an open immersion $i : X \to \tilde{X}$ such that $pi = f$.*

The factorization $f = pi$ is called a *compactification of f*.

The following theorem is known as a *global duality theorem for proper morphisms*.

(1.2.20) Theorem. *Let $p : Y \to X$ be a proper morphism. Then the derived functor $\underline{R}^+ p_* : D_{qc}^+(Y) \to D_{qc}^+(X)$ has the right adjoint $p^! : D_{qc}^+(X) \to D_{qc}^+(Y)$.*

Let $\mathrm{Fin}(X)$ denote the category of X-schemes of finite type. A morphism $f : Y \to Y'$ in $\mathrm{Fin}(X)$ has a compactification $f = pi$ by (1.2.19). We define

$$f^! := i^* \circ p^! : D_{qc}^+(Y') \to D_{qc}^+(Y),$$

where $p^!$ is the right adjoint of $\underline{R}^+ p_*$. The functors $f^!$ have the following properties.

(1.2.21) Proposition. *With the above notation, the following hold:*

(a) *The definition of $f^!$ is independent of a compactification $f = pi$.*

(b) *For any two morphisms f, g in $\mathrm{Fin}(X)$ we have $(g \circ f)^! = f^! \circ g^!$, provided $g \circ f$ is defined.*

(c) *If $h : Y \to Y'$ is a smooth morphism in $\mathrm{Fin}(X)$, of relative dimension d, then $h^!$ is isomorphic to the functor $h^\sharp$, where*

$$h^\sharp(\mathcal{F}) = h^* \mathcal{F} \otimes_{\mathcal{O}_Y}^L \omega_{Y'/Y}[d].$$

(d) *If $g : Y \to Y'$ is a finite morphism in $\mathrm{Fin}(X)$, then $g^!$ is isomorphic to $g^\sharp$, where*

$$g^\sharp(\mathcal{F}) = \bar{g}^* \underline{R} \, \mathrm{Hom}_{\mathcal{O}_{Y'}}^\bullet(g_* \mathcal{O}_Y, \mathcal{F}),$$

where $\bar{g} : (Y, \mathcal{O}_Y) \to (Y', g_* \mathcal{O}_Y)$ is the canonical morphism of ringed spaces associated to g.

(e) Let $f : Y \to Y'$ be a morphism from $\mathrm{Fin}(X)$, *and let* $g : Z' \to Y'$ *be a flat morphism of Noetherian schemes. Let* $Z = Y \times_{Y'} Z'$ *with the commutative square*

$$
\begin{array}{ccc}
Z & \stackrel{f'}{\to} & Z' \\
\downarrow g' & & \downarrow g \\
Y & \stackrel{f}{\to} & Y'
\end{array}.
$$

Then we have a canonical isomorphism $(g')^* \circ f^! = (f')^! \circ g^*$.

(f) Let $f : X \to Y$ *be a morphism of finite type. Let* $\mathcal{I}_Y^{\bullet}$ *be a dualizing complex on* Y. *Then* $\mathcal{I}_X^{\bullet} := f^!(\mathcal{I}_Y^{\bullet})$ *is a dualizing complex on* X.

(1.2.22) Theorem (Duality for Proper Morphisms). *Let* $p : Y \to X$ *be a proper morphism between Noetherian schemes,* $F_{\bullet} \in D_{qc}^-(Y)$, $G^{\bullet} \in D_{qc}^+(X)$. *Then there is an isomorphism*

$$
\theta_p : \underline{R}p_* \underline{R} \operatorname{Hom}_{\mathcal{O}_Y}^{\bullet}(F_{\bullet}, p^! G^{\bullet}) \cong \underline{R} \operatorname{Hom}_{\mathcal{O}_X}^{\bullet}(\underline{R}p_* F_{\bullet}, G^{\bullet}).
$$

(1.2.23) Remark. *The proof of Theorem (1.2.22) is the main subject of Hartshorne's lecture notes [H2]. The existence of the adjoint $p^!$ is proven in the appendix by Deligne. In our application we will use only the case when p is projective, and therefore one needs only the contents of first three chapters of [H2]. The proofs in [H2] are rather complicated, and the signs are not always correct. These questions are addressed in recent lecture notes of Brian Conrad [C], where the fully rigorous version of the duality theorem is developed. An alternative approach based on techniques from algebraic topology was developed by Neeman [Ne]. Still, this approach depends on unbounded derived functors and techniques of Thomason [TT]. The reader should also compare the notes [Ha7] of Hashimoto for fuller treatment of this material.*

One of the applications of duality for proper morphisms is the notion of rational singularities. We follow the approach of Kempf from [KKMSD].

(1.2.24) Corollary. *Let* $p : Y \to X$ *be a proper morphism of smooth varieties. Denote by* m, n *the dimensions of* X, Y *respectively. Let* $\mathcal{L}$ *be an invertible sheaf on* Y. *Assume that* $R^i f_*(\omega_Y \otimes \mathcal{L}^{-1}) = 0$ *for* $i > 0$. *Then there are natural isomorphisms of sheaves on* X *given by*

$$
R^i p_* \mathcal{L} = \underline{\operatorname{Ext}}_{\mathcal{O}_X}^{n-m+i}(p_*(\omega_Y \otimes \mathcal{L}^{-1}), \omega_X).
$$

Proof. We apply Theorem (1.2.22) to the morphism f, and to the sheafs $\mathcal{F} = \omega_Y \otimes \mathcal{L}^{-1}$, $\mathcal{G} = \omega_X$. Then we take the cohomology sheaves of complexes on both sides. ∎

Let X be a smooth scheme of dimension m. A quasicoherent sheaf $\mathcal{F}$ on X is *Cohen–Macaulay of pure dimension k* if $\underline{\mathrm{Ext}}_{\mathcal{O}_X}^{m-j}(\mathcal{F}, \omega_X)$ is zero unless $j = k$ is the dimension of the support of $\mathcal{F}$.

(1.2.25) Remark. *The above definition is related to the definition of Cohen–Macaulay modules. Assume that X is an affine smooth variety. Let A be the coordinate ring of X. Then we have $\omega_X = \mathcal{O}_X = \tilde{A}$. For a sheaf $\mathcal{F} = \tilde{M}$ the condition on vanishing of $\underline{\mathrm{Ext}}_{\mathcal{O}_X}^{m-j}(\mathcal{F}, \omega_X)$ is equivalent to the vanishing of modules $\mathrm{Ext}_A^{m-j}(M, A)$. This translates to the depth condition by Theorem (1.2.1), which, by Proposition (1.2.6) and the following definition, is equivalent to the Cohen–Macaulayness of the module M.*

(1.2.26) Corollary. *Let $p : Y \to X$ be a proper map of smooth varieties. Assume that $R^i f_*(\omega_Y \otimes \mathcal{L}^{-1}) = 0$ for $i > 0$ and additionally $R^i p_* \mathcal{L} = 0$ for $i > 0$. Then the sheaves $p_*(\omega_Y \otimes \mathcal{L}^{-1})$, $p_* \mathcal{L}$ are Cohen–Macaulay of pure dimension y. In fact the duality $\mathcal{F} \to \underline{\mathrm{Ext}}^{m-n}(-, \omega_X)$ interchanges them.*

We refer to the duality $\mathcal{F} \mapsto \underline{\mathrm{Ext}}^{m-n}(-, \omega_X)$ as the Ext *duality*. It is an exact involution on the category of Cohen–Macaulay sheaves supported in $p(Y)$.

(1.2.27) Definition. *Let $f : Z \to Y$ be a proper birational morphism, with Z smooth. We call such f a resolution of singularities. The resolution f is a rational resolution if the following conditions are satisfied:*

(a) Y is normal, i.e., the natural map $\mathcal{O}_Y \to f_* \mathcal{O}_Z$ is an isomorphism,
(b) $R^i f_* \mathcal{O}_Z = 0$ for $i > 0$,
(c) $R^i f_* \omega_Z = 0$ for $i > 0$.

Condition (c) is not needed over a field of characteristic 0 because of the following relative version of the Kodaira vanishing theorem.

(1.2.28) Theorem (Grauert–Riemenschneider, [GR], [Ke4]). *Let $\mathbf{K}$ be a map of characteristic 0. Let $f : Z \to Y$ be a proper map of smooth varieties defined over $\mathbf{K}$, with $k = \dim Z$, $n = \dim Y$. Then*

$$R^i f_* \omega_Z = 0$$

for $i > k - n$.

(1.2.29) Proposition. *Let* $f : Z \to Y$ *be a resolution of singularities. Assume that condition (c) from (1.2.27) holds. Assume that* Y *can be embedded in a smooth variety* X. *Then the resolution* f *is a rational resolution if and only if the following conditions are satisfied:*

  (d) $\mathcal{O}_Y$ *is Cohen–Macaulay.*
  (e) The natural morphism $f_* \omega_Z \to \omega_Y$ *(where* $\omega_Y = \underline{\operatorname{Ext}}^{m-n}_{\mathcal{O}_X}(\mathcal{O}_Y, \mathcal{O}_X)$ *is a dualizing sheaf on* Y) *is an isomorphism.*

Proof. Let $j : Y \to X$ be an embedding. Let $g = j \circ f$. Condition (c) is satisfied, so by (1.2.24) we get the isomorphisms

$$j_* R^i f_* \mathcal{O}_Z = \underline{\operatorname{Ext}}^{m-n+i}_{\mathcal{O}_X}(g_* \omega_Z, \omega_X).$$

Let us assume that conditions (a) and (b) are satisfied. Condition (b) implies that $g_* \omega_Z$ or $f_* \omega_Z$ are Cohen–Macaulay. If conditions (a) and (b) are satisfied, we have that $\mathcal{O}_Y = f_* \mathcal{O}_X$ is the Ext-dual of the Cohen–Macaulay sheaf $g_* \omega_Z$. Thus $\mathcal{O}_Y$ is Cohen–Macaulay. The homomorphism in (e) is the Ext-dual of the isomorphism $\mathcal{O}_Y \to f_* \mathcal{O}_Z$. Thus (d) and (e) hold.

Assume that (d) and (e) are true. Then $\mathcal{O}_Y$ is Cohen–Macaulay with Ext-dual Cohen–Macaulay sheaf ω_Y by (d). Using (e), we see that $f_* \omega_Z$ is Cohen–Macaulay. The isomorphisms

$$j_* R^i f_* \mathcal{O}_Z = \underline{\operatorname{Ext}}^{m-n+i}_{\mathcal{O}_X}(g_* \omega_Z, \omega_X)$$

imply condition (b) and that $f_* \mathcal{O}_Z$ is the Ext-dual of $f_* \omega_Z$. Further, the Ext-dual of the homomorphism $\mathcal{O}_Y \to f_* \mathcal{O}_Z$ of Cohen–Macaulay sheaves is an isomorphism by (e). This implies that the homomorphism $\mathcal{O}_Y \to f_* \mathcal{O}_Z$ is an isomorphism. This implies (a), and the proposition is proven. ∎

(1.2.30) Remark. *Let us assume that we work over a field* **K** *of characteristic zero. By [GR] the sheaf* $f_* \omega_Z$ *does not depend on the choice of desingularization* Z, *only on the variety* Y. *Therefore conditions (d) and (e) are just conditions on the variety* Y. *The assumption of embeddability in a smooth variety is locally satisfied for any variety* Y.

(1.2.31) Definition. *The variety* Y *defined over a field* **K** *of characteristic zero has rational singularities if one of the equivalent conditions holds:*

  (a) Conditions (d) and (e) are (locally) satisfied.
  (b) There exists a desingularization $f : Z \to Y$ *which is a rational resolution.*
  (c) Every desingularization $f : Z \to Y$ *is a rational resolution.*

We finish this section with a nice criterion, due to Hinich [Hi], for a scheme to have rational singularities and be Gorenstein.

(1.2.32) Proposition. *Let* $f : X \to Y$ *be a proper birational morphism of finite type schemes over a field* **K**, *with X smooth and Y normal. Suppose that there exists a morphism of sheaves* $\phi : \mathcal{O}_X \to \omega_X$ *such that* $f_*(\phi) :$ $f_*(\mathcal{O}_X) \to f_*(\omega_X)$ *is an isomorphism. Then Y is Gorenstein and it has rational singularities.*

Proof. Consider the diagram

$$
\begin{array}{ccc}
X & \xrightarrow{\ f\ } & Y \\
& \searrow p \qquad \nearrow q & \\
& \text{Spec } \mathbf{K} &
\end{array}
$$

By (1.2.21) (b), $p^!(\mathbf{K}) = f^! q^!(\mathbf{K})$. Applying (1.2.21) (c) to $p : X \to$ Spec (**K**), one obtains $p^!(\mathbf{K}) = \underline{L}p^*(\mathbf{K}) \otimes^L \omega_X[n] = \omega_X[n]$, where $n = \dim X = \dim Y$. Denoting $\omega_Y := q^!(\mathbf{K})[-n]$, we have $f^!(\omega_Y) = \omega_X$. By (1.2.21) (f), ω_Y is the dualizing complex on Y.

Applying the duality theorem (1.2.22) to the morphism f and complexes ω_X, ω_Y, we get

$$
\underline{R}f_*\underline{R}\operatorname{Hom}_X(\omega_X, \omega_X) = \underline{R}\operatorname{Hom}_Y(\underline{R}f_*(\omega_X), \omega_Y). \qquad (*)
$$

Since $H^i(\underline{R}\operatorname{Hom}_X(\omega_X, \omega_X)) = \mathcal{O}_X$ for $i = 0$ and 0 for $i > 0$, the left hand side can be identified with $\underline{R}f_*\mathcal{O}_X$.

By the Grauert–Riemenschneider theorem (1.2.28), $R^i f_*(\omega_X) = 0$ for $i > 0$. Since f is proper, birational, X is smooth, and Y is normal, we have $f_*\mathcal{O}_X = \mathcal{O}_Y$. Therefore by the assumption of the proposition we have the isomorphism

$$
\alpha : \underline{R}f_*(\omega_X) \to \mathcal{O}_Y,
$$

$\alpha = f(\phi)^{-1}$, and $(*)$ gives the isomorphism

$$
\beta : \underline{R}f_*(\mathcal{O}_X) \to \omega_Y.
$$

Recall that we have a morphism $\phi : \mathcal{O}_X \to \omega_X$. It induces the morphism $\psi = \alpha\phi\beta^{-1} : \omega_Y \to \mathcal{O}_Y$. By assumption of the proposition we know that ψ induces an isomorphism $H^0(\psi)$ in zero cohomology, since $R^0 f_*(\phi) = f_*(\phi)$.

Let $i : U \to Y$ be an open immersion. Then by (1.2.21) (f), the complex $i^*(\omega_Y)$ is a dualizing complex on U, and $i^*(\psi)$ induces an isomorphism in zero cohomology. Since being Gorenstein is a local property, it is enough to prove the assertion in the case when $Y = \text{Spec}(R)$ for some commutative Noetherian ring R.

We can therefore work in the category of R-modules.

We are given a morphism $\psi : \omega_R \to R$ in $D^+(R)$ which induces an isomorphism $H^0(\psi) : H^0(\omega_R) \to R$. By definition ψ is represented by a diagram

$$\omega_R \xleftarrow{s} C^{\bullet} \xrightarrow{\Psi} R,$$

where s is a quasiisomorphism. The morphism Ψ is given by a diagram

$$
\begin{array}{ccccccccc}
\cdots & \to & C^{-1} & \xrightarrow{d^{-1}} & C^0 & \xrightarrow{d^0} & C^1 & \to & \cdots \\
 & & \downarrow & & \downarrow \Psi^0 & & \downarrow & & \\
\cdots & \to & 0 & \to & R & \to & 0 & \to & \cdots
\end{array}
$$

with $\Psi^0 d^{-1} = 0$.

Since $H^0(\Psi) : H^0(C^{\bullet}) \to R$ is an isomorphism, there exists a cycle $z \in C^0$ such that $\Psi^0(z) = 1$. Therefore the R-module map $a \mapsto az$ from R to C^0 extends to a morphism $\sigma : R \to C^{\bullet}$ of complexes. One obviously has $\Psi \sigma = 1_R$. Thus $C^{\bullet} = R \oplus \text{Ker}(\Psi)$. Since $C^{\bullet}$ has a finite injective dimension (it is a dualizing complex), R also has a finite injective dimension. Therefore R is Gorenstein.

Now, since Y is Gorenstein, ω_Y has only one nonzero cohomology, so $\psi : \omega_Y \to \mathcal{O}_Y$ is an isomorphism. We conclude that $R^i f_*(\mathcal{O}_X) = H^i(\omega_Y) = 0$ for $i > 0$. ∎

1.3. Determinants of Complexes

In this section we collect the facts we need about determinants of complexes of vector spaces and modules.

Let us start with the complex of vector spaces

$$V_{\bullet} : 0 \to V_n \xrightarrow{d_n} V_{n-1} \xrightarrow{d_{n-1}} \cdots \to V_{m+1} \xrightarrow{d_{m+1}} V_m$$

over a field $\mathbf{K}$.

For a vector space of dimension n we define its determinant to be the one dimensional vector space $\det(V) := \bigwedge^n V$. Similarly we define the inverse of the determinant of V by setting $\det(V)^{-1} := \bigwedge^n (V^*)$. We define the determinant of a complex $V_{\bullet}$ to be a one dimensional vector space

$$\det(V_{\bullet}) = \bigotimes_{i=m}^{n} \det(V_i)^{(-1)^i}.$$

(1.3.1) Proposition. *Let*

$$0 \to V' \to V \to V'' \to 0$$

be an exact sequence of complexes. Then we have a canonical isomorphism

$$\det(V') \otimes \det(V'') \to \det(V).$$

Proof. Let $\{u_1, \ldots, u_m\}$ be a basis of V'. Let $\{v_1, \ldots, v_n\}$ be a basis of V''. Denote by g the linear map from V' to V, and let f denote the linear map from V to V''. We choose the elements $w_1, \ldots, w_n$ in V so $f(w_i) = v_i$ for $i = 1, \ldots, n$. We define the isomorphism

$$j : \det(V') \otimes \det(V'') \to \det(V)$$

by setting $j(u_1 \wedge \ldots \wedge u_m \otimes v_1 \wedge \ldots \wedge v_n) := g(u_1) \wedge \ldots \wedge g(u_m) \wedge w_1 \wedge \ldots \wedge w_n$. One checks directly that this isomorphism does not depend on the choice of vectors $w_1, \ldots, w_n$ or on the choice of bases $\{u_1, \ldots, u_m\}$, $\{v_1, \ldots, v_n\}$. ∎

(1.3.2) Proposition. *Let $V_\bullet$ be a complex of vector spaces. Let $H(V_\bullet)$ be a complex of vector spaces whose i-th term is $H_i(V_\bullet)$, with zero differential. Then we have a canonical isomorphism*

$$\mathrm{Eu}_d : \det(V_\bullet) = \det(H(V_\bullet)).$$

Proof. Decompose the complex $V_\bullet$ to short exact sequences

$$0 \to \mathrm{Ker}(d_i) \to V_i \to \mathrm{Im}(d_i) \to 0,$$

$$0 \to \mathrm{Im}(d_{i+1}) \to \mathrm{Ker}(d_i) \to H_i(V_\bullet) \to 0,$$

and use isomorphisms from Proposition (1.3.1) ∎

We have the following properties of determinants of complexes of vector spaces. They easily follow from definitions.

(1.3.3) Proposition.
  (a) *Let $V[i]_\bullet$ be a complex $V_\bullet$ shifted to the left (i.e. $V[i]_j := V_{i+j}$). Then $\det(V[i]_\bullet) = \det(V_\bullet)^{(-1)^i}$.*
  (b) *Let*

$$0 \to V'_\bullet \to V_\bullet \to V''_\bullet \to 0$$

be an exact sequence of complexes of vector spaces. Then $\det(V_\bullet) =$
$\det(V_\bullet') \otimes \det(V_\bullet'')$.

(c) If the complex $V_\bullet$ *is exact, we have the isomorphism* $\mathrm{Eu}_d : \det(V_\bullet) \to \mathbf{K}$.

Assume that $(V_\bullet, v)$ is a *based exact complex*, i.e., $V_\bullet$ is exact and v denotes the choice of bases $\{v_1^{(i)}, \ldots, v_{n_i}^{(i)}\}$ of vector spaces V_i. Then we have a natural basis element (denoted also v) in one dimensional vector space $\bigotimes_i \det(V_i)^{(-1)^i}$ given by the tensor product of volume elements $(v_1^{(i)} \wedge v_{n_i}^{(i)})$ and their duals. We define a determinant of the based complex $(V_\bullet, v)$ to be a nonzero scalar

$$\det(V_\bullet, v) = \mathrm{Eu}_d(v).$$

Let R be a commutative domain, and let

$$F_\bullet : F_n \overset{d_n}{\to} F_{n-1} \to \ldots \overset{d_{m+1}}{\to} F_m$$

be a complex of free R-modules. Assume that $F_\bullet$ is generically exact. This means that, tensoring with the field of fractions $\mathbf{K} := R_{(0)}$, we get an exact sequence of vector spaces. Let us fix bases $\{f_1^{(i)}, \ldots, f_{n_i}^{(i)}\}$ in the R-modules F_i. Then we can talk about the element $\det(F_\bullet, f) \in \mathbf{K}^*$, where f is the corresponding volume element. For another choice of bases (over R) in F_i corresponding to a volume element f' we see that $\det(F_\bullet, f') = \det(F_\bullet, f) u$ where u is a unit in R. Thus a determinant of a generically exact complex of free modules is well defined as an element of $\mathbf{K}^*/R^*$. We want to investigate the numerator and denominator of this function.

Notice that the construction of the determinant of a complex of free modules commutes with the localization. The properties (1.3.3) of determinants of complexes are also true for generically exact complexes of free modules over R.

Let us assume that R is a unique factorization domain. We can write

$$\det(F_\bullet) = \prod_{f \in \mathrm{Irr}(R)} f^{\mathrm{ord}_f(\det(F_\bullet))}.$$

where $\mathrm{Irr}(R)$ denotes the set of irreducible elements in R.

In order to understand the determinant of $F_\bullet$ it is enough to understand the numbers $\mathrm{ord}_f(\det(F_\bullet))$

Let us fix the irreducible element f. We recall that the ideal (f) is prime and that the localization $R_{(f)}$ is a discrete valuation ring. For each finitely generated R-module M we define its f-multiplicity $\mathrm{mult}_f(M)$ as the length of the localization $M \otimes R_{(f)}$. The multiplicity of a finitely generated R-module

M is finite if and only if the localization $M_{(f)}$ is annihilated by some power of f.

(1.3.4) Theorem. *Let us assume that R is a unique factorization domain. Let $F_\bullet$ be a complex of free R-modules that is generically exact. Then the order $\mathrm{ord}_f(\det(F_\bullet))$ of the irreducible element f in the determinant of $F_\bullet$ is given by the formula*

$$\mathrm{ord}_f(\det(F_\bullet)) = \sum_i (-1)^i \mathrm{mult}_f(H_i(F_\bullet)).$$

Proof. By localizing we may assume that the ring R is a discrete valuation ring, so we need just to calculate the determinant of a free complex over such ring. However, each homology module $H_i(F_\bullet)$ is torsion, because $F_\bullet$ is generically exact. Therefore each $H_i(F_\bullet)$ is a direct sum of cyclic modules $R/(f^j)$. Let $G^{(i)} : 0 \to G_1^{(i)} \to G_0^{(i)}$ be a free resolution of $H_i(F_\bullet)$ (which is a direct sum of complexes $0 \to R \xrightarrow{f^j} R$).

We use the decreasing induction on j, starting with $j = n+1$ to define complexes $F_\bullet^{(j)}$ such that

$$H_i(F_\bullet^{(j)}) = \begin{cases} H_i(F_\bullet) & \text{if } i < j, \\ 0 & \text{otherwise.} \end{cases}$$

The complex $F_\bullet^{(n+1)} = F_\bullet$. Suppose we constructed the complex $F_\bullet^{(j+1)}$. Its top nonzero homology is $H_j(F_\bullet^{(j+1)}) = H_j(F_\bullet)$. Therefore we can construct a map $\psi_j : G^{(j)}[j]_\bullet \to F_\bullet^{(j+1)}$ lifting the identity map on homology in degree j. We define $F_\bullet^{(j)}$ to be the cone of the map ψ_j. The complex $F_\bullet^{(m-1)}$ will be exact, and thus its determinant will be a unit in R. Using multiplicativity of the determinant with respect to short exact sequences and using exact sequences associated to the cone construction, we see that it is enough to prove the statement of the theorem for the complexes $G_\bullet^{(j)}$. This means it is enough to check the statement for each summand $0 \to R \xrightarrow{f^j} R$. Here both sides of the equality give j, so Theorem (1.3.4) is proved. ∎

(1.3.5) Corollary. *Assume that $m = 0$ and the complex $F_\bullet$ is acyclic in codimension 2, i.e., all modules $H_i(F_\bullet)$ are supported in codimension 2. Then the determinant of $F_\bullet$ is the greatest common divisor of the maximal minors of the map d_1.*

Proof. Using the statement (1.3.4), we see that the only factors that can occur in $\det(F_\bullet)$ are the equations of codimension 1 components of the support of

$H_0(F_\bullet)$. This means after localizing at (f) the complex will be acyclic. This again reduces the statement to the case of complexes of length 1 resolving a torsion module, in which case we can check it directly. ∎

(1.3.6) Remark. *The last statement is closely related to the so-called first structure theorem of Buchsbaum and Eisenbud [BE3]. It states that if* rank $(d_i) = r_i$, *then there exists a sequence of maps* $a_i : R \to \bigwedge^{r_i} F_{i-1}$ *such that* $\bigwedge^{r_i} d_i = a_i a_{i+1}^*$. *In fact one knows (Theorem 3, Chapter 7.) that a complex satisfying the assumptions of (1.3.5) always satisfies this theorem. One knows also that for $i > 1$ the ideal generated by entries of a_i has depth ≥ 2. This means that the determinant of $F_\bullet$ is the entry of a_1. But the ideal of maximal minors of d_1 equals $a_1 I(a_2)$, where $I(a_2)$ is the ideal of entries of a_2, which has depth ≥ 2. Thus in this case a_1 is the greatest common divisor of maximal minors of d_1.*

2

Schur Functors and Schur Complexes

In this chapter we develop the representation theory of general linear groups. We follow the approach from [ABW2] based on the explicit characteristic free definition of Schur and Weyl functors. This approach is sufficient for our goals, and it seems to be easier to grasp for the reader not familiar with representation theory. In section 2.1 we define Schur and Weyl functors and prove the standard basis theorems. In section 2.2 we discuss the connection of Schur functors with highest weight theory, and provide the alternate definition using Young symmetrizers in characteristic 0. In section 2.3 we derive various formulas from representation theory, including the Littlewood–Richardson rule and Cauchy formulas. Finally in section 2.4 we give the definition and basic properties of Schur complexes.

2.1. Schur Functors and Weyl Functors

Let us fix a free module E of dimension n over a commutative ring $\mathbf{K}$. Let $\lambda = (\lambda_1, \ldots, \lambda_s)$ be a partition of a number m. We consider the module

$$L_\lambda E = \bigwedge^{\lambda_1} E \otimes \bigwedge^{\lambda_2} E \otimes \ldots \otimes \bigwedge^{\lambda_s} E / R(\lambda, E),$$

where the submodule $R(\lambda, E)$ is the sum of submodules:

$$\bigwedge^{\lambda_1} E \otimes \ldots \otimes \bigwedge^{\lambda_{a-1}} E \otimes R_{a,a+1}(E) \otimes \bigwedge^{\lambda_{a+2}} E \otimes \ldots \otimes \bigwedge^{\lambda_s} E$$

for $1 \leq a \leq s - 1$, where $R_{a,a+1}(E)$ is the submodule spanned by the images of the following maps $\theta(\lambda, a, u, v; E)$ with $u + v < \lambda_{a+1}$:

$$\bigwedge^u E \otimes \bigwedge^{\lambda_a - u + \lambda_{a+1} - v} E \otimes \bigwedge^v E$$
$$\downarrow 1 \otimes \Delta \otimes 1$$
$$\bigwedge^u E \otimes \bigwedge^{\lambda_a - u} E \otimes \bigwedge^{\lambda_{a+1} - v} E \otimes \bigwedge^v E$$
$$\downarrow m_{12} \otimes m_{34}$$
$$\bigwedge^{\lambda_a} E \otimes \bigwedge^{\lambda_{a+1}} E.$$

Let $e_1, \ldots, e_n$ will be a fixed ordered basis of E. We introduce the ordered set $[1, n] = \{1, 2, \ldots, n\}$, which is the set indexing our basis.

We refer to section 1.1.2. for notions related to tableaux used in this section.

Let T be a tableau of shape λ with the entries in $[1, n]$. We associate to T the element in $L_\lambda E$ which is a coset of the tensor

$$e_{T(1,1)} \wedge \ldots \wedge e_{T(1,\lambda_1)} \otimes e_{T(2,1)} \wedge \ldots \wedge e_{T(2,\lambda_2)}$$
$$\otimes \ldots \otimes e_{T(s,1)} \wedge \ldots \wedge e_{T(s,\lambda_s)}$$

in $L_\lambda E$. In the sequel we will identify these two objects and we will call both of them the tableaux of shape λ corresponding to the basis $\{e_1, e_2, \ldots, e_n\}$.

(2.1.1) Remark. *It is convenient to think about the relations $R(\lambda, E)$ in graphical terms using the Young frames. Let us define the Young scheme of shape λ to be the Young frame of shape λ with some boxes empty and some filled. We associate to the map $\theta(\lambda, a, u, v; E)$ its Young scheme which is empty in all rows except the a-th and $(a + 1)$-st, and its restriction to these rows is*

with u empty boxes, followed by $\lambda_a - u$ filled boxes in the a-th row, and $\lambda_{a+1} - v$ filled boxes, followed by v empty boxes in the $(a + 1)$-st row. Notice that the condition $u + v < \lambda_{a+1}$ assures that there will be at least one column in this frame with two boxes filled. The image of typical element $U_1 \otimes \ldots \otimes U_{a-1} \otimes V_1 \otimes V_2 \otimes V_3 \otimes U_{a+2} \otimes \ldots \otimes U_s$, where $U_j = e_{T(j,1)} \wedge \ldots \wedge e_{T(j,\lambda_j)}$, $V_1 = e_{x_1} \wedge \ldots \wedge e_{x_u}$, $V_2 = e_{y_1} \wedge \ldots \wedge e_{y_{\lambda_a + \lambda_{a+1} - u - v}}$, $V_3 = e_{z_1} \wedge \ldots \wedge e_{z_v}$, is a sum of tableaux, where we put in each tableau element $T(j, t)$ in the t-th box in the j-th row for $j \neq a, a + 1$. In the a-th and $(a + 1)$st row we put $x_1, \ldots, x_u$ in the empty u boxes in the a-th row, put $z_1, \ldots, z_v$ in the empty v boxes in the $(a + 1)$st row, and shuffle the elements $y_1, \ldots, y_{\lambda_a + \lambda_{a+1} - u - v}$ between the filled boxes in the a-th and $(a + 1)$st rows, with the appropriate signs coming from exterior diagonal.

(2.1.2) Example. *Take $\lambda = (3, 3)$, $u = v = 1$. The corresponding Young scheme is*

Take $x_1 = 1$, $z_1 = 6$, $\{y_1, y_2, y_3, y_4\} = \{2, 3, 4, 5\}$. The image of the corresponding vector by $\theta(\lambda, 1, u, v; E)$ is

$$
\begin{array}{|c|c|c|}\hline 1&2&3\\\hline 4&5&6\\\hline\end{array}
-
\begin{array}{|c|c|c|}\hline 1&2&4\\\hline 3&5&6\\\hline\end{array}
+
\begin{array}{|c|c|c|}\hline 1&2&5\\\hline 3&4&6\\\hline\end{array}
+
\begin{array}{|c|c|c|}\hline 1&3&4\\\hline 2&5&6\\\hline\end{array}
-
\begin{array}{|c|c|c|}\hline 1&3&5\\\hline 2&4&6\\\hline\end{array}
+
\begin{array}{|c|c|c|}\hline 1&4&5\\\hline 2&3&6\\\hline\end{array}.
$$

(2.1.3) Example.

   (a) *If $\lambda = (t)$ then $L_\lambda E = \bigwedge^t E$. Indeed, by definition $L_\lambda E = \bigwedge^t E / R((t), E)$, but $R((t), E) = 0$, since the partition (t) has only one part.*

   (b) *If $\lambda = (1^t)$ then $L_\lambda E = S_t E$. Indeed, the relations $R_{a,a+1}(E)$ express the symmetry between the a-th and $(a+1)$st row.*

   (c) *Let $\lambda = (2, 1)$. In that case there is only one choice of u, v, namely $u = v = 0$. The corresponding Young scheme is*

The image of $\theta(1, 0, 0; E)$ on the typical element $e_{y_1} \wedge e_{y_2} \wedge e_{y_3}$ is

$$
\begin{array}{|c|c|}\hline y_1&y_2\\\hline y_3\\\cline{1-1}\end{array}
-
\begin{array}{|c|c|}\hline y_1&y_3\\\hline y_2\\\cline{1-1}\end{array}
+
\begin{array}{|c|c|}\hline y_2&y_3\\\hline y_1\\\cline{1-1}\end{array}.
$$

We conclude that $L_{2,1} E$ is the cokernel of the diagonal map

$$
\Delta : \bigwedge^3 E \to \bigwedge^2 E \otimes E.
$$

   (d) *Let $\lambda = (2, 2)$. There are three types of relations $\theta(1, u, v)$. The choices of u, v are $(u, v) = (0, 0), (1, 0), (0, 1)$. The corresponding Young schemes are*

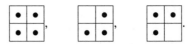

It is easy to see that the relations coming from $\theta(1, 1, 0; E)$ are the consequences of two other types of relations. Therefore we have two types of relations. For the map $\theta(1, 0, 0; E)$ the image of the typical element $e_{y_1} \wedge e_{y_2} \wedge e_{y_3} \wedge e_{y_4}$ is

$$
\begin{array}{|c|c|}\hline y_1&y_2\\\hline y_3&y_4\\\hline\end{array}
-
\begin{array}{|c|c|}\hline y_1&y_3\\\hline y_2&y_4\\\hline\end{array}
+
\begin{array}{|c|c|}\hline y_1&y_4\\\hline y_2&y_3\\\hline\end{array}
+
\begin{array}{|c|c|}\hline y_2&y_3\\\hline y_1&y_4\\\hline\end{array}
-
\begin{array}{|c|c|}\hline y_2&y_4\\\hline y_1&y_3\\\hline\end{array}
+
\begin{array}{|c|c|}\hline y_3&y_4\\\hline y_1&y_2\\\hline\end{array}.
$$

For the relation $\theta(1, 0, 1; E)$ the image of the typical element $e_{y_1} \wedge$
$e_{y_2} \wedge e_{y_3} \otimes e_z$ is

$$\begin{array}{|c|c|} \hline y_1 & y_2 \\ \hline y_3 & z \\ \hline \end{array} - \begin{array}{|c|c|} \hline y_1 & y_3 \\ \hline y_2 & z \\ \hline \end{array} + \begin{array}{|c|c|} \hline y_2 & y_3 \\ \hline y_1 & z \\ \hline \end{array}.$$

We conclude that $L_{2,2}E$ is the factor of $\bigwedge^2 E \otimes \bigwedge^2 E$ by the images
of two maps $\bigwedge^3 E \otimes E \to \bigwedge^2 E \otimes \bigwedge^2 E$ (corresponding to $u = 0$,
$v = 1$) and the diagonal $\bigwedge^4 E \to \bigwedge^2 E \otimes \bigwedge^2 E$ (corresponding to
$u = v = 0$).

(e) *Let $\lambda = (3, 2)$. There are three choices of pairs u, v: $(u, v) = (0, 0)$,*
$(1, 0)$, $(0, 1)$. The Young schemes are

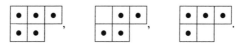

It follows that $L_{3,2}E$ is a factor of the module $\bigwedge^3 E \otimes \bigwedge^2 E$ by the
images of three maps: $\bigwedge^4 E \otimes E \to \bigwedge^3 E \otimes \bigwedge^2 E$ (corresponding to
$u = 0, v = 1$), $E \otimes \bigwedge^4 E \to \bigwedge^3 E \otimes \bigwedge^2 E$ (corresponding to $u=1$,
$v = 0$), and $\bigwedge^5 E \to \bigwedge^3 E \otimes \bigwedge^2 E$ (corresponding to $u = v = 0$).

(f) *We will show below that for two rowed partitions there are two ways of*
choosing smaller set of relations $\theta(1, u, v; E)$ that still suffice to define
the Schur functor. One choice is to take all pairs (u, v) with $u = 0$. The
other choice is to take all pairs (u, v) with one overlap, i.e. all pairs
(u, v) for which the Young scheme has exactly one column with two
filled boxes (equivalently, $u + v = \lambda_2 - 1$).

(g) *Let $\lambda = (2, 2, 1)$. We have two types of relations corresponding to the*
first pair of rows (described in example (d)), and one type correspond-
ing to the second and third rows (described in example (c)). Choosing
the relations with one overlap the Young schemes are

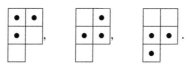

(h) *Let λ be a hook i.e. a partition of the form $\lambda = (p, 1^{q-1})$. Graphically,*

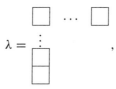

with p boxes in the first row and q boxes in the first column. The relations between two rows of length 1 express the symmetry (cf. example (b)). There is only one type of relations corresponding to the first two rows, for the pair $u = v = 0$. It follows that the Schur functor $L_{(p,1^{q-1})}E$ is the cokernel of the composition map

$$\bigwedge^{p+1} E \otimes S_{q-2}E \xrightarrow{\Delta \otimes 1} \bigwedge^{p} E \otimes E \otimes S_{q-2}E \xrightarrow{1 \otimes m} \bigwedge^{p} E \otimes S_{q-1}E.$$

We recall from section 1.1.2 that a tableau T is *standard* if the numbers in each row of T form an increasing sequence and the numbers in each column of T form a nondecreasing sequence. This notion plays a key role in representation theory thanks to the following

(2.1.4) Proposition. *Let E be a free $\mathbf{K}$-module of dimension n. Let $e_1, \ldots, e_n$ be a basis of E. The set $\mathrm{ST}(\lambda, [1, n])$ of standard tableaux of shape λ with entries from $[1, n]$ form a basis of $L_\lambda E$. In particular, $L_\lambda E$ is also a free module.*

Proof of Proposition (2.1.4). First we prove that the standard tableaux generate $L_\lambda E$. It is clear that the set $\mathrm{RST}(\lambda, [1, n])$ of row standard tableaux with entries from $[1, n]$ generate $L_\lambda E$. Let us order the set of such tableaux by the order $\preceq$ defined in section 1.1.2. We will prove that if the tableau T is not standard then we can express it modulo $R(\lambda, E)$ as a combination of earlier tableaux. Let us assume first that T has two rows. Since T is not standard, we can find w for which $T(1, w) > T(2, w)$. We consider the map $\theta(\lambda, 1, u, v; E)$ for $u = w - 1$ and $v = \lambda_2 - w$. The key observation is that image of the tensor $V_1 \otimes V_2 \otimes V_3$, where

$$V_1 = e_{T(1,1)} \wedge e_{T(1,2)} \wedge \ldots \wedge e_{T(1,w-1)},$$
$$V_2 = e_{T(1,w)} \wedge \ldots \wedge e_{T(1,\lambda_1)} \wedge e_{T(2,1)} \wedge \ldots \wedge e_{T(2,w)},$$
$$V_3 = e_{T(2,w+1)} \wedge \ldots \wedge e_{T(2,\lambda_2)},$$

contains the tableau T with the coefficient 1, and all the other tableaux occurring in this image are earlier than T in the order $\preceq$. Indeed, in all summands other than T we shuffle the smaller numbers from the second row to the first one, replacing bigger numbers. Therefore T can be expressed modulo $R(\lambda, E)$ as a combination of earlier tableaux.

Let us consider the general case. If T is not standard, then we can find such a and w that $T(a, w) > T(a + 1, w)$. Now we apply the previous argument to the tableau S which consists of the a-th and $(a+1)$st rows of T. Notice that the relations $R(\lambda, E)$ we are using do not do anything to the other rows of T, so we can express T as a sum of earlier tableaux in the order $\preceq$.

It remains to prove that the standard tableaux are linearly independent in $L_\lambda E$. Consider a map

$$\phi_\lambda : \overset{\lambda_1}{\bigwedge} E \otimes \overset{\lambda_2}{\bigwedge} E \otimes \ldots \otimes \overset{\lambda_s}{\bigwedge} E \overset{\alpha}{\longrightarrow} \otimes_{(i,j)\in\lambda} E(i, j)$$
$$\overset{\beta}{\longrightarrow} S_{\lambda_1'} E \otimes S_{\lambda_2'} E \otimes \ldots \otimes S_{\lambda_t'} E,$$

where α is the tensor product of exterior diagonals

$$\Delta : \overset{\lambda_j}{\bigwedge} E \to E(j, 1) \otimes E(j, 2) \otimes \ldots \otimes E(j, \lambda_j)$$

and β is the tensor product of multiplications

$$m : E(1, i) \otimes E(2, i) \otimes \ldots \otimes E(\lambda_i', i) \to S_{\lambda_i'} E.$$

If we imagine the copies of E correspond to the boxes of λ with $\bigwedge^{\lambda_j} E$ corresponding to the boxes in the first row and $S_{\lambda_i'} E$ corresponding to boxes in the i-th column of λ, we can think of the image $\phi_\lambda(T)$ of the tableau T as first shuffling (with signs) the terms of T in each row, and then multiplying the terms in each column of the tableaux we obtain. The map ϕ_λ is called the *Schur map associated to the partition λ*.

(2.1.5) Example.
(a) Let $\lambda = (2, 2)$. Then the map

$$\phi_\lambda : \overset{2}{\bigwedge} E \otimes \overset{2}{\bigwedge} E \longrightarrow S_2 E \otimes S_2 E$$

is given by the formula

$$\phi_\lambda(x \wedge y \otimes u \wedge v) = xu \otimes yv - yu \otimes xv - xv \otimes yu + yv \otimes xu.$$

(b) It is useful to think about the map ϕ_λ in graphical terms as follows. We consider the case $\lambda = (3, 2)$, but the general case will become clear. The tensor product $\bigwedge^3 E \otimes \bigwedge^2 E$ has a basis corresponding to the set $\mathrm{RST}(\lambda, [1, n])$. It can be thought of as a set of standard tableaux of shapes (3) and (2), corresponding to rows of λ. The tensor product $S_2 E \otimes S_2 E \otimes E$ has a basis consisting of triples of costandard

tableaux of shapes (2), (2), (1), corresponding to columns of λ*. The map* ϕ_λ *acts according to the scheme*

and the image of a tableau T is the sum (with signs) of tableaux obtained from T by shuffling each of its rows.

The role of the map ϕ_λ is explained in the next statement.

(2.1.6) Proposition. *The image* $\phi_\lambda(R(\lambda, E))$ *equals 0.*

Proof. We want to show that the spaces

$$\overset{\lambda_1}{\bigwedge} E \otimes \ldots \otimes \overset{\lambda_{a-1}}{\bigwedge} E \otimes R_{a,a+1}(E) \otimes \overset{\lambda_{a+2}}{\bigwedge} E \otimes \ldots \otimes \overset{\lambda_s}{\bigwedge} E$$

are in the kernel of ϕ_λ. Since $R_{a,a+1}$ is the span of images of the maps of type $\theta(\lambda, a, u, v; E)$, we choose one such map (i.e., we choose a and u, v such that $u + v < \lambda_{a+1}$).

Let us consider the element

$$U = U_1 \otimes U_2 \otimes \ldots \otimes U_{a-1} \otimes U' \otimes U_{a+2} \otimes \ldots \otimes U_s,$$

where $U_i = x_{i,1} \wedge x_{i,2} \wedge \ldots \wedge x_{i,\lambda_i} \in \bigwedge^{\lambda_i} E$ and U' is the image under $\theta(\lambda, a, u, v; E)$ of the element $x_1 \wedge \ldots \wedge x_u \otimes y_1 \wedge \ldots \wedge y_{\lambda_a + \lambda_{a+1} - u - v} \otimes z_1 \wedge \ldots \wedge z_v$. Let us consider the tensor $\phi_\lambda U$. It is formally a sum of tensors in $S_{\lambda'_1} E \otimes S_{\lambda'_2} E \otimes \ldots \otimes S_{\lambda'_t} E$ each of which is the tensor product of products of $x_{i,j}$'s, x_j's, y_m's, and z_p's shuffled in some way. Let us write our image formally as such a sum by writing products in each $S_{\lambda'_j} E$ in the order they get multiplied by β. Let us consider a summand T in our sum. Since $u + v < \lambda_{a+1}$, two of the y_m's have to occur in the same symmetric power $S_{\lambda'_j} E$. Let us choose the smallest such j. Let y_b and y_c occur in $S_{\lambda'_j} E$. Let us consider another summand T' in our sum $\phi_\lambda U$ which differs from T by changing places of y_b and y_c when applying the map $1 \otimes \Delta \otimes 1$ from $\theta(\lambda, a, u, v; E)$. We can easily check that the correspondence $T \mapsto T'$ defines an involution on the summands in ϕ_λ. Moreover, each pair of such summands cancels out in

$S_{\lambda_1'} E \otimes S_{\lambda_2'} E \otimes \ldots \otimes S_{\lambda_t'} E$, because they come with different signs and the product $y_b y_c$ is symmetric. This means $\phi_\lambda U = 0$. ∎

(2.1.7) Example. *Let us take* $\lambda = (3, 3)$, $u = v = 1$. *Consider* $U = x \otimes y_1 \wedge y_2 \wedge y_3 \wedge y_4 \otimes z$. *Then if* $T = xy_1 \otimes y_3 y_4 \otimes y_2 z$ *then* $T' = xy_1 \otimes y_4 y_3 \otimes y_2 z$. *Now* T *occurs as a summand in* $\phi_\lambda(x \wedge y_3 \wedge y_2 \otimes y_1 \wedge y_4 \wedge z)$, *and* T' *occurs as a summand in* $\phi_\lambda(x \wedge y_4 \wedge y_2 \otimes y_1 \wedge y_3 \wedge z)$. *One checks easily that* T *and* T' *cancel out.*

Proposition (2.1.6) means that ϕ_λ induces a surjective map from $L_\lambda E$ to Im ϕ_λ. Next we will show that the map ϕ_λ maps standard tableaux to linearly independent elements of $S_{\lambda_1'} E \otimes S_{\lambda_2'} E \otimes \ldots \otimes S_{\lambda_t'} E$. This will prove (2.1.4), and at the same time it will prove that $L_\lambda E = \text{Im } \phi_\lambda$.

In order to see that the images of standard tableaux are linearly independent, we notice that the module $S_{\lambda_1'} E \otimes S_{\lambda_2'} E \otimes \ldots \otimes S_{\lambda_t'} E$ has a basis corresponding naturally to the set $\text{RST}(\lambda', [1, n]')$ of row costandard tableaux of shape λ'. We order this set by the order $\preceq$ defined in section 1.1.2.

If T is a standard tableau of shape λ, then the smallest element (with respect to the order $\preceq$) occurring in $\phi_\lambda(T)$ is

$$e_{T(1,1)} \cdots e_{T(\lambda_1',1)} \otimes e_{T(1,2)} \cdots e_{T(\lambda_2',2)} \otimes \ldots \otimes e_{T(1,t)} \cdots e_{T(\lambda_t',t)}.$$

Indeed, if in applying the map α we make an exchange of elements in some row, we put bigger elements to the earlier columns, so we get the later (with respect to $\preceq$) elements. Moreover, one sees instantly that

$$e_{T(1,1)} \cdots e_{T(\lambda_1',1)} \otimes e_{T(1,2)} \cdots e_{T(\lambda_2',2)} \otimes \ldots \otimes e_{T(1,t)} \cdots e_{T(\lambda_t',t)}$$

occurs in $\alpha(T)$ with coefficient 1. It is also obvious that the initial elements

$$e_{T(1,1)} \cdots e_{T(\lambda_1',1)} \otimes e_{T(1,2)} \cdots e_{T(\lambda_2',2)} \otimes \ldots \otimes e_{T(1,t)} \cdots e_{T(\lambda_t',t)}$$

are different for different standard tableaux T. This proves that the images $\phi_\lambda T$ of standard tableaux T are linearly independent.●

Since the exterior and symmetric powers are $\text{GL}(E)$-modules, and the diagonal and multiplication maps are $\text{GL}(E)$-equivariant, it is clear that the group $\text{GL}(E)$ acts on $L_\lambda E$ in a natural way. The space $L_\lambda E$ becomes a $\text{GL}(E)$-module, which is called *the Schur module corresponding to the partition* λ.

The notion of a Schur module can be generalized to skew partitions. Let λ/μ be a skew partition. We define the Schur map

$$\phi_{\lambda/\mu} : \overset{\lambda_1-\mu_1}{\bigwedge} E \otimes \overset{\lambda_2-\mu_2}{\bigwedge} E \otimes \ldots \otimes \overset{\lambda_s-\mu_s}{\bigwedge} E$$
$$\to S_{\lambda_1'-\mu_1'} E \otimes \ldots \otimes S_{\lambda_t'-\mu_t'} E$$

as a composition

$$\phi_{\lambda/\mu} : \overset{\lambda_1-\mu_1}{\bigwedge} E \otimes \overset{\lambda_2-\mu_2}{\bigwedge} E \otimes \ldots \otimes \overset{\lambda_s-\mu_s}{\bigwedge} E \overset{\alpha}{\longrightarrow} \otimes_{(i,j)\in\lambda/\mu} E(i,j)$$

$$\overset{\beta}{\longrightarrow} S_{\lambda_1'-\mu_1'} E \otimes S_{\lambda_2'-\mu_2'} E \otimes \ldots \otimes S_{\lambda_t'-\mu_t'} E,$$

where α is the tensor product of exterior diagonals

$$\Delta : \overset{\lambda_j-\mu_j}{\bigwedge} E \to E(j, \mu_j+1) \otimes E(j, \mu_j+2) \otimes \ldots \otimes E(j, \lambda_j)$$

and β is the tensor product of multiplications

$$m : E(\mu_i'+1, i) \otimes E(\mu_i'+2, i) \otimes \ldots \otimes E(\lambda_i', i) \to S_{\lambda_i'} E.$$

If we imagine the copies of E correspond to the boxes in a skew Young frame λ/μ, with $\bigwedge^{\lambda_j-\mu_j} E$ corresponding to the boxes in the j-th row, and $S_{\lambda_i'-\mu_i'} E$ corresponding to boxes in the i-th column, we can think of the image $\phi_{\lambda/\mu}(T)$ of a tableau T as first shuffling (with signs) the terms of T in each row, and then multiplying the terms of each column of each summand.

(2.1.8) Example. *Let $\lambda = (3, 2)$, $\mu = (1)$. The tensor product $\bigwedge^2 E \otimes \bigwedge^2 E$ has a basis corresponding to the set $\mathrm{RST}(\lambda/\mu, [1, n])$. It can be thought of as a pair of standard tableaux of shapes (2) and (2) (corresponding to the rows of $(3, 2)/(1)$). The tensor product $E \otimes S_2 E \otimes E$ has a basis consisting of triples of costandard tableaux of shapes $(1), (2), (1)$ (corresponding to columns of $(3, 2)/(1)$). The map $\phi_{(3,2)/(1)}$ acts according to the scheme*

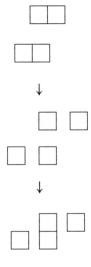

We define the skew Schur module $L_{\lambda/\mu}E$ to be the image of $\phi_{\lambda/\mu}$. The description of the relations and of the standard basis of skew Schur modules is the same as for the Schur modules.

(2.1.9) Proposition.

(a) $L_{\lambda/\mu}E = \bigwedge^{\lambda_1-\mu_1} E \otimes \bigwedge^{\lambda_2-\mu_2} E \otimes \ldots \otimes \bigwedge^{\lambda_s-\mu_s} E / R(\lambda/\mu, E)$
where $R(\lambda/\mu, E)$ is spanned by the subspaces:

$$\overset{\lambda_1-\mu_1}{\bigwedge} E \otimes \ldots \otimes \overset{\lambda_{a-1}-\mu_{a-1}}{\bigwedge} E \otimes R_{a,a+1}(E) \otimes \overset{\lambda_{a+2}-\mu_{a+2}}{\bigwedge} E$$
$$\otimes \ldots \otimes \overset{\lambda_s-\mu_s}{\bigwedge} E$$

for $1 \le a \le s - 1$, where $R_{a,a+1}(E)$ is the vector space spanned by the images of the following maps $\theta(\lambda/\mu, a, u, v; E)$ with $u + v < \lambda_{a+1} - \mu_a$:

$$\bigwedge^u E \otimes \bigwedge^{\lambda_a-\mu_a-u+\lambda_{a+1}-\mu_{a+1}-v} E \otimes \bigwedge^v E$$
$$\downarrow {\scriptstyle 1\otimes\Delta\otimes1}$$
$$\bigwedge^u E \otimes \bigwedge^{\lambda_a-\mu_a-u} E \otimes \bigwedge^{\lambda_{a+1}-\mu_{a+1}-v} E \otimes \bigwedge^v E$$
$$\downarrow {\scriptstyle m_{12}\otimes m_{34}}$$
$$\bigwedge^{\lambda_a-\mu_a} E \otimes \bigwedge^{\lambda_{a+1}-\mu_{a+1}} E.$$

(b) The standard tableaux of shape λ/μ form a basis of $L_{\lambda/\mu}$.

Proof. First of all one can check by direct calculation that the map $\phi_{\lambda/\mu}$ sends the elements from $R(\lambda/\mu, E)$ to zero. This is done in the same way as in the case $\mu = \emptyset$, so we leave it to the reader. Then it is enough to check two things:

(1) The standard tableaux of shape λ/μ generate the factor

$$\overset{\lambda_1-\mu_1}{\bigwedge} E \otimes \overset{\lambda_2-\mu_2}{\bigwedge} E \otimes \ldots \otimes \overset{\lambda_s-\mu_s}{\bigwedge} E / R(\lambda/\mu, E).$$

(2) The images $\phi_{\lambda/\mu}(T)$ for the standard tableaux T of shape λ/μ are linearly independent.

Facts (1) and (2) allow us to identify the image of $\phi_{\lambda/\mu}$ with the factor

$$\overset{\lambda_1-\mu_1}{\bigwedge} E \otimes \overset{\lambda_2-\mu_2}{\bigwedge} E \otimes \ldots \otimes \overset{\lambda_s-\mu_s}{\bigwedge} E / R(\lambda/\mu, E).$$

The proof of (1) and (2) is exactly the same as for Schur functors, so we leave it to the reader as an exercise. ∎

We can associate to each of the relations $\theta(\lambda/\mu, a, u, v; E)$ its Young scheme, as we did in the case $\mu = \emptyset$. The only difference is that there are missing boxes. Again the condition $u + v < \lambda_{a+1} - \mu_a$ assures at least one overlap in the Young scheme of the relation.

(2.1.10) Example. *Let us take $\lambda = (4, 3)$, $\mu = (1, 0)$. The Young scheme of the relation $\theta((4, 3)/(1, 0), 1, 1, 0; E)$ is*

(2.1.11) Remark. *We can interpret the relations $R(\lambda/\mu, E)$ in terms of tableaux as we did in the case $\mu = \emptyset$. Again we start with the case of two rows. The image of typical element $V_1 \otimes V_2 \otimes V_3$ with $V_1 = e_{x_1} \wedge \ldots \wedge e_{x_u}$, $V_2 = e_{y_1} \wedge \ldots \wedge e_{y_{\lambda_a - \mu_a + \lambda_{a+1} - \mu_{a+1} - u - v}}$, $V_3 = e_{z_1} \wedge \ldots \wedge e_{z_v}$ is a sum of tableaux, where we put in each tableau $x_1, \ldots, x_u$ in the u empty boxes in the first row, put $z_1, \ldots, z_v$ in the v empty boxes in the second row, and shuffle the elements $y_1, \ldots, y_{\lambda_a + \lambda_{a+1} - u - v}$ between the filled boxes in the first and second rows, with the appropriate signs coming from exterior diagonal. If the number of parts of λ is bigger than 2, the relations $R(\lambda/\mu, E)$ can be interpreted in terms of tableaux as follows. Fix a, u, v. The Young scheme of the map $\theta(\lambda/\mu, a, u, v; E)$ has all the boxes empty except the a-th and $(a + 1)$st, where the scheme is the same as in the case of two rows. The image of the typical element $U_1 \otimes \ldots \otimes U_{a-1} \otimes V_1 \otimes V_2 \otimes V_3 \otimes U_{a+2} \otimes \ldots \otimes U_s$, where $U_j = e_{i(j,1)} \wedge \ldots \wedge e_{i(j,\lambda_j - \mu_j)}$, $V_1 = e_{x_1} \wedge \ldots \wedge e_{x_u}$, $V_2 = e_{y_1} \wedge \ldots \wedge e_{y_{\lambda_a - \mu_a + \lambda_{a+1} - \mu_{a+1} - u - v}}$, and $V_3 = e_{z_1} \wedge \ldots \wedge e_{z_v}$, is the same as in the case of two rows, except that in each summand we put element $i(j, t)$ in the t-th box in the j-th row for $j \neq a, a + 1$.*

(2.1.12) Example. *Let $\lambda = (3, 2)$, $\mu = (1, 0)$, $u = v = 0$. The corresponding Young scheme is*

Take $\{y_1, y_2, y_3, y_4\} = \{2, 3, 4, 5\}$. The image of the corresponding vector $e_2 \wedge e_3 \wedge e_4 \wedge e_5$ by $\theta((3, 2)/(1, 0), 1, 0, 0; E)$ is

$$
\begin{array}{cc} 2 & 3 \\ \hline 4 & 5 \end{array}
-
\begin{array}{cc} 2 & 4 \\ \hline 3 & 5 \end{array}
+
\begin{array}{cc} 2 & 5 \\ \hline 3 & 4 \end{array}
+
\begin{array}{cc} 3 & 4 \\ \hline 2 & 5 \end{array}
-
\begin{array}{cc} 3 & 5 \\ \hline 2 & 4 \end{array}
+
\begin{array}{cc} 4 & 5 \\ \hline 2 & 3 \end{array}.
$$

(2.1.13) Example.

(a) *Consider the skew shape* $\lambda = (3, 2)$, $\mu = (1, 0)$. *The only possible relation in this case is* $\theta((3, 2)/(1, 0), 1, 0, 0; E)$. *Its Young scheme is*

The Schur module $L_{(3,2)/(1,0)}E$ *is a factor of* $\bigwedge^2 E \otimes \bigwedge^2 E$ *by the image of* $\bigwedge^4 E$ *embedded by the exterior diagonal.*

(b) *Take* $\lambda = (4, 2)$, $\mu = (2, 0)$. *In this case there are no possible relations, and* $L_{(4,2)/(2,0)}E$ *is just the tensor product* $\bigwedge^2 E \otimes \bigwedge^2 E$.

(c) *The previous example generalizes as follows. Assume that the skew Young frame* λ/μ *is disconnected, i.e., it can be written as a union of two skew Young frames* $\lambda(1)/\mu(1)$ *and* $\lambda(2)/\mu(2)$ *with no boxes in the same row or column. Then* $L_{\lambda/\mu}E = L_{\lambda(1)/\mu(1)}E \otimes L_{\lambda(2)/\mu(2)}E$.

Proof of (c). The relations of type $\theta(\lambda/\mu, a, u, v; E)$ between two rows of $\lambda(i)/\mu(i)$ $(i = 1, 2)$ are the same as the relations between corresponding rows of λ/μ. Since λ/μ is disconnected, there are no relations between rows of $\lambda(1)/\mu(1)$ and of $\lambda(2)/\mu(2)$. ∎

We conclude this section with the definition of the skew Weyl modules $K_{\lambda/\mu}$. They are the duals of the skew Schur modules.

To define the modules $K_{\lambda/\mu}E$ we take the definition of $L_{\lambda/\mu}E$, but instead of exterior powers we use divided powers, and instead of symmetric powers we use exterior powers. Thus we define the *Weyl map*

$$\psi_{\lambda/\mu} : D_{\lambda_1-\mu_1} E \otimes D_{\lambda_2-\mu_2} E \otimes \ldots \otimes D_{\lambda_s-\mu_s} E$$
$$\to \overset{\lambda'_1-\mu'_1}{\bigwedge} E \otimes \ldots \otimes \overset{\lambda'_t-\mu'_t}{\bigwedge} E$$

as a composition

$$\psi_{\lambda/\mu} : D_{\lambda_1-\mu_1} E \otimes \ldots \otimes D_{\lambda_s-\mu_s} E \overset{\alpha}{\longrightarrow} \bigotimes_{(i,j)\in\lambda-\mu} E(i, j)$$
$$\overset{\beta}{\longrightarrow} \overset{\lambda'_1-\mu'_1}{\bigwedge} E \otimes \ldots \otimes \overset{\lambda'_t-\mu'_t}{\bigwedge} E,$$

where α is the tensor product of divided diagonals

$$\Delta : D_{\lambda_j-\mu_j} E \to E(j, \mu_j + 1) \otimes E(j, \mu_j + 2) \otimes \ldots \otimes E(j, \lambda_j),$$

and β is the tensor product of multiplications

$$m : E(\mu'_i + 1, i) \otimes E(\mu'_i + 2, i) \otimes \ldots \otimes E(\lambda'_i, i) \to \bigwedge^{\lambda'_i - \mu'_i} E.$$

We define *the skew Weyl module* $K_{\lambda/\mu}E$ to be the image of $\psi_{\lambda/\mu}$. The description of the relations and of the standard basis of skew Weyl modules is analogous to that for the skew Schur modules. Before we state it, let us define the tableaux. As before we fix an ordered basis $e_1, \ldots, e_n$ of E. Let us choose the tensors $U_j \in D_{\lambda_j - \mu_j}$, $U_j = e_1^{r(j,1)} \ldots e_n^{r(j,n)}$, where of course $r(j, 1) + \ldots + r(j, n) = \lambda_j - \mu_j$. Then the image $\psi_{\lambda/\mu}(U_1 \otimes \ldots \otimes U_s)$ will be denoted by a tableau T of shape λ/μ which in the j-th row has $r(j, 1)$ 1's, $r(j, 2)$ 2's, $\ldots$, $r(j, n)$ n's. The order of these elements will not matter, because we will assume each tableau to be symmetric in the symbols in every row. We will identify the tableau T with the tensor $\psi_{\lambda/\mu}(U_1 \otimes \ldots \otimes U_s)$.

(2.1.14) Example. *Take* $\lambda = (3, 2)$, $\mu = \emptyset$. *The tableau*

$$T = \begin{array}{|c|c|c|} \hline 1 & 1 & 2 \\ \hline 2 & 2 \\ \cline{1-2} \end{array}$$

is identified with the image $\psi_{(3,2)}(e_1^{(2)}e_2 \otimes e_2^{(2)})$.

(2.1.15) Proposition.

(a) $K_{\lambda/\mu}E = D_{\lambda_1 - \mu_1}E \otimes D_{\lambda_2 - \mu_2}E \otimes \ldots \otimes D_{\lambda_s - \mu_s}E / U(\lambda/\mu, E)$, *where* $U(\lambda/\mu, E)$ *is the sum of subspaces*

$$D_{\lambda_1 - \mu_1}E \otimes \ldots \otimes D_{\lambda_{a-1} - \mu_{a-1}}E \otimes U_{a,a+1}(E) \otimes$$
$$D_{\lambda_{a+2} - \mu_{a+2}}E \otimes \ldots \otimes D_{\lambda_s - \mu_s}E$$

for $1 \le a \le s - 1$, *where* $U_{a,a+1}(E)$ *is the module spanned by the images of the following maps* $\theta'(\lambda/\mu, a, u, v; E)$ *with* $u + v < \lambda_{a+1} - \mu_a$:

$$D_u E \otimes D_{\lambda_a - \mu_a - u + \lambda_{a+1} - \mu_{a+1} - v}E \otimes D_v E$$
$$\downarrow 1 \otimes \Delta \otimes 1$$
$$D_u E \otimes D_{\lambda_a - \mu_a - u}E \otimes D_{\lambda_{a+1} - \mu_{a+1} - v}E \otimes D_v E$$
$$\downarrow m_{12} \otimes m_{34}$$
$$D_{\lambda_a - \mu_a}E \otimes D_{\lambda_{a+1} - \mu_{a+1}}E.$$

(b) *The costandard tableaux* $ST(\lambda, \mu, [1, n]')$ *of shape* λ/μ *(cf. section 1.1.2) with the entries from* $[1', n']$ *form a basis of* $K_{\lambda/\mu}E$.

Proof. For the remainder of the section we write j instead of j' for $j' \epsilon [1, n]$. First of all, one can check by direct calculation that the map $\psi_{\lambda/\mu}$ sends the elements from $U(\lambda/\mu, E)$ to zero. This is done in the same way as in the case of Schur modules.

Then it is enough to check two things:

(1) The costandard tableaux of shape λ/μ generate the factor

$$D_{\lambda_1 - \mu_1} E \otimes D_{\lambda_2 - \mu_2} E \otimes \ldots \otimes D_{\lambda_s - \mu_s} E / U(\lambda/\mu, E).$$

(2) The images $\psi_{\lambda/\mu}(T)$ for the costandard tableaux T of shape λ/μ are linearly independent.

Facts (1) and (2) allow us to identify the image of $\psi_{\lambda/\mu}$ with the factor

$$D_{\lambda_1 - \mu_1} E \otimes D_{\lambda_2 - \mu_2} E \otimes \ldots \otimes D_{\lambda_s - \mu_s} E / U(\lambda/\mu, E).$$

The proof of (2) is exactly the same as for Schur modules, so we leave it to the reader. However, we prove fact (1) because the proof in the case of Weyl modules is slightly different. The reason is that the map $\theta'(\lambda/\mu, a, u, v; E)$ involves the multiplication in the divided powers algebra, which involves some integer coefficients.

It is clear that the row costandard tableaux $RST(\lambda/\mu, [1, n]')$ with entries from $[1, n]$ generate $K_{\lambda/\mu} E$. Let us order the set of such tableaux by the order $\preceq$ defined in section 1.1.2. We will prove that if the tableau T is not costandard, then we can express it modulo $U(\lambda/\mu, E)$ as a combination of earlier tableaux. Let us assume first that T has two rows. Since T is not costandard, we can find w for which $T(1, w) \geq T(2, w)$. Let us find the smallest w with this property. Let w' be the biggest number for which $T(2, w') = T(2, w)$. We consider the map $\theta'(\lambda/\mu, a, u, v, E)$ for $u = w - \mu_1 - 1$ and $v = \lambda_2 - w'$. The key observation is that image of the tensor $V_1 \otimes V_2 \otimes V_3$ where

$$V_1 = e_{T(1,\mu_1+1)} \cup e_{T(1,\mu_1+2)} \cup \ldots \cup e_{T(1,w-1)},$$
$$V_2 = e_{T(1,w)} \cup \ldots \cup e_{T(1,\lambda_1)} \cup e_{T(2,\mu_2+1)} \cup \ldots \cup e_{T(2,w')},$$
$$V_3 = e_{T(2,w'+1)} \cup \ldots \cup e_{T(2,\lambda_2)},$$

where $\cup$ indicates we take the corresponding tensor in the divided power, contains the tableau T with the coefficient 1, and all the other tableaux occurring in this image are earlier than T. The coefficients of multiplication in the divided algebra do not spoil anything, because by choice of w and w' we have $T(1, w - 1) < T(1, w)$ and $T(2, w') < T(2, w' + 1)$. Therefore T can be expressed modulo $U(\lambda/\mu, E)$ as a combination of earlier tableaux.

Let us consider the general case. If T is not costandard, then we can find a and w such that $T(a, w) \geq T(a + 1, w)$. Now we apply the previous argument to the tableau S which consists of the a-th and $(a+1)$st rows of T. Notice that the relations $U(\lambda/\mu, E)$ we are using do nothing to the other rows of T, so in the same way we can express T as a sum of earlier tableaux. ∎

(2.1.16) Example. *Let us standardize the tableau*

$$
\begin{array}{|c|c|c|}
\hline
1 & 4 & 5 \\
\hline
2 & 3 & 3 \\
\hline
\end{array}
$$

in $K_{(3,3)}E$. We have to use the relation $\theta'(\lambda, a, u, v; E)$ with $u = 1, v = 0$:

$$
\begin{array}{|c|c|c|}
\hline
1 & 4 & 5 \\
\hline
2 & 3 & 3 \\
\hline
\end{array}
= -
\begin{array}{|c|c|c|}
\hline
1 & 2 & 3 \\
\hline
3 & 4 & 5 \\
\hline
\end{array}
-
\begin{array}{|c|c|c|}
\hline
1 & 2 & 4 \\
\hline
3 & 3 & 5 \\
\hline
\end{array}
-
\begin{array}{|c|c|c|}
\hline
1 & 2 & 5 \\
\hline
3 & 3 & 4 \\
\hline
\end{array}
-
\begin{array}{|c|c|c|}
\hline
1 & 3 & 3 \\
\hline
2 & 4 & 5 \\
\hline
\end{array}
$$

$$
-
\begin{array}{|c|c|c|}
\hline
1 & 3 & 4 \\
\hline
2 & 3 & 5 \\
\hline
\end{array}
-
\begin{array}{|c|c|c|}
\hline
1 & 3 & 5 \\
\hline
2 & 3 & 4 \\
\hline
\end{array}.
$$

All the tableaux on the right hand side except the third one and the last one are standard, and we have

$$
\begin{array}{|c|c|c|}
\hline
1 & 2 & 5 \\
\hline
3 & 3 & 4 \\
\hline
\end{array}
= -
\begin{array}{|c|c|c|}
\hline
1 & 2 & 3 \\
\hline
3 & 4 & 5 \\
\hline
\end{array}
-
\begin{array}{|c|c|c|}
\hline
1 & 2 & 4 \\
\hline
3 & 3 & 5 \\
\hline
\end{array}.
$$

Similarly,

$$
\begin{array}{|c|c|c|}
\hline
1 & 3 & 5 \\
\hline
2 & 3 & 4 \\
\hline
\end{array}
= -
\begin{array}{|c|c|c|}
\hline
1 & 2 & 3 \\
\hline
3 & 4 & 5 \\
\hline
\end{array}
- 2
\begin{array}{|c|c|c|}
\hline
1 & 3 & 3 \\
\hline
2 & 4 & 5 \\
\hline
\end{array}
-
\begin{array}{|c|c|c|}
\hline
1 & 3 & 4 \\
\hline
2 & 3 & 5 \\
\hline
\end{array}.
$$

Putting all these expressions together, we get

$$
\begin{array}{|c|c|c|}
\hline
1 & 4 & 5 \\
\hline
2 & 3 & 3 \\
\hline
\end{array}
=
\begin{array}{|c|c|c|}
\hline
1 & 2 & 3 \\
\hline
3 & 4 & 5 \\
\hline
\end{array}
+
\begin{array}{|c|c|c|}
\hline
1 & 3 & 3 \\
\hline
2 & 4 & 5 \\
\hline
\end{array}.
$$

In the case when $\mu = 0$, i.e. in the case of the partition, we denote $K_{\lambda/\mu}$ by K_λ and we call it *a Weyl module*.

(2.1.17) Example.

(a) *For $\lambda = (t)$, $K_\lambda E = D_t E$. Indeed, by definition $K_\lambda E = D_t E/U((t), E)$, but $U((t), E) = 0$, since the partition (t) has only one part.*

(b) *For $\lambda = (1^t)$, $K_\lambda E = \bigwedge^t E$. Indeed, the map ψ_λ is onto in this case. The relations $\theta'(a, u, v; E)$ between two rows of length 1 correspond to $u = v = 0$, and they express antisymmetry in the rows.*

(c) Let $\lambda = (2, 1)$. The only relation that occurs in $K_{(2,1)}E$ is $\theta'(1, 0, 0; E)$. The corresponding Young scheme is

The image by $\theta'(1, 0, 0; E)$ of the typical element $e_{y_1} \cup e_{y_2} \cup e_{y_3}$ is

$$\begin{array}{|c|c|}\hline y_1 & y_2 \\\hline y_3 \\\cline{1-1}\end{array} + \begin{array}{|c|c|}\hline y_1 & y_3 \\\hline y_2 \\\cline{1-1}\end{array} + \begin{array}{|c|c|}\hline y_2 & y_3 \\\hline y_1 \\\cline{1-1}\end{array}$$

if y_1, y_2, y_3 are all different,

$$\begin{array}{|c|c|}\hline y_1 & y_1 \\\hline y_3 \\\cline{1-1}\end{array} + \begin{array}{|c|c|}\hline y_1 & y_3 \\\hline y_1 \\\cline{1-1}\end{array}$$

if $y_1 = y_2 \neq y_3$, and

$$\begin{array}{|c|c|}\hline y_1 & y_1 \\\hline y_1 \\\cline{1-1}\end{array}$$

if $y_1 = y_2 = y_3$.

(d) Let $\lambda = (2, 2)$. There are three types of relations $\theta'(1, u, v; E)$. The possible pairs (u, v) are $(0, 0)$, $(1, 0)$, $(0, 1)$. The corresponding Young schemes are

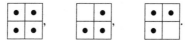

Therefore we have three types of relations. For the map $\theta'(1, 0, 0; E)$ the image of typical element $e_{y_1} \cup e_{y_2} \cup e_{y_3} \cup e_{y_4}$ is

$$\begin{array}{|c|c|}\hline y_1 & y_2 \\\hline y_3 & y_4 \\\hline\end{array} + \begin{array}{|c|c|}\hline y_1 & y_3 \\\hline y_2 & y_4 \\\hline\end{array} + \begin{array}{|c|c|}\hline y_1 & y_4 \\\hline y_2 & y_3 \\\hline\end{array} + \begin{array}{|c|c|}\hline y_2 & y_3 \\\hline y_1 & y_4 \\\hline\end{array} + \begin{array}{|c|c|}\hline y_2 & y_4 \\\hline y_1 & y_3 \\\hline\end{array} + \begin{array}{|c|c|}\hline y_3 & y_4 \\\hline y_1 & y_2 \\\hline\end{array}$$

when all numbers y_i are different, with easy adjustments when repetitions occur. For the map $\theta'(1, 0, 1; E)$ the image of the typical element $e_{y_1} \cup e_{y_2} \cup e_{y_3} \otimes e_z$ is

$$\begin{array}{|c|c|}\hline y_1 & y_2 \\\hline y_3 & z \\\hline\end{array} + \begin{array}{|c|c|}\hline y_1 & y_3 \\\hline y_2 & z \\\hline\end{array} + \begin{array}{|c|c|}\hline y_2 & y_3 \\\hline y_1 & z \\\hline\end{array}$$

if all numbers are different, with easy adjustments when repetitions occur. Finally, for the map $\theta'(1, 1, 0; E)$ the image of typical element

$e_x \otimes e_{y_1} \cup e_{y_2} \cup e_{y_3}$ is

$$
\begin{array}{|c|c|}\hline x & y_1 \\\hline y_2 & y_3 \\\hline\end{array}
+
\begin{array}{|c|c|}\hline x & y_2 \\\hline y_1 & y_3 \\\hline\end{array}
+
\begin{array}{|c|c|}\hline x & y_3 \\\hline y_1 & y_2 \\\hline\end{array}
$$

if all numbers are different, with easy adjustments when repetitions occur. We note a slight difference between the cases of Schur and Weyl modules. For Schur modules we could eliminate the relation $\theta'(1, 0, 0; E)$. This is impossible for Weyl modules. Indeed, let us consider the case when all four numbers are the same and they equal y. The relations coming from $\theta'(1, 1, 0; E)$ and $\theta'(1, 0, 1; E)$ give the relation

$$
2\begin{array}{|c|c|}\hline y & y \\\hline y & y \\\hline\end{array} = 0
$$

in $K_{2,2}E$. To get the relation

$$
\begin{array}{|c|c|}\hline y & y \\\hline y & y \\\hline\end{array} = 0
$$

we need the relation $\theta'(1, 0, 0; E)$.

(e) *Let $\lambda = (3, 2)$. There are three choices of pairs u, v. $(u, v) = (0, 0)$, $(1, 0)$, $(0, 1)$. The Young schemes are*

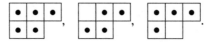

It follows that $K_{3,2}E$ is a factor of the module $D_3E \otimes D_2E$ by the images of three maps: $D_4E \otimes E \to D_3E \otimes D_2E$ (corresponding to $u = 0$, $v = 1$), $E \otimes D_4E \to D_3E \otimes D_2E$ (corresponding to $u = 1$, $v = 0$), and $D_5E \to D_3E \otimes D_2E$ (corresponding to $u = v = 0$).

(f) *We will show below that for two rowed partitions one can choose a smaller set of relations $\theta(1, u, v; E)$ with $u = 0$ that still suffice to define the Weyl functor. Example (c) above shows that the other choice that worked for Schur functors (choosing relations with one overlap) does not define a Weyl functor.*

(g) *Let $\lambda = (2, 2, 1)$. We have three types of relations corresponding to the first pair of rows (described in example (d)), and one type corresponding to the second and third row (described in example (c)). The Young schemes are*

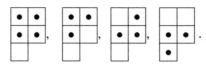

(h) Let λ be a hook, i.e. a partition of the form $\lambda = (p, 1^{q-1})$. *Graphically*

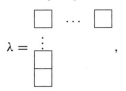

$$\lambda = \quad ,$$

with p boxes in the first row and q boxes in the first column. The relations between two rows of length 1 express the antisymmetry (cf. example (b)). There is only one type of relations corresponding to the first two rows, for the pair u = v = 0. It follows that the Weyl functor $K_{(p,1^{q-1})}E$ *is the cokernel of the map*

$$D_{p+1}E \otimes \overset{q-2}{\bigwedge} E \xrightarrow{\Delta \otimes 1} D_p E \otimes E \otimes \overset{q-2}{\bigwedge} E \xrightarrow{1 \otimes m} D_p E \otimes \overset{q-1}{\bigwedge} E.$$

We finish this section by stating the obvious functoriality property of all above constructions.

(2.1.18) Proposition.

    (a) The constructions of modules $L_{\lambda/\mu}E$, $K_{\lambda/\mu}E$ *are functorial with respect to the free module E. They define the endofunctors of the category of free **K**-modules. We refer to these functors as Schur functors and Weyl functors respectively.*

    (b) The functors $L_{\lambda/\mu}$, $K_{\lambda/\mu}$ *are polynomial, homogeneous of degree* $|\lambda/\mu|$.

    (c) We have the functorial isomorphisms $K_{\lambda/\mu}E = (L_{\lambda'/\mu'}E^*)^*$.

Proof. We start with the proof of (a). The modules $L_{\lambda/\mu}E$, $K_{\lambda/\mu}E$ are defined as images of natural transformations $\phi_{\lambda/\mu}$, $\psi_{\lambda/\mu}$ of endofunctors of a category of free **K**-modules. They are therefore functors themselves. Their values are again in the category of free **K**-modules by the standard basis theorem. Part (b) follows because the exterior, symmetric, and divided powers are homogeneous polynomial functors. Statement (c) is a consequence of (1.1.7), because that statement implies that the map $\psi_{\lambda'/\mu'}$ is the dual of the map $\phi_{\lambda/\mu}$ for E^*. ∎

2.2. Schur Functors and Highest Weight Theory

The modules $L_\lambda E$ play a crucial role in the representation theory of the general linear group. In this section we describe this connection.

We assume that the commutative ring **K** is an infinite field of arbitrary characteristic.

Let us denote by $\mathbf{T}$ (by $\mathbf{U}$) the subgroup of $GL(E)$ of all diagonal matrices (all upper triangular matrices with 1's on the diagonal) with respect to a fixed basis $e_1, \ldots, e_n$ of E.

We recall that a *rational representation* of $GL(E)$ is a vector space V together with a homomorphism of algebraic groups $\rho : GL(E) \to GL(V)$. In terms of coordinate rings this means that we have a homomorphism

$$\hat{\rho} : \mathbf{K}[GL(V)] \longrightarrow \mathbf{K}[GL(E)]$$

satisfying the conditions dual to the conditions for homomorphism. A rational representation V is called *polynomial* if ρ extends to the algebraic map $\rho' : \text{End}_{\mathbf{K}}(E) \to \text{End}_{\mathbf{K}}(V)$.

(2.2.1) Proposition. *Let V be a rational representation of $GL(E)$, and let $\bigwedge^n E$ be a determinant representation of $GL(E)$. Then for $m \gg 0$ the representation $V \otimes_{\mathbf{K}} (\bigwedge^n E)^{\otimes m}$ is polynomial.*

Proof. Indeed, for each k, l $(1 \le k, l \le \dim V)$ the image $\hat{\rho}(Y_{k,l})$ is an element of $\mathbf{K}[GL(E)] = \mathbf{K}[\{X_{i,j}\}_{1 \le i,j \le n}, T]/(T \det(X_{i,j}) - 1)$, where $X_{i,j}$ is the (i, j)-th entry function on GL(E). Multiplying the representation ρ by the determinant $\bigwedge^n E$ means that the corresponding image $\hat{\rho}(Y_{k,l})$ will be multiplied by $\det(X_{i,j})$. In this way, by multiplying by a sufficiently high power of $\det(X_{i,j})$ we can clear the denominators of all elements $\rho(Y_{i,j})$. ∎

The following fact is well known in representation theory (cf. [B, section 8]).

(2.2.2) Proposition.
  (a) *Every character $\chi : \mathbf{T} \to GL(1) = \mathbf{K}^*$ is of the form $(t_1, \ldots, t_n) \mapsto t_1^{\chi_1} t_2^{\chi_2} \ldots t_n^{\chi_n}$ for some integers $\chi_1, \chi_2, \ldots, \chi_n$. Here we denote by $(t_1, \ldots, t_n)$ the diagonal $n \times n$ matrix with the entries $t_1, \ldots, t_n$.*
  (b) *Every rational representation V of $GL(E)$ has a decomposition*

$$V = \bigoplus_{\chi \in \text{char}(\mathbf{T})} V_\chi,$$

where $V_\chi = \{v \in V \,|\, \rho(t)v = \chi(t)v\}$ for all $t \in \mathbf{T}$.

The characters of $\mathbf{T}$ are called *weights*. The subspace V_χ of V is called the *weight space of V corresponding to the weight χ*. We denote by ϵ_i the weight $\epsilon_i(t_1, \ldots, t_n) = t_i$.

The next step is to investigate how the elements of $\mathbf{U}$ change weights of vectors from V. We denote by $A_{i,j}(x)$ the elementary endomorphism $A_{i,j}(x)(e_s) = e_s + x\delta_{sj}e_i$.

(2.2.3) Proposition.

(a) *Let V be a polynomial representation of* GL(E), *and W a subrepresentation. Let $v \in V_\chi$. Then there exists a natural number r and elements $v_0, \ldots, v_r$ in V such that for every $x \in \mathbf{K}$*

$$A_{i,j}(x)v = v_0 + xv_1 + \ldots + x^r v_r.$$

Moreover, the vector v_s is a weight vector of weight $\chi + s(\epsilon_i - \epsilon_j)$ and if $v \in W$, then $v_0, \ldots, v_r \in W$.

(b) *Every nonzero polynomial representation V of* GL(E) *contains a nonzero $\mathbf{U}$-invariant weight vector.*

Proof. The existence of the vectors v_i follows at once from the fact that the representation V is a polynomial representation. To calculate the weight of v_i we notice that if $t = (t_1, \ldots, t_n)$ is a diagonal matrix, then

$$t A_{i,j}(x) = A_{i,j}(xt_i t_j^{-1})t.$$

Applying both sides to v yields that the weight of v_s is $\chi + s(\epsilon_i - \epsilon_j)$. To prove the last statement we notice that we can use the assumption that $\mathbf{K}$ is infinite. Then we can find $r + 1$ values $x_1, \ldots, x_{r+1}$ for which the corresponding Vandermonde determinant is nonzero. This means that if W is a subrepresentation and the vectors $A_{i,j}(x_1)v, \ldots, A_{i,j}(x_{r+1})v$ are in W, then $v_0, \ldots, v_{r+1}$ are also in W. This completes the proof of the first part.

To prove the second part we order the weights $\chi = (\chi_1, \chi_2, \ldots, \chi_n)$ lexicographically with respect to the sequence $\chi_1, \chi_2, \ldots, \chi_n$. If V is a finite dimensional rational representation of GL(E), then there exists the earliest weight χ in this order for which $V_\chi \neq 0$. By part (a) we see that every element of V_χ is a $\mathbf{U}$-invariant. Indeed, if $v \in V_\chi$ and $i < j$, then the weights of elements v_s from part (a) for $s > 0$ are earlier than χ. This means that $v_s = 0$ for $s > 0$, so $A_{i,j}(x)v = v$. Since the elements $A_{i,j}(x)$ $(i < j, x \in k)$ generate $\mathbf{U}$, v is a $\mathbf{U}$-invariant. ∎

(2.2.4) Example. *The canonical tableau*

$$c_\lambda(i, j) = j$$

for all $(i, j) \in \lambda$ is a $\mathbf{U}$-invariant in $L_\lambda E$.

The following fact is crucial for the whole theory.

(2.2.5) Lemma. *The vector space $(L_\lambda E)^U$ of U-invariants in $L_\lambda E$ is one dimensional (i.e., it is spanned by c_λ).*

Proof. Let U^- be the subgroup of lower triangular matrices with 1's on the diagonal in $GL(E)$. We will actually prove the opposite result stating that the space of U^--invariants in $L_\lambda E$ is one dimensional, spanned by the anticanonical tableau $\hat{c}_\lambda$ defined by

$$\hat{c}_\lambda(i, j) = n - \lambda_i + j.$$

We start with two combinatorial results.

Let T be a standard tableau, and let us take $1 \leq i < j \leq n$. Define the tableau $S_i^j(T)$ to be the tableau obtained from T by replacing i by j in every row in T containing i but not j. Let $h_i^j(T)$ denote the number of rows in T containing i but not j. The tableau $S_i^j(T)$ does not have to be standard. In some situations it turns out $S_i^j(T)$ is standard.

(2.2.6) Lemma. *Let us fix $1 \leq i < j \leq n$. Assume that T is a standard tableau of shape λ, with entries from $[1, n]$. Assume that every row of T containing an integer $\leq i$ contains also all integers $i, i + 1, \ldots, j - 1$. Then $S_i^j(T)$ is standard, and T is determined by $h_i^j(T)$ and $S_i^j(T)$.*

Proof. From our assumptions it follows that the rows of T not containing i must follow the rows of T containing i. Thus T is obtained from $S_i^j(T)$ by replacing j by i in the first $h_i^j(T)$ rows that contain j but not i.

To show that $S_i^j(T)$ is standard it is enough to do it in the case when T has two rows. The action of S_i^j on a row satisfying our assumption can at most replace the entry p with $p + 1$ for $i \leq p < j$. Therefore the only possible violation of standardness in $S_i^j(T)$ arises when i occurs in the first row of T and, for some p with $i \leq p < j$, p occurs in the same positions in the first and second row of T. Since the first row of T must contain $i, i + 1, \ldots, p$ and since the entries in the second row are strictly increasing and T is standard, i must occur in the second row of T in the same position as in the first row. If j does not occur in the second row of T, then p in the second row will be replaced by $p + 1$ and standardness will be preserved. But if j occurs in the second row, then it must occur exactly $j - i$ positions after i, since all integers $i + 1, \ldots, j - 1$ must occur there by assumption. However the first row of T

must have an element $> j$ that is $j - i$ positions after i, because it contains $i + 1, \ldots j - 1$ but not j. This contradicts the standardness of T. ∎

We can pass from any standard tableau T to the anticanonical tableau by applying the composite operator

$$S_{n-1}^n S_{n-2}^n S_{n-2}^{n-1} \ldots S_1^3 S_1^2.$$

Denote by h_i^j the number of substitutions of j for i made by the application of S_i^j in that sequence.

(2.2.7) Corollary. *The standard tableau T of shape λ is determined by the numbers h_i^j defined above.*

Proof. The corollary follows from Lemma (2.2.6) by induction. ∎

Now we conclude the proof of (2.2.5). For $1 \le i < j \le n$ and for $x \in \mathbf{K}$ we denote by $A_{j,i}(x)$ the matrix from $\mathbf{U}^-$ which has entries on the diagonal equal to 1, the entry in the (j, i)th position equal to x, and all other entries equal to 0. Consider the expansion from Proposition (2.2.3). It is clear that for $v = T$ we have $v_r = S_i^j(T)$ and that $r = h_i^j(T)$.

Consider a nonzero linear combination

$$y = \sum_{s=1}^m a_s T_s$$

of standard, distinct tableaux T_s with all $a_s \ne 0$. To prove (2.2.5) it is enough to show that the anticanonical tableau $\hat{c}_\lambda$ is contained in the span of $\mathbf{U}^- y$.

Consider $A_{j,i}(x)(y)$. This can be expanded as a polynomial in x, as in (2.2.3). The expansion gives

$$A_{j,i}(x)(y) = x^h \sum_{h_i^j(T_s)=h} a_s S_i^j(T_s) + \{\text{terms of lower degree in } x\}.$$

The element

$$y' = \sum_{h_i^j(T_s)=h} a_s S_i^j(T_s)$$

is nonzero by (2.2.6) and by the assumption $a_s \ne 0$. The element y' is contained in the linear span of $\mathbf{U}^- y$ by the Vandermonde determinant argument used in the proof of Proposition (2.2.3). ●

The statement (2.2.5) has important consequences.

(2.2.8) Proposition. *The submodule $M_\lambda E$ of $L_{\lambda'} E$ generated by the canonical tableau c_λ is an irreducible $GL(E)$-module. Every irreducible polynomial representation of $GL(E)$ is isomorphic to $M_\lambda E$ for some λ.*

Proof. A nonzero submodule W of $M_\lambda E$ contains a U-invariant element, so it contains c_λ. This means that $W = M_\lambda E$. This proves the first statement. To prove the second part of the proposition let us consider the irreducible polynomial representation V of $GL(E)$. By the second part of (2.2.3) we see that V contains for some weight λ a nonzero U-invariant vector v_λ of weight λ. We will show that V and M_λ are isomorphic. Let us consider the submodule W in $V \oplus M_\lambda$ generated by $(v_\lambda, c_{\lambda'})$. We consider two projections $p : W \to V$ and $q : W \to M_\lambda$. Both maps p, q are nonzero, so they are epimorphisms. Let us assume that one of those maps, say p, has nonzero kernel. This means that the module $\mathrm{Ker}\, p$ is isomorphic to M_λ, so $W = V \oplus M_\lambda$. This is however impossible, because the weight space W_λ is one dimensional. Indeed, if U^- denotes the group of lower triangular matrices with 1's on the diagonal, then the subset $U^- TU$ is a Zariski dense subset of $GL(E)$. Therefore W is spanned by $U^- TU(v_\lambda, c_{\lambda'})$, which is the span of $U^-(v_\lambda, c_{\lambda'})$. By part (b) of (2.2.3) we see that the only weight vector in the last span is $(v_\lambda, c_{\lambda'})$. The same reasoning works for the projection q. Thus p and q are isomorphisms and we are done. ∎

If $\mathbf{K}$ is a field of characteristic zero, then it is well known (see [Hu1] for the proof) that $GL(E)$ is linearly reductive, i.e., all finite dimensional representations are direct sums of simple ones. Then it follows instantly from (2.2.7) that $L_\lambda E$ is irreducible. Let us state these facts.

(2.2.9) Theorem. *All irreducible rational representations of $GL(E)$ are isomorphic to $M_\lambda E \otimes_\mathbf{K} (\bigwedge^n E)^{\otimes m}$ for some partition $\lambda = (\lambda_1, \ldots, \lambda_{n-1})$ and $m \in \mathbf{Z}$. This correspondence is bijective.*

(2.2.10) Theorem. *Assume that $\mathbf{K}$ is a field of characteristic zero.*

  (a) We have $M_\lambda E = L_{\lambda'} E$.
  (b) Every rational representation of $GL(E)$ is a direct sum of irreducibles.

Theorem (2.2.9) is the main statement of the highest weight theory. The irreducible representations are parametrized by the sequences $(\lambda_1, \ldots, \lambda_n)$ with $\lambda_i \in \mathbf{Z}, \lambda_1 \geq \ldots \geq \lambda_n$. These are *dominant integral weights* for the group $GL(E)$. They can be defined as the weights whose values on the roots of $GL(E)$

are integral, and whose values on positive roots are nonnegative. The *integral weights* are the weights whose values on the roots of $GL(E)$ are integral.

Let V be a rational representation of $GL(E)$. We define the character of V

$$\text{char}(V) = \sum \dim V_\chi \, x_1^{\chi_1} \dots x_n^{\chi_n},$$

where $x_1, \dots, x_n$ are indeterminates. Obviously we have

$$\text{char}(V \oplus V') = \text{char}(V) + \text{char}(V'), \text{char}(V \otimes V') = \text{char}(V)\text{char}(V').$$

(2.2.11) Remark.

(a) *When* **K** *is a field of characteristic 0, then every representation is determined by its character. This follows easily from linear reductivity (2.2.10) (b) and from (2.2.5).*

(b) *The function* char(V) *is a symmetric function of* $x_1, \dots, x_n$, *because for each* $\chi_1, \dots, \chi_n$ *and for each permutation* $\sigma \in \Sigma_n$ *we have* $\dim V_{\chi_1,\dots,\chi_n} = \dim V_{\chi_{\sigma(1)},\dots,\chi_{\sigma(n)}}$. *Indeed, the permutation matrix in* $GL(N)$ *corresponding to* σ *carries one space isomorphically into another.*

(c) *The function* char($L_\lambda E$) *is called the Schur function (cf. [MD, chapter I]). The Schur functions play an important role in combinatorics.*

(d) *The correspondence* $E \mapsto L_\lambda E$ *gives rise to a functor from the category* Vect$_\mathbf{K}$ *to itself. We will refer to it as a Schur functor.*

We finish our discussion with the simple example showing that (2.2.10) is false in positive characteristics.

(2.2.12) Example. *Let* char $\mathbf{K} = p > 0$. *Let* $\lambda = (1^p)$. *Then* $L_\lambda E = S_p E$, *but* $M_{\lambda'} E$ *is the span of the elements* e_i^p *for* $1 \le i \le n$. *This subspace is* $GL(E)$*-invariant, because in characteristic* p *we have* $(x + y)^p = x^p + y^p$.

For the remainder of this section we assume that the commutative ring **K** has characteristic 0, i.e., it is a **Q**-algebra. We give an alternate description of Schur modules using Young idempotents (cf. [DC]).

Consider the natural action of the symmetric group Σ_m on the tensor product $E^{\otimes m}$ given by

$$\sigma(v_1 \otimes \dots \otimes v_m) = v_{\sigma^{-1}(1)} \otimes \dots \otimes v_{\sigma^{-1}(m)}.$$

Let λ be a partition of m and let D be a tableau of shape λ, with entries from $[1, m]$ of weight (1^m) (i.e. with distinct entries). We define the Young symmetrizer $e(D) \in K[\Sigma_m]$ as follows. Denote by $R(D)$ (by $C(D)$) the subgroup of Σ_m of permutations preserving the rows (columns) of D. We set

$$e(D) = \sum_{\tau \in R(D)} \sum_{\sigma \in C(D)} \text{sgn}(\sigma)\, \tau\sigma.$$

We define the representation of $GL(E)$ depending on the tableau D by

$$L_D(E) = e(D)E^{\otimes m}.$$

(2.2.13) Lemma.
  (a) *If D, D' are the tableaux of the same shape λ, then $L_D(E)$ and $L_{D'}(E)$ are isomorphic as $GL(E)$-modules.*
  (b) *If D is a tableau of shape λ, then $L_D(E)$ is isomorphic to the Schur module $L_\lambda(E)$.*

Proof. We start with part (a). Let σ be a permutation such that $\sigma(D) = D'$, i.e., for every $(i, j) \in D(\lambda)$ we have $D'(i, j) = \sigma(D(i, j))$. Then we have $R(D') = \sigma R(D)\sigma^{-1}$, $C(D') = \sigma C(D)\sigma^{-1}$, and therefore $e(D') = \sigma e(D)\sigma^{-1}$, which implies $L_{D'}(E) = \sigma(L_D(E))$. Indeed, the isomorphism is given by acting by σ on a tensor.

To prove (b) let us choose D which is a row standard tableau, minimal with respect to the order $\preceq$ defined in section 1.1.2 or, more precisely, $D(i, j) = \lambda_1 + \ldots + \lambda_{i-1} + j$.

We can identify representations $L_\lambda E$ and $L_D(E)$ as follows. $L_\lambda E$ can be interpreted as $\text{Im } \phi_\lambda$. We embed $\text{Im } \phi_\lambda$ into the tensor product $\otimes_{(i,j)\in D(\lambda)} E$ by symmetrizing in each column. This can be done, because in characteristic 0 the symmetric power can be identified with the set of symmetric tensors (which in a characteristic free way is isomorphic to the divided power). Call the symmetrization map η. Then the image $\eta\phi_\lambda(v)$ can be identified with $e(D)(\alpha(v))$, where α is the product of exterior diagonals used to define ϕ_λ. ∎

The approach based on the use of the action of Σ_m on $E^{\otimes m}$, due to Schur, can also give the irreducibility of Schur modules in characteristic 0. We sketch the basic steps in the proof. The interested reader may consult [DC].

The actions of Σ_m and of $GL(E)$ on $E^{\otimes m}$ commute. Let us denote the span of all endomorphisms of type $g^{\otimes m}$ ($g \in GL(E)$) in $\text{End}_K(E^{\otimes m})$ by $S(m, E)$. We also denote by $\Sigma(m, E)$ the endomorphisms of $E^{\otimes m}$ that are induced by the elements of the group ring $K[\Sigma_m]$.

(2.2.14) Lemma (Schur Commutation Lemma). *The algebras $S(m, E)$ and $\Sigma(m, E)$ are their own commutants in* $\mathrm{End}_\mathbf{K}(E^{\otimes m})$. *More precisely,*

$$S(m, E) = \{h \in \mathrm{End}_\mathbf{K}(E^{\otimes m}) \mid gh = hg \text{ for all } g \in \Sigma(m, E)\},$$
$$\Sigma(m, E) = \{h \in \mathrm{End}_\mathbf{K}(E^{\otimes m}) \mid gh = hg \text{ for all } g \in S(m, E)\}.$$

The algebra $\Sigma(m, E)$ is semisimple by Maschke's theorem. This means the action of $S(m, E)$ on $E^{\otimes m}$ is also semisimple and the Schur modules are precisely the irreducible representations. This follows from the following facts, proven in [DC], in the appendix on non-commutative algebra, part IV.

(2.2.15) Proposition. *Let B be a semisimple subalgebra in the matrix algebra $M_n(\mathbf{K})$. Let C be the commutant of B, i.e.,*

$$C = \{x \in M_N(\mathbf{K}) \mid xy = yx \text{ for all } y \in B\}.$$

Then the subalgebra C is also semisimple. Denote by V_C the vector space $V = \mathbf{K}^N$ with the structure of a C-module. Every simple module in V_C is isomorphic to a module bV where b is an element of a minimal left ideal in B. If b, b' generate the same left ideal in B, then bV_C and $b'V_C$ are isomorphic as C-modules.

(2.2.16) Proposition. *The isomorphism classes of minimal left ideals in $K[\Sigma_m]$ are in one to one correspondence with partitions of m. The correspondence is given by associating to λ the left ideal generated by $e(D)$, where D is an arbitrary tableau of shape λ with entries from $[1, m]$, with distinct entries.*

2.3. Properties of Schur Functors. Cauchy Formulas, Littlewood–Richardson Rule, and Plethysm

In this section we discuss some formulas from the representation theory of general linear groups. They will be used in the calculations involving vector bundles in chapters 6 through 9. We try to give both characteristic 0 and characteristic free statements.

We start with the direct sum decompositions. Let E and F be two free modules over a commutative ring $\mathbf{K}$. Our formulas express the modules $L_{\lambda/\mu}(E \oplus F)$ and $K_{\lambda/\mu}(E \oplus F)$ through the corresponding modules for E and F.

(2.3.1) Proposition.

(a) *There is a* $\mathrm{GL}(E) \times \mathrm{GL}(F)$-*equivariant filtration on* $L_{\lambda/\mu}(E \oplus F)$ *with the associated graded object*

$$\bigoplus_{\nu|\mu\subset\nu\subset\lambda} L_{\lambda/\nu}E \otimes L_{\nu/\mu}F.$$

If **K** *is a commutative ring of characteristic* 0, *then we have a* $\mathrm{GL}(E) \times \mathrm{GL}(F)$-*equivariant isomorphism*

$$L_{\lambda/\mu}(E \oplus F) = \bigoplus_{\nu|\mu\subset\nu\subset\lambda} L_{\lambda/\nu}E \otimes L_{\nu/\mu}F.$$

(b) *There is a* $\mathrm{GL}(E) \times \mathrm{GL}(F)$-*equivariant filtration on* $K_{\lambda/\mu}(E \oplus F)$ *with the associated graded object*

$$\bigoplus_{\nu|\mu\subset\nu\subset\lambda} K_{\lambda/\nu}E \otimes K_{\nu/\mu}F.$$

If **K** *is a commutative ring of characteristic* 0, *then we have a* $\mathrm{GL}(E) \times \mathrm{GL}(F)$-*equivariant isomorphism*

$$K_{\lambda/\mu}(E \oplus F) = \bigoplus_{\nu|\mu\subset\nu\subset\lambda} K_{\lambda/\nu}E \otimes K_{\nu/\mu}F.$$

Proof. Since the proofs of both parts of the proposition are the same, we will just prove part (a). We observe that it is enough to prove the first statement, since from it one deduces that the characters of the left and right hand sides of the second formula are the same.

Let us denote dim $E = n$, dim $F = m$. Let us choose the bases $e_1, \ldots, e_n$ of E, $f_1, \ldots, f_m$ of F. Then $e_1, \ldots, e_n, f_1, \ldots f_m$ is the basis of $E \oplus F$, and we order it so $f_1 < \ldots < f_m < e_1 < \ldots < e_n$. We can consider the tableaux corresponding to this basis. For each such tableau T we define its F-part $f(T)$ to be the sequence $a_1, \ldots, a_s$, where a_i is the number of elements f_j in the i-th row of T. We order sequences $f(T)$ lexicographically and denote this order by $\preceq$. Now for each ν such that $\mu \subset \nu \subset \lambda$ we define the subspace $\mathcal{F}_\nu$ to be the span of all tableaux T such that $(\nu_1 - \mu_1, \ldots, \nu_s - \mu_s) \preceq f(T)$. It is clear that $\mathcal{F}_\nu$ is a $\mathrm{GL}(E) \times \mathrm{GL}(F)$-submodule. We also order all possible partitions ν by saying that $\nu \preceq \xi$ if $(\xi_1, \ldots, \xi_s) \preceq (\nu_1, \ldots, \nu_s)$. It is clear by definition that if $\xi \preceq \nu$ then $\mathcal{F}_\xi \subset \mathcal{F}_\nu$.

Claim. $\mathcal{F}_\nu / \sum_{\xi \prec \nu} \mathcal{F}_\xi$ *is a factor of* $L_{\nu/\mu}F \otimes L_{\lambda/\nu}E$.

It is obvious from the definition that the factor $\mathcal{F}_\nu / \sum_{\xi \prec \nu} \mathcal{F}_\xi$ is spanned by the tableaux T such that $f(T) = (\nu_1 - \mu_1, \ldots, \nu_s - \mu_s)$. Each tableau T such that $f(T) = (\nu_1 - \mu_1, \ldots, \nu_s - \mu_s)$ can be considered as a pair of tableaux: the tableau $T(1)$ of shape ν/μ with the entries from the set $\{f_1, \ldots, f_m\}$, and the tableau $T(2)$ of shape λ/ν with the entries from the set $\{e_1, \ldots e_n\}$. We will denote such T by $(T(1), T(2))$.

We show that the standard relations defining $L_{\nu/\mu} F \otimes L_{\lambda/\nu} E$ are satisfied in $\mathcal{F}_\nu / \sum_{\xi < \nu} \mathcal{F}_\xi$.

Let us consider the element in $\mathcal{F}_\nu / \sum_{\xi < \nu} \mathcal{F}_\xi$ which is the relation of type $\theta(\lambda/\nu, a, u, v; E)$ (cf. section 2.1) on the tableau $T(2)$. More precisely, let us denote by $T(1)_j$ the j-th row of $T(1)$, and let us choose the rows $T(2)_j$ for j different than a, $a+1$, and finally let us choose $V_1 \in \bigwedge^u E$, $V_2 \in \bigwedge^{\lambda_a - \nu_a + \lambda_{a+1} - \nu_{a+1} - u - v} E$ and $V_3 \in \bigwedge^v E$. We consider the relation which shuffles the entries of V_2 into the last $\lambda_a - \nu_a - u$ spots in the a-th row of λ/ν and the first $\lambda_{a+1} - \nu_{a+1} - v$ spots in the $(a+1)$st row of λ/ν and leaves all other entries in their spots. Let us call this relation R_1. We want to show that this relation is zero in $\mathcal{F}_\nu / \sum_{\xi < \nu} \mathcal{F}_\xi$. Let us consider the relation R_2 of type $\theta(\lambda/\mu, a, \nu_a - \mu_a + u, v; E \oplus F)$ which shuffles the entries of $V_2 \wedge T(1)_a$. This relation, being the defining relation of $L_{\lambda/\mu}(E \oplus F)$, is identically zero in $\mathcal{F}_\nu / \sum_{\xi < \nu} \mathcal{F}_\xi$. The relation R_2 has more summands than R_1. However, all the summands occurring in R_2 and not in R_1 involve shuffling some basis elements from F into the earlier rows of λ/μ. Such elements are automatically contained in $\sum_{\xi < \nu} \mathcal{F}_\xi$, so they are automatically zero in $\mathcal{F}_\nu / \sum_{\xi < \nu} \mathcal{F}_\xi$. Similarly we deal with the relations on the F-side. This proves our claim.

Now the statement of the proposition follows, because by the standard basis theorem the left and right hand sides of the second formula have the same dimension, so in fact

$$\mathcal{F}_\nu / \sum_{\xi < \nu} \mathcal{F}_\xi = L_{\nu/\mu} F \otimes L_{\lambda/\nu} E.$$

This completes the proof of (2.3.1). ∎

Next we discuss the Cauchy formulas. Let E and F be two free modules. We are interested in the modules $S_m(E \otimes F)$ and $\bigwedge^m(E \otimes F)$. We want to express these modules in terms of Schur and Weyl modules.

(2.3.2) Theorem.

(a) *There is a natural filtration on $S_m(E \otimes F)$ whose associated graded object is*

$$\bigoplus_{|\lambda|=m} L_\lambda E \otimes L_\lambda F.$$

(b) There is a natural filtration on $\bigwedge^m(E \otimes F)$ whose associated graded object is

$$\bigoplus_{|\lambda|=m} L_\lambda E \otimes K_\lambda F.$$

We will prove part (a) of this theorem in chapter 3. We also give there the description of the filtration giving part (b). For the proof we refer to [ABW2], or exercises 4, 5, 6 in chapter 3. Now we state the characteristic zero consequence.

(2.3.3) Corollary. *Let **K** be a commutative ring of characteristic 0. We have natural isomorphisms*

$$S_m(E \otimes F) = \bigoplus_{|\lambda|=m} L_\lambda E \otimes L_\lambda F,$$

$$\bigwedge^m(E \otimes F) = \bigoplus_{|\lambda|=m} L_\lambda E \otimes L_{\lambda'} F.$$

Proof. Without loss of generality we can assume that **K** is a field. Indeed, if the theorem is true over the field **Q** of rational numbers, then the result extends to any commutative ring of characteristic 0 by base change.

By Theorem (2.3.2) the characters of both sides of our formulas are the same (we notice that since char $\mathbf{K} = 0$, $K_\lambda F = L_{\lambda'} F$). Using the remark (2.2.11) (a), we get our statement. ∎

The formulas from Corollary (2.3.3) are special cases of the *problem of outer plethysm*. The general problem is to find the multiplicities $v(\mu, \nu, \lambda)$ in the decomposition

$$L_\lambda(E \otimes F) = \bigoplus_{|\mu|=|\nu|=|\lambda|} v(\mu, \nu, \lambda)\, L_\mu E \otimes L_\nu F.$$

This is a very difficult problem, solved in very few cases. Notice that substituting in (2.3.3) (a) $F \otimes G$ for F, we see that

$$S_m(E \otimes F \otimes G) = \bigoplus_{|\lambda|=|\mu|=|\nu|=m} v(\mu, \nu, \lambda)\, L_\lambda E \otimes L_\mu F \otimes L_\nu G,$$

so the multiplicities $v(\mu, \nu, \lambda)$ are symmetric in λ, μ, and ν.

This interpretation explains why the problem of outer plethysm is so complicated. Let us consider the action of the group SL $(E) \times$ SL$(F) \times$ SL(G) on $E \otimes F \otimes G$. The dimension of $E \otimes F \otimes G$ is in general bigger than the dimension of the acting group SL$(E) \times$ SL$(F) \times$ SL(G). This means that the structure of the ring of invariants is very complicated. Finding the expression for the Hilbert function of the ring of invariants is, however, equivalent to finding some of the multiplicities $v(\mu, v, \lambda)$. This does not exclude the existence of a combinatorial formula for $v(\mu, v, \lambda)$, but it means that such a formula will not lead to an easy calculation of our multiplicities.

Next we state the Littlewood–Richardson rule. It describes a decomposition of the tensor product of Schur functors into Schur functors.

Let $\mathbf{K}$ be a commutative ring of characteristic 0. Let λ, μ be two partitions. In the case where $\mathbf{K}$ is a field we have by (2.2.10)

$$L_\lambda E \otimes L_\mu E = \bigoplus_{|v|=|\lambda|+|\mu|} u(\lambda, \mu; v) L_v E,$$

where $u(\lambda, \mu; v)$ are some multiplicities. This decomposition carries over to the case of an arbitrary ring $\mathbf{K}$ of characteristic 0. Indeed, the explicit isomorphism over $\mathbf{Q}$ remains an isomorphism when tensored with $\mathbf{K}$.

The Littlewood–Richardson rule gives a beautiful combinatorial description of these multiplicities.

In order to state the rule, we need one combinatorial notion.

A word $w = w_1 \ldots w_t$, with $w_1, \ldots, w_t$ being positive integers, is *a lattice permutation* if for each $s(1 \leq s \leq t)$ and each positive integer i, the number of occurrences of i in $w_1, \ldots, w_s$ is not smaller than the number of occurrences of $i + 1$.

Let T be a tableau of skew shape v/λ. From such T we form a word $w(T)$ by reading T column by column, starting in each column with the lowest entry. In other words,

$$w(T) = (T(v_1', 1), T(v_1' - 1, 1), \ldots, T(v_1' - \lambda_1' + 1, 1),$$
$$T(v_2', 2), \ldots, T(v_s' - \lambda_s' + 1, s)).$$

We say that the tableau T *satisfies the condition* LP if the word $w(T)$ is a *lattice permutation*. Let us denote by $P(\lambda, \mu; v)$ the set of all standard tableaux of shape v/λ of weight μ' satisfying the condition LP. Then we have

(2.3.4) Theorem (Littlewood–Richardson Rule).

$$u(\lambda, \mu; v) = \text{card } P(\lambda, \mu; v).$$

Again there is a characteristic free statement of the Littlewood–Richardson rule involving filtrations (cf. [Bo2]), but we will not need it in the applications. Since we will use the rule sporadically, we give a combinatorial proof as a series of exercises at the end of this chapter. The reader might also look for the proof in [MD, chapter I]. MacDonald proves the statement about the symmetric functions, but we notice that his statement means that the representations $L_\lambda E \otimes L_\mu E$ and $\bigoplus_{|\nu|=|\lambda|+|\mu|} u(\lambda, \mu; \nu) L_\nu E$, with $u(\lambda, \mu; \nu)$ defined by (2.3.4), have the same characters.

Let us state two important special cases of the Littlewood–Richardson rule, known as *Pieri's formulas*. We recall that a skew partition λ/μ is called *a vertical strip* if it contains at most one box in each row, i.e., $\lambda_i \le \mu_i + 1$ for all i. We denote the set of all vertical strips by VS. Similarly, the skew partition λ/μ is *a horizontal strip* if it contains at most one box in each column, i.e. when λ'/μ' is a vertical strip. We denote the set of all horizontal strips by HS.

(2.3.5) Corollary (Pieri's Formulas). *Let* **K** *be a commutative ring of characteristic* 0. *Then we have the natural isomorphisms*

$$L_\lambda E \otimes S_j E = \bigoplus_{\{\nu | \lambda \subset \nu, |\nu/\lambda|=j, \nu/\lambda \in \mathrm{VS}\}} L_\nu E,$$

$$L_\lambda E \otimes \bigwedge^{j} E = \bigoplus_{\{\nu | \lambda \subset \nu, |\nu/\lambda|=j, \nu/\lambda \in \mathrm{HS}\}} L_\nu E.$$

The Littlewood–Richardson rule has an analogue for skew Schur functors. Assuming that **K** has characteristic 0, we can write

$$L_{\nu/\mu} E = \bigoplus_{|\lambda|=|\nu/\mu|} w(\nu/\mu; \lambda) L_\lambda E$$

Then we have

(2.3.6) Theorem (Littlewood–Richardson Rule for Skew Schur Functors).

$$w(\nu/\mu; \lambda) = u(\lambda, \mu; \nu) = \mathrm{card}\ P(\lambda, \mu; \nu).$$

Again we refer for the proof of this fact to [MD, chapter I]. The characteristic free statement involving filtrations is also true (cf. [Bo2]).

We state the analogues of Pieri's formulas for skew shapes.

(2.3.7) Corollary. *Let* **K** *be a commutative ring of characteristic* 0.

(a)

$$L_{\nu/(1^j)}E = \bigoplus_{\{\lambda \mid \lambda \subset \nu, \, |\nu/\lambda|=j, \, \nu/\lambda \in \mathrm{VS}\}} L_\lambda E.$$

(b)

$$L_{\nu/(j)}E = \bigoplus_{\{\lambda \mid \lambda \subset \nu, \, |\nu/\lambda|=j, \, \nu/\lambda \in \mathrm{HS}\}} L_\lambda E.$$

We conclude this section with a brief discussion of the *problem of inner plethysm*. The general problem is to decompose the functor $L_\lambda(L_\mu E)$ into Schur functors. This problem is probably even more difficult than the outer plethysm. To see why, let us look at the situation from the point of view of invariant theory. We look at the special case $\lambda = (1^m)$, $\mu = (1^n)$. Decomposing $S_m(S_n E)$ into Schur functors involves the formula for the dimension of the homogeneous components of the ring of invariants of $\mathrm{SL}(E)$ acting on $S_n E$. Such rings are extremely complicated, at least according to nineteenth century invariant theorists.

In view of this remark it is logical to expect nice formulas for $S_n(S_2 E)$ and $S_n(\bigwedge^2 E)$, because the action of $\mathrm{GL}(E)$ on $S_2 E$ or $\bigwedge^2 E$ has finitely many orbits, so the rings of $\mathrm{SL}(E)$-invariants are very simple. Such formulas are given in the next proposition.

(2.3.8) Proposition. *Let* **K** *be a commutative ring of characteristic* 0.

(a)

$$S_m(S_2 E) = \bigoplus_{|\lambda|=2m, \, \lambda'_i \text{ even for all } i} L_\lambda E,$$

(b)

$$S_m\left(\overset{2}{\bigwedge} E \right) = \bigoplus_{|\lambda|=2m, \, \lambda_i \text{ even for all } i} L_\lambda E.$$

Proof. It is enough to prove the proposition when **K** is a field. The corresponding formulas for characters are given in [MD, chapter I]. However, it is convenient for later applications to give a proof based on U-invariants. Let us fix an ordered basis $e_1, \ldots, e_n$ of the vector space E. We first prove the formula (a). We identify the symmetric algebra $S = \mathrm{Sym}(S_2 E)$ with the polynomial ring $\mathbf{K}[X_{i,j}]$, where $X_{i,j} = e_i e_j$. We consider the generic $n \times n$ symmetric matrix $X = (X_{i,j})_{1 \le i, j \le n}$ over S. For each r, $1 \le r \le n$, we choose the $r \times r$ minor of X by deleting rows and columns with numbers $> r$. Let g_r be the determinant of this minor. Then an easy calculation shows that g_r is an U-invariant of weight $(2^r, 0^{n-r})$ in $S_r(S_2 E)$. Let λ be a partition of $2m$ such that $\lambda'_j = 2\mu_j$ for $1 \le j \le n$. This means that for each $i > 0$, $\lambda_{2i-1} = \lambda_{2i}$. Therefore the product $g_\lambda = \prod_{i>0} g_{\lambda_{2i}}$ is a nonzero U-invariant of weight λ' in $S_m(S_2 E)$. This means that the left hand side of (a) contains the right hand side of (a). Now we prove (a) by induction on $n = \dim E$. For $n = 1$ the formula is obvious. Let us assume that the formula (a) is true for $\dim E = n$. We will prove that the left and right hand sides have the same dimension for $\dim E = n + 1$. In view of the above construction, it is enough to prove (a).

To prove the statement about the dimensions, let us consider the space $E \oplus \mathbf{K}$ of dimension $n + 1$ over $\mathbf{K}$. We will actually prove that the left and right hand sides of (a) for $E \oplus \mathbf{K}$ are isomorphic as $GL(E)$-modules. The left hand side decomposes in the following way: $S_2(E \oplus \mathbf{K}) = S_2 E \oplus E \oplus \mathbf{K}$. Therefore

$$S_m(S_2(E \oplus \mathbf{K})) = \bigoplus_{i+j+l=m} S_i(S_2 E) \otimes S_j E \otimes S_l \mathbf{K}$$

$$= \bigoplus_{i+j \le m} S_i(S_2 E) \otimes S_j E.$$

Applying the formula (a) for $i \le n$, we get

$$S_m(S_2(E \oplus \mathbf{K})) = \bigoplus_{i+j \le m} \bigoplus_{|\nu|=2i, \ \nu'_l \ even \ for \ all \ l} L_\nu E \otimes S_j E.$$

On the other hand, for the right hand side we get

$$\bigoplus_{|\lambda|=2m, \ \lambda'_i \ even \ for \ all \ i} L_\lambda(E \oplus \mathbf{K}) = \bigoplus_{|\lambda|=2m, \ \lambda'_i \ even \ for \ all \ i} \bigoplus_{\mu \subset \lambda, \ \lambda/\mu \in VS} L_\mu E.$$

It remains to show that for any partition α, $L_\alpha E$ occurs in the right hand sides of both formulas with the same multiplicity. Let us fix m and α. The multiplicity of $L_\alpha E$ in the first formula is equal to the cardinality of the set

$$A_1^\alpha = \{(\nu, i, j) \mid |\nu| = 2i, \ \nu'_l \ is \ even \ for \ all \ l, \ \nu \subset \alpha, \ \alpha/\nu \in VS, \ |\alpha/\nu|$$
$$= j, \ i + j \le m\}.$$

The multiplicity of $L_\alpha E$ in the second formula is equal to the cardinality of the set

$$A_2^\alpha = \{\lambda \mid |\lambda| = 2m, \ \lambda'_l \ even \ for \ all \ l, \ \alpha \subset \lambda, \ \lambda/\alpha \in VS\}.$$

To finish the proof of (a) it is enough to construct a bijection h from A_1^α to A_2^α. Let $(\nu, i, j) \in A_1^\alpha$. We construct $\lambda \in A_2^\alpha$ as follows. Let us define the numbers a_j $(0 \le j)$. If α_j' is even, then $\alpha_j' - \nu_j' = 2a_j$. If α_j' is odd, then $\alpha_j' - \nu_j' = 2a_j + 1$. We also define $a_{-1} = 0$. Let us construct the partition β by adding to the j-th column of α $2a_{j-1}$ boxes if α_j is even and $1 + 2a_{j-1}$ boxes if α_j is odd. Then β is a partition of $2i + 2j$ (we added j boxes to α) with each β_j' even. We construct $\lambda = h(\nu, i, j)$ by adding to the first column of β $2(m - i - j)$ boxes. The reader will check easily that the map h defines a bijection from A_1^α to A_2^α. This proves the statement (a).

We prove (b) using the same technique. We identify $S = \mathrm{Sym}(\bigwedge^2 E)$ with the polynomial ring in the variables $X_{i,j} = e_i \wedge e_j$. Then we consider the $n \times n$ skew symmetric generic matrix $X = (X_{i,j})_{1 \le i, j \le n}$ over S. For each even number $2r$, $0 \le 2r \le n$, we define the element g_{2r} to be the Pfaffian of the skew symmetric $2r \times 2r$ matrix obtained from X by deleting rows and columns with numbers $> 2r$. We see easily that for each $2r$ the element g_{2r} is a $\mathbf{U}$-invariant of the weight $(1^{2r}, 0^{n-2r})$. Now for a partition λ with all λ_i even we find that $g_\lambda = \prod_i g_{\lambda_i}$ is a nonzero $\mathbf{U}$-invariant of weight λ'. This shows that the right hand side of (b) is contained in the left hand side. Then we can finish the proof of (b) with a similar (but easier) reasoning to the one in the proof of (a). $\blacksquare$

It turns out that the companion formulas for $\bigwedge^m (S_2 E)$ and $\bigwedge^m (\bigwedge^2 E)$ also can be easily obtained. Let us recall that every partition λ can be written in the hook notation as $\lambda = (a_1, \ldots a_r | b_1, \ldots, b_r)$ (cf. section 1.1.2). Let us denote by $Q_1(m)$ the set of partitions λ of m for which $a_i = b_i + 1$ for each i. Similarly we denote by $Q_{-1}(m)$ the set of partitions λ of m for which $b_i = a_i + 1$ for each i. Then we have

(2.3.9) Proposition. *Let* $\mathbf{K}$ *be a commutative ring of characteristic 0.*

(a)

$$\bigwedge^m (S_2 E) = \bigoplus_{\lambda \in Q_{-1}(2m)} L_\lambda E.$$

(b)

$$\bigwedge^m \left(\bigwedge^2 E \right) = \bigoplus_{\lambda \in Q_1(2m)} L_\lambda E.$$

Proof. The corresponding formulas for characters are given in [MD, chapter I]. Let us just indicate the nonzero $\mathbf{U}$-invariants in $\bigwedge^m (S_2 E)$ and in $\bigwedge^m (\bigwedge^2 E)$.

The proof of (2.3.9) can be finished by an argument similar to the one in the proof of (2.3.8).

Let λ be a partition from $Q_{-1}(2m)$ which can be written in the hook notation as $\lambda = (a_1, \ldots a_r | a_1 + 1, \ldots, a_r + 1)$. Let us denote by $X_{i,j}$ the element $e_i e_j$ of $S_2 E$. Then we define $g_j = X_{j,j} \wedge X_{j,j+1} \wedge \ldots \wedge X_{j,j+a_i-1}$ and $g_\lambda = g_1 \wedge g_2 \wedge \ldots \wedge g_r$. Then one checks easily that g_λ is a U-invariant of weight λ'.

Similarly for $\bigwedge^m (\bigwedge^2 E)$. We fix $\lambda \in Q_1(2m)$ such that λ can be written in the hook notation $\lambda = (a_1 + 1, \ldots a_r + 1 | a_1, \ldots, a_r)$. We denote by $X_{i,j}$ the element $e_i \wedge e_j$ from $\bigwedge^2 E$. Then we define $g_j = X_{j,j+1} \wedge X_{j,j+2} \wedge \ldots \wedge X_{j,j+a_i}$ and $g_\lambda = g_1 \wedge g_2 \wedge \ldots \wedge g_r$. It is easy to check that g_λ is a nonzero U-invariant of weight λ'. ∎

Finally let us mention that all the formulas proven in this section are functorial, so they extend to vector bundles.

2.4. The Schur Complexes

In this section we review the theory of Schur complexes. First we deal with the case of a general $\mathbf{Z}_2$-graded module and define $\mathbf{Z}_2$-graded Schur modules, which are common generalizations of Schur and Weyl modules.

We apply this definition to complexes. Over a field of characteristic zero the Schur complexes obtained in this way have many nice acyclicity properties. We review the main properties of these complexes. Then we discuss the special case of complexes of length 1. It turns out that in this case the acyclicity properties are true in characteristic free settings.

The Schur complexes will be used in section 6.2 when proving the properties of determinantal ideals in positive characteristic and in the discussion of the differentials in resolutions of determinantal ideals.

We work in the category of $\mathbf{Z}_2$-graded free modules over a commutative ring $\mathbf{K}$. The objects of our category are $\mathbf{Z}_2$-graded modules $\Phi = F_0 \oplus F_1$ where both F_0, F_1 are free $\mathbf{K}$-modules. The maps are all $\mathbf{K}$-linear maps of degree 0.

Our theory associates to the $\mathbf{Z}_2$-graded module Φ the family of $\mathbf{Z}_2$-graded modules $L_\lambda \Phi$. They are a common generalization of Schur and Weyl modules. For $F_1 = 0$ we have $L_\lambda \Phi = L_\lambda F_0$, and for $F_0 = 0$ we have $L_\lambda \Phi = K_\lambda F_1[|\lambda|]$, where the bracket denotes shift in homological degree.

The strategy of our approach is similar when defining Schur functors. First we define the exterior and symmetric powers of Φ and then we imitate the definition from section 2.1.

The *i-th exterior power* $\bigwedge^i \Phi$ is a $\mathbf{Z}_2$-graded module defined in the following way.

Consider the i-fold tensor product $\Phi^{\otimes i}$. The permutation $\sigma \in \Sigma_i$ acts of $\Phi^{\otimes i}$ in the following way:

$$\sigma(v_1 \otimes \ldots \otimes v_i) = \pm v_{\sigma^{-1}(1)} \otimes \ldots \otimes v_{\sigma^{-1}(i)},$$

where v_i are homogeneous elements from Φ and the sign $\pm$ is determined by the rule that exchanging the elements v and w contributes the sign $(-1)^{\deg(v)\deg(w)}$. This means that $\pm = (-1)^N$, where $N = \sum_{(i,j)\in \text{Inv}(\sigma)} \deg(v_i)\deg(v_j)$, where we sum over inversions of σ. This formula defines a ($\mathbf{Z}_2$-graded) action of Σ_i on $\Phi^{\otimes i}$.

We define the i-th exterior power of Φ as the subset of antisymmetric elements in $\Phi^{\otimes i}$ with respect to this action of Σ_i. For elements of degree 0 this means antisymmetry of elements, but for degree 1 elements this means symmetry. More precisely, $\bigwedge^i \Phi$ is a $\mathbf{Z}$-graded vector space whose t-th graded piece is

$$\left(\bigwedge^i \Phi \right)_t = D_t F_1 \otimes \bigwedge^{i-t} F_0.$$

The *i-th symmetric power* $S_i \Phi$ is a $\mathbf{Z}$-graded module defined as a factor of $\Phi^{\otimes i}$ by the span of all elements $v - \sigma(v)$ ($v \in \Phi^{\otimes i}$, $\sigma \in \Sigma_i$).

The t-th graded piece of $S_t \Phi$ is

$$(S_i \Phi)_t = \bigwedge^t F_1 \otimes S_{i-t} F_0.$$

(2.4.1) Proposition.

(a) *There exist natural maps of $\mathbf{Z}$-graded modules*

$$\Delta : \bigwedge^{i+j} \Phi \to \bigwedge^i \Phi \otimes \bigwedge^j \Phi$$

whose components are given by the products of exterior and divided diagonals.

(b) *There exist natural maps of $\mathbf{Z}$-graded modules*

$$m : S_i \Phi \otimes S_j \Phi \to S_{i+j} \Phi$$

whose components are given by the products of exterior and symmetric multiplications.

(c) *There exist natural maps of $\mathbf{Z}$-graded modules*

$$m : \bigwedge^i \Phi \otimes \bigwedge^j \Phi \to \bigwedge^{i+j} \Phi$$

whose components are given by the products of exterior and divided multiplications.

In the case when **K** *is a field, all above maps are* $\mathrm{GL}(F_0) \times \mathrm{GL}(F_1)$-*equivariant.*

Proof. We will just define the maps from parts (a), (b), (c) of the proposition.

We start with part (a). Choose the pair of indices a, b such that $0 \le a \le i$, $0 \le b \le j$. We define the component

$$\Delta_{a,b} : \left(\bigwedge^{i+j} \Phi \right)_{a+b} \to \left(\bigwedge^i \Phi \right)_a \otimes \left(\bigwedge^j \Phi \right)_b$$

as the following map:

$$D_{a+b} F_1 \otimes \bigwedge^{i+j-a-b} F_0$$
$$\downarrow \Delta \otimes \Delta$$
$$D_a F_1 \otimes D_b F_1 \otimes \bigwedge^{i-a} F_0 \otimes \bigwedge^{j-b} F_0$$
$$\downarrow t_{23}$$
$$D_a F_1 \otimes \bigwedge^{i-a} F_0 \otimes D_b F_1 \otimes \bigwedge^{j-b} F_0,$$

where t_{23} is the map exchanging the second and third positions in the tensor product.

Now we proceed with part (b) of the proposition. The component

$$m_{a,b} : (S_i \Phi)_a \otimes (S_j \Phi)_b \to (S_{i+j} \Phi)_{a+b}$$

is defined as a composition

$$\bigwedge^a F_1 \otimes S_{i-a} F_0 \otimes \bigwedge^b F_1 \otimes S_{j-b} F_0$$
$$\downarrow t_{23}$$
$$\bigwedge^a F_1 \otimes \bigwedge^b F_1 \otimes S_{i-a} F_0 \otimes S_{j-b} F_0$$
$$\downarrow m \otimes m$$
$$\bigwedge^{a+b} F_1 \otimes S_{i+j-a-b} F_0.$$

Finally, the component

$$m_{a,b} : \left(\bigwedge^i \Phi \right)_a \otimes \left(\bigwedge^j \Phi \right)_b \to \left(\bigwedge^{i+j} \Phi \right)_{a+b}$$

is defined as a composition

$$D_a F_1 \otimes \bigwedge^{i-a} F_0 \otimes D_b F_1 \otimes \bigwedge^{j-b} F_0$$
$$\downarrow t_{23}$$
$$D_a F_1 \otimes D_b F_1 \otimes \bigwedge^{i-a} F_0 \otimes \bigwedge^{j-b} F_0$$
$$\downarrow m \otimes m$$
$$D_{a+b} F_1 \otimes \bigwedge^{i+j-a-a} F_0.$$

This completes the proof of the proposition. ∎

We call the maps Δ and m from Proposition (2.4.1) the diagonal and multiplication maps.

We could also define the diagonal maps on symmetric powers, and the multiplication maps on exterior powers, but they will not be needed in our application.

Now we are ready to define the $\mathbf{Z}_2$-graded Schur modules.

Let $\Phi = F_0 \oplus F_1$ be as above, and let λ be a partition. For two partitions $\mu \subset \lambda$ we define

$$L_{\lambda/\mu}\Phi = \overset{\lambda_1-\mu_1}{\bigwedge} \Phi \otimes \overset{\lambda_2-\mu_2}{\bigwedge} \Phi \otimes \ldots \otimes \overset{\lambda_s-\mu_s}{\bigwedge} \Phi / R(\lambda/\mu, \Phi),$$

where $R(\lambda/\mu, \Phi)$ is the sum of submodules:

$$\overset{\lambda_1-\mu_1}{\bigwedge} \Phi \otimes \ldots \otimes \overset{\lambda_{a-1}-\mu_{a-1}}{\bigwedge} \Phi \otimes R_{a,a+1}(\Phi) \otimes \overset{\lambda_{a+2}-\mu_{a+2}}{\bigwedge} \Phi$$
$$\otimes \ldots \otimes \overset{\lambda_s-\mu_s}{\bigwedge} \Phi$$

for $1 \leq a \leq s-1$, where $R_{a,a+1}(\Phi)$ is the submodule spanned by the images of the following maps $\Theta(a, u, v, \Phi)$:

$$\overset{u}{\bigwedge} \Phi \otimes \overset{\lambda_a-\mu_a-u+\lambda_{a+1}-\mu_{a+1}-v}{\bigwedge} \Phi \otimes \overset{v}{\bigwedge} \Phi$$
$$\downarrow \, 1 \otimes \Delta \otimes 1$$
$$\overset{u}{\bigwedge} \Phi \otimes \overset{\lambda_a-\mu_a-u}{\bigwedge} \Phi \otimes \overset{\lambda_{a+1}-\mu_{a+1}-v}{\bigwedge} \Phi \otimes \overset{v}{\bigwedge} \Phi$$
$$\downarrow \, m_{12} \otimes m_{34}$$
$$\overset{\lambda_a-\mu_a}{\bigwedge} \Phi \otimes \overset{\lambda_{a+1}-\mu_{a+1}}{\bigwedge} \Phi$$

for $u + v < \lambda_{a+1} - \mu_a$.

The next step is the description of the standard basis in $L_{\lambda/\mu}\Phi$.

Let $f_1, \ldots, f_m$ and $g_1, \ldots, g_n$ be fixed bases in F_0 and F_1 respectively. We consider the $\mathbf{Z}_2$-graded set $A = (A_0, A_1)$ where $A_0 = [1, m]$, $A_1 = [1, n]$.

Let us recall from section 1.1.2 that a $\mathbf{Z}_2$-graded tableau of shape λ/μ with values in A is a map $T : D(\lambda/\mu) \to A$.

We define the map $\varphi : A \to F_0 \oplus F_1$ by setting $\varphi(i) = f_i$ for $i \in A_0$ and $\varphi(j) = g_j$ for $j \in A_1$.

Let T be a $\mathbf{Z}_2$-graded tableau of shape λ with the values in A. We associate to T the element in $L_\lambda \Phi$ which is a coset of the tensor

$$\varphi(T(1, \mu_1 + 1)) \wedge \ldots \wedge \varphi(T(1, \lambda_1)) \otimes \varphi(T(2, \mu_2 + 1)) \wedge \ldots$$
$$\wedge \varphi(T(2, \lambda_2)) \otimes \ldots \otimes \varphi(T(s, \mu_s + 1)) \wedge \ldots \wedge \varphi(T(s, \lambda_s)).$$

In the sequel we will identify these two objects and we will call both of them the ($\mathbf{Z}_2$-graded) tableaux of shape λ/μ.

Let us order the set A by an arbitrary order $\lhd$. Let us recall that in section 1.1 we defined a standard $\mathbf{Z}_2$-graded tableau relative to the order $\lhd$ as a tableau satisfying the conditions

(1) $T(u, v) \lhd T(u, v + 1)$ with equality possible when $T(u, v) \in A_1$,
(2) $T(u, v) \lhd T(u + 1, v)$ with equality possible when $T(u, v) \in A_0$.

We have the following generalization of Proposition (2.1.9) (b).

(2.4.2) Proposition. *Let us fix the order $\lhd$ on A. The standard $\mathbf{Z}_2$-graded tableaux of shape λ/μ with values in A form a basis of the module $L_{\lambda/\mu}\Phi$.*

Proof of proposition (2.4.2). First we prove that the standard $\mathbf{Z}_2$-graded tableaux generate $L_{\lambda/\mu}\Phi$. It is clear that the row standard $\mathbf{Z}_2$-graded tableaux generate $L_{\lambda/\mu}\Phi$. Let us order the set of such tableaux by the order $\preceq$ defined in section 1.1.1. We will prove that if the $\mathbf{Z}_2$-graded tableau T is not standard, then we can express it modulo $R(\lambda/\mu, \Phi)$ as a combination of earlier $\mathbf{Z}_2$-graded tableaux. Since T is not standard, we can find a and w for which $T(a, w) (T(a + 1, w)$ with possible equality if $T(a, w) \in A_1$. Let w' be such an index that $T(a + 1, w) = T(a + 1, w + 1) = \cdots = T(a + 1, w') \lhd T(a + 1, w' + 1)$. We consider the map $\Theta(a, u, v; \Phi)$ for $u = w - \mu_a - 1$ and $v = \lambda_{a+1} - w'$. The key observation is that the image of the tensor $U_1 \otimes \ldots \otimes U_{a-1} \otimes V_1 \otimes V_2 \otimes V_3 \otimes U_{a+2} \otimes \ldots \otimes U_s$, where

$$U_j = \varphi(T(j, \mu_j + 1)) \wedge \ldots \wedge \varphi(T(j, \lambda_j - \mu_j))$$

for $j \neq a, a + 1$, and where

$$V_1 = \varphi(T(a, \mu_a + 1)) \wedge \varphi(T(a, \mu_a + 2)) \wedge \ldots \wedge \varphi(T(a, w - 1)),$$
$$V_2 = \varphi(T(a, w)) \wedge \ldots \wedge \varphi(T(a, \lambda_a - \mu_a)) \wedge \varphi(T(a + 1, \mu_{a+1} + 1))$$
$$\wedge \ldots \wedge \varphi(T(a + 1, w')),$$
$$V_3 = \varphi(T(a + 1, w' + 1)) \wedge \ldots \wedge \varphi(T(a + 1, \lambda_{a+1} - \mu_{a+1})),$$

contains the tableau T with the coefficient 1, and all the other tableaux occurring in this image are earlier than T.

It remains to prove that the standard tableaux are linearly independent in $L_{\lambda/\mu}\Phi$. Let us consider a map

$$H_{\lambda/\mu} : \overset{\lambda_1 - \mu_1}{\bigwedge} \Phi \otimes \overset{\lambda_2 - \mu_2}{\bigwedge} \Phi \otimes \ldots \otimes \overset{\lambda_s - \mu_s}{\bigwedge} \Phi \overset{\alpha}{\longrightarrow} \otimes_{(i,j)\in D(\lambda/\mu)} \Phi(i, j)$$
$$\overset{\beta}{\longrightarrow} S_{\lambda'_1 - \mu'_1} \Phi \otimes S_{\lambda'_2 - \mu'_2} \Phi \otimes \ldots \otimes S_{\lambda'_t - \mu'_t} \Phi,$$

where α is the tensor product of exterior diagonals

$$\delta : \bigwedge^{\lambda_j - \mu_j} \Phi \to \Phi(j, \mu_j + 1) \otimes \Phi(j, \mu_j + 2) \otimes \ldots \otimes \Phi(j, \lambda_j - \mu_j)$$

and β is the tensor product of multiplications

$$m : \Phi(\mu_i' + 1, i) \otimes \Phi(\mu_i' + 2, i) \otimes \ldots \otimes \Phi(\lambda_i', i) \to S_{\lambda_i' - \mu_i'} \Phi.$$

The map $H_{\lambda/\mu}$ is called the $\mathbf{Z}_2$-graded Schur map associated to the partition λ/μ.

The straightforward calculation (compare [ABW2, II.2]) shows that

(2.4.3) Proposition. *The image* $H_\lambda(R(\lambda/\mu, \Phi))$ *equals* 0.

This means that $H_{\lambda/\mu}$ induces a surjective map from $L_{\lambda/\mu}\Phi$ to Im $H_{\lambda/\mu}$. Now we will show that the map $H_{\lambda/\mu}$ maps standard tableaux to linearly independent elements of

$$S_{\lambda_1' - \mu_1'} \Phi \otimes S_{\lambda_2' - \mu_2'} \Phi \otimes \ldots \otimes S_{\lambda_t' - \mu_t'} \Phi.$$

This will show (2.4.2), and at the same time it will prove that $L_{\lambda/\mu}\Phi = $ Im $H_{\lambda/\mu}$.

A typical basis element of $S_t \Phi$ is $\varphi(s_1) \ldots \varphi(s_t)$ where $\varphi(s_1) \lhd \varphi(s_2) \lhd \ldots \lhd \varphi(s_t)$ with equality $\varphi(s_i) = \varphi(s_{i+1})$ allowed only when $s_i \in A_0$. We order these elements lexicographically with respect to the sequence $(s_1, \ldots, s_t)$. We denote this order by $\ll$. If $w = w_1 \otimes w_2 \otimes \ldots \otimes w_t$ and $w = w_1' \otimes w_2' \otimes \ldots \otimes w_t'$ are two tensor products of such elements in $S_{\lambda_1' - \mu_1'} E \otimes S_{\lambda_2' \mu_2'} E \otimes \ldots \otimes S_{\lambda_t' - \mu_t'} E$, we say that $w \ll w'$ iff $w_j \ll w_j'$ for the smallest j for which $w_j \neq w_j'$.

If T is a standard $\mathbf{Z}_2$-graded tableau of shape λ/μ, then the smallest element (with respect to the order we just defined) occurring in $H_{\lambda/\mu}(T)$ is

$$\varphi(T(\mu_1' + 1, 1)) \ldots \varphi(T(\lambda_1', 1)) \otimes \ldots \otimes \varphi(T(\mu_t' + 1, t)) \ldots \varphi(T(\lambda_t', t)).$$

Indeed, if, in applying the map α, we make the exchange of elements in some row, then we put bigger elements in earlier columns, so we get the earlier (with respect to $\ll$) elements. Moreover, it follows easily from the definitions that

$$\varphi(T(\mu_1' + 1, 1)) \ldots \varphi(T(\lambda_1', 1)) \otimes \ldots \otimes \varphi(T(\mu_t' + 1, t)) \ldots \varphi(T(\lambda_t', t))$$

occurs in $\alpha(T)$ with coefficient 1. It is also obvious that the elements

$$\varphi(T(\mu_1' + 1, 1)) \ldots \varphi(T(\lambda_1', 1)) \otimes \ldots \otimes \varphi(T(\mu_t' + 1, t)) \ldots \varphi(T(\lambda_t', t))$$

are different for different standard tableaux T. This proves that the images $H_{\lambda/\mu}T$ of standard $\mathbf{Z}_2$-graded tableaux T are linearly independent.●

(2.4.4) Example.

(a) *The module $L_{2,1}\Phi$ is by definition the cokernel of the diagonal map $\Delta : \bigwedge^3 \Phi \to \bigwedge^2 \Phi \otimes \Phi$. It is a direct sum*

$$(L_{2,1}\Phi)_3 \oplus (L_{2,1}\Phi)_2 \oplus (L_{2,1}\Phi)_1 \oplus (L_{2,1}\Phi)_0.$$

There are two basic descriptions of the graded components $(L_{2,1}\Phi)_i$. If we choose the order $\lhd$ so $F_0 \lhd F_1$, then $(L_{2,1}\Phi)_3 = K_{2,1}F_1$, $(L_{2,1}\Phi)_2 = F_1 \otimes F_1 \otimes F_0$, $(L_{2,1}\Phi)_1$ has a filtration with associated graded object $F_1 \otimes \bigwedge^2 F_0 \oplus F_1 \otimes S_2F_0$, and $(L_{2,1}\Phi)_0 = L_{2,1}F_0$. If we set $F_1 \lhd F_0$, we get $(L_{2,1}\Phi)_3 = K_{2,1}F_1$, $(L_{2,1}\Phi)_2$ has a filtration with associated graded object $D_2F_1 \otimes F_0 \oplus \bigwedge^2 F_1 \otimes F_0$, $(L_{2,1}\Phi)_1 = F_1 \otimes F_0 \otimes F_0$, and $(L_{2,1}\Phi)_0 = L_{2,1}F_0$.

(b) *The module $L_{2,2}\Phi$ is by definition the factor of $\bigwedge^2 \Phi \otimes \bigwedge^2 \Phi$ divided by the following images of maps $\Theta(1, u, v; \Phi)$: $\bigwedge^3 \Phi \otimes \Phi \to \bigwedge^2 \Phi \otimes \bigwedge^2 \Phi$ (corresponding to $u = 0, v = 1$) and $\bigwedge^4 \Phi \to \bigwedge^2 \Phi \otimes \bigwedge^2 \Phi$ (corresponding to $u = v = 0$). It is a direct sum*

$$(L_{2,2}\Phi)_4 \oplus (L_{2,2}\Phi)_3 \oplus (L_{2,2}\Phi)_2 \oplus (L_{2,2}\Phi)_1 \oplus (L_{2,2}\Phi)_0.$$

The graded components have the following descriptions, the same for both possible orders: $(L_{2,2}\Phi)_4 = K_{2,2}F_1$, $(L_{2,2}\Phi)_3 = K_{2,1}F_1 \otimes F_0$, $(L_{2,2}\Phi)_1 = L_{2,1}F_0 \otimes F_1$ $(L_{2,2}\Phi)_0 = L_{2,2}F_0$. Here we use the isomorphisms $K_{2,2/1}F_1 \cong K_{2,1}F_1$, $L_{2,2/1}F_0 \cong L_{2,1}F_0$. The middle component has similar description whether we use the order $F_0 \lhd F_1$ or $F_1 \lhd F_0$. The module $(L_{2,2}\Phi)_2$ has a filtration with the associated graded object $D_2F_1 \otimes \bigwedge^2 F_0 \oplus \bigwedge^2 F_1 \otimes S_2F_0$.

As a consequence of (2.4.2) we prove the following properties of $\mathbf{Z}_2$-graded Schur modules.

(2.4.5) Theorem. *The Schur modules $L_{\lambda/\mu}\Phi$ have the following properties:*

(a) *The t-th term $(L_{\lambda/\mu}\Phi)_t$ has a natural filtration with the associated graded object*

$$\bigoplus_{|\nu|=|\lambda|-t} K_{\lambda/\nu}F_1 \otimes L_{\nu/\mu}F_0.$$

(b) *The t-th term* $(L_{\lambda/\mu}\Phi)_t$ *has a natural filtration with the associated graded object*

$$\bigoplus_{|v|=t} K_{v/\mu}F_1 \otimes L_{\lambda/v}F_0.$$

Proof. We start with (a). Let us choose the order $\lhd$ by setting $A_0 \lhd A_1$ and $i \lhd j$ if and only if $i < j$ for $i, j \in A_s$ for $s = 0, 1$.

Let us order all sequences $(u_1, \ldots, u_s)$ by saying that $(u_1, \ldots, u_s) \preceq (v_1, \ldots, v_s)$ if $u_j > v_j$ for the smallest j for which $u_j \neq v_j$. For each v we define $(\mathcal{F}_{\leq v}\Phi)_s$ as the span of the images of

$$\left(\overset{\lambda_1-\mu_1}{\bigwedge}\Phi\right)_{u_1} \otimes \cdots \otimes \left(\overset{\lambda_s-\mu_s}{\bigwedge}\Phi\right)_{u_s},$$

where the sequence $(u_1, \ldots, u_s)$ is $\preceq$ than $(v_1, \ldots, v_s)$. We notice that the proof of (2.4.2) implies that if we take the tableau T from $(\mathcal{F}_{\leq v}\Phi)_s$ and we standardize it, we express it as a linear combination of earlier standard tableaux from $(\mathcal{F}_{\leq v}\Phi)_s$. Let us order all partitions v by the order $\preceq$, and let us consider the factor

$$(\mathcal{F}_{\leq v}\Phi)_s / \sum_{\eta < v}(\mathcal{F}_{\leq \eta}\Phi)_s.$$

This space is generated by all the $\mathbf{Z}_2$-graded tableaux T of shape λ/μ where elements from A_0 occupy places from v/μ and elements from A_1 occupy places from λ/v. In fact, by analyzing the relations $R(\lambda, \Phi)$ it is easy to see that the relations $R(v/\mu, F_0)$ in the F_0-part of the tableau and the relations for $U(\lambda/vF_1)$ in the F_1-part of the tableau are satisfied in $(\mathcal{F}_{\leq v}\Phi)_s / \sum_{\eta<v}(\mathcal{F}_{\leq \eta}\Phi)_s$.

It follows that $(\mathcal{F}_{\leq v}\Phi)_s / \sum_{\eta<v}(\mathcal{F}_{\leq \eta}\Phi)$ is a factor of $K_{\lambda/v}F_1 \otimes L_{v/\mu}F_0$. In fact these two modules are equal, because by (2.4.2)

$$\dim (L_{\lambda/\mu}\Phi)_s = \sum_{|v|=|\lambda|-s} \dim K_{\lambda/v}F_1 \otimes L_{v/\mu}F_0.$$

This proves the first part of (a).

To prove (b) we use the same argument as above, choosing the order $\lhd$ in such way that the elements of A_1 are earlier then the elements of A_0. $\blacksquare$

The $\mathbf{Z}_2$-graded Schur modules have also nice properties with respect to direct sum decomposition

(2.4.6) Theorem. *Let* $\Phi = F_0 \oplus F_1$, $\Psi = G_0 \oplus G_1$ *be two* $\mathbf{Z}_2$-*graded modules. Then there is a functorial (in Φ and Ψ) filtration on $L_{\lambda/\mu}(\Phi \oplus \Psi)$ with*

the associated graded object

$$\bigoplus_{\mu \subset \nu \subset \lambda} L_{\lambda/\nu}\Phi \otimes L_{\nu/\mu}\Psi.$$

Proof. The proof is the repetition of the analogous fact for Schur modules. We use the order on the basis of $\Phi \oplus \Psi$ in which the basis of Φ precedes the basis of Ψ. The standard basis theorem gives us the bijections between the bases of the left and right sides of our formula, but we can also define the filtration by ordering summands according to ν as we did in proving (2.3.1). ∎

In the next part of the section we assume that characteristic of **K** is zero (i.e. that **K** is a **Q**-algebra). We work in the category of chain complexes of free modules over **K**. A typical object in this category will be written

$$\Phi : 0 \to \Phi_n \to \Phi_{n-1} \to \ldots \to \Phi_1 \to \Phi_0 \to 0.$$

The permutation group Σ_m acts on the tensor product $\Phi^{\otimes m}$. In order for the action to commute with the differentials, we need to use the $\mathbf{Z}_2$-graded action. We introduce a $\mathbf{Z}_2$-graduation on Φ by setting $F_0 = \bigoplus_{i\ even} \Phi_i$, $F_1 = \bigoplus_{i\ odd} \Phi_i$. Then our differential has degree 1. The action of Σ_m on $\Phi^{\otimes m}$ commutes with the differential on $\Phi^{\otimes m}$.

We can now apply the machinery of (2.4.1)–(2.4.6) to define *the Schur complexes* $L_{\lambda/\mu}\Phi$. All above constructions commute with a differential induced from Φ.

This theory is satisfactory only in characteristic zero. The point is that when making such definitions one has in mind the following acyclicity properties.

(2.4.7) Proposition. *Let* **K** *be a commutative ring of characteristic* 0.

   (a) *Let* Φ *be an exact complex of free* **K**-*modules. Then the complex* $L_{\lambda/\mu}\Phi$ *is exact unless* $\lambda = \mu$.
   (b) *Let* $\Phi \cong \Phi_0 \oplus \Psi$ *with* Ψ *exact. Then the natural inclusion* $L_{\lambda/\mu}\Phi_0 \to L_{\lambda/\mu}\Phi$ *is a quasiisomorphism.*

Proof. To prove (a) we observe that by localization we can assume that **K** is local. Every exact complex of free modules over a local commutative ring splits, so it is enough to prove the statement for a split exact complex Φ.

A split exact complex is a direct sum of complexes $\mathbf{K} \overset{\mathrm{id}}{\to} \mathbf{K}$ with shifted homological degree. Using (2.4.6) we reduce to Φ being such a complex.

Notice that the complex $L_{\lambda/\mu}\Phi$ will be identically zero if the skew diagram λ/μ contains the 2×2 rectangle. Let us assume it does not. Moreover, let us assume that the diagram $D(\lambda/\mu)$ is connected. If not, the Schur complex will be a tensor product of Schur complexes associated to connected components of $D(\lambda/\mu)$ so it is enough to show exactness for these. Such shapes are called skew strips. They are essentially the sequences of horizontal and vertical strips joined at the corners. Let us choose bases for components of Φ. The corresponding Z_2-graded set is $A = A_0 \cup A_1$ where A_0 and A_1 have both one element. Let us denote $A_0 = \{0\}$, $A_1 = \{1\}$. Consider the standard tableaux of the shape λ/μ with values in A. We see that if a box of λ/μ has a neighbor to the left, it is forced to be filled with 1, and if it has a neighbor below it, it is forced to be filled with 0. But now we see that all boxes of λ/μ are filled, except for the one at the left lower end, which might be filled with 0 or 1. Using (2.4.5) we deduce that the complex $L_{\lambda/\mu}\Phi$ is isomorphic (up to shift in grading) to Φ and is therefore exact. $\blacksquare$

(2.4.8) Remarks.
   (a) *The complexes $L_\lambda\Phi$ can be also defined using Young idempotents.*
   (b) *The analogues of the Littlewood–Richardson rule and the Cauchy formulas are also true for complexes $L_\lambda\Phi$.*
   (c) *The theory of Schur complexes in characteristic zero was first described in an unpublished paper [Ni] of Nielsen.*

(2.4.9) Example. *Proposition (2.4.7) is not true over an arbitrary ring* **K**. *Consider the complex*

$$\Phi : 0 \to \Phi_2 \to \Phi_1 \to \Phi_0$$

of length 2. The complex $\bigwedge^2 \Phi$ has the nonzero terms

$$0 \to \bigwedge^2 \Phi_2 \to \Phi_2 \otimes \Phi_1 \to \Phi_2 \otimes \Phi_0 \oplus D_2\Phi_1 \to \Phi_1 \otimes \Phi_0 \to \bigwedge^2 \Phi_0.$$

Let us assume now that $\Phi_0 = 0$ but that the complex Φ is exact. Our complex reduces to

$$0 \to \bigwedge^2 \Phi_2 \to \Phi_2 \otimes \Phi_1 \to D_2\Phi_1.$$

The right hand map cannot be an epimorphism in characteristic 2, because it does not cover the image of the map $D_2\Phi_2 \to D_2\Phi_1$, which is the second divided power applied to the original differential in Φ.

The example above shows that for characteristic free theory one needs bigger complexes. Such theory can be developed for general polynomial functors; cf. [TW].

However, it turns out that for complexes concentrated in degrees 1 and 0 the Schur complexes have right acyclicity properties in characteristic free setting.

Let $\Phi : \Phi_1 \to \Phi_0$ be a linear map of free **K**-modules. We start with the definition of differentials on complexes $\bigwedge^m \Phi$ and $S_m \Phi$. The differential on $\bigwedge^m \Phi$ can be defined as the composition

$$D_i \Phi_1 \otimes \overset{m-i}{\bigwedge} \Phi_0 \to D_{i-1}\Phi_1 \otimes \Phi_1 \otimes \overset{m-i}{\bigwedge} \Phi_0 \to D_{i-1}\Phi_1 \otimes$$
$$\Phi_0 \otimes \overset{m-i}{\bigwedge} \Phi_0 \to D_{i-1}\Phi_1 \otimes \overset{m-i+1}{\bigwedge} \Phi_0,$$

where the first map is the diagonal on Φ_1, the second is the differential of Φ tensored with identity components, and the third one is the exterior multiplication on Φ_0.

Similarly, the differential on $S_m \Phi$ can be defined as the composition

$$\overset{i}{\bigwedge} \Phi_1 \otimes S_{m-i}\Phi_0 \to \overset{i-1}{\bigwedge} \Phi_1 \otimes \Phi_1 \otimes S_{m-i}\Phi_0 \to \overset{i-1}{\bigwedge} \Phi_1 \otimes$$
$$\Phi_0 \otimes S_{m-i}\Phi_0 \to \overset{i-1}{\bigwedge} \Phi_1 \otimes S_{m-i+1}\Phi_0,$$

where the first map is the exterior diagonal on Φ_1, the second is the differential of Φ tensored with identity components, and the third is the multiplication on Φ_0.

Then the construction of modules $L_{\lambda/\mu}\Phi$ commutes with the differentials and thus leads to (characteristic free) Schur complexes. We denote them again by $L_{\lambda/\mu}\Phi$. If characteristic of the field **K** is zero and Φ is concentrated in degrees 0 and 1, the two definitions overlap.

Let us collect all properties of characteristic free Schur complexes we need.

(2.4.10) Theorem. *Let* **K** *be an arbitrary commutative ring. Assume that* Φ *is concentrated in degrees 0 and 1, i.e.* $\Phi_i = 0$ *for* $i > 1$. *We denote* $\Phi_0 := F_0$, $\Phi_1 := F_1$.

    (a) The t-th component of $L_{\lambda/\mu}\Phi$ *has a natural filtration with associated graded object*

$$\bigoplus_{|v|=|\lambda|-t} K_{\lambda/v} F_1 \otimes L_{v/\mu} F_0.$$

(b) *The t-th component of $L_{\lambda/\mu}\Phi$ has a natural filtration with associated graded object*

$$\bigoplus_{|v|=|\mu|+t} K_{v/\mu}F_1 \otimes L_{\lambda/v}F_0.$$

(c) *The complex $L_{\lambda/\mu}\Phi$ has for each s a natural subcomplex $L_{\lambda/\mu}\Phi(s)$ whose t-th term has a natural filtration with the associated graded object*

$$\bigoplus_{|v|=|\lambda|-t, v_1 \geq s} K_{\lambda/v}F_1 \otimes L_{v/\mu}F_0.$$

(d) *If Φ is exact, then $L_{\lambda/\mu}\Phi$ is exact for $\lambda \neq \mu$.*

(e) *If $\Phi = \Phi' \oplus \Psi$ with Ψ exact, then the natural embedding $L_{\lambda/\mu}\Phi' \to L_{\lambda/\mu}\Phi$ is a quasiisomorphism.*

Proof. Parts (a) and (b) follow from (2.4.5). To prove part (c) we choose the ordering on the basis of Φ so the basis of Φ_0 preceeds the basis of Φ_1. We define the subcomplex $L_{\lambda/\mu}\Phi(s)$ to be the span of tableaux having $\geq s$ elements from Φ_0 in the fist row of the diagram. It is clear we get a subcomplex, and the rest of (c) follows from (2.5.2), because this subcomplex is a part of filtration whose associated graded is described in (a).

Part (e) follows from part (d) using Theorem (2.4.6), To prove part (d) we notice that, using Theorem (2.4.6), it is enough to show that the complexes $L_{\lambda/\mu}\Phi$ are exact for $\lambda \neq \mu$ and $\Phi : \mathbf{K} \to \mathbf{K}$ with the differential equal to identity. Notice that such a complex will be identically zero if the skew diagram λ/μ contains the 2×2 rectangle. Let us assume it does not. Moreover, let us assume that the diagram $D(\lambda/\mu)$ is connected. If not, the Schur complex will be a tensor product of Schur complexes associated to connected components of $D(\lambda/\mu)$, so it is enough to show exactness for these. Such shapes are called *skew strips*. They are essentially sequences of horizontal and vertical strips joined at the corners. Let us choose bases for components of Φ. The corresponding $\mathbf{Z}_2$-graded set is $A = A_0 \cup A_1$, where A_0 and A_1 each have one element. Let us denote $A_0 = \{0\}$, $A_1 = \{1\}$. Consider the standard tableaux of the shape λ/μ with values in A. We see that if a box of λ/μ has a neighbor to the left, it is forced to be filled with 1, and if it has a neighbor below it, it is forced to be filled with 0. But now we see that all boxes of λ/μ are filled, except for the one at the left lower end, which might be filled with 0 or 1. This means the complex $L_{\lambda/\mu}\Phi$ is isomorphic (up to shift in grading) to Φ and is therefore exact. ∎

(2.4.11) Remark. *We will use part (c) of Theorem (2.4.10) in section 6.2, where the complexes $L_{\lambda/\mu}\Phi(s)$ will be used to calculate some sheaf cohomology.*

Exercises for Chapter 2

Definition of Schur Modules

1. Let $\lambda = (\lambda_1, \ldots, \lambda_s)$ be a partition.
   (a) (Towber [To]) Prove that to define the Schur module $L_\lambda E$ it is enough to take the relations $\theta(a, u, v; E)$ with $u = 0$, $v < \lambda_{a+1}$.
   (b) Prove that if **K** is a commutative ring of characteristic 0, then to define the Schur module $L_\lambda E$ it is enough to take the relations $\theta(a, u, v; E)$ with $u = 0$, $v = \lambda_{a+1} - 1$ for $a = 1, \ldots, s - 1$.

2. Let **K** be a commutative ring. Let E be a free **K**-module of dimension n. Consider a complex of free modules

$$K_m(E) : 0 \to \overset{m}{\bigwedge} E \to \ldots \to \overset{m-i}{\bigwedge} E \otimes S_i E \to \overset{m-i-1}{\bigwedge} E \otimes S_{i+1} E$$
$$\to \ldots \to S_m E \to 0$$

   with the differential $\partial_i : \bigwedge^{m-i} E \otimes S_i E \to \bigwedge^{m-i-1} E \otimes S_{i+1} E$ defined as a composition

$$\overset{m-i}{\bigwedge} E \otimes S_i E \xrightarrow{\Delta \otimes 1} \overset{m-i-1}{\bigwedge} E \otimes E \otimes S_i E \xrightarrow{1 \otimes m} \overset{m-i-1}{\bigwedge} E \otimes S_{i+1} E.$$

   Show that for $m > 0$ the complex $K_m(E)$ is split exact and that for $i = 1, \ldots, m$ the module of cycles of $K_m(E)$ at $\bigwedge^{m-i} E \otimes S_i E$ is isomorphic to $L_{m-i+1,1^{i-1}} E$.

3. ([AB1]) Let λ, μ be two-rowed partitions. Assume that λ/μ has t overlaps, i.e. that $D(\lambda/\mu)$ has t columns of length 2. Assume $\lambda_1 > \lambda_2$, $\mu_1 > \mu_2$. Define $\lambda(1) = (\lambda_1 - 1, \lambda_2)$, $\mu(1) = (\mu_1 - 1, \mu_2)$. The shape $\lambda(1)/\mu(1)$ has rows of the same length as λ/μ, but the first row is shifted by one place to the left, so we have t+1 overlaps. Show that there is a natural epimorphism

$$\pi(\lambda, \mu) : L_{\lambda/\mu} E \to L_{\lambda(1)/\mu(1)} E.$$

   Show that the kernel of $\pi(\lambda, \mu)$ is isomorphic to $L_{\lambda_1 - \mu_2, \lambda_2 - \mu_1} E$ (with the convention that Ker $\pi(\lambda, \mu) = 0$ if $\lambda_2 < \mu_1$). Formulate and prove the analogous result for skew Weyl modules.

Schur and Weyl Modules in Positive Characteristic

4. Define two morphisms $j_d : S_d E \to D_d E, i_d : D_d E \to S_d E$ by formulas

$$i_d(e_1^{a_1} \ldots e_n^{a_n}) = \frac{n!}{(a_1)! \ldots (a_n)!} e_1^{(a_1)} \ldots e_n^{(a_n)},$$

$$j_d(e_1^{(a_1)} \ldots e_n^{(a_n)}) = (a_1)! \ldots (a_n)! e_1^{a_1} \ldots e_n^{a_n}.$$

Prove that i_d, j_d define GL(E)-equivariant maps and that the compositions $i_d j_d = n!(\mathrm{id}_{S_d E})$, $j_d i_d = n!(\mathrm{id}_{D_d E})$.

5. Let $\lambda = (\lambda_1, \ldots, \lambda_s)$ be a partition.

   (a) Define a morphism

   $$\tilde{j}_\lambda : \overset{\lambda_1}{\bigwedge} E \otimes \ldots \otimes \overset{\lambda_s}{\bigwedge} E \overset{\phi_\lambda}{\longrightarrow} S_{\lambda'_1} E \otimes \ldots \otimes S_{\lambda'_t} E$$
   $$\overset{j_{\lambda'_1} \otimes \ldots \otimes j_{\lambda'_s}}{\longrightarrow} D_{\lambda'_1} E \otimes \ldots \otimes D_{\lambda'_t} E.$$

   Prove that $\tilde{j}_\lambda$ factors to give an equivariant map

   $$j_\lambda : L_\lambda E \to K_\lambda E.$$

   (b) Define a morphism

   $$\tilde{i}_\lambda : D_{\lambda_1} E \otimes \ldots \otimes D_{\lambda_s} E \overset{\psi_\lambda}{\longrightarrow} \overset{\lambda'_1}{\bigwedge} E \otimes \ldots \otimes \overset{\lambda'_t}{\bigwedge} E.$$

   Prove that $\tilde{i}_\lambda$ factors to give an equivariant map

   $$i_\lambda : K_{\lambda'} E \to L_\lambda E.$$

   (c) Prove that $i_\lambda j_{\lambda'} = h_\lambda(\mathrm{id}_{L_{\lambda'} E})$, $j_\lambda i_{\lambda'} = h_\lambda(\mathrm{id}_{K_\lambda E})$, where $h_\lambda = \prod_{(x,y) \in \lambda} h_\lambda(x, y)$, where $h_\lambda(x, y) = \lambda_y - x + \lambda'_x - y + 1$ is the hook length of a hook in λ with the corner at (x, y).

   (d) Deduce that over a field of characteristic $p > 0$ the module $L_\lambda E$ is irreducible as long as λ does not contain a box (x, y) such that $h_\lambda(x, y)$ is divisible by p.

6. We call a partition λ *p-regular* if it does not contain p rows of the same length. Otherwise we call λ *p-singular*.

   (a) For a partition λ define $p^i \lambda = (p^i \lambda_1, \ldots, p^i \lambda_s)$. Prove that every partition can be written uniquely as $\lambda = \sum_{j \geq 0} p^j \lambda(j)$ where $\lambda(j)$ are p-regular partitions. We call this decomposition a *p-adic decomposition* of λ.

(b) Let $|\lambda| = n = \dim E$. Prove that the module $M_\lambda E$ contains a nonzero element of weight (1^n) if λ is p-regular.

(c) Let $\mathbf{K}$ be an infinite field of characteristic $p > 0$. We define the Frobenius functor $Fr : Vect_\mathbf{K} \to Vect_\mathbf{K}$ by setting $Fr(E) = E$ and $Fr(\phi) = \phi^{(p)}$, where $\phi^{(p)}$ is a matrix $(\phi_{i,j}^p)_{1 \le i \le m, 1 \le j \le n}$. For a functor $H : Vect_\mathbf{K} \to Vect_\mathbf{K}$ we denote $H^{(i)} = H \circ Fr^i$. This means that we take "the same" functor as H on object, but when evaluating $H \circ Fr^i$ on a linear map η we raise every entry of the matrix $H(\eta)$ to the power p^i.

(d) Define a $\mathbf{U} - p$-invariant in $L_\lambda E$ to be a vector $v \in L_\lambda E$ such that for the generic matrix $t := \mathrm{id} + \sum_{i<j} t_{i,j} E_{i,j}$ from $\mathbf{U}$

$$tv = v + \sum_\alpha \prod_{i<j} t_{i,j}^{p\alpha_{i,j}} v_\alpha.$$

Prove that if v is a $\mathbf{U} - p$-invariant, then all vectors v_α are also $\mathbf{U} - p$-invariants. Prove that if λ is p-regular, then the only $\mathbf{U} - p$-invariant in $L_\lambda E$ is the canonical tableau.

(e) (Steinberg theorem) Let λ be a partition, $\lambda = \sum_{j \ge 0} p^j \lambda(j)$ its p-adic decomposition. Prove that

$$M_\lambda E = \bigotimes_{j \ge 0} M_{\lambda(j)}^{(j)} E.$$

Deduce that the reverse implication in (b) is also true.

7. Let $\mathbf{K}$ be an infinite field of characteristic $p > 0$.

(a) Prove that the exterior power $\bigwedge^i E$ is an irreducible representation of $GL(E)$.

(b) Consider the symmetric power $S_d E$. Assume $p^i \le d < p^{i+1}$. The sequence $\underline{d} = (d_0, \ldots, d_i)$ is a *p-adic representation* of d if $d = d_0 + d_1 p + \ldots + d_i p^i$. Assume that $n = \dim E \ge d$. For any p-adic representation $\underline{d}$ of d we denote by $N_{\underline{d}}$ the $GL(E)$-submodule of $S_d E$ generated by the weight vector of weight $(1^{d_0}, p^{d_1}, \ldots, (p^i)^{d_i})$. Describe $N_{\underline{d}}$, and prove that these are the only equivariant subspaces in $S_d E$.

(c) Let $\underline{d}, \underline{e}$ be two p-adic representations of the same number d. We say that $\underline{e}$ is a *refinement* of $\underline{d}$ if it can be obtained from $\underline{d}$ by several steps, each of which involves decreasing d_j by 1 and simultaneously increasing d_{j-1} by p (for some $j = 1, \ldots, i$). This defines a partial order on the set of p-adic representations of d, denoted $\underline{d} \subset \underline{e}$. Prove that $\underline{d} \subset \underline{e}$ if and only if $N_{\underline{d}} \subset N_{\underline{e}}$.

(d) Let $\underline{d}$ be a p-adic representation of d. We define the partition $\lambda(p, \underline{d})$ where $\lambda_v = \sum_{j=0}^{i} m_v(d_j)p^j$, where the numbers $m_v(d)$ are defined as follows

$$m_v(d) = \begin{cases} p - 1 & \text{if } d \geq (v + 1)(p - 1), \\ d - v(p - 1) & \text{if } v(p - 1) < d < (v + 1)(p - 1), \\ 0 & \text{otherwise.} \end{cases}$$

Prove that for each p-adic representation $\underline{d}$ of d the module $N_{\underline{d}} / \sum_{\underline{e} \subset \underline{d}} N_{\underline{e}}$ is an irreducible representation of $GL(E)$ of highest weight $\lambda(p, \underline{d})$.

8. Let **K** be a field. Consider a vector space E of dimension n. For a partition λ of m we denote by S^λ the weight space of $L_\lambda E$ of weight (1^m). We can think of S^λ as a span of tableaux of shape λ of weight (1^m) modulo the usual standard relations. The module S^λ is called a *Specht module* corresponding to the partition λ.
   (a) Prove that S^λ has the natural structure of a Σ_m-module.
   (b) Let **K** be a field of characteristic 0. Prove that the modules S^λ give a complete set of nonisomorphic irreducible Σ_m-modules.
   (c) Let **K** be a field of characteristic $p > 0$. Let λ be a partition of m. Let M^λ be the weight space of $M_\lambda E$ of weight (1^m). The module M^λ is $\neq 0$ if and only if λ is p-regular. It is proven in $[jm]$ that the representations M^λ give a complete set of isomorphism classes of irreducible representations of S_m.

9. Let **K** be a field of characteristic 3. Let E be a vector space of dimension n. Consider the Schur functors of degree 5.
   (a) Prove that $L_{(3,1,1)}E, L_{(4,1)}E, L_{(5)}E$ are irreducible,
   (b) Prove the following exact sequences, which imply the composition series of the remaining Schur functors:

$$0 \rightarrow M_{(5)}E \rightarrow L_{(1^5)}E \rightarrow M_{(2^2,1)}E \rightarrow 0,$$
$$0 \rightarrow M_{(4,1)}E \rightarrow L_{(2,1^3)}E \rightarrow M_{(3,2)}E \rightarrow 0,$$
$$0 \rightarrow M_{(3,2)}E \rightarrow L_{(2,2,1)}E \rightarrow M_{(1^5)}E \rightarrow 0,$$
$$0 \rightarrow M_{(2,2,1)}E \rightarrow L_{(3,2)}E \rightarrow M_{(2,1^3)}E \rightarrow 0.$$

Littlewood–Richardson Rule

10. Use the Littlewood–Richardson rule to find the multiplicities of irreducible representations $L_\lambda E$ in the tensor products $L_{2,1}E \otimes L_{2,1}E$, $L_{3,1}E \otimes L_{2,1}E$.

11. Let λ, μ be two rectangular partitions, i.e. $\lambda = (l^s)$, $\mu = (m^t)$. Prove that the tensor product $L_\lambda E \otimes L_\mu E$ is multiplicity free, i.e. that for each ν the multiplicity $u(\lambda, \mu; \nu)$ equals 0 or 1. Characterize the partitions ν such that $L_\nu E$ occurs in $L_\lambda E \otimes L_\mu E$.

12. Let $\nu = (m^t)$ be a rectangular partition. Show that $L_\nu E$ occurs in the tensor product $L_\lambda E \otimes L_\mu E$ if and only if λ and μ can be fitted together to fill the rectangle ν, i.e. when for $\lambda = (\lambda_1, \ldots, \lambda_t)$, $\mu = (\mu_1, \ldots, \mu_t)$ we have $\lambda_i + \mu_{t+1-i} = m$ for $1 \le i \le t$. Show that if λ and μ can be fitted together to fill the rectangle ν, then the multiplicity of $L_\nu E$ in $L_\lambda E \otimes L_\mu E$ is equal to 1.

13. Let us fix five numbers a, b, c, d, e with $a \ge b \ge c, d \ge e, a + b + c = d + e$. Denote by $t(m)$ the multiplicity of $L_{d^m, e^m} E$ in $L_{a^m} E \otimes L_{b^m} E \otimes L_{c^m} E$. Prove that $t(m) = \binom{t(1)+m-2}{m-1}$.

14. Let us fix five numbers a, b, c, d, e with $a \ge b \ge c \ge d$, $a + b + c + d = 2e$. Denote by $t(m)$ the multiplicity of $L_{e^{2m}} E$ in $L_{a^m} E \otimes L_{b^m} E \otimes L_{c^m} E \otimes L_{d^m} E$. Prove that $t(m) = \binom{t(1)+m-2}{m-1}$.

15. Let us fix seven numbers : $a_1, a_2, b_1, b_2, c_1, c_2, d$. Assume $a_1 + b_1 + c_1 + a_2 + b_2 + c_2 = 3d$. Denote by $t(m)$ the multiplicity of $L_{d^{3m}} E$ in $L_{a_1^m, a_2^m} E \otimes L_{b_1^m, b_2^m} E \otimes L_{c_1^m, c_2^m} E$. Prove that $t(m) = \binom{t(1)+m-2}{m-1}$.

16. Let $\lambda = (\lambda_1, \ldots, \lambda_s)$, $\mu = (\mu_1, \ldots, \mu_t)$ be two partitions. Let $\nu = (\nu_1, \ldots, \nu_{s+t})$ be a partition resulting from permuting the sequence $(\lambda_1, \ldots, \lambda_s, \mu_1, \ldots, \mu_t)$ to be nonincreasing. Let $\pi \in \Sigma_{s+t}$ be the resulting permutation, i.e. $\lambda_i = \nu_{\pi(i)}$ for $1 \le i \le s$, $\mu_j = \nu_{s+j}$ for $1 \le j \le t$. Prove that the morphism

$$\hat{m}(\lambda, \mu) : \overset{\lambda_1}{\bigwedge} E \otimes \ldots \otimes \overset{\lambda_s}{\bigwedge} E \otimes \overset{\mu_1}{\bigwedge} E \otimes \ldots \otimes \overset{\mu_t}{\bigwedge} E$$
$$\to \overset{\nu_1}{\bigwedge} E \otimes \ldots \otimes \overset{\nu_{s+t}}{\bigwedge} E$$

given by permuting the factors according to the permutation π factors to give an equivariant epimorphism

$$m(\lambda, \mu) : L_\lambda E \otimes L_\mu E \to L_\nu E.$$

Use the Littlewood–Richardson rule to show that the multiplicity of $L_\nu E$ in $L_\lambda E \otimes L_\mu E$ is equal to 1. Let us order the partitions lexicographically. Prove that for all partitions η such that $L_\eta E$ occurs in $L_\lambda E \otimes L_\mu E$ we have $\eta \ge \nu$. Sometimes the factor $L_\nu E$ is called the *Cartan piece* of the tensor product $L_\lambda E \otimes L_\mu E$.

17. Let $\lambda = (\lambda_1, \ldots, \lambda_s)$ and $\mu = (\mu_1, \ldots, \mu_s)$ be two partitions. Let ν be a partition $\nu = (\lambda_1 + \mu_1, \ldots, \lambda_s + \mu_s)$. Define the map

$$\hat{\Delta}(\lambda, \mu): \overset{\lambda_1+\mu_1}{\bigwedge} E \otimes \ldots \otimes \overset{\lambda_s+\mu_s}{\bigwedge} E \to \overset{\lambda_1}{\bigwedge} E \otimes \ldots \otimes \overset{\lambda_s}{\bigwedge} E \otimes \overset{\mu_1}{\bigwedge} E$$

$$\otimes \ldots \otimes \overset{\mu_s}{\bigwedge} E.$$

to be the tensor product of the diagonals followed by a permutation of factors. Prove that $\hat{\Delta}(\lambda, \mu)$ factors to give an equivariant map

$$\Delta(\lambda, \mu): L_\nu E \to L_\lambda E \otimes L_\mu E.$$

Use the Littlewood–Richardson rule to show that the multiplicity of $L_\nu E$ in $L_\lambda E \otimes L_\mu E$ is equal to 1. Let us order the partitions lexicographically. Prove that for all partitions η such that $L_\eta E$ occurs in $L_\lambda E \otimes L_\mu E$ we have $\eta \le \nu$.

Schur Functors and Duality

18. Let E be a vector space of dimension n.
    (a) Prove the canonical isomorphisms

    $$L_{\lambda_1,\ldots,\lambda_s} E^* = L_{n-\lambda_s,\ldots,n-\lambda_1} E \otimes \left(\overset{n}{\bigwedge} E^* \right)^{\otimes s},$$

    (b) Prove the canonical isomorphism

    $$K_{\lambda_1,\ldots,\lambda_n} E^* = K_{-\lambda_n,\ldots,-\lambda_1} E.$$

19. Let E be a vector space of dimension n. Use duality and the Littlewood–Richardson rule to decompose $\bigwedge^i E \otimes \bigwedge^j E^*$ to the irreducible highest weight representations as a $GL(E)$-module.

Acyclicity Properties of Schur Complexes

20. Let R be a commutative ring, let M be an R-module, and let

$$F_1 \to F_0 \to M \to 0$$

be a presentation of M. Let Φ be a complex $F_1 \to F_0$. We define the module $L_\lambda M$ to be

$$L_\lambda M = H_0(L_\lambda \Phi).$$

Prove that this definition does not depend on the choice of the presentation Φ.

21. Let M be an R-module of projective dimension 1 with a free resolution

$$0 \to F_1 \to F_0 \to M \to 0.$$

Let Φ be the complex $F_1 \to F_0$. Prove that if λ is a partition of d and M is a $(d\text{–}1)$st syzygy, then $L_\lambda(\Phi)$ gives a free resolution of $L_\lambda M$.

22. Let $\Phi : F_1 \to F_0$ be a linear map of free **K**-modules. For each partition ν the complex $L_\lambda \Phi$ has a subcomplex $\mathcal{X}_{<\nu}$ whose chains have a filtration with associated graded

$$\bigoplus_{\mu < \nu} K_{\lambda/\mu} F_1 \otimes L_\mu F_0.$$

Similarly, $L_\lambda \Phi$ has a subcomplex $\mathcal{X}_{\leq \nu}$ whose chains have a filtration with associated graded

$$\bigoplus_{\mu \leq \nu} K_{\lambda/\mu} F_1 \otimes L_\mu F_0.$$

3

Grassmannians and Flag Varieties

In this chapter we discuss the properties of Grassmannians and flag varieties. They can be defined as homogeneous spaces $\mathbf{G}/\mathbf{P}$ for the general linear group $\mathbf{G}$ and a parabolic subgroup $\mathbf{P}$, but we take a direct approach.

In section 3.1 we define the Plücker embedding and prove the quadratic relations satisfied by the image of the flag variety by the Plücker embedding in the product of projective spaces. They turn out to be the shuffling relations defining the Schur functors. This leads to the identification of the multihomogeneous components of the coordinate rings of flag varieties and Schur modules.

In section 3.2 we describe the standard coverings of flag varieties by affine spaces. We apply these coverings to prove the Cauchy formula stated in section 2.2.

In section 3.3 we define tautological vector bundles on flag varieties and define the flag varieties in a relative setting. We also prove an important result about a Koszul complex resolving the coordinate sheaf of tautological bundle on a Grassmannian. This is a prototype of the technique we will develop in the later chapters.

3.1. The Plücker Embeddings

Let E be a vector space of dimension n over a field $\mathbf{K}$, and let r be an integer, $0 < r \le n$. We consider the set Grass (r, E) of r-dimensional subspaces of E. Our first objective is to define the structure of a projective variety on Grass(r, E). Let $e_1, \ldots, e_n$ be a fixed basis of E.

For a subspace $R \subset E$ of dimension r in E we choose a basis $z_1, \ldots, z_r$ of R. We write each z_i as a linear combination of basis vectors from E:

$$
\begin{aligned}
z_1 &= z_{11}e_1 + z_{12}e_2 + \cdots + z_{1n}e_n, \\
z_2 &= z_{21}e_1 + z_{22}e_2 + \cdots + z_{2n}e_n, \\
&\vdots \\
z_r &= z_{r1}e_1 + z_{r2}e_2 + \cdots + z_{rn}e_n.
\end{aligned}
$$

We denote by Z the $r \times n$ matrix (z_{ij}). For each r-tuple $(i_1, \ldots, i_r)$, we consider the $r \times r$ submatrix $Z(i_1, \ldots, i_r)$ of Z whose j-th column is the i_j-th column of Z. We define $p(i_1, \ldots, i_r) := \det Z(i_1, \ldots, i_r)$. We call $p(i_1, \ldots, i_r)$ the Plücker coordinates associated to the basis $z_1, \ldots, z_r$. If $z'_1, \ldots, z'_r$ is another basis of R, then there exists a unique $r \times r$ nonsingular matrix $A = (a_{ij})$ such that

$$
\begin{aligned}
z'_1 &= a_{11}z_1 &+& a_{12}z_2 &+& \cdots &+& a_{1r}z_r, \\
z'_2 &= a_{21}z_1 &+& a_{22}z_2 &+& \cdots &+& a_{2r}z_r, \\
&\vdots \\
z'_r &= a_{r1}z_1 &+& a_{r2}z_2 &+& \cdots &+& a_{rr}z_r.
\end{aligned}
$$

This means that if $p'(i_1, \ldots, i_r)$ denote the Plücker coordinates associated to the basis $z'_1, \ldots, z'_r$, we have

$$
p'(i_1, \ldots, i_r) = \det(A) \, p(i_1, \ldots, i_r).
$$

We define the Plücker embedding

$$
\pi : \mathrm{Grass}(r, E) \longrightarrow P\left(\bigwedge^r E \right)
$$

sending the subspace R to the point with homogeneous coordinates $p(i_1, \ldots, i_r)$ $(1 \le i_1 < \ldots < i_r \le n)$. The formula above shows that π is well defined (i.e., it does not depend on the choice of basis $z_1, \ldots, z_r$).

(3.1.1) Proposition. *The map* $\pi : \mathrm{Grass}(r, E) \longrightarrow P(\bigwedge^r E)$ *is an embedding of* $\mathrm{Grass}(r, E)$ *onto a closed algebraic subset of* $P(\bigwedge^r E)$.

Proof. First we prove that π is injective. Let R, R' be two subspaces from $\mathrm{Grass}(r, E)$. Let us assume that $\dim(R \cap R') = i < r$. Then we can choose a basis $e_1, \ldots, e_n$ of E such that

(1) $e_1, \ldots, e_i$ is a basis of $R \cap R'$,
(2) $e_1, \ldots, e_r$ is a basis of R,
(3) $e_1, \ldots, e_i, e_{r+1}, \ldots, e_{2r-i}$ is a basis of R'.

We see that $p(1, \ldots, r)$ is 0 on $\pi(R')$ and $\neq 0$ on $\pi(R)$, and that $p(1, \ldots, i, r+1, \ldots, 2r-i)$ is 0 on $\pi(R)$ and $\neq 0$ on $\pi(R')$. This proves that $\pi(R) \neq \pi(R')$.

Let us define $A(r) = \bigoplus_{s \ge 0} A_s(r)$ the graded coordinate ring of the closure of $\pi(\mathrm{Grass}(r, E))$. We proceed to describe the relations satisfied by Plücker coordinates in $A(r)$.

(3.1.2) Proposition. *The Plücker coordinates in $A(r)$ satisfy the following relations:*

$$\sum_{\beta} \mathrm{sgn}(\beta)\, p(i_1, \ldots, i_u, j_{\beta(1)}, \ldots, j_{\beta(r-u)}) p(j_{\beta(r-u+1)}, \ldots,$$

$$j_{\beta(2r-u-v)}, l_1, \ldots, l_v), = 0$$

where we sum over all permutations β of $\{1, \ldots, 2r - u - v\}$ such that $\beta(1) < \beta(2) < \ldots < \beta(r - u)$ and $\beta(r - u + 1) < \beta(r - u + 2) < \ldots < \beta(2r - u - v)$. We denote the left hand sides of these relations by $R(i_1, \ldots, i_u; l_1, \ldots, l_v; j_1, \ldots, j_{2r-u-v})$.

Proof of Proposition (3.1.2). Let $\mathrm{GL}(E)$ be the general linear group of E. Let us fix an r-dimensional subspace R in E. We define the subgroup $P_r = \{g \in \mathrm{GL}(E) | g(R) = R\}$. The group $\mathrm{GL}(E)$ acts transitively on $\mathrm{Grass}(r, E)$, so we can identify $\mathrm{Grass}(r, E)$ with $\mathrm{GL}(E)/P_r$ via the map $g \mapsto g(R)$. To prove (3.1.2) we notice that the span of the relations $R(i_1, \ldots, i_u; l_1, \ldots, l_v; j_1, \ldots, j_{2r-u-v})$ is $\mathrm{GL}(E)$-invariant. Indeed, we can identify the **K**-span of Plücker coordinates $p(i_1, \ldots, i_r)$ with $\bigwedge^r E^*$. Under this identification the coordinate $p(i_1, \ldots, i_r)$ corresponds to the tensor $e_{i_1}^* \wedge \ldots \wedge e_{i_r}^*$, where $e_1^*, \ldots, e_n^*$ denotes the dual basis to $e_1, \ldots, e_n$.

The relations in (3.1.2) can now be defined in terms of exterior diagonals and exterior multiplications. They span the image of the following map:

$$\bigwedge\nolimits^u E^* \otimes \bigwedge\nolimits^{2r-u-v} E^* \otimes \bigwedge\nolimits^v E^*$$

$$\downarrow {\scriptstyle 1 \otimes \Delta \otimes 1}$$

$$\bigwedge\nolimits^u E^* \otimes \bigwedge\nolimits^{r-u} E^* \otimes \bigwedge\nolimits^{r-v} E^* \otimes \bigwedge\nolimits^v E^*$$

$$\downarrow {\scriptstyle m_{12} \otimes m_{34}}$$

$$\bigwedge\nolimits^r E^* \otimes \bigwedge\nolimits^r E^*.$$

To prove our proposition it is enough to show that the relations in (3.1.2) vanish on one element of $\mathrm{Grass}(r, E)$, say $R = \mathbf{K}e_1 + \cdots + \mathbf{K}e_r$. But one sees immediately that all Plücker coordinates except $p(1, \ldots, r)$ vanish on R, so the summation in (3.1.2) evaluated on R has to be equal to 0. This concludes the proof. ■

(3.1.3) Remark. *Notice that the relations in (3.1.2) are familiar. They are the special case of the relations $\theta(\lambda, a, u, v; E^*)$ for $\lambda_a = \lambda_{a+1} = r$. This means that $A_s(r)$ is a factor of the Schur functor $L_{(r^s)}E^*$. In the next statement we will show that both spaces are isomorphic.*

(3.1.4) Proposition. *The graded component $A_s(r)$ of the homogeneous coordinate ring of $\mathrm{Grass}(r, E)$ is isomorphic to $L_{r^s}E^*$.*

Proof. Proposition (3.1.2) and Remark (3.1.3) imply that for each s there is a GL(E)-equivariant epimorphism $L_{r^s} E^* \to A_s(r)$. Denote by V_s its kernel. Applying (2.2.3) (b) and (2.2.5), we see that if V_s is nonzero, it has to contain the canonical tableau. However, our identifications are such that the canonical tableau is the s-th power of the Plücker coordinate $p(1, \ldots, r)$. This would mean that $p(1, \ldots, r)$ vanishes on Grass(r, E), which is a contradiction. Therefore $V_s = 0$ for all positive s. ∎

. Now we can conclude the proof of Proposition (3.1.1), i.e. prove that $\pi(\text{Grass}(r, E))$ is closed. Let us suppose otherwise. This means that the closure of $\pi(\text{Grass}(r, E))$ contains a proper invariant closed subset. Let J be its defining ideal in $A(r)$. The ideal J is a nonzero radical homogeneous GL(E)-invariant ideal in $A(r)$. Let s be such that J_s is nonzero. Then J_s has to contain the canonical tableau. But this is a power of a Plücker coordinate, and J is radical, so the Plücker coordinate is in J. By GL(E)-invariance we conclude that J is equal to the maximal ideal in $A(r)$. This is a contradiction, because the maximal ideal defines an empty subset of the projective space. Proposition (3.1.1) is proved. ∎

Let T be a tableau of shape r^s. We associate to this tableau the product of s Plücker coordinates $p(T(1, 1), \ldots, T(1, r))p(T(2, 1), \ldots, T(2, r)) \ldots$ $p(T(s, 1), \ldots, T(s, r))$. We will also call this product a tableau of shape r^s. Proposition (2.1.4) allows us to describe the basis of $A_s(r)$.

(3.1.5) Proposition. *The set of standard tableaux of shape r^s forms a basis of $A_s(r)$.*

We extend the above results to the flag varieties. Define *the full flag variety*

$$\text{Flag}(E) = \{(R_1, \ldots, R_{n-1}) \in \text{Grass}(1, E) \times \ldots \times \text{Grass}(n-1, E)|$$
$$R_1 \subset R_2 \subset \ldots \subset R_{n-1}\}.$$

Let

$$\pi : \text{Flag}(E) \longrightarrow P\left(\overset{1}{\bigwedge} E\right) \times P\left(\overset{2}{\bigwedge} E\right) \times \ldots \times P\left(\overset{n-1}{\bigwedge} E\right)$$

be the product of Plücker embeddings. The map π is easily seen to be injective. We will describe the (multigraded) coordinate ring

$$A(1, 2, \ldots, n-1) = \bigoplus_{s_1, \ldots, s_{n-1} \geq 0} A(1, 2, \ldots, n-1)_{s_1, \ldots, s_{n-1}}$$

of Flag(E). The relations between Plücker coordinates of various sizes turn
out to be the following.

(3.1.6) Proposition. *Let $0 \leq a_2 \leq a_1 \leq n - 1$, and let u, v be two numbers
such that $u + v < a_2$. Then the following relations are satisfied by Plücker
coordinates of sizes a_1, a_2 in $A(1, 2, \ldots, n - 1)$:*

$$\sum_{\beta} \mathrm{sgn}(\beta) \, p(i_1, \ldots, i_u, j_{\beta(1)}, \ldots, j_{\beta(a_1-u)}) p(j_{\beta(a_1-u+1)}, \ldots,$$

$$j_{\beta(a_1+a_2-u-v)}, l_1, \ldots, l_v), = 0$$

*where we sum over all permutations β of $\{1, \ldots, a_1 + a_2 - u - v\}$ such that
$\beta(1) < \beta(2) < \ldots < \beta(a_1 - u)$ and $\beta(a_1 - u + 1) < \beta(a_1 - u + 2) < \ldots < \beta(a_1 + a_2 - u - v)$.*

Proof. Let us fix a flag $R = (R_1, \ldots, R_{n-1})$ in E, say $R_i = \mathbf{K}e_1 + \cdots + \mathbf{K}e_i$.
Let $\mathbf{B} \subset \mathrm{GL}(E)$ be the subgroup of all elements g of $\mathrm{GL}(E)$ stabilizing R,
i.e. such that $g(R_i) \subset R_i$ for $1 \leq i \leq n - 1$. The subgroup $\mathbf{B}$ can be identified with the set of upper triangular matrices. Since $\mathrm{GL}(E)$ acts transitively on Flag(E), we can identify Flag(E) with $\mathrm{GL}(E)/\mathbf{B}$ via the map
$g \mapsto (g(R_1), \ldots, g(R_{n-1}))$. Next, we observe that the $\mathbf{K}$-span of the relations in (3.1.6) is $\mathrm{GL}(E)$-invariant. Indeed, for given a_1, a_2, u, v they span
the image of the following map:

$$\bigwedge^u E^* \otimes \bigwedge^{a_1-u+a_2-v} E^* \otimes \bigwedge^v E^*$$

$$\downarrow {\scriptstyle 1 \otimes \Delta \otimes 1}$$

$$\bigwedge^u E^* \otimes \bigwedge^{a_1-u} E^* \otimes \bigwedge^{a_2-v} E^* \otimes \bigwedge^v E^*$$

$$\downarrow {\scriptstyle m_{12} \otimes m_{34}}$$

$$\bigwedge^{a_1} E^* \otimes \bigwedge^{a_2} E^*,$$

where we again identify the span of Plücker coordinates of size r with $\bigwedge^r E^*$.
Now it is enough to show that the relations in (3.1.6) vanish at one point of
Flag(E), say $R = (R_1, \ldots, R_{n-1})$ where $R_i = \mathbf{K}e_1 + \ldots + \mathbf{K}e_i$. This however is obvious, since all Plücker coordinates of size r ($1 \leq r \leq n - 1$) except
$p(1, \ldots, r)$ vanish on R. ∎

(3.1.7) Remark. *The relations in (3.1.6) can be recognized as the relations $\theta(a, u, v; E^*)$ defined in section 2.1, corresponding to two rows of a
Schur functor of length a_1 and a_2. This means that the space $A(1, 2, \ldots,
n - 1)_{s_1,\ldots,s_{n-1}}$ is a factor of $L_{(n-1)^{s_{n-1}},(n-2)^{s_{n-2}},\ldots 1^{s_1}} E^*$. In fact in the next state-
ment we show it is an isomorphism.*

(3.1.8) Proposition.

   (a) *The multigraded component $A(1, 2, \ldots, n - 1)_{s_1, \ldots, s_{n-1}}$ is isomorphic to $L_{(n-1)^{s_{n-1}}, (n-2)^{s_{n-2}}, \ldots, 1^{s_1}}(E^*)$.*

   (b) *The map π is an embedding of* Flag(E) *onto a closed set of $P(\bigwedge^1 E) \times \ldots \times P(\bigwedge^{n-1} E)$ whose defining ideal is generated by the relations in (3.1.6).*

Proof. Let us start with (a). We denote by $V_{s_{n-1}, \ldots, s_1}$ the kernel of the epimorphism from $L_{(n-1)^{s_{n-1}}, (n-2)^{s_{n-2}}, \ldots, 1^{s_1}}(E^*)$ onto $A(1, 2, \ldots, n - 1)_{s_1, \ldots, s_{n-1}}$, the existence of which follows from (3.1.6). This is a GL(E)-submodule, so if it is nonzero, it has to contain a **U**-invariant. But by (2.2.5) the only **U**-invariant in $L_{(n-1)^{s_{n-1}}, (n-2)^{s_{n-2}}, \ldots, 1^{s_1}}(E^*)$ is the canonical tableau, which is a product of powers of Plücker coordinates $p(1, \ldots, r)$ for various r. The canonical tableau cannot therefore vanish on $\pi(\mathrm{Flag}(E))$. This contradiction shows that $V_{s_{n-1}, \ldots, s_1} = 0$ for all $s_{n-1}, \ldots, s_1$, and (a) is proven.

To prove (b), let us assume that it is false. Then the ring $A(1, 2, \ldots, n - 1)$ contains a nonzero GL(E)-invariant prime ideal J (the defining ideal of some nonempty closed invariant irreducible subset in the closure of $\pi(\mathrm{Flag}(E))$). The ideal J has to contain some nonzero **U**-invariant which is a canonical tableau, i.e. the product of Plücker coordinates. Therefore one of the Plücker coordinates $p(1, \ldots, r)$ is in J. By the invariance of J we see that it contains all Plücker coordinates of some size r. This is a contradiction, since J corresponds to a nonempty subset of $P(\bigwedge^1 E) \times \ldots \times P(\bigwedge^{n-1} E)$. ∎

We conclude this section with a brief discussion of partial flag varieties. The proofs of the results are the same as for the full flag variety, so we only give the statements.

For every sequence $(b) = (b_1, \ldots, b_t)$ with $1 \leq b_1 < \ldots < b_t \leq n - 1$ we can define *the partial flag variety*

$$\mathrm{Flag}(b_1, \ldots, b_t; E) = \{(R_{b_1}, \ldots, R_{b_t}) \in \mathrm{Grass}(b_1, E) \times \ldots \times$$
$$\mathrm{Grass}(b_t, E) | R_{b_1} \subset \ldots \subset R_{b_t}\}.$$

(3.1.9) Proposition.

   (a) *The partial flag variety* Flag$(b_1, \ldots, b_t; E)$ *can be identified with the homogeneous space* GL$(E)/\mathbf{P}_{(b)}$, *where $\mathbf{P}_{(b)}$ is a subgroup of elements in* GL(E) *stabilizing a fixed flag in* Flag $(b_1, \ldots b_t; E)$.

   (b) *The product π of Plücker embeddings sends the variety* Flag$(b_1, \ldots, b_t; E)$ *onto a closed subset of $P(\bigwedge^{b_1} E) \times \ldots \times P(\bigwedge^{b_t} E)$. We denote $A(b_1, \ldots, b_t)$ the coordinate ring of $\pi(\mathrm{Flag}(b_1, \ldots, b_t; E))$,*

(c) The multigraded component $A(b_1, \ldots, b_t)_{s_1,\ldots,s_t}$ is isomorphic to the Schur module $L_{b_1^{s_t},\ldots,b_1^{s_1}}(E^*)$,

(d) The defining ideal of $\pi(\mathrm{Flag}(b_1, \ldots, b_t; E))$ is generated by relations of the type in (3.1.6).

3.2. The Standard Open Coverings of Flag Manifolds and the Straightening Law

In this section we introduce the standard open coverings of the flag varieties. We prove that the sets from these coverings are affine spaces. We apply these coverings to prove the characteristic free version of the Cauchy formula stated in the first part of (2.3.2).

Let us consider the Plücker embedding

$$\pi : \mathrm{Grass}(r, E) \longrightarrow P(\overset{r}{\bigwedge} E),$$

where E is an n-dimensional vector space over a field $\mathbf{K}$. Let $I = \{i_1, \ldots, i_r\}$ be a subset of $[1, n]$ of cardinality r. We consider the open covering of $\mathrm{Grass}(r, E)$ by the subsets

$$U_I = \{R \in \mathrm{Grass}(r, E) \mid p(i_1, \ldots, i_r)(R) \neq 0\}.$$

Obviously the sets U_I form an open covering of $\mathrm{Grass}(r, E)$.

For a given I we denote by π_I the projection of E onto $\mathbf{K}e_{i_1} + \ldots + \mathbf{K}e_{i_r}$ sending e_j with $j \notin I$ to 0. Then $R \in U_I$, if and only if $\pi_I|R$ is an isomorphism. This means that we can choose the unique basis $z_1, \ldots, z_r$ of R such that

$$
\begin{aligned}
z_1 &= \textstyle\sum_{j \notin I} z_{1j} e_j &+\ & e_{i_1}, \\
z_2 &= \textstyle\sum_{j \notin I} z_{2j} e_j &+\ & e_{i_2}, \\
&\ \ \vdots & & \\
z_r &= \textstyle\sum_{j \notin I} z_{rj} e_j &+\ & e_{i_r}.
\end{aligned}
$$

Conversely, any subspace R spanned by vectors of the above form (which are always linearly independent) is an element of U_I. This reasoning proves

(3.2.1) Proposition.

(a) The subset U_I of $\mathrm{Grass}(r, E)$ can be identified with the affine space of dimension $r(n - r)$. The coordinate functions in this space are the functions z_{ij} with $1 \le i \le r$, $j \notin I$.

(b) The Grassmannian $\mathrm{Grass}(r, E)$ is a rational nonsingular variety of dimension $r(n - r)$.

Let us choose $I_0 = \{n - r + 1, \ldots, n\}$. Recall that the coordinate ring of U_{I_0} is an algebra generated over $\mathbf{K}$ by the functions $\hat{p}(i_1, \ldots, i_r) = p(i_1, \ldots, i_r)p(n - r + 1, \ldots, n)^{-1}$. We want to see what basis in the polynomial ring in $r(n - r)$ variables we obtain by restricting the standard products of Plücker coordinates $p(i_1, \ldots, i_r)$ to U_{I_0}. First of all let us notice that our matrix Z from section 3.1 is specialized in U_{I_0} to

$$\begin{pmatrix} z_{11} & \cdots & z_{1,n-r} & 1 & 0 & \cdots & 0 \\ z_{21} & \cdots & z_{2,n-r} & 0 & 1 & \cdots & 0 \\ \vdots & & \vdots & \vdots & \vdots & & \vdots \\ z_{r1} & \cdots & z_{r,n-r} & 0 & 0 & \cdots & 1 \end{pmatrix}.$$

Let us consider the function $\hat{p}(i_1, \ldots, i_r)$. It is the $r \times r$ minor of the above matrix corresponding to columns $i_1, \ldots, i_r$. Let us assume that $1 \leq i_1 < \ldots < i_s \leq n - r < i_{s+1} < \ldots < i_r \leq n$. Then it follows from the Laplace expansion that on U_{I_0} the coordinate $p(i_1, \ldots, i_r)$ equals, up to sign, to the $s \times s$ minor of the matrix

$$\begin{pmatrix} z_{11} & \cdots & z_{1,n-r} \\ z_{21} & \cdots & z_{2,n-r} \\ \vdots & & \vdots \\ z_{r1} & \cdots & z_{r,n-r} \end{pmatrix} \tag{$*$}$$

corresponding to the rows from the set $[1, r] \setminus \{i_{s+1} - (n - r), \ldots, i_r - (n - r)\}$ and the columns $i_1, \ldots, i_s$.

This means that the product of Plücker coordinates on $\mathrm{Grass}(r, E)$ restricts to the product of minors of various sizes of the matrix $(*)$ on U_{I_0}. Let us check how to express the standardness condition in this setting.

Let us consider two Plücker coordinates $p(i_1, \ldots, i_r)$, $p(j_1, \ldots, j_r)$ such that the product $p(i_1, \ldots, i_r)p(j_1, \ldots, j_r)$ is standard (i.e. $i_1 < \ldots < i_r$, $j_1 < \ldots < j_r$, $i_s \leq j_s$ for $1 \leq s \leq r$). Let us assume that $1 \leq i_1 < \ldots < i_s \leq n - r < i_{s+1} < \ldots < i_r \leq n$, $1 \leq j_1 < \ldots < j_t \leq n - r < j_{t+1} < \ldots < j_r \leq n$. This means that $s \geq t$ and that the tableau

i_1	i_2	$\cdots$	i_t		$\cdots$	i_s
j_1	j_2	$\cdots$	j_t			

is standard. Our condition also means that if $(u_1, \ldots, u_s)$ denotes the sequence of elements of $[1, r] \setminus \{i_{s+1} - (n - r), \ldots, i_r - (n - r)\}$ written in increasing order, and $(v_1, \ldots, v_t)$ denotes the sequence of elements of

$[1, r] \setminus \{j_{t+1} - (n - r), \ldots, j_r - (n - r)\}$ written in increasing order, the tableau

u_s	$\ldots$		$\ldots$	u_1
v_t	$\ldots$	v_1		

satisfies the conditions $u_s > \ldots > u_1$, $v_t > \ldots > v_1$, $u_{s-j} \geq v_{t-j}$ for $1 \leq j \leq t$. This means that when we reverse the order of numbers in this tableau, we will get the standardeness condition for both tableaux.

Recalling the definition of the standard tableau from section 1.A. we see that it is natural to call the double tableau

$$\left(\begin{array}{|c|c|c|c|c|} \hline u_s & \ldots & & \ldots & u_1 \\ \hline v_t & \ldots & v_1 & & \\ \hline \end{array} \quad \begin{array}{|c|c|c|c|c|} \hline i_1 & i_2 & \ldots & i_t & \ldots & i_s \\ \hline j_1 & j_2 & \ldots & j_t & & \\ \hline \end{array} \right)$$

a standard double tableau.

(3.2.2) Proposition. *The standard double tableaux form a basis of the coordinate ring* $\mathbf{K}[U_{I_0}]$ *(which by (3.2.1) is the polynomial ring in the variables* z_{ij} *for* $1 \leq i \leq r$, $1 \leq j \leq n - r$*).*

Proof. The coordinate ring $\mathbf{K}[U_{I_0}]$ is generated by functions $\hat{p}(i_1, \ldots, i_r)$. The standard products of Plücker coordinates restrict to the standard double tableaux. If a product of functions $\hat{p}(i_1, \ldots, i_r)$ is a double tableau which is not standard, we can express as a linear combinations of standard products of Plücker coordinates, which in turn will express our double tableau as a linear combination of standard double tableaux. This shows that standard double tableaux span $\mathbf{K}[U_{I_0}]$.

To show that the standard double tableaux are linearly independent on U_{I_0}, we notice that if some linear combination of standard double tableaux (say in degree s) is zero, then by multiplying by some power of $p(n - r + 1, \ldots, n)$ we can extend each summand to Grass(r, E). Each summand extends to a standard tableau, since adding at the end the row $(n - r + 1, \ldots, n)$ does not affect standardness. Thus we get a nontrivial linear combination of standard tableaux on Grass(r, E) which vanishes on U_{I_0}. This is a contradiction. ■

Let us denote by F, G respectively the space of rows and columns of the matrix (∗). The polynomial ring $\mathbf{K}[U_{I_0}]$ is identified with Sym$(F \otimes G)$. We want to consider more closely the meaning of Plücker relations in terms of double tableaux. Let us consider the nonstandard tableau

$$T = \begin{array}{|c|c|c|c|c|c|} \hline i_1 & i_2 & \cdots & i_a & \cdots & i_r \\ \hline j_1 & j_2 & \cdots & j_a & \cdots & j_r \\ \hline \end{array}.$$

More precisely, let us assume that $i_1 < \ldots < i_r$, $j_1 < \ldots < j_r$, $i_a > j_a$, and $i_a \leq n - r$. To express T as a sum of earlier tableaux we use the relation

$$R(i_1, \ldots, i_{a-1}; j_{a+1}, \ldots, j_r; j_1, \ldots, j_a, i_a, \ldots, i_r)$$

from (3.1.2). Let us suppose that after the restriction to U_{I_0}, $p(i_1, \ldots, i_r)$ becomes an $s \times s$ minor of $(*)$ and $p(j_1, \ldots, j_r)$ becomes a $t \times t$ minor of $(*)$. This means that $1 \leq i_1 < \ldots < i_s \leq n - r < i_{s+1} < \ldots < i_r \leq n$, $1 \leq j_1 < \ldots < j_t \leq n - r < j_{t+1} < \ldots < j_r \leq n$. We also assume that $s \leq t$.

We notice that in the summands of $R(i_1, \ldots, i_{a-1}; j_{a+1}, \ldots, j_r; j_1, \ldots, j_a, i_a, \ldots, i_r)$ the number of entries that are $\leq n - r$ in the first row is $\geq s$. The summands in $R(i_1, \ldots, i_{a-1}; j_{a+1}, \ldots, j_r; j_1, \ldots, j_a, i_a, \ldots, i_r)$ where the number of entries that are $\leq n - r$ in the first row is equal to s correspond to shuffling $j_1, \ldots, j_a$ with $i_a, \ldots, i_s$. In all those summands the numbers bigger than $n - r$ stay fixed. This combination of products of minors of size s and t ($s \geq t$) is, in terms of double tableaux, the relation of the type $\theta(a, u, v; F)$ on the left side of the double tableau with its right side fixed. Our relation says that this combination belongs to the linear span of products of minors of size $s + i$ multiplied by minors of size $t - i$ for various $i > 0$. This establishes the following proposition.

(3.2.3) Proposition. *Let us consider the composition* Θ *given by*

$$\bigwedge{}^u F \otimes \bigwedge{}^{s+t-u-v} F \otimes \bigwedge{}^v F \otimes \bigwedge{}^s G \otimes \bigwedge{}^t G$$
$$\downarrow {\scriptstyle 1 \otimes \Delta \otimes 1 \otimes 1 \otimes 1}$$
$$\bigwedge{}^u F \otimes \bigwedge{}^{s-u} F \otimes \bigwedge{}^{t-v} F \otimes \bigwedge{}^v F \otimes \bigwedge{}^s G \otimes \bigwedge{}^t G$$
$$\downarrow {\scriptstyle m_{12} \otimes m_{34} \otimes 1 \otimes 1}$$
$$\bigwedge{}^s F \otimes \bigwedge{}^t F \otimes \bigwedge{}^s G \otimes \bigwedge{}^t G$$
$$\downarrow {\scriptstyle 1 \otimes t_{23} \otimes 1}$$
$$\bigwedge{}^s F \otimes \bigwedge{}^s G \otimes \bigwedge{}^t F \otimes \bigwedge{}^t G$$
$$\downarrow {\scriptstyle \zeta_s \otimes \zeta_t}$$
$$S_s(F \otimes G) \otimes S_t(F \otimes G)$$
$$\downarrow {\scriptstyle m}$$
$$S_{s+t}(F \otimes G),$$

where $\zeta_s : \bigwedge{}^s F \otimes \bigwedge{}^s G \longrightarrow S_s(F \otimes G)$ *sends* $f_{i_1} \wedge f_{i_2} \wedge \ldots \wedge f_{i_s} \otimes g_{j_1} \wedge g_{j_2} \wedge \ldots \wedge g_{j_s}$ *to the minor of the matrix* $Z' = (z_{i,j})_{i,j}(f_i \otimes g_j)_{i,j}$ *corresponding to rows* $i_1, \ldots, i_s$ *and columns* $j_1, \ldots, j_s$. *Then each element in*

Im Θ *is a linear combination of products of* $(s+i) \times (s+i)$ *minors and* $(t-i) \times (t-i)$ *minors for* $i > 0$.

(3.2.4) Example.

$$\left(\boxed{\begin{array}{ccc}3&2&1\\ \hline 4\end{array}}\ \boxed{\begin{array}{ccc}1&2&3\\ \hline 4\end{array}}\right) - \left(\boxed{\begin{array}{ccc}4&2&1\\ \hline 3\end{array}}\ \boxed{\begin{array}{ccc}1&2&3\\ \hline 4\end{array}}\right)$$

$$+ \left(\boxed{\begin{array}{ccc}4&3&1\\ \hline 2\end{array}}\ \boxed{\begin{array}{ccc}1&2&3\\ \hline 4\end{array}}\right) - \left(\boxed{\begin{array}{ccc}4&3&2\\ \hline 1\end{array}}\ \boxed{\begin{array}{ccc}1&2&3\\ \hline 4\end{array}}\right)$$

$$= \left(\boxed{3\,2\,1\,4}\ \boxed{1\,2\,3\,4}\right),$$

$$\left(\boxed{\begin{array}{ccc}5&4&3\\ \hline 2&1\end{array}}\ \boxed{\begin{array}{ccc}1&2&3\\ \hline 4&5\end{array}}\right) - \left(\boxed{\begin{array}{ccc}5&4&2\\ \hline 3&1\end{array}}\ \boxed{\begin{array}{ccc}1&2&3\\ \hline 4&5\end{array}}\right) \left(\boxed{\begin{array}{ccc}5&3&2\\ \hline 4&1\end{array}}\ \boxed{\begin{array}{ccc}1&2&3\\ \hline 4&5\end{array}}\right)$$

$$- \left(\boxed{\begin{array}{ccc}4&3&2\\ \hline 5&1\end{array}}\ \boxed{\begin{array}{ccc}1&2&3\\ \hline 4&5\end{array}}\right)$$

$$= \left(\boxed{\begin{array}{cccc}5&4&3&2\\ \hline 1\end{array}}\ \boxed{\begin{array}{cccc}1&2&3&4\\ \hline 5\end{array}}\right) - \left(\boxed{\begin{array}{cccc}5&4&3&2\\ \hline 1\end{array}}\ \boxed{\begin{array}{cccc}1&2&3&5\\ \hline 4\end{array}}\right).$$

For each partition $\lambda = (\lambda_1, \ldots, \lambda_t)$ let us consider the maps

$$\zeta_\lambda : \bigwedge^{\lambda_1} F \otimes \bigwedge^{\lambda_1} G \otimes \bigwedge^{\lambda_2} F \otimes \bigwedge^{\lambda_2} G \otimes \ldots \otimes \bigwedge^{\lambda_t} F \otimes \bigwedge^{\lambda_t} G$$
$$\longrightarrow \mathrm{Sym}_{|\lambda|}(F \otimes G),$$

where $\zeta_\lambda = \zeta_{\lambda_1} \zeta_{\lambda_2} \cdots \zeta_{\lambda_t}$.

Now we are ready to restate the characteristic free version of the Cauchy formula.

(3.2.5) Theorem. *The symmetric power* $S_m(F \otimes G)$ *has a natural* $\mathrm{GL}(F) \times \mathrm{GL}(G)$-*invariant filtration whose associated graded object equals* $\bigoplus_{|\lambda|=m} L_\lambda F \otimes L_\lambda G$.

Proof. Let us order all the partitions λ of weight m by the order $\leq$ (cf. section 1.1). We define

$$\Xi_\lambda = \sum_{\mu \leq \lambda} \mathrm{Im}\ \zeta_\mu, \qquad \Xi'_\lambda = \sum_{\mu < \lambda} \mathrm{Im}\ \zeta_\mu$$

Then the relations in (3.2.3) and the relations we get from them by changing the roles of F and G show that $\Xi_\lambda / \Xi_\lambda'$ is a factor of $L_\lambda F \otimes L_\lambda G$. In fact these two spaces have to be equal, because by (3.2.2) the dimensions of $S_m(F \otimes G)$ and of $\bigoplus_{|\lambda|=m} L_\lambda F \otimes L_\lambda G$ are equal. ∎

(3.2.6) Remark. *This theorem is sometimes referred to as the straightening law, because it gives an algorithm to "straighten" nonstandard tableaux so that they become linear combinations of standard tableaux. The statement is classical (see, for example, the book of Turnbull [T]). The first modern proof was given by Doubilet, Rota, and Stein ([DRS]). The proof using Plücker relations comes from [DEP1].*

The proof of second part of (2.3.2) is based on an analogue of the straightening law. In order to find the filtration on $\bigwedge^\bullet (F \otimes G)$ we need first to define the analogues of the maps ζ_s. We define the map

$$\zeta_s' : \overset{s}{\bigwedge} F \otimes D_s G \longrightarrow \overset{s}{\bigwedge}(F \otimes G)$$

by sending $f_1 \wedge f_2 \wedge \ldots \wedge f_s \otimes g_1^{(i_1)} \ldots g_t^{(i_t)}$, $i_1 + \ldots + i_t = s$, to the element

$$\sum_\beta \operatorname{sgn}(\beta)\,(f_1 \otimes g_{\beta(1)}) \wedge \ldots \wedge (f_s \otimes g_{\beta(s)}),$$

where we sum over all $\beta \in \Sigma_s / \Sigma_{i_1} \times \ldots \times \Sigma_{i_t}$. Then for any partition $\lambda = (\lambda_1, \ldots, \lambda_s)$ of m we define

$$\zeta_\lambda' : \overset{\lambda_1}{\bigwedge} F \otimes D_{\lambda_1} G \otimes \overset{\lambda_2}{\bigwedge} F \otimes D_{\lambda_2} G \otimes \ldots \otimes \overset{\lambda_s}{\bigwedge} F \otimes D_{\lambda_s} G$$

$$\longrightarrow \overset{m}{\bigwedge}(F \otimes G), \ \zeta_\lambda' = \zeta_{\lambda_1}' \wedge \ldots \wedge \zeta_{\lambda_s}'$$

Finally we define $Z_\lambda = \sum_{\mu \leq \lambda} \operatorname{Im} \zeta_\mu'$, $Z_\lambda' = \sum_{\mu < \lambda} \operatorname{Im} \zeta_\mu'$. Then one proves that $Z_\lambda / Z_\lambda' = L_\lambda F \otimes K_\lambda G$. This implies imediately

(3.2.7) Proposition. *For each partition λ the factor Z_λ / Z_λ' is isomorphic to $L_\lambda F \otimes K_\lambda G$.*

Proof. See [ABW2], section III. ∎

We finish this section with the definition of standard open covering of the flag variety. Let us consider the Plücker embedding

$$\pi : \operatorname{Flag}(E) \longrightarrow P\left(\overset{1}{\bigwedge} E\right) \times \ldots \times P\left(\overset{n-1}{\bigwedge} E\right)$$

For each permutation $\sigma \in \Sigma_n$ we define

$$U_\sigma = \mathrm{Flag}(E) \cap (U_{\sigma(1)} \times \ldots \times U_{\sigma(1),\ldots,\sigma(i)} \times \ldots \times U_{\sigma(1),\ldots,\sigma(n-1)}).$$

Then we have the properties analogous to the Plücker embedding of the Grassmannians:

(3.2.8) Proposition.
  (a) *The sets U_σ form an open covering of the flag variety. Each of the sets U_σ is isomorphic to the affine space of dimension $\binom{n}{2}$,*
  (b) *The variety $\mathrm{Flag}(E)$ is rational and nonsingular of dimension $\binom{n}{2}$.*

Proof. The second part follows at once from the first. To prove part (a), note that all U_σ are open by definition. Let us show that they form a covering of $\mathrm{Flag}(E)$. Let $R = (R_1, \ldots, R_{n-1})$ belong to $\mathrm{Flag}(E)$. Then we can choose a basis $x_1, \ldots, x_n$ of E such that $R_i = \mathbf{K}x_1 + \ldots + \mathbf{K}x_i$ for all i. The Plücker coordinates of size i are the coordinates of the tensor $x_1 \wedge \ldots \wedge x_i$ written as a combination of tensors $e_{j_1} \wedge \ldots \wedge e_{j_i}$.

Now we prove by reverse induction on i that we can choose the sequence of subsets $J_1, \ldots J_{n-1}$ such that $J_i \subset J_{i+1}$ and $p(J_i)(R) \neq 0$ for all i.

First of all, we can, choose J_{n-1}, since $x_1 \wedge \ldots \wedge x_{n-1}$ is nonzero. Suppose we have chosen $J_{n-1}, \ldots, J_{i+1}$. Then since, in the expansion of the tensor $(x_1 \wedge \ldots \wedge x_i) \wedge x_{i+1}$ in the standard basis, $e_{J_{i+1}}$ occurs with a nonzero coefficient, we see that $x_1 \wedge \ldots \wedge x_i$ has to contain with nonzero coefficient some basis element e_I with $I \subset J_{i+1}$. This proves the existence of J_i. Finally we define $\sigma(i) = J_i \setminus J_{i-1}$. Then by construction $R \in U_\sigma$.

To prove the second statement of part (a) of the proposition, let us fix a permutation σ. For any flag R from U_σ we will choose a special set of vectors $x_1, \ldots, x_{n-1}$ such that $R_i = \mathbf{K}x_1 + \ldots + \mathbf{K}x_i$ for all i. We choose x_1 to be the unique vector from R_1 in which $e_{\sigma(1)}$ occurs with coefficient 1. Since other vectors x_i can be choosen modulo x_1, we can assume that $e_{\sigma(1)}$ does not occur in their expansion. We choose x_2 to be the only vector from R_2 which does not contain $e_{\sigma(1)}$ and in which $e_{\sigma(2)}$ occurs with coefficient 1. Continuing like that, we choose x_i to be the only vector in R_i whose expansion does not contain $e_{\sigma(1)}, \ldots e_{\sigma(i-1)}$ and contains $e_{\sigma(i)}$ with coefficient 1. Notice that for each R from U_σ we can choose the sequence x_i in a unique way. By construction we can write

$$x_i = e_{\sigma(i)} + \sum_{\sigma^{-1}(j) > i} z_{ij} e_j.$$

One sees easily that the functions z_{ij} are the coordinate functions on U_σ. ∎

A similar covering can be constructed for any partial flag variety $\text{Flag}(b_1, \ldots, b_t; E)$ Let me just mention that the open subsets in this covering are indexed by the sequences $(J_1, \ldots, J_t)$ where J_i is a subset of $[1, n]$ with b_i elements and $J_i \subset J_{i+1}$ for all i.

The indexing set can be identified with $\Sigma_n / \Sigma_{b_1} \times \ldots \times \Sigma_{b_t}$. The permutation whose coset corresponds to $(J_1, \ldots, J_t)$ is constructed by listing (each part in increasing order) $J_1, J_2 \setminus J_1, \ldots, J_t \setminus J_{t-1}$. We therefore denote sets in our coverings by U_σ ($\sigma \in \Sigma_n / \Sigma_{b_1} \times \ldots \times \Sigma_{b_t}$). The proofs are the same as for full flag varieties, so we just state the results.

(3.2.9) Proposition.

(a) *The sets* U_σ *(*$\sigma \in \Sigma_n / \Sigma_{b_1} \times \ldots \times \Sigma_{b_t}$*) form an open covering of the flag variety* $\text{Flag}(b_1, \ldots, b_t; E)$. *Each of the sets* U_σ *is isomorphic to the affine space of dimension* $\sum_{1 \leq i < j \leq t+1} (b_i - b_{i-1})(b_j - b_{j-1})$, *with the convention* $b_0 = 0$, $b_{t+1} = n$.

(b) *The variety* $\text{Flag}(E)$ *is rational, nonsingular of dimension* $\sum_{1 \leq i < j \leq t+1} (b_i - b_{i-1})(b_j - b_{j-1})$.

3.3. The Homogeneous Vector Bundles on Flag Manifolds

First let us recall some basic definitions. By a vector bundle of dimension n over a scheme X we mean a scheme $\mathcal{E}$ equipped with the morphism $h : \mathcal{E} \longrightarrow X$ for which there exists a Zariski open covering $X = \bigcup_{\alpha \in A} X_\alpha$ such that

(1) for each α there exists an isomorphism $r_\alpha : h^{-1}(X_\alpha) \longrightarrow X_\alpha \times \mathbf{K}^n$ such that the composition of r_α with the first projection equals h;

(2) for each α, β the map

$$r_{\alpha\beta} : (X_\alpha \cap X_\beta) \times \mathbf{K}^n \longrightarrow (X_\alpha \cap X_\beta) \times \mathbf{K}^n$$

defined as $r_{\alpha\beta} = r_\beta (r_\alpha)^{-1}$ is linear on the fibers, i.e., there is a regular function $\phi_{\alpha\beta} : X_\alpha \cap X_\beta \to GL(k^n)$ such that

$$r_{\alpha\beta}(x, y) = (x, \phi_{\alpha\beta}(x)y).$$

We call $\mathcal{E}$ the *total space* of the vector bundle, and the map h the *structure map* of the vector bundle $\mathcal{E}$.

We define an n-dimensional trivial vector bundle over X to be $\mathcal{E} = \mathbf{K}^n \times X$ with $h(x, y) = x$.

For a morphism $f : Y \to X$ of schemes and a vector bundle $\mathcal{E}$ on X we define the induced bundle

$$f^*(\mathcal{E}) = \mathcal{E} \times_X Y$$

with the obvious structure map.

The subject of this section is the construction and properties of some important vector bundles on Grassmannians and flag varieties. Let us start with the Grassmannian $\text{Grass}(r, E)$. We consider the following variety:

$$\mathcal{R} = \{(x, R) \in E \times \text{Grass}(r, E) | x \in R\}.$$

Denote by $p : \mathcal{R} \longrightarrow \text{Grass}(r, E)$ the restriction of the second projection to $\mathcal{R}$.

(3.3.1) Proposition. *The morphism p defines on $\mathcal{R}$ the structure of a vector bundle over* $\text{Grass}(r, E)$.

Proof. We will show that the standard covering $\{U_I\}$ defined in section 3.2 satisfies the conditions of our definition. Let us fix a subset I and the subspace R from U_I. Choose a basis

$$
\begin{aligned}
z_1 &= z_{11}e_1 + z_{12}e_2 + \cdots + z_{1n}e_n, \\
z_2 &= z_{21}e_1 + z_{22}e_2 + \cdots + z_{2n}e_n, \\
&\vdots \\
z_r &= z_{r1}e_1 + z_{r2}e_2 + \cdots + z_{rn}e_n
\end{aligned}
$$

of R. As in the previous section, we form a matrix

$$
Z = \begin{pmatrix}
z_{11} & z_{12} & \cdots & z_{1n} \\
z_{21} & z_{22} & \cdots & z_{2n} \\
\vdots & \vdots & & \vdots \\
z_{r1} & z_{r2} & \cdots & z_{rn}
\end{pmatrix},
$$

and by Z_I we denote the minor of Z corresponding to the columns from I. Since R is in U_I, the matrix Z_I is nonsingular. The matrix $Z_I^{-1}Z$ has in the columns from I all entries equal to 0 or 1 (they are consecutive columns of identity matrix). We denote by W_I the matrix $Z_I^{-1}Z$ with the columns from I deleted.

We define the map $r_I : p^{-1}(U_I) \to U_I \times \mathbf{K}^{r(n-r)}$ by setting

$$r_I(x, R) = (x, W_I).$$

It is clear that r_I is an isomorphism. It remains to show that point (2) of the definition of the vector bundle is satisfied. Let us assume the subspace R is in $U_I \cap U_J$. Then for any $x \in R$ we have $r_I(x, R) = (x, W_I)$ and $r_J(x, R) = (x, W_J)$. The map $r_J r_I^{-1}$ is linear on each fiber, because it can be expressed as a composition

$$W_I \to Z_I^{-1}Z \to Z_J^{-1}Z \to W_J$$

of linear operations on matrices. Each entry of Z is a ratio of two Plücker coordinates, so it is a regular function on $U_I \cap U_J$. So are $\det(Z_I)^{-1}$ and $\det(Z_J)^{-1}$. Therefore each entry of W_J is a regular function on $U_I \cap U_J$. This completes the proof of Proposition (3.3.1). ∎

The bundle $\mathcal{R}$ is called *a tautological subbbundle* of the trivial bundle $E \times \text{Grass}(r, E)$. It plays an important role because of the following universal property.

(3.3.2) Proposition. *Let X be a scheme, and let $\mathcal{E}$ be an r-dimensional vector bundle which is a subbundle of n-dimensional trivial bundle $E \times X$. Then there exists a map $f : X \to \text{Grass}(r, E)$ such that $\mathcal{E} = f^*(\mathcal{R})$.*

Proof. Let $i : \mathcal{E} \to E \times X$ be the embedding of $\mathcal{E}$ as a subbundle. For each $x \in X$ the fiber $\mathcal{E}_x$ becomes a subspace of E. We define the map $f : X \to \text{Grass}(r, E)$ by setting $f(x) = \mathcal{E}_x$. The reader can easily verify that $f^*(\mathcal{R}) = \mathcal{E}$. ∎

The bundle $\mathcal{R}$ is by definition a subbundle of the trivial bundle $E \times \text{Grass}(r, E)$. We will denote the factorbundle $E \times \text{Grass}(r, E)/\mathcal{R}$ by $\mathcal{Q}$. We call this bundle *the tautological factorbundle of the trivial bundle $E \times \text{Grass}(r, E)$*. Over $\text{Grass}(r, E)$ we have an exact sequence of vector bundles

$$0 \longrightarrow \mathcal{R} \longrightarrow E \times \text{Grass}(r, E) \longrightarrow \mathcal{Q} \longrightarrow 0$$

We call this sequence *a tautological sequence* on the Grassmannian $\text{Grass}(r, E)$.

Next we deal with the equations defining $\mathcal{R}$ as a subvariety in $E \times \text{Grass}(r, E)$. The Koszul complex of sheaves that arises is central to the techniques of calculating syzygies we develop. Since the map $p : \mathcal{R} \to \text{Grass}(r, E)$ is the restriction of the projection from $E \times \text{Grass}(r, E)$ onto $\text{Grass}(r, E)$, we also denote this projection by p.

(3.3.3) Proposition.
  (a) *The structure sheaf $\mathcal{O}_\mathcal{R}$ has a locally free resolution as an $\mathcal{O}_{E \times \text{Grass}(r, E)}$-module given by the Koszul complex*

$$\mathcal{K}_\bullet(\mathcal{Q}^*) : 0 \to \bigwedge^{n-r}(p^*\mathcal{Q}^*) \to \ldots \to \bigwedge^{2}(p^*\mathcal{Q}^*)$$
$$\to p^*(\mathcal{Q}^*) \to \mathcal{O}_{E \times \text{Grass}(r, E)}.$$

The differentials in this complex are homogeneous of degree 1 *in the coordinate functions on E.*

(b) *The direct image* $p_*\mathcal{O}_\mathcal{R}$ *can be identified with the sheaf of algebras* $\mathrm{Sym}(\mathcal{R}^*)$ *on* $\mathrm{Grass}(r, E)$.

Proof. We use the notation from the proof of (3.3.2). Let $a_1, \ldots, a_n$ be the coordinate functions on E. We write the equations of $\mathcal{R}|U_I$ in $E \times U_I$. The point (x, R) from $E \times U_I$ is in $\mathcal{R}|U_I$ if and only if x belongs to the subspace generated by the row vectors of the matrix $Z_I^{-1}Z$. Writing $x = \sum_{j=1}^n a_j e_j$, we see that this means that $x = \sum_{u=1}^r a_{i_u} W_u$, where W_u is the u-th row of $Z_I^{-1}Z$. This gives us the set of $n - r$ equations

$$a_j = a_{i_1} z'_{1j} + \cdots + a_{i_r} z'_{rj} \qquad \text{for} \quad j \notin I, \tag{*}$$

where z'_{uj} denote the entries of $Z_I^{-1}Z$ from the columns not in I, which are the coordinates in U_I. The coordinate ring $\mathbf{K}[E \times U_I]$ is a polynomial ring in the variables a_i, z'_{uj}. The factor of $\mathbf{K}[E \times U_I]$ by the equations (*) is the polynomial ring in a_i $(i \in I)$ and z'_{uj}. This means that the elements (*) are the regular sequence and they generate the prime ideal in $\mathbf{K}[E \times U_I]$. The resolution of $\mathcal{O}_{\mathcal{R}|U_I}$ as $\mathcal{O}_{E \times U_I}$-module is the Koszul complex on the equations (*).

It remains to show that these Koszul complexes patch together to give the complex (3.3.3) (a). It is enough to show that the equations (*) patch together as a cosection of $p^*\mathcal{Q}^*$, or equivalently, as a section of $p^*\mathcal{Q}$.

The bundle $p^*\mathcal{Q}$ has an obvious section s given by the map

$$(x, R) \mapsto (\bar{x}, (x, R)),$$

where $\bar{x}$ is the coset of x in E/R. Let us write this section in local coordinates on $E \times U_I$. The bundle $\mathcal{Q}$ can be trivialized over the covering $\{U_I\}$ by defining $r' : \mathcal{Q}|U_I \to \mathbf{K}^{n-r} \times U_I$, $r'(x, R) = ((a_j - \sum_{u=1}^r a_{i_u} z'_{uj})_{j \notin I}, R)$. Therefore over U_I

$$s(x, R) = \left(\left(a_j - \sum_{u=1}^r a_{i_u} z'_{uj} \right)_{j \notin I}, (x, R) \right)$$

This proves that vanishing of s locally gives the equations (*). The first statement of Proposition (3.3.3) (a) follows. Now the second statement of (a) follows instantly from (*).

To prove (b) we observe that p is an affine map ([H1, chapter II, exercise 5.17]), so the direct image functor p_* is exact ([H1, chapter II, exercise 8.2]). We apply the functor p_* to the complex $\mathcal{K}_\bullet(\mathcal{Q}^*)$. By the projection formula

([H1, chapter II, exercise 5.1]) we have

$$p_*\left(\bigwedge^j (p^* Q^*)\right) = \bigwedge^j Q^* \otimes \mathrm{Sym}(E^*) \otimes \mathcal{O}_{\mathrm{Grass}(r,E)}.$$

The direct image of the cosection s is easily identified with the map

$$Q^* \otimes \mathrm{Sym}(E^*) \otimes \mathcal{O}_{\mathrm{Grass}(r,E)} \longrightarrow \mathrm{Sym}(E^*) \otimes \mathcal{O}_{\mathrm{Grass}(r,E)}$$

induced by the inclusion $Q^* \hookrightarrow E^* \otimes \mathcal{O}_{\mathrm{Grass}(r,E)}$. The statement follows. ■

(3.3.4) Remark. *The complex (3.3.3) (a) is a prototype of similar Koszul complexes associated to vector bundles that will be considered in chapter 5. They play the key role in the geometric method for constructing syzygies.*

The tangent bundle on the Grassmannian can be conveniently expressed through tautological bundles.

(3.3.5) Proposition.

(a) *The tangent bundle $T_{\mathrm{Grass}(r,E)}$ can be canonically identified as*

$$T_{\mathrm{Grass}(r,E)} = \mathrm{Hom}_{\mathcal{O}_{\mathrm{Grass}(r,E)}}(\mathcal{R}, Q) = \mathcal{R}^* \otimes Q.$$

(b) *The bundle of differentials $\Omega_{\mathrm{Grass}(r,E)}$ can be canonically identified with $\mathcal{R} \otimes Q^*$.*

(c) *The canonical bundle $K_{\mathrm{Grass}(r,E)}$ on $\mathrm{Grass}(r, E)$ can be canonically identified with $(\bigwedge^r \mathcal{R})^{\otimes (n-r)} \otimes (\bigwedge^{(n-r)} Q^*)^{\otimes r}$.*

Proof. It is enough to prove part (a). Let us consider the tautological sequence

$$0 \to \mathcal{R} \to p^* \mathcal{E} \to Q \to 0.$$

We consider the Plücker embedding $\pi : \mathrm{Grass}(r, E) \to P = P(\bigwedge^r E)$. By [H1, chapter II, exercise 8.17] we see that we have the following exact sequences of bundles:

$$0 \to T_{\mathrm{Grass}(r,E)} \to T_P|_{\mathrm{Grass}(r,E)} \to (\mathcal{I}/\mathcal{I}^2)^* \to 0$$

where $\mathcal{I}$ is the sheaf of ideals of $\mathrm{Grass}(r, E)$ in P. By [H1, chapter II, 8.20.1] the bundle $T_{P(\bigwedge^r E)}$ can be identified with $\bigwedge^r E \otimes \mathcal{O}_P(1)/\mathcal{O}_P$. However, $\mathcal{O}_P(1)|_{\mathrm{Grass}(r,E)}$ equals $\bigwedge^r \mathcal{R}^*$, so $T_{\mathrm{Grass}(r,E)}$ is a subbundle of $(\bigwedge^r E \otimes \bigwedge^r \mathcal{R}^*)/(\bigwedge^r \mathcal{R} \otimes \bigwedge^r \mathcal{R}^*)$. Notice that there exists a natural subbundle $\bigwedge^{r-1} \mathcal{R} \otimes Q \otimes \bigwedge^r \mathcal{R}^* = Q \otimes \mathcal{R}^*$ of the right dimension in that bundle. We will show that it can be canonically identified with $T_{\mathrm{Grass}(r,E)}$. Indeed, let us calculate the fiber of the tangent bundle to the Grassmannian. Recall that the Plücker embedding identifies the affine cone Y over the Grassmannian with the set of totally decomposable tensors in $\bigwedge^r E$. Let $e_1, \ldots, e_n$ be a basis in E. Let us consider the point $x^0 = e_1 \wedge \ldots \wedge e_r$ in this cone. Then the tangent space at x^0 to the cone Y is the subspace T^0 of $\bigwedge^r E$ spanned by $e_1 \wedge \ldots \wedge e_r$ and all vectors $e_1 \wedge \ldots \hat{e}_i \wedge \ldots \wedge e_r \wedge e_j$ for all $1 \le i \le r$, $j > r$. This identifies the tangent space to the Grassmannian at x^0 with the factor of T^0 by $e_1 \wedge \ldots \wedge e_r$. However this is also a fiber of the bundle $\bigwedge^{r-1} \mathcal{R} \otimes \mathcal{Q} \otimes \bigwedge^r \mathcal{R}^*$ under similar identification, because at x^0, the vectors $e_1, \ldots, e_r$ are the generators of the corresponding subspace R. ∎

The tautological bundles can be defined over any flag variety. We start with the full flag variety Flag(E). Let us denote by π the natural embedding of Flag(E) into Grass($1, E$) $\times \ldots \times$ Grass($n - 1, E$). For each i we denote by $\mathcal{R}_i$ the tautological bundle on Grass(i, E). We use the same notation for the bundle on Grass($1, E$) $\times \ldots \times$ Grass($n - 1, E$) induced by $\mathcal{R}_i$ and for the bundle $(\pi)^*(\mathcal{R}_i)$ on Flag(E). This defines on Flag(E) the family of bundles $\mathcal{R}_i$ such that $\mathcal{R}_1 \subset \mathcal{R}_2 \subset \ldots \subset \mathcal{R}_{n-1} \subset E \times$ Flag(E). We call $\mathcal{R}_i$ the *i-dimensional tautological subbundle* on Flag(E). The factor $E/\mathcal{R}_i$ is denoted by $\mathcal{Q}_{n-i}$. This bundle is called *the $(n - i)$-dimensional tautological factor bundle* on Flag(E). We also introduce the line bundles $\mathcal{L}_i = \mathcal{R}_i/\mathcal{R}_{i-1}$.

Finally we deal with an arbitrary partial flag variety Flag($b_1, \ldots, b_t; E$). It is clear form the previous discussion that on Flag($b_1, \ldots, b_t; E$) we can define the tautological subbundles $\mathcal{R}_{b_j}$ for $1 \le j \le t$. Obviously we have $\mathcal{R}_{b_1} \subset \mathcal{R}_{b_2} \subset \ldots \subset \mathcal{R}_{b_t}$. We denote by $\mathcal{Q}_{n-b_j}$ the corresponding factor bundles.

We conclude this section with a brief discusion of the relative Grassmannians and flag varieties.

Consider a scheme X and a vector bundle $\mathcal{E}$ of dimension n over X with the structure map h. There exists an open covering $X = \bigcup_{\alpha \in \mathcal{A}} X_\alpha$ and the trivialization maps $r_\alpha : h^{-1}(X_\alpha) \longrightarrow X_\alpha \times k^n$ satisfying the conditions (1) and (2) of the definition of the vector bundle above.

For a number r satisfying $1 \le r \le n - 1$ we construct the scheme Grass $(r, \mathcal{E})$. For each α we consider the scheme $X_\alpha \times$ Grass($r, \mathbf{K}^n$). We patch these schemes in the following way. For two indices α and β we patch $X_\alpha \times$ Grass($r, \mathbf{K}^n$) and $X_\beta \times$ Grass($r, \mathbf{K}^n$) along $X_\alpha \cap X_\beta \times$ Grass($r, \mathbf{K}^n$) by identifying the point (x, R) from $X_\alpha \times$ Grass($r, \mathbf{K}^n$) with $x \in X_\alpha \cap X_\beta$ with $(x, \phi_{\alpha\beta}R)$ in $X_\beta \times$ Grass($r, \mathbf{K}^n$). The resulting scheme is the relative Grassmannian Grass($r, \mathcal{E}$). It is clear that the first projections from $X_\alpha \times$ Grass($r, \mathbf{K}^n$) to X_α patch into the map Grass($r, \mathcal{E}$) $\longrightarrow X$, which we will call the structure map and we will again denote by h. By definition Grass($r, \mathcal{E}$) is a locally trivial fibration over X with the fiber Grass($r, \mathbf{K}^n$).

In a similar way we construct the spaces $\text{Flag}(\mathcal{E})$ and more generally $\text{Flag}(b_1, \ldots, b_t; \mathcal{E})$. They are locally trivial fibrations over X with the fibers $\text{Flag}(\mathbf{K}^n)$ and $\text{Flag}(b_1, \ldots, b_t; \mathbf{K}^n)$ respectively. The reader should be able to fill the details of these constructions. Each relative flag variety associated to the vector bundle $\mathcal{E}$ comes equipped with the structure map to X. All these maps will be denoted by the same letter as the structure map of $\mathcal{E}$—in this case by h. It will be always clear from the context which map h we have in mind.

The definitions of tautological bundles carry over to the relative case. Indeed, let us consider the relative Grassmannian $\text{Grass}(r, \mathcal{E})$. Let us consider the structure map $h : \text{Grass}(r, \mathcal{E}) \to X$ and the bundle $h^*\mathcal{E}$ over $\text{Grass}(r, \mathcal{E})$. We define the tautological subbundle $\mathcal{R}$ as follows. The preimage $h^{-1}X_\alpha$ in $h^*\mathcal{E}$ is the set of triples (x, R, y) with $x \in X$, $R \subset \mathbf{K}^n$, $y \in \mathbf{K}^n$. Then we define the subset $\mathcal{R}_\alpha$ to be the set of triples (x, R, y) such that $y \in R$. The condition $y \in R$ is independent of the choice of basis in $\mathbf{K}^n$, so the subsets $\mathcal{R}_\alpha$ patch together. The resulting subset $\mathcal{R}$ is easily seen to be a subbundle in $h^*\mathcal{E}$. We call $\mathcal{R}$ a tautological subbundle of $h^*\mathcal{E}$. Indeed, we can check this locally, and over each X_α we are dealing with the product of X_α with the Grassmannian $\text{Grass}(r, \mathbf{K}^n)$, so the claim follows from previous considerations.

The tautological factor bundle $\mathcal{Q}$ over $\text{Grass}(r, \mathcal{E})$ is by definition $h^*\mathcal{E}/\mathcal{R}$.

Similarly one can define the tautological subbundles over the flag varieties $\text{Flag}(\mathcal{E})$ and $\text{Flag}(b_1, \ldots, b_t; \mathcal{E})$. We denote them by $\mathcal{R}_i$ and $\mathcal{R}_{b_i}$ respectively. The corresponding tautological factor bundles will be denoted by $\mathcal{Q}_{n-i}$ and $\mathcal{Q}_{n-b_i}$ respectively. As before, the bundle $\mathcal{R}_i/\mathcal{R}_{i-1}$ on $\text{Flag}(\mathcal{E})$ will be denoted by $\mathcal{L}_i$.

Exercises for Chapter 3

Isotropic Grassmannians

1. Let F be a vector space of dimension $2n$ with a nondegenerate skew-symmetric bilinear form $\langle \ , \ \rangle$. Let $e_1, \ldots e_n, \bar{e}_n, \ldots, \bar{e}_1$ be a symplectic basis in F. Let r be a number such that $1 \leq r \leq n$. Consider the cone X over $\text{Grass}(r, F)$ embedded into the projective space by the Plücker embedding. Let $Y \subset X$ be the cone over the set $\text{IGrass}(r, F)$ of isotropic subspaces in F. The coordinate ring $A = K[X]$ can be identified with $\bigoplus_{s \geq 0} L_{r^s} F^*$. Consider the linear relations of the form

$$t \wedge f_1 \wedge \ldots \wedge f_{r-2}$$

where $t = \sum_{i=1}^n e_i^* \wedge \bar{e}_i^* \in \bigwedge^2 F^*$ is a tensor corresponding to the form $\langle \ , \ \rangle$ and $f_1, \ldots, f_{r-2}$ are vectors from F^*. Prove that these relations vanish identically on Y and that they define Y set-theoretically.

2. Let F be a vector space of dimension $2n + 1$ with a nondegenerate symmetric bilinear form $\langle \ , \ \rangle$. Let $e_0, e_1, \ldots e_n, \bar{e}_n, \ldots, \bar{e}_1$ be a basis in F such that $\langle e_i, \bar{e}_i \rangle = 1$, $\langle e_0, e_0 \rangle = 1$ with the subspaces generated by $e_i, \bar{e}_i$ orthogonal to the space generated by e_0 and to each other for different values of i. Let r be a number such that $1 \leq r \leq n$. Consider the cone X over $\text{Grass}(r, F)$ embedded into the projective space by the Plücker embedding. Let $Y \subset X$ be the cone over the set $\text{IGrass}(r, F)$ of isotropic subspaces in F. The coordinate ring $A = K[X]$ can be identified with $\bigoplus_{s \geq 0} L_{r^s} F^*$. Consider the quadratic relations of the form

$$\sum_j (t_1^j \wedge f_1 \wedge \ldots \wedge f_{r-1})(t_2^j \wedge f_1' \wedge \ldots \wedge f_{r-1}')$$

where $t = e_0^{*2} + \sum_{i=1}^n (e_i^*)(\bar{e}_i^*) \in S_2 F^*$ is a tensor corresponding to the form $\langle \ , \ \rangle$, and denote

$$\Delta(t) = \sum_j t_j^1 \otimes t_j^2 \in F^* \otimes F^*.$$

The elements $f_1, \ldots, f_{r-1}, f_1', \ldots, f_{r-1}'$ are the vectors from F^*. Prove that these relations vanish identically on Y and that they define Y set-theoretically.

3. Let F be a vector space of dimension $2n$ with a nondegenerate symmetric bilinear form $\langle \ , \ \rangle$. Let $e_1, \ldots e_n, \bar{e}_n, \ldots, \bar{e}_1$ be a basis in F such that $\langle e_i, \bar{e}_i \rangle = 1$ with the subspaces generated by $e_i, \bar{e}_i$ orthogonal to each other for different values of i. Let r be a number such that $1 \leq r \leq n$. Consider the cone X over $\text{Grass}(r, F)$ embedded into the projective space by the Plücker embedding. Let $Y \subset X$ be the cone over the set $\text{IGrass}(r, F)$ of isotropic subspaces in F. The coordinate ring $A = K[X]$ can be identified with $\bigoplus_{s \geq 0} L_{r^s} F^*$. Consider quadratic relations of the form

$$\sum_j (t_1^j \wedge f_1 \wedge \ldots \wedge f_{r-1})(t_2^j \wedge f_1' \wedge \ldots \wedge f_{r-1}'),$$

where $t = \sum_{i=1}^n (e_i^*)(\bar{e}_i^*) \in S_2 F^*$ is a tensor corresponding to the form $\langle \ , \ \rangle$, and we denote

$$\Delta(t) = \sum_j t_j^1 \otimes t_j^2 \in F^* \otimes F^*.$$

The elements $f_1, \ldots, f_{r-1}, f_1', \ldots, f_{r-1}'$ are the vectors from F^*. Prove that these relations vanish identically on Y. Prove that if $1 \leq r \leq n - 1$ then these relations define Y set-theoretically.

Proof of Cauchy Formula for $\bigwedge^\bullet(F \otimes G)$ ([AB1]).

4. For a composition $\lambda = (\lambda_1, \ldots, \lambda_s)$ with $k = \lambda_1 + \ldots + \lambda_s$, define

$$D^\lambda(E) = D_{\lambda_1} E \otimes \ldots \otimes D_{\lambda_s} E.$$

Let W be a homogeneous representation of $GL(E)$ of degree k. Prove that the space of equivariant maps $\operatorname{Hom}_{GL(E)}(D^\lambda(E), W)$ is isomorphic to the weight space W_λ of W. Thus the representations $D^\lambda(E)$ are the projective objects in the category $C(k, E)$ of homogeneous representations of $GL(E)$ of degree k. Conclude that every object W in $C(k, E)$ has a left resolution whose terms are direct sums of representations $D^\lambda(E)$.

5. Assume that $\dim E \geq k$. Prove that there exist functors $T_k : C(k, E) \to C(k, E)$ sending $D^\lambda(E)$ into

$$\Lambda^\lambda(E) = \overset{\lambda_1}{\bigwedge} E \otimes \ldots \otimes \overset{\lambda_s}{\bigwedge} E$$

and sending diagonal (respectively, multiplication) maps into diagonal (respectively, multiplication) maps. The functors T_k are involutions on the category $C(k, E)$. We call the functors T_k the Schur–Weyl duality functors. Prove that T_k are exact and that $T_k(K_\lambda E) = L_\lambda E$, $T_m(L_\lambda E) = K_\lambda E$.

6. Let F, G be two vector spaces of dimensions m, n respectively. Consider the category $C(k, F, G)$ of representations of $GL(F) \times GL(G)$ of homogeneous degree k on both coordinates. Prove that there exists an endofunctor $\operatorname{Id} \otimes T_k$ of $C(k, F, G)$ sending $D^\lambda(F) \otimes D^\mu(G)$ to $D^\lambda(F) \otimes \Lambda^\mu(G)$. Prove that $T_k(S_k(F \otimes G)) = \bigwedge^k(F \otimes G)$. Derive the statement of Theorem (2.3.2) (b).

Combinatorial Proof of Littlewood–Richardson Rule

7. (Schensted process.) Let us consider a standard tableau $T \in ST(\lambda, [1, n])$ and a number $p \in [1, n]$ (treated as a standard tableau of shape (1)). We define the tableau $(p \to T)$ of a shape μ such that $\mu/\lambda = (1)$ as follows. We enter p into the first row of T at the place of the smallest element p_1 such that $p_1 \geq p$. If such p_1 does not exist, we add a box to λ at the end of the first row of T and write p in that box. Next we repeat the procedure with p_1 and the second row of T, and so on. Prove that the tableau $(p \to T)$ is standard.

Example.

$$
2 \rightarrow
\begin{array}{|c|c|c|c|}
\hline
1 & 2 & 3 & 6 \\
\hline
1 & 3 & 4 \\
\cline{1-3}
5 \\
\cline{1-1}
\end{array}
\quad = \quad
\begin{array}{|c|c|c|c|}
\hline
1 & 2 & 3 & 6 \\
\hline
1 & 2 & 4 \\
\cline{1-3}
3 \\
\cline{1-1}
5 \\
\cline{1-1}
\end{array}
\quad .
$$

8. (Reverse Schensted process.) Let U be a standard tableau, $U \in$ $ST(\mu, [1, n])$. Let r be a number filling an extremal box x in U. We take out r from U and insert it into the previous row of U at the place x_1 of the largest element r_1 such that $r_1 \leq r$. We continue the process with r_1, x_1, and the previous row of U, and so on. At the end we get a tableau $T := (U \rightarrow r)$ of shape $\lambda = \mu \setminus x$ and a number from $[1, n]$ that was bumped from the first row of the tableau.
   (a) Prove that T is standard.
   (b) Prove that $((p \rightarrow T) \rightarrow r) = T$ where r is the added box in $(p \rightarrow T)$.
   (c) Prove that Schensted process and reverse Schensted process define bijections

$$
ST(\lambda, [1, n]) \times [1, n] \cong \bigcup_{\mu; \ \mu/\lambda=(1)} ST(\mu, [1, n])
$$

   where the union on the right hand side is disjoint.

Example.

$$
\begin{array}{|c|c|c|c|}
\hline
1 & 2 & 3 & 6 \\
\hline
1 & 2 & 4 \\
\cline{1-3}
3 \\
\cline{1-1}
5 \\
\cline{1-1}
\end{array}
\quad \rightarrow 5 =
\begin{array}{|c|c|c|c|}
\hline
1 & 2 & 3 & 6 \\
\hline
1 & 3 & 4 \\
\cline{1-3}
5 \\
\cline{1-1}
\end{array}
\quad .
$$

9. Consider two partitions λ, μ. Let us consider the sets $\mathcal{A} = ST(\lambda, [1, n]) \times ST(\mu, [1, n])$ and $\mathcal{B} = \bigcup_\nu ST(\nu, [1, n]) \times P(\lambda, \mu; \nu)$, where $P(\lambda, \mu; \nu)$ is the set used in the description of Littlewood–Richardson rule. More precisely, $P(\lambda, \mu; \nu)$ is the set of all tableaux of the shape ν/λ of weight μ' satisfying the condition LP defined in section 2.3. We define the weight of the pair (S, T) to be the sum of weights of S and T. The weight of the pair (U, V) is defined to be the weight of U. We define the map $\alpha : \mathcal{A} \rightarrow \mathcal{B}$ as follows. Let $(S, T) \in \mathcal{A}$. We define the word $v(T) = (T(1, \mu_1), T(1, \mu_1 - 1), \ldots, T(1, 1), T(2, \mu_2), \ldots)$. Let $v(T) =$

$(v_1, v_2, \ldots)$. We define the tableau U of shape v to be

$$U = (v_t \to \ldots \to (v_2 \to (v_1 \to S))\ldots).$$

We also define the tableau V of shape v/λ as follows. We insert the number s into the box $x \in v/\lambda$ if a box x was added to the shape of U when inserting an element from the s-th column of T.

Example.

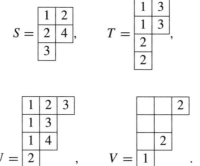

Prove that $(U, V) \in \mathcal{B}$. Conclude that α is a weight preserving map from $\mathcal{A}$ to $\mathcal{B}$.

10. Let $(U, V) \in \mathcal{B}$, and let us assume that it comes from the part of $\mathcal{B}$ corresponding to the shape v. We bump out from U, using the reverse Schensted process, boxes from v/λ. We start with the boxes corresponding to the highest number in V, starting with the lowest one. We get S as a remaining standard tableau of shape λ. We construct T form the elements we bump out. We collect the elements corresponding to boxes with number i in V into the i-th column of T (the element that comes out first is the last element in the i-th column). Prove that the map $\beta : \mathcal{B} \to \mathcal{A}$ defined in this way is a well-defined map which is the inverse of the map α constructed in exercise 9. Conclude that there is a weight preserving bijection from $\mathcal{A}$ to $\mathcal{B}$.

11. Prove that the characters of the representations occurring on both sides of Littlewood–Richardson rule (2.3.4) are the same, and thus, over a field of characteristic zero, the two representations are isomorphic.

12. (Robinson–Schensted correspondence). Fix a natural number n. For a partition λ of n let $ST(\lambda, [1, n])^0$ denote the set of standard tableaux of shape λ of weight (1^n), i.e. where every number occurs once. Let $\mathcal{A} = \bigcup_{|\lambda|=n} ST(\lambda, [1, n])^0 \times ST(\lambda, [1, n])^0$, where the union is disjoint. Let $(S, T) \in \mathcal{A}$. The permutation $\sigma := \gamma(S, T)$ is defined inductively. The value $\sigma(n)$ is the number bumped out from S when we perform the reverse Schensted process to the box in S which is the same as the box in T containing the number n. After completing this process we have still a pair of diagrams (S', T') of the same shape. We define $\sigma(n - 1)$ to be the number bumped out of S' by reverse Schensted process applied to the extremal box in S that corresponds to the box occupied by the number $n - 1$ in T', etc. Prove that the map $\gamma : \mathcal{A} \to \Sigma_n$ defined in this way is a bijection.

Canonical Bundles on Flag Varieties

13. Let us consider the variety $\mathrm{Flag}(b_1, \ldots, b_t; E)$. Let $u_i := b_i - b_{i-1}$ (with the convention $b_0 := 0$, $b_{t+1} := n$) for $i = 1, \ldots, t + 1$. Consider the line bundle $\mathcal{L}(v_1^{u_1}, \ldots, v_{t+1}^{u_{t+1}})$ where $v_i = u_1 + \ldots + u_{i-1} - u_{i+1} - \ldots - u_{t+1}$. Prove that $\mathcal{L}(v_1^{u_1}, \ldots, v_{t+1}^{u_{t+1}})$ is the pull-back of a line bundle $\mathcal{K}(b_1, \ldots, b_t)$ on $\mathrm{Flag}(b_1, \ldots, b_t; E)$. Prove that $\mathcal{K}(b_1, \ldots, b_t)$ is the canonical bundle on $\mathrm{Flag}(b_1, \ldots, b_t; E)$ by showing that it is the only homogeneous line bundle on $\mathrm{Flag}(b_1, \ldots, b_t; E)$ whose top cohomology group is a trivial representation of $\mathrm{GL}(E)$.

4

Bott's Theorem

This chapter is devoted to the discussion of several versions of Bott's theorem on cohomology of line bundles on homogeneous spaces $\mathbf{G}/\mathbf{B}$. We also prove some corresponding results on cohomology of some vector bundles on homogeneous spaces $\mathbf{G}/\mathbf{P}$. They follow by applying the spectral sequence of the composition to a line bundle on $\mathbf{G}/\mathbf{B}$.

Throughout the chapter we prove all results for general linear groups and state them for arbitrary reductive groups. This allows us to avoid technicalities of general theory while preserving the basic ideas of the proof.

In section 4.1 we classify line bundles on flag varieties. They correspond to integral weights for the general linear group. We also state Bott's theorem for the general linear groups. We prove related results on cohomology of some vector bundles on Grassmannians and partial flag varieties. Finally we state Kempf's vanishing theorem.

In section 4.2 we prove Bott's theorem for the general linear group. The proof follows closely the approach of Demazure [D].

In section 4.3 we state Bott's theorem for an arbitrary reductive group. We also give some explicit calculations of cohomology on homogeneous spaces $\mathbf{G}/\mathbf{P}$ where $\mathbf{G}$ is a classical group and $\mathbf{P}$ is a maximal parabolic subgroup. They will be needed in chapter 8.

4.1. The Formulation of Bott's Theorem for the General Linear Group

Throughout this chapter, unless stated otherwise, we work over an algebraically closed field $\mathbf{K}$ of characteristic 0.

Let X be a nonsingular projective variety, and let $\mathcal{E}$ be a vector bundle of dimension n over X. We denote by $h : \mathcal{E} \to X$ the structural morphism of $\mathcal{E}$. In chapter 3 we introduced the flag variety $\mathrm{Flag}_X(\mathcal{E})$ associated to $\mathcal{E}$. The structure morphism from $\mathrm{Flag}_X(\mathcal{E})$ to X is again denoted by h.

The object of Bott's theorem is to calculate the higher direct images $\mathcal{R}^i h_*(\mathcal{L})$ for every line bundle $\mathcal{L}$ on $\mathrm{Flag}_X(\mathcal{E})$. In order to formulate this result we need to classify such bundles. The classification is based on the following result.

Let X and $\mathcal{E}$ be as above. Let $Y = P_X(\mathcal{E})$ be the projectivised bundle, and let $h : Y \to X$ be the structural map. Identifying Y with $\mathrm{Grass}_X(n-1, \mathcal{E})$, we have the tautological subbundle $\mathcal{R}$ of dimension $n-1$ in $h^*(\mathcal{E})$ and the tautological factor bundle $\mathcal{Q}$, which is a line bundle.

We also denote $\mathcal{Q}$ by $\mathcal{O}_Y(1)$. Then for any line bundle $\mathcal{L}$ on Y we can define $\mathcal{L}(m) := \mathcal{L} \otimes \mathcal{O}_Y(1)^{\otimes m}$ for any $m \in \mathbf{Z}$.

Then we have

(4.1.1) Proposition. *Every line bundle $\mathcal{L}$ on Y is of the form $h^*(\mathcal{M})(m)$ for some $m \in \mathbf{Z}$ and some line bundle $\mathcal{M}$ on X. This identification induces the isomorphism* $\mathrm{Pic}\,(P(\mathcal{E})) = \mathrm{Pic}(X) \times \mathbf{Z}$.

Proof. Let $X = \bigcup_{\alpha \in A} X_\alpha$ be the open covering of X such that for each α, $\mathcal{E}|X_\alpha$ is trivial. Let $r_\alpha : h^{-1}(X_\alpha) \longrightarrow X_\alpha \times k^n$ be the trivialization functions. We recall the following standard result: ∎

(4.1.2) Proposition. *Let Y be a Noetherian nonsingular quasiprojective variety over $\mathbf{K}$. We consider the variety $Y \times P^n$ and denote by p_1, p_2 the projection from $Y \times P^n$ on Y and P^n respectively. Then every line bundle $\mathcal{L}$ over $Y \times P^n$ is of the form $p_1^*(\mathcal{M}) \otimes p_2^*(\mathcal{O}(m))$ for some line bundle $\mathcal{M}$ on Y and $m \in \mathbf{Z}$. Moreover, this identification induces the isomorphism* $\mathrm{Pic}(Y \times P^n)/p_1^*(\mathrm{Pic}(Y)) = \mathbf{Z}$.

Proof. This follows from [H1, exercise 6.1, p. 146] together with [H1, Corollary 6.16, p. 145].

Now we apply this proposition to the subsets X_α. It follows that for each α there exists a line bundle $\mathcal{M}_\alpha$ on X_α and an integer m_α such that

$$\mathcal{L}|h^{-1}(X_\alpha) = (r_\alpha)^*(p_1^*(\mathcal{M}_\alpha) \otimes p_2^*(\mathcal{O}(m_\alpha))).$$

Considering the restrictions to $X_\alpha \cap X_\beta$, we see that for all α and β we have $m_\alpha = m_\beta = m$. Let us consider $\mathcal{L}' = \mathcal{L} \otimes \mathcal{O}_Y(-m)$. For each α

$$\mathcal{L}'|h^{-1}(X_\alpha) = (r_\alpha)^*(p_1^*(\mathcal{M}_\alpha)) = h^*(\mathcal{M}_\alpha).$$

By restricting to $X_\alpha \cap X_\beta$ we also see that $\mathcal{M}_\alpha|X_\alpha \cap X_\beta = \mathcal{M}_\beta|X_\alpha \cap X_\beta$. Therefore the line bundles $\mathcal{M}_\alpha$ patch together into the line bundle $\mathcal{M}$ on X, and $\mathcal{L}' = h^*(\mathcal{M})$, which proves Proposition (4.1.1). ∎

Let us come back to flag variety. Let us recall that it was equipped with the collection of tautological bundles $\mathcal{R}_i$ of dimension i for $i = 1, 2, \ldots$ $n - 1$. For $\alpha = (\alpha_1 \ldots, \alpha_n) \in \mathbf{Z}^n$ we define the line bundle $\mathcal{L}(\alpha)$ on $\mathrm{Flag}(\mathcal{E})$ as follows:

$$\mathcal{L}(\alpha) = (\mathcal{R}_1)^{-\alpha_1} \otimes (\mathcal{R}_2/\mathcal{R}_1)^{-\alpha_2} \otimes \ldots \otimes (\mathcal{E}/\mathcal{R}_{n-1})^{-\alpha_n}.$$

We will call the elements of $\mathbf{Z}^n$ the *weights*, and so the bundle $\mathcal{L}(\alpha)$ will be called the line bundle corresponding to the weight α.

Now we can classify the line bundles on $\mathrm{Flag}_X(\mathcal{E})$.

(4.1.3) Proposition. *Every line bundle $\mathcal{L}$ on $\mathrm{Flag}_X(\mathcal{E})$ is of the form*

$$\mathcal{L} = \mathcal{L}(\alpha) \otimes h^*(\mathcal{M})$$

for some $\alpha \in \mathbf{Z}^n$ and a line bundle $\mathcal{M}$ on X. This identification establishes the isomorphism $\mathrm{Pic}(\mathrm{Flag}_X(\mathcal{E}))/h^*(\mathrm{Pic}(X)) = \mathbf{Z}^n/\mathbf{Z}(1, 1 \ldots, 1)$.

Proof. This follows by induction using the sequence of fibrations

$$\mathrm{Flag}(\mathcal{E}) \to \mathrm{Flag}(2, 3, \ldots, n - 1; \mathcal{E}) \to \ldots$$
$$\to \mathrm{Flag}(n - 2, n - 1; \mathcal{E}) \to \mathrm{Flag}(n - 1, \mathcal{E}) \to X,$$

each of which is a projectivized vector bundle. Indeed, the fibration

$$\mathrm{Flag}_X(i, i + 1, \ldots, n - 1; \mathcal{E}) \to \mathrm{Flag}_X(i + 1, \ldots, n - 1; \mathcal{E})$$

can be identified with the projectivization $P_{\mathrm{Flag}_X(i+1,\ldots,n-1;\mathcal{E})}(\mathcal{R}_{i+1})$ of the bundle $\mathcal{R}_{i+1}$ on $\mathrm{Flag}_X(i + 1, \ldots, n - 1; \mathcal{E})$. The repeated application of (4.1.1) gives the first part of our result, and the identification of $\mathrm{Pic}(\mathrm{Flag}_X(\mathcal{E}))/h^*$ $(\mathrm{Pic}(X))$ with $\mathbf{Z}^{n-1}$. The second part follows after we observe that $\mathcal{L}(1, 1, \ldots, 1) = h^*(\bigwedge^n \mathcal{E})$. ∎

Our goal is to compute the higher direct images of the line bundles $\mathcal{L}(\alpha)$ for $\alpha \in \mathbf{Z}^n$.

Before formulating Bott's theorem we need some notation. Let $\rho = (n - 1, n - 2, \ldots, 0)$. The permutation group Σ_n on n letters acts on the set $\mathbf{Z}^n$ in the natural way:

$$\sigma((\alpha_1, \ldots \alpha_n)) = (\alpha_{\sigma(1)}, \ldots, \alpha_{\sigma(n)}).$$

We define the dotted action of Σ_n on $\mathbf{Z}^n$ in the following way:

$$\sigma^{\cdot}(\alpha) = \sigma(\alpha + \rho) - \rho.$$

The dotted action is central for the formulation of Bott's theorem.

(4.1.4) Theorem (Bott). *Let $X, \mathcal{E}, h :$ $\mathrm{Flag}_X(\mathcal{E}) \to X$ be as above. Let $\alpha \in \mathbf{Z}^n$ be a weight, and let $\mathcal{L}(\alpha)$ be the corresponding line bundle over $\mathrm{Flag}_X(\mathcal{E})$. Then one of two mutually exclusive possibilities occurs:*

> *(1) There exists $\sigma \in \Sigma_n$, $\sigma \neq 1$, such that $\sigma \cdot(\alpha) = \alpha$. Then the higher direct images $\mathcal{R}^i h_* \mathcal{L}(\alpha)$ are zero for $i \geq 0$;*
> *(2) There exists a unique $\sigma \in \Sigma_n$ such that $\sigma \cdot(\alpha) := (\beta)$ is nonincreasing. In this case all higher direct images $\mathcal{R}^i h_* \mathcal{L}(\alpha)$ are zero for $i \neq l(\sigma)$, and*

$$\mathcal{R}^{l(\sigma)} h_* \mathcal{L}(\alpha) = K_\beta \mathcal{E}^*,$$

where $K_\beta \mathcal{E}^*$ is defined as a Weyl module tensored with an appropriate power of determinant bundle, i.e. as $K_{\beta_1 - \beta_n, \ldots, \beta_{n-1} - \beta_n} \mathcal{E}^* \otimes \bigwedge^n \mathcal{E}^{\otimes(-\beta_n)}$.

There is another way to state this result in form of an algorithm.

(4.1.5) Remark (Bott's Algorithm). *Let $\alpha = (\alpha_1, \ldots, \alpha_n)$. The permutation $\sigma_i = (i, i+1)$ acts on the set of weights in the following way:*

$$\sigma_i \cdot \alpha = (\alpha_1, \ldots, \alpha_{i-1}, \alpha_{i+1} - 1, \alpha_i + 1, \alpha_{i+2}, \ldots, \alpha_n). \qquad (*)$$

If α is a nonincreasing, then $\mathcal{R}^0 h_* \mathcal{L}(a) = K_{(\alpha)} \mathcal{E}^*$ and $\mathcal{R}^i h_* \mathcal{L}(\alpha) = 0$ for $i > 0$. If α is not a partition, then we start to apply the exchanges of type $(*)$, trying to move bigger number to the right past the smaller number. Two possibilities can occur:

> *(1) We get at some point $\alpha_{i+1} = \alpha_i + 1$ when the exchange of type $(*)$ leads to the same sequence. In this case $\mathcal{R}^i h_* \mathcal{L}(\alpha) = 0$ for all $i \geq 0$.*
> *(2) After applying say j exchanges, we transform α into a nonincreasing sequence β. Then $\mathcal{R}^i h_* \mathcal{L}(\alpha) = 0$ for $i \neq j$ and $\mathcal{R}^j h_* \mathcal{L}(\alpha) = K_\beta \mathcal{E}^*$.*

(4.1.6) Example. *Let us take $n = 4$.*

> *(a) $\alpha = (3, 2, 5, 1)$. We apply the exchange $\sigma_2^\bullet$ to get $(3, 4, 3, 1)$. Then we see that possibility (1) occurs, so all higher direct images of $\mathcal{L}(\alpha)$ are 0.*
> *(b) $\alpha = (3, 2, 6, 1)$. We apply the exchange $\sigma_2^\bullet$ to get $(3, 5, 3, 1)$ and then we apply $\sigma_1^\bullet$ to get $(4, 4, 3, 1)$. Therefore the only nonzero higher direct image is $\mathcal{R}^2 h_* \mathcal{L}(3, 2, 6, 1) = K_{(4,4,3,1)} \mathcal{E}^*$.*

Let us notice the special case of Theorem (4.1.4) which occurs when the space X is a point. The vector bundle $\mathcal{E}$ reduces to the vector space of dimension n, and the relative flag variety $\mathrm{Flag}(\mathcal{E})$ becomes the usual flag variety $\mathrm{Flag}(E)$. The higher direct images are replaced by the cohomology groups. Let us state this special case explicitly.

(4.1.7) Corollary. *Let $\alpha = (\alpha_1, \ldots, \alpha_n)$ be a weight, and let $\mathcal{L}(\alpha)$ be the corresponding line bundle on $\mathrm{Flag}(E)$. Then one of two mutually exclusive possibilities occurs:*

(1) *There exists $\sigma \in \Sigma_n$, $\sigma \neq 1$ such that $\sigma \cdot (\alpha) = \alpha$. Then the cohomology groups $H^i(\mathrm{Flag}(E), \mathcal{L}(\alpha))$ are zero for $i \geq 0$.*

(2) *There exists a unique $\sigma \in \Sigma_n$ such that $\sigma \cdot (\alpha) := (\beta)$ is nonincreasing). In this case all higher direct images $H^i(\mathrm{Flag}(E), \mathcal{L}(\alpha))$ are zero for $i \neq l(\sigma)$, and*

$$H^{l(\sigma)}(\mathrm{Flag}(E), \mathcal{L}(\alpha)) = K_\beta E^*.$$

We shall also need the companion statement for partial flag varieties. It enables us to calculate the cohomology groups of bundles which are the images of line bundles on the full flag variety.

Let us consider the partial flag variety $\mathrm{Flag}(b_1, \ldots, b_t; \mathcal{E})$. The weight α is called (b)-dominant if $\alpha_i \geq \alpha_{i+1}$ provided i is not equal to one of b_j's. For such α we define the weights α^j of length $b_{j+1} - b_j$ for $j = 0, \ldots, t$ by taking $\alpha^j = (\alpha_{b_j+1}, \ldots, \alpha_{b_{j+1}})$. The weights α^j are nonincreasing because α is (b)-dominant, so for each j we can define the Weyl functor $K_{\alpha^j}(\mathcal{R}_{b_{j+1}}/\mathcal{R}_{b_j})^*$.

For a (b)-dominant α we can consider the vector bundle

$$\mathcal{V}(\alpha) = \bigotimes_{j=0}^{t} K_{\alpha^j}(\mathcal{R}_{b_{j+1}}/\mathcal{R}_{b_j})^*$$

on $\mathrm{Flag}(b_1, \ldots, b_t; \mathcal{E})$. Then we have the following version of Bott's theorem.

(4.1.8) Theorem (Bott's Theorem for Partial Flag Varieties). *Let $h : \mathrm{Flag}(b_1, \ldots, b_t; \mathcal{E}) \to X$ be the structure map for the partial flag variety. Let α be a (b)-dominant weight, and let $\mathcal{V}(\alpha)$ be the bundle defined above. Then one of two mutually exclusive possibilities occurs:*

(1) *There exists $\sigma \in \Sigma_n$, $\sigma \neq 1$ such that $\sigma \cdot (\alpha) = \alpha$. Then the higher direct images $\mathcal{R}^i h_* \mathcal{V}(\alpha)$ are zero for $i \geq 0$.*

(2) *There exists a unique $\sigma \in \Sigma_n$ such that $\sigma \cdot (\alpha) := (\beta)$ is a partition (i.e. is nonincreasing). In this case all higher direct images $\mathcal{R}^i h_* \mathcal{V}(\alpha)$ are zero for $i \neq l(\sigma)$, and*

$$\mathcal{R}^{l(\sigma)} h_* \mathcal{V}(\alpha) = K_\beta \mathcal{E}^*.$$

Proof. We deduce this result from Theorem (4.1.4). Let us consider the natural projection $h' : \text{Flag}(\mathcal{E}) \to \text{Flag}(b_1, \ldots, b_t; \mathcal{E})$. We notice that this map makes $\text{Flag}(\mathcal{E})$ the relative product of flag varieties for the bundles $\mathcal{R}_{b_{j+1}} / \mathcal{R}_{b_j}$ for $j = 0, \ldots, t$ over $\text{Flag}(b_1, \ldots, b_t; \mathcal{E})$. Indeed, the fiber of the map h' over the fixed flag of type (b) consists of various ways of completing this flag to a full flag. Such a completion is just a choice of full flags in each of the factors $\mathcal{R}_{b_{j+1}} / \mathcal{R}_{b_j}$.

This means we can apply Bott's theorem (4.1.4) to the bundle $\mathcal{L}(\alpha)$. The direct images $\mathcal{R}^j h'_* \mathcal{L}(\alpha)$ are therefore calculated by applying Bott's algorithm (4.1.5) to each segment α^j of α. Since α is assumed to be (b)-dominant, there are no exchanges, and we see that

$$\mathcal{R}^j h'_* \mathcal{L}(\alpha) = 0$$

for $j > 0$, and

$$\mathcal{R}^0 h'_* \mathcal{L}(\alpha) = \mathcal{V}(\alpha).$$

Applying Bott's theorem to the composition hh' and to the line bundle $\mathcal{L}(\alpha)$, we see that the higher direct images $\mathcal{R}^j(h\,h')_* \mathcal{L}(\alpha)$ are calculated according to Bott's algorithm. Now the spectral sequence of the composition degenerates and gives us the statement of (4.1.8). ∎

The important special case occurs when $t = 1$. Then $b_1 = r$, and we are dealing with the Grassmannian $\text{Grass}(r, \mathcal{E})$. Let us assume that the weight α is (b)-dominant. This means that $\alpha_1 \geq \ldots \geq \alpha_r$ and $\alpha_{r+1} \geq \ldots \geq \alpha_n$. We can consider two weights $\beta = (\alpha_1 \ldots, \alpha_r)$ and $\gamma = (\alpha_{r+1}, \ldots, \alpha_n)$. Then

$$\mathcal{V}(\alpha) = K_\beta \mathcal{R}^* \otimes K_\gamma \mathcal{Q}^*.$$

Therefore we get the following result.

(4.1.9) Corollary. *Let $h : \text{Grass}(r, \mathcal{E}) \to X$ be the structure map. We consider the weight α satisfying $\alpha_i \geq \alpha_{i+1}$ for $i \neq r$ and the corresponding vector bundle $\mathcal{V}(\alpha)$ over $\text{Grass}(r, \mathcal{E})$ defined above. Then one of two mutually exclusive possibilities occurs:*

(1) There exists $\sigma \in \Sigma_n$, $\sigma \neq 1$, such that $\sigma^{\cdot}(\alpha) = \alpha$. Then the higher direct images $\mathcal{R}^i h_* \mathcal{V}(\alpha)$ are zero for $i \geq 0$.

(2) There exists a unique $\sigma \in \Sigma_n$ such that $\sigma^{\cdot}(\alpha) := (\beta)$ is a partition (i.e. is nonincreasing). In this case all higher direct images $\mathcal{R}^i h_* \mathcal{V}(\alpha)$ are zero for $i \neq l(\sigma)$, and

$$\mathcal{R}^{l(\sigma)} h_* \mathcal{V}(\alpha) = K_\beta \mathcal{E}^*.$$

We notice that (4.1.8) and (4.1.9) can be specialized to usual flag varieties and Grassmannians in the style of Corollary (4.1.7). We leave the formulation to the reader.

We conclude this section by stating Kempf's vanishing theorem. Among the line bundles $\mathcal{L}(\alpha)$ there are those for which $\mathcal{R}^0 h_* \mathcal{L}(\alpha) \neq 0$ and $\mathcal{R}^i h_* \mathcal{L}(\alpha) = 0$ for $i > 0$. This happens for dominant weights α, i.e. such sequences $\alpha = (\alpha_1, \ldots, \alpha_n)$ that $\alpha_1 \geq \ldots \geq \alpha_n$. These bundles are important because their sections give the homogeneous components of the coordinate ring of flag varieties. It turns out that for such bundles the statement of Bott's theorem is true in a characteristic free setting.

Before we state the result we need one more piece of notation. Let $\alpha \in \mathbf{Z}^n$ be a dominant integral weight. We denote $\bar{\alpha} = (\alpha_1 - \alpha_n, \ldots, \alpha_{n-1} - \alpha_n, 0)$. By definition the weight $\bar{\alpha}$ is a partition.

(4.1.10) Theorem (Kempf). *Let K be a field of characteristic $p > 0$. Let $X, \mathcal{E}, h : \mathrm{Flag}_X(\mathcal{E}) \to X$ be as above. For a dominant integral weight $\alpha \in \mathbf{Z}^n$ let $\mathcal{L}(\alpha)$ be the corresponding line bundle on $\mathrm{Flag}_X(\mathcal{E})$. Then*

$$\mathcal{R}^0 h_* \mathcal{L}(\alpha) = L_{\bar{\alpha}} \mathcal{E}^* \otimes \left(\bigwedge^n \mathcal{E}^* \right)^{\otimes \alpha_n} \quad and \quad \mathcal{R}^i h_* \mathcal{L}(\alpha) = 0 \quad for \ i > 0.$$

(4.1.11) Remark. *One cannot write $K_{\bar{\alpha}} \mathcal{E}^*$ instead of $L_{\bar{\alpha}} \mathcal{E}^*$ in the above result, because these modules are isomorphic only in characteristic 0. In fact, the sections of line bundles on flag varieties are isomorphic to the multigraded components of the coordinate rings of flag varieties, which, as we saw in chapter 3, are canonically isomorphic to Schur functors, not Weyl functors.*

For the proof of Kempf's theorem the reader may consult Jantzen's book [Ja].

Finally we state the characteristic free analogue of (4.1.9).

(4.1.12) Corollary. *Let $h : \mathrm{Grass}_X(\mathcal{E}) \to X$ be a structure map. Consider the integral dominant weight α and the corresponding vector bundle $\mathcal{V}(\alpha)$ on*

Grass($\mathcal{E}$) defined in (4.1.9). Then

$$\mathcal{R}^0 h_* \mathcal{V}(\alpha) = L_{\tilde{\alpha}'} \mathcal{E}^* \otimes \left(\bigwedge^n \mathcal{E}^* \right)^{\otimes \alpha_n} \quad and \quad \mathcal{R}^i h_* \mathcal{V}(\alpha) = 0 \quad for \ i > 0.$$

The corollary follows from (4.1.10) in the same way (4.1.9) followed from (4.1.4).

4.2. The Proof of Bott's Theorem for the General Linear Group

We start with some results on P^1 bundles.

Let X be a nonsingular variety, and $\mathcal{T}$ a vector bundle of dimension 2 over X. Let $P = P_X(\mathcal{T})$, $\hat{P} = P_X(\mathcal{T}^*)$. We denote by h the projection $h : P \to X$, $\hat{h} : \hat{P} \to X$. Notice that because $\mathcal{T}$ has dimension 2, we have a canonical isomorphism $\mathcal{T}^* = \mathcal{T} \otimes \bigwedge^2 \mathcal{T}^*$. Therefore there exists an isomorphism $\epsilon : P \to \hat{P}$ such that the diagram

$$
\begin{array}{ccc}
P & \xrightarrow{\epsilon} & \hat{P} \\
 & h \searrow \quad \swarrow \hat{h} & \\
 & X &
\end{array}
$$

is commutative. Indeed, it is true in general that for any vector bundle $\mathcal{T}$ over X and any line bundle $\mathcal{L}$ over X the varieties $P_X(\mathcal{T})$ and $P_X(\mathcal{T} \otimes \mathcal{L})$ are canonically isomorphic. The variety $P_X(\mathcal{T})$ parametrizes one dimensional quotients of $\mathcal{T}$, and the variety $P_X \mathcal{T} \otimes \mathcal{L}$ does likewise for $\mathcal{T} \otimes \mathcal{L}$, so tensoring by $\mathcal{L}$ sets up our isomorphism.

If $0 \to \mathcal{R} \to h^* \mathcal{T} \to \mathcal{Q} \to 0$ is the tautological sequence on P, then $0 \to \mathcal{Q}^* \to \hat{h} \mathcal{T}^* \to \mathcal{R}^* \to 0$ is the tautological sequence on $\hat{P}$. It is clear that $\epsilon^* \mathcal{R} = \mathcal{Q}^*$ and $\epsilon^* \mathcal{R}^* = \mathcal{Q}$. Notice also that the isomorphism ϵ is functorial in X and $\mathcal{T}$. The bundles $\mathcal{Q}$ and $\mathcal{R}^*$ can be identified with the twisting bundle $\mathcal{O}(1)$. We also recall that because of (3.3.5) the relative bundle of differentials $\Omega_{P/X}$ is canonically isomorphic to $\mathcal{R} \otimes \mathcal{Q}^*$. In terms of the twisting bundle we have

$$\Omega_{P/X} = h^* \left(\bigwedge^2 \mathcal{T}^* \right)(-2).$$

Let $\mathcal{L}$ be a line bundle over P. We say that the *degree* of $\mathcal{L}$ (with respect to h) equals d if $\mathcal{L} = \mathcal{O}(d) \otimes h^*(\mathcal{M})$ for some line bundle $\mathcal{M}$ on X. Such d always exists and is unique by (4.1.2). The degree has also the following properties:

(1) $d(\Omega_{P/X}) = -2$;
(2) $d(\mathcal{L}) = 0$ if and only if $\mathcal{L} = h^* h_*(\mathcal{L})$.

Let us define $\Omega^{(d)}$ for $d \in \mathbf{Z}$ as follows:

$$\Omega^{(d)} = \begin{cases} S_d(\Omega_{P/X}) & \text{if } d \geq 0, \\ S_{-d}(T_{P/X}) & \text{if } d < 0. \end{cases}$$

For any line bundle $\mathcal{L}$ of degree d on P we define its dual

$$\mathcal{L}^{\vee} = \mathcal{L}^* \otimes \Omega^{(-d)}.$$

Notice that $d(\mathcal{L}^{\vee}) = d(\mathcal{L})$.

(4.2.1) Lemma. *There exists a canonical pairing $h_*(\mathcal{L}) \otimes h_*(\mathcal{L}^{\vee}) \to \mathcal{O}_X$, functorial with respect to change of base and isomorphisms of pairs $h : P \to X$, which induces a duality between $h_*(\mathcal{L})$ and $h_*(\mathcal{L}^{\vee})$.*

Proof. Let $\mathcal{M} = h_*(\mathcal{L}(-d))$. Then we can identify $\mathcal{L}$ with $h^*(\mathcal{M})(d)$ and $\mathcal{L}^{\vee}$ with $h^*(\mathcal{M}^*)(d) \otimes h^*((\bigwedge^2 T)^d)$. Indeed, using the formula for the sheaf of relative differentials given above, we see that $\Omega_{P/X}^{-d} = h^*(\bigwedge^2 T)^d(2d)$.

We conclude by Serre's theorem ([H1], Theorem III.5.1) that if $d \geq 0$ we have

$$h_*(\mathcal{L}) = \mathcal{M} \otimes S_d(T^*),$$

$$h_*(\mathcal{L}^{\vee}) = \mathcal{M}^* \otimes S_d(T^*) \otimes \left(\overset{2}{\bigwedge} T \right)^d,$$

and for $d < 0$ both direct images are zero.

We can define our pairing to be the product of the canonical pairing $\mathcal{M} \otimes \mathcal{M}^* \to \mathcal{O}_X$ with the pairing

$$S_d(T^*) \otimes S_d(T^*) \to \left(\overset{2}{\bigwedge} T^* \right)^d$$

composed with the pairing with $(\bigwedge^2 T)^d$.

Notice that the last map is defined only in characteristic zero. It is a composition of the Cauchy embedding $S_d(T^*) \otimes S_d(T^*) \to S_d(T^* \otimes T^*)$ followed by the d-th symmetric power of the exterior multiplication. The natural characteristic free map would involve divided powers instead of symmetric powers. This is the only place in the proof of Bott's theorem where we use the characteristic assumption. ∎

The pairing from Lemma (4.2.1) is called *Demazure pairing*.

(4.2.2) Proposition. *Let $h : P \to X$ be as above. Let us consider the map $g : X \to Y$, where Y is another variety. For any line bundle $\mathcal{L}$ on P such that*

$d = d(\mathcal{L}) \geq -1$ *and for each $n \geq 0$ we have the canonical isomorphisms on Y*

$$\mathcal{R}^n(gh)_*(\mathcal{L}) = \mathcal{R}^{n+1}(gh)_*(\mathcal{L} \otimes \Omega^{(d+1)})$$

for all n (this includes $\mathcal{R}^0(gh)_*(\mathcal{L} \otimes \Omega^{(d+1)}) = 0$).

Proof. First we notice that $d(\mathcal{L} \otimes \Omega^{(d+1)}) = -d(\mathcal{L}) - 2$. By Serre's theorem, $\mathcal{R}^j h_*(\mathcal{L}) = 0$ for $j \neq 0$, and $\mathcal{R}^j h_*(\mathcal{L} \otimes \Omega^{(d+1)}) = 0$ for $j \neq 1$. The spectral sequence of a composition ([Go, chapter II, section 7.3]) gives

$$\mathcal{R}^n(gh)_*\mathcal{L} = \mathcal{R}^n g_*(h_*\mathcal{L}),$$

$$\mathcal{R}^{n+1}(gh)_*(\mathcal{L} \otimes \Omega^{(d+1)}) = \mathcal{R}^n g_*(\mathcal{R}^1 h_*(\mathcal{L} \otimes \Omega^{(d+1)})).$$

We notice that Serre's duality for h gives us a pairing

$$\mathcal{R}^1 h_*(\mathcal{L} \otimes \Omega^{(d+1)}) \otimes h_*(\mathcal{L}^\vee) \to \mathcal{O}_X.$$

Compairing this with Demazure's pairing, we get a canonical isomorphism

$$\mathcal{R}^1 h_*(\mathcal{L} \otimes \Omega^{(d+1)}) = h_*\mathcal{L},$$

which, together with the spectral sequence of the composition, gives the statement of the proposition. ∎

Proof of Theorem (4.1.4). We introduce the following notation: $\mathcal{F} = \mathrm{Flag}_X$ $(\mathcal{E})$, $\mathcal{F}_i = \mathrm{Flag}_X(1, 2, \ldots, i-1, i+1, \ldots, n-1; \mathcal{E})$ for $i = 1, 2, \ldots, n-1$. We denote by h_i the natural projection $h_i : \mathcal{F} \to \mathcal{F}_i$. Notice that for each i, $\mathcal{F}$ can be identified with the P^1 bundle $P_{\mathcal{F}_i}(\mathcal{R}_{i+1}/\mathcal{R}_{i-1})$. We will apply the Demazure duality to the maps h_i. For the bundle $\mathcal{L}(\alpha)$ on $\mathrm{Flag}_X(\mathcal{E})$ let $d_i(\mathcal{L})$ denote the degree of $\mathcal{L}(a)$ with respect to h_i. We also denote the bundle of relative differentials $\Omega_{\mathcal{F}/\mathcal{F}_i}$ by Ω_i. We also set $\Omega = \Omega_{\mathcal{F}/X}^{\dim \mathcal{F}/X}$, the relative canonical bundle on $\mathcal{F}$. Finally, we denote by $\epsilon_i = (0, 0, \ldots, 0, 1, 0, \ldots, 0)$ (1 in the i-th place) the i-th basis vector in $\mathbf{Z}^n$.

(4.2.3) Proposition.
    (a) $\Omega = \mathcal{L}(\beta)$ where $\beta = (-n+1, -n+3, \ldots, n-1)$.
    (b) $\Omega_i = \mathcal{L}(\beta^i)$ where $\beta^i = \epsilon_i - \epsilon_{i+1}$.
    (c) $d_i(\mathcal{L}(\alpha)) = \alpha_i - \alpha_{i+1}$.

Proof. Let us start with (b). On the projective line bundle $P_{\mathcal{F}_i}(\mathcal{R}_{i+1}/\mathcal{R}_{i-1})$ the tautological sequence is

$$0 \to \mathcal{R}_i/\mathcal{R}_{i-1} \to \mathcal{R}_{i+1}/\mathcal{R}_{i-1} \to \mathcal{R}_{i+1}/\mathcal{R}_i \to 0,$$

and therefore

$$\Omega_i = \mathcal{R}_i/\mathcal{R}_{i-1} \otimes (\mathcal{R}_{i+1}/\mathcal{R}_i)^* = \mathcal{L}(\epsilon_i - \epsilon_{i+1}).$$

To see (c) let us notice that for the projective line bundle $P_{\mathcal{F}_i}(\mathcal{R}_{i+1}/\mathcal{R}i)$ we have $\mathcal{R} = \mathcal{R}_i/\mathcal{R}_{i-1} = \mathcal{L}(-\epsilon_i)$, $\mathcal{Q} = \mathcal{R}_{i+1}/\mathcal{R}_i = \mathcal{L}(\epsilon_{i+1})$. It follows from (4.1.1) and the definition of the degree that $d_i(\mathcal{L}(\epsilon_i)) = 1$ and $d_i(\mathcal{L}(\epsilon_{i+1})) = -1$. Moreover, for $k \neq i, k \neq i + 1$ the bundle $\mathcal{L}(\epsilon_k)$ is induced from $\mathcal{F}_i$, so $d_i(\mathcal{L}(\epsilon_k)) = 0$. This proves (c).

To prove (a) we use induction on n. For $n = 2$ it follows from (3.3.5). To make the inductive step, let us consider the diagram

$$\mathcal{F} \xrightarrow{q} P_X(1; \mathcal{E}) = P \to X.$$

We know by (3.3.5) that $\Omega^{n-1}_{P/X} = \bigwedge^n \mathcal{E}_P^*(-n)$. By induction

$$\Omega^{\dim \mathcal{F}/P}_{\mathcal{F}/P} = \mathcal{L}(0, -(n - 2), \dots, n - 2).$$

Therefore

$$\Omega = \Omega^{\dim \mathcal{F}/P}_{\mathcal{F}/P} \otimes q^*(\Omega^{n-1}_{P/X}) = \mathcal{L}(-n, -(n - 2), \dots, n - 2) \otimes h^*\left(\bigwedge^n \mathcal{E}^*\right).$$

Taking into account that $h^*(\bigwedge^n \mathcal{E}^*) = \mathcal{L}(1, 1, \dots, 1)$, we get (a). ∎

Let us consider the line bundle $\mathcal{L}(\alpha)$. Let us assume first that possibility (1) of Theorem (4.1.4) occurs. Therefore there exists a permutation σ such that $\sigma(\alpha + \rho) = \alpha + \rho$. This means that the weight $\alpha + \rho$ has two repeated entries. Let us suppose they occur in places i and j with $i < j$. We proceed by induction on $j - i$. If $j - i = 1$, we see that $\alpha_{i+1} = \alpha_i + 1$. This means that the degree $d_i(\mathcal{L})(a) = -1$. This means that $\mathcal{L}(\alpha) = \mathcal{L}(\alpha) \otimes \Omega_i^{d_i(\mathcal{L}(\alpha))+1}$. By Proposition (4.2.2) applied to h_i we see that

$$\mathcal{R}^u h_* \mathcal{L}(\alpha) = \mathcal{R}^{u+1} h_* \mathcal{L}(\alpha)$$

for each $u \geq 0$. This means all these direct images have to be zero, because for $u > \dim \mathcal{F}/X$ they are zero by the relative version of the Grothendieck theorem ([H1, chapter III, Corollary 11.2]).

Let us assume that $j > i + 1$ and that theorem is proven for smaller $j - i$. We take $\beta = \sigma_i(\alpha)$, so $\beta = (\alpha_1, \dots, \alpha_{i-1}, \alpha_{i+1} - 1, \alpha_i + 1, \alpha_{i+2}, \dots, \alpha_n)$. Notice that β still satisfies our property, but now the pair (i, j) is changed to $(i + 1, j)$. This means by induction that $\mathcal{R}^u h_* \mathcal{L}(\beta) = 0$ for all $u \geq 0$. Now we again use the map h_i. We notice that $\mathcal{L}(\beta) = \mathcal{L}(\alpha) \otimes \Omega_i^{d_i(\mathcal{L}(\alpha))+1}$. and vice versa $\mathcal{L}(\alpha) = \mathcal{L}(\beta) \otimes \Omega_i^{d_i(\mathcal{L}(\beta))+1}$. Since one of the line bundles $\mathcal{L}(\alpha)$, $\mathcal{L}(\beta)$ has degree d_i which is ≥ -1, we can apply Proposition (4.2.2) to this bundle.

We get either

$$\mathcal{R}^u h_* \mathcal{L}(\alpha) = \mathcal{R}^{u+1} h_* \mathcal{L}(\beta)$$

for all u or

$$\mathcal{R}^u h_* \mathcal{L}(\alpha) = \mathcal{R}^{u-1} h_* \mathcal{L}(\beta)$$

for all u. In both cases all higher direct images of $\mathcal{R}^u h_* \mathcal{L}(\alpha)$ are 0.

Let us assume now that possibility (2) of (4.1.4) occurs. This means that the weight $\alpha + \rho$ has no repeated entries. Therefore there exists a unique permutation σ such that $\sigma(\alpha + \rho)$ is a strictly decreasing sequence, so $\sigma \cdot \alpha$ is a partition β. Let us write a reduced expression

$$\sigma = \sigma_{v_1} \dots \sigma_{v_l},$$

where $l = l(\sigma)$. We define the weights $\beta^s = \sigma_{v_{l-s+1}} \dots \sigma_{v_l}(\alpha)$ for $s = 1, \dots, l$. Obviously $\beta^l = \beta$. We also set $\beta^0 = \alpha$. We apply Proposition (4.2.2) to $p_{v_{l-s}}$ and to $\mathcal{L}(\beta^{s+1})$. Since $\sigma_{v_1} \dots \sigma_{v_l}$ is a reduced expression, we have $d_{l_s}(\mathcal{L}(\beta^{s+1})) \geq 0$, so the proposition applies. We get

$$\mathcal{R}^{u+1} h_* \mathcal{L}(\beta^s) = \mathcal{R}^u h_* \mathcal{L}(\beta^{s+1})$$

for all u and s. Putting these equalities together, we get

$$\mathcal{R}^{u+l} h_* \mathcal{L}(\alpha) = \mathcal{R}^u h_* \mathcal{L}(\beta)$$

for all u. Therefore we are reduced to calculating the cohomology of the bundles $\mathcal{L}(\beta)$ where β is a partition.

First we show that for such bundles $\mathcal{R}^u h_* \mathcal{L}(\beta) = 0$ for $u > 0$. In order to do this we use the permutation $\tau(i) = (n + 1 - i)$. Clearly $l(\tau) = \binom{n}{2}$. Let us choose the reduced expression

$$\tau = \sigma_1(\sigma_2 \sigma_1) \dots (\sigma_{n-2} \dots \sigma_1)(\sigma_{n-1} \dots \sigma_1).$$

Using Proposition (4.2.2) repeatedly, as above, we get

$$\mathcal{R}^u h_* \mathcal{L}(\beta) = \mathcal{R}^{u+\binom{n}{2}} h_* \mathcal{L}(\tau \cdot (\beta))$$

for each u. Since $\binom{n}{2} = \dim \mathcal{F}/X$, we get $\mathcal{R}^u h_* \mathcal{L}(\beta) = 0$ for $u > 0$.

It remains to identify the direct images $h_* \mathcal{L}(\beta)$ for every partition β. To do this we first notice that the question is local in X. Therefore we can assume that the bundle $\mathcal{E}$ is trivial, so $\mathcal{E} = X \times E$ for some vector space E of dimension n over k. This means that we can identify Flag$(\mathcal{E})$ with $X \times$ Flag(E). The map h is just the first projection. The direct image $h_* \mathcal{L}(\beta)$ becomes $\mathcal{O}_X \otimes H^0(\text{Flag}(E), \mathcal{L}(\beta))$. To conclude the proof we have to show that for every nonincreasing sequence $\beta = (\beta_1, \dots, \beta_n)$ the space of sections

$H^0(\text{Flag}(E), \mathcal{L}(\beta))$ is isomorphic to $K_{(\beta_1,\ldots,\beta_n)}E^*$. Notice that we can also assume that $\beta_n = 0$, because adding 1 to each coordinate corresponds to tensoring our bundle by $\bigwedge^n E^*$, and the same is true for the functors $K_{(\beta_1,\ldots,\beta_n)}E^*$.

Pick $\beta = (\beta_1,\ldots,\beta_n)$. Let $\{b_1,\ldots,b_t\} = \{1 \le i \le n-1 \mid \beta_i > \beta_{i+1}\}$. Consider the graded ring

$$S(\beta) = \bigoplus_{n \ge 0} K_{n\beta_1,\ldots,n\beta_n} E^*.$$

This is a homogeneous coordinate ring of the flag variety $\text{Flag}(b_1,\ldots,b_t; E)$ embedded in a projective space $\mathbf{P}(L_{\beta'} E^*)$ by using the line bundle $\mathcal{L}(\beta)$ on $\text{Flag}(b_1,\ldots,b_t; E)$.

(4.2.4) Lemma. *The ring $S(\beta)$ is a domain.*

Proof. Assume $S(\beta)$ is not a domain. Then the zero divisors of $S(\beta)$ are the union of finitely many prime ideals $P_1,\ldots,P_m$ in $S(\beta)$. The group $\text{GL}(E)$ acts on $S(\beta)$. Since $\text{GL}(E)$ is connected, it follows that all ideals P_i are equivariant. This means P_1 has to contain a U-invariant. But P_1 is prime, so it has to contain a U-invariant in degree 1 – the canonical tableau. Since the representation $L_{\beta'} E^*$ is irreducible, P_1 contains all elements of degree 1. This is a contradiction proving that $S(\beta)$ is a domain. ∎

By ([H1, chapter II, Exercise 5.14) we see that the normalization of $S(\beta)$

$$\overline{S(\beta)} = \bigoplus_{n \ge 0} H^0(\text{Flag}(b_1,\ldots,b_t; E), \mathcal{L}(n\beta)).$$

Moreover, the same exercise shows that for $n \gg 0$ we have

$$H^0(\text{Flag}(b_1,\ldots,b_t; E), \mathcal{L}(n\beta)) = K_{n\beta_1,\ldots,n\beta_n} E^*. \qquad (**)$$

Now it is easy to finish the proof. We will show in fact that $S(\beta)$ is integrally closed.

Without loss of generality we can assume that β is not a multiple of another weight. For such β we will show that $(**)$ holds for every $n > 0$. By (2.2.3) it is enough to show that every U-invariant in $\overline{S(\beta)}$ is a power of a canonical tableau c_β. Let x be a U-invariant in $\overline{S(\beta)}$. Then a high enough power x^m is the power of the canonical tableau c_β, because $\overline{S(\beta)}/S(\beta)$ has to have finite length since it is supported at the origin. Let $x^m = c_\beta^l$. If m divides l, we are done, because $S(\beta)$ is a domain, so x has to be the power of canonical tableau. If not, then the weight of x is not an integral weight. The proof of Theorem (4.1.4) is complete.●

4.3. Bott's Theorem for General Reductive Groups

In this section we assume that the reader is familiar with the basic notions concerning reductive groups and root systems. We state here Bott's theorem for reductive groups. The results of this section will be used only in Chapter 8.

We start with recalling some standard notation. Let $\mathbf{K}$ be an algebraically closed field, and let $\mathbf{G}$ be a reductive linear group over $\mathbf{K}$. Let $\mathbf{T}$ be a maximal torus, and $\mathbf{B}$ a Borel subgroup containing $\mathbf{T}$. We denote by Φ the root system associated to the pair $\mathbf{G}, \mathbf{T}$. This is by definition a finite set of vectors in $\mathrm{Hom}_{\mathbf{K}}(\mathrm{Lie}(\mathbf{T}), \mathbf{K})$, where $\mathrm{Lie}(\mathbf{T})$ is the Lie algebra of the torus $\mathbf{T}$. The choice of $\mathbf{B}$ determines the subset Φ^+ of positive roots in Φ. The space $\mathrm{Hom}_{\mathbf{K}}(\mathrm{Lie}(\mathbf{T}), \mathbf{K})$ is equipped with a nondegenerate scalar product $(\ ,\ )$. We define, for $\alpha, \beta \in \mathrm{Hom}_{\mathbf{K}}(\mathrm{Lie}(\mathbf{T}), \mathbf{K})$, $\langle \beta, \alpha \rangle = 2(\beta, \alpha)/(\alpha, \alpha)$. We denote by Λ the lattice of integral weights in $\mathrm{Hom}_{\mathbf{K}}(\mathrm{Lie}(\mathbf{T}), \mathbf{K})$:

$$\Lambda = \{\gamma \in \mathrm{Hom}_{\mathbf{K}}(\mathrm{Lie}(\mathbf{T}), \mathbf{K}) \mid \forall \alpha \in \Phi, \ \langle \gamma, \alpha \rangle \in \mathbf{Z}\}.$$

The lattice Λ contains the cone Λ^+ of dominant integral weights,

$$\Lambda^+ = \{\gamma \in \Lambda \mid \forall \alpha \in \Phi^+, \ \langle \gamma, \alpha \rangle \in \mathbf{Z}^+\}.$$

Bott's theorem gives a rule for calculating cohomology groups of the line bundles on the homogeneous space $\mathbf{G}/\mathbf{B}$. Such bundles are described by weights. Indeed, for each character γ of $\mathbf{T}$ we define a one dimensional rational $\mathbf{B}$-module $V(\gamma)$ by letting the unipotent radical $\mathbf{U}$ of $\mathbf{B}$ act trivially on $V(\gamma)$ and the torus $\mathbf{T}$ act by the character γ, and by letting

$$\mathcal{L}(\gamma) = \mathbf{G} \times_{\mathbf{B}} V(\gamma),$$

where for any rational $\mathbf{B}$-module V we denote by $\mathbf{G} \times_{\mathbf{B}} V$ the product $\mathbf{G} \times V$ divided by the equivalence relation $(g, v) \sim (gb, b^{-1}v)$ for $b \in \mathbf{B}$.

We can identify the group of characters of $\mathbf{T}$ with the additive subgroup in Λ by associating to each character its derivative at identity.

The Weyl group W of $\mathbf{G}$ acts naturally on weights. Let $\rho = \frac{1}{2}\sum_{\alpha>0}\alpha$ be half of the sum of positive roots. We define the dotted action of W on weights

$$\sigma^{\cdot}(\gamma) = \sigma(\gamma + \rho) - \rho.$$

Let us recall that the irreducible representations of $\mathbf{G}$ correspond to the dominant integral weights. For a dominant integral weight β we denote by V_β the irreducible $\mathbf{G}$-module of highest weight β.

(4.3.1) Theorem (Bott). *Let* $\mathbf{G}, \mathbf{T}, \mathbf{B}, W, \Phi$ *be as above. Let* γ *be an integral weight, and let* $\mathcal{L}(\gamma)$ *be the corresponding line bundle over* $\mathbf{G}/\mathbf{B}$. *Then one* *of two mutually exclusive possibilities occurs:*

(1) *There exists* $\sigma \in W$, $\sigma \neq 1$, *such that* $\sigma \cdot (\gamma) = \gamma$. *Then the cohomology groups* $H^i(\mathbf{G}/\mathbf{B}, \mathcal{L}(\gamma))$ *are zero for* $i \geq 0$.

(2) *There exists a unique* $\sigma \in W$ *such that* $\sigma \cdot (\gamma) := (\alpha)$ *is a dominant integral weight. In this case all cohomology groups* $H^i(\mathbf{G}/\mathbf{B}, \mathcal{L}(\gamma))$ *are zero for* $i \neq l(\sigma)$, *and*

$$H^{l(\sigma)}(\mathbf{G}/\mathbf{B}, \mathcal{L}(\gamma)) = V_\alpha.$$

(4.3.2) Remark. *The proof in the general case follows the same scheme as in the case of the general linear group. The role of the flag varieties* Flag $(1, 2, \ldots, i-1, i+1, \ldots, n-1; \mathcal{E})$ *is played by the homogeneous space* $\mathbf{G}/\mathbf{P}_\alpha$, *where* $\mathbf{P}_\alpha$ *is a parabolic subgroup corresponding to a simple root* α.

(4.3.3) Examples.

(a) *Let us fix a vector space* E *of dimension* n, *and let us consider the general linear group* $\mathbf{G} = \mathrm{GL}(E)$. *Then we can choose the maximal torus* T *to be the subgroup of diagonal matrices, and the Borel subgroup* $\mathbf{B}$ *to be the subgroup of upper triangular matrices. The homogeneous space* $\mathbf{G}/\mathbf{B}$ *can be identified with* $\mathrm{Flag}(E)$, *the set* $\Lambda = \mathbf{Z}^n$, *and the Weyl group is isomorphic to* Σ_n. *The statement of Bott's theorem reduces to Corollary (4.1.2).*

(b) *Let* F *be a vector space of dimension* $2n + 1$ *with a nondegenerate symmetric bilinear form* $\langle \ , \ \rangle$. *Let us take* $\mathbf{G} = \mathrm{SO}(F)$ *to be the special orthogonal group. The lattice* $\Lambda = \mathbf{Z}^n$. *The Weyl group* W *is a hyperoctahedral group acting on* Λ *by signed permutations. The half sum of the positive roots is* $\rho = (\frac{2n-1}{2}, \frac{2n-3}{2}, \ldots, \frac{1}{2})$

(c) *Let* F *be a vector space of dimension* $2n$ *with a nondegenerate skew symmetric bilinear form* $\langle \ , \ \rangle$ *on* F. *Let us take* $\mathbf{G} = \mathrm{Sp}(F)$ *to be the symplectic group associated to* F. *The lattice* $\Lambda = \mathbf{Z}^n$. *The Weyl group* W *is a hyperoctahedral group acting on* Λ *by signed permutations. The half sum of the positive roots is* $\rho = (n, n-1, \ldots, 1)$

(d) *Let* F *be a vector space of dimension* $2n$ *with a nondegenerate symmetric bilinear form* $\langle \ , \ \rangle$. *Let us take* $\mathbf{G} = \mathrm{SO}(F)$ *to be the special orthogonal group. The lattice* $\Lambda = \mathbf{Z}^n$. *The Weyl group* W *is a subgroup of hyperoctahedral group acting on* Λ *by signed permutations with even number of sign changes. The half sum of the positive roots is* $\rho = (n-1, n-2, \ldots, 0)$.

We proceed with some explicit calculations on homogeneous spaces $\mathbf{G}/\mathbf{P}$ where $\mathbf{P}$ is a maximal parabolic subgroup in $\mathbf{G}$. We will need these kinds of calculations in chapter 8. We limit ourselves to some examples related to classical groups. They can be viewed as analogues of (4.1.9). We recall that by the general theory ([Hu2], [Bou]) there are (up to conjugation) finitely many types of parabolic subgroups in $\mathbf{G}$, and they correspond to subsets of simple roots.

We start with the symplectic group. Let F be a vector space of even dimension $2n$ with a nondegenerate skew symmetric form $\langle \ , \ \rangle$. The simple roots of Φ are $\alpha_j = \epsilon_j - \epsilon_{j+1}$ for $j = 1, \ldots, n-1$ and $\alpha_n = 2\epsilon_n$.

Next we give a concrete description of the homogeneous spaces $\mathbf{G}/\mathbf{B}$. Let us recall that a subspace $R \subset F$ is *isotropic* if the restriction of $\langle \ , \ \rangle$ to R is zero.

We consider the set

$$\mathrm{IFlag}(F) = \{(R_1, \ldots, R_n) \in \mathrm{Flag}(1, 2, \ldots, n; F) \mid R_n \text{ is isotropic}\}.$$

The group $\mathrm{Sp}(F)$ acts on the set $\mathrm{IFlag}(F)$ transitively. To see this, observe that for a flag $(R_1 \ldots, R_n) \in \mathrm{IFlag}(F)$ we can choose a symplectic basis $e_1, \ldots, e_n, \bar{e}_n, \ldots, \bar{e}_1$ of F so $e_1, \ldots, e_i$ is a basis of R_i ($1 \le i \le n$). Since the symplectic group operates transitively on symplectic bases, we are done.

The space $\mathrm{IFlag}(F)$ can be identified with the homogeneous space $\mathbf{G}/\mathbf{H}$, where $\mathbf{H}$ is a subgroup of elements in $\mathrm{Sp}(F)$ stabilizing the flag $(R_1, \ldots, R_n)$. By Borel's theorem there exists a Borel subgroup $\mathbf{B}$ contained in $\mathbf{H}$. Since every parabolic subgroup of a connected reductive group is connected ([Hu2]), we have $\mathbf{H} = \mathbf{B}$.

This realization of the homogeneous space $\mathbf{G}/\mathbf{B}$ allows us to develop relative theory in the same spirit as in section 3.3. We just give the definitions, leaving the details to the reader.

Let $\mathcal{F}$ be a symplectic vector bundle over a scheme X, i.e. a vector bundle $\mathcal{F}$ equipped with a map $\langle \ , \ \rangle : \bigwedge^2 \mathcal{F} \to \mathcal{O}_X$ for which the restriction to each fiber gives a nondegenerate skew symmetric form on it. We can construct a relative isotropic flag variety $\mathrm{IFlag}(\mathcal{F})$ with the structure map $p : \mathrm{IFlag}(\mathcal{F}) \to X$. The statement of Bott's theorem is true in a relative version with higher direct images replacing cohomology groups. We leave the formulation of this result to the reader.

For each $j = 1, \ldots, n$ we consider the maximal parabolic subgoup $\mathbf{P}_j$ in $\mathbf{G} = \mathrm{Sp}(2n)$ which corresponds to the subset of all simple roots except α_j. The space $\mathbf{G}/\mathbf{P}_j$ can be identified with the isotropic Grassmannian $\mathrm{IGrass}(j, F)$ of isotropic subspaces of dimension j in F. This is a closed subset in $\mathrm{Grass}(F)$, so we can talk about the tautological subbundle $\mathcal{R}_j$ on $\mathrm{IGrass}(j, F)$. For any isotropic subspace from $\text{IGrass}(j, F)$ we define the orthogonal complement

$$R^\vee = \{x \in F \mid \forall y \in R \; \langle x, y \rangle = 0\}.$$

The space $R^\vee$ contains R and has dimension $2n - j$. The correspondence $R \mapsto R^\vee$ defines a tautological bundle $\mathcal{R}_j^\vee$ of dimension $2n - j$ on IGrass (j, F). We have the inclusions of bundles on $\text{IGrass}(j, F)$

$$\mathcal{R}_j \subset \mathcal{R}_j^\vee \subset F \times \text{IGrass}(j, F).$$

The bundle $\mathcal{R}_j^\vee/\mathcal{R}_j$ is a symplectic bundle. Indeed, we have a map of vector bundles

$$\overset{2}{\bigwedge} \mathcal{R}_j^\vee/\mathcal{R}_j \to \mathcal{O}_{\text{IGrass}(j,F)},$$

which on each fiber is induced by the form $\langle \;,\; \rangle$, so it is nondegenerate.

This means that for each dominant weight $\mu = (\mu_1, \ldots, \mu_{n-j})$ for the root system of type C_{n-j} we can talk about the bundle $V_\mu(\mathcal{R}_j^\vee/\mathcal{R}_j)$. Its fiber over a point corresponding to the isotropic space R is $V_\mu(R^\vee/R)$.

(4.3.4) Corollary. *Let us consider the vector bundle* $\mathcal{V}_{\beta,\mu} = K_\beta \mathcal{R}_j \otimes V_\mu$ $(\mathcal{R}_j^\vee/\mathcal{R}_j)$ *over* $\text{IGrass}(r, F)$, *where* $\beta = (\beta_1, \ldots, \beta_j)$ *is a dominant integral weight for the root system of type* A_{j-1} *and* $\mu = (\mu_1, \ldots, \mu_{n-j})$ *is the integral dominant weight for the root system of type* C_{n-j}. *Let us consider the weight*

$$\gamma = (-\beta_j, \ldots, -\beta_1, \mu_1, \ldots, \mu_{n-j}).$$

Then one of the mutually exclusive possibilities occurs:

> (1) *There exists* $\sigma \in W$, $\sigma \neq 1$, *such that* $\sigma(\gamma) = \gamma$. *Then all cohomology groups* $H^i(\text{IGrass}(j, F), \mathcal{V}_{\beta,\mu})$ *are* 0 *for* $i \geq 0$,
> (2) *There exists unique* $\sigma \in W$ *such that* $\sigma \cdot (\gamma) := \alpha$ *is a dominant integral weight for the root system of type* C_n. *Then all cohomology groups* $H^i(\text{IGrass}(j, F), \mathcal{V}_{\beta,\mu})$ *are* 0 *for* $i \neq l(\sigma)$, *and*

$$H^{l(\sigma)}(\text{IGrass}(j, F), \mathcal{V}_{\beta,\mu}) = V_\alpha(F).$$

Proof. Let us consider the projection $p : \mathbf{G/B} \to \mathbf{G/P}_j$. Identifying, as above, $\mathbf{G/B}$ with the space of isotropic flags and $\mathbf{G/P}$ with the isotropic Grassmannian, we see that the fiber over a point corresponding to a subspace R is $p^{-1}(R) = \text{Flag}(R) \times \text{IFlag}(R^\vee/R)$. Therefore we can identify $\mathbf{G/B}$ with the relative variety $\text{Flag}(\mathcal{R}_j) \times \text{IFlag}(\mathcal{R}_j^\vee/\mathcal{R}_j)$. Consider the line bundle $\mathcal{L}(\gamma)$ on $\mathbf{G/B}$. Using Bott's theorem in relative situation for the types A_{j-1} and C_{n-j}

we see that $\mathcal{R}^i p_*(\mathcal{L}(\gamma)) = 0$ for $i > 0$ and $\mathcal{R}^0\mathcal{L}(\gamma) = V_{\beta,\mu}$. Now our statement follows from Bott's theorem (4.3.1) and the spectral sequence of the composition. ∎

(4.3.5) Remark. *In the above corollary and in the following calculations we adopt the convention that for a vector space E of dimension m, $K_{(\beta_1,\ldots,\beta_m)}E^* \cong K_{(-\beta_m,\ldots,-\beta_1)}E$. Thus the above proposition also allows the calculation of the cohomology groups $H^i(\mathrm{IGrass}(j, F), K_\beta \mathcal{R}_j^* \otimes V_\alpha(\mathcal{R}_j^\vee/\mathcal{R}_j))$.*

Let us look more closely at the isotropic Grassmannian $\mathrm{IGrass}(j, F)$ as a subset of the Grassmannian Grass (j, F).

(4.3.6) Proposition. *The isotropic Grassmannian $\mathrm{IGrass}(j, F)$ is locally a complete intersection in $\mathrm{Grass}(j, F)$. The structure sheaf of $\mathrm{IGrass}(j, F)$ can be resolved by locally free sheaves over $\mathrm{Grass}(j, F)$ by means if the Koszul complex*

$$0 \to \bigwedge^{\binom{j}{2}}\left(\bigwedge^2 \mathcal{R}_j\right) \to \ldots \to \bigwedge^2 \mathcal{R}_j \xrightarrow{\psi} \mathcal{O}_{\mathrm{Grass}(j,F)}.$$

Proof. Since F is a symplectic space, we have the following map of locally free sheaves over $\mathrm{Grass}(j, F)$:

$$\bigwedge^2 \mathcal{R}_j \to \bigwedge^2 F \times \mathcal{O}_{\mathrm{Grass}(j,F)} \to \mathcal{O}_{\mathrm{Grass}(j,F)}$$

with the left map coming from tautological inclusion and the right one induced by the form $\langle\ ,\ \rangle$. The composition gives us the cosection ψ of $\bigwedge^2 \mathcal{R}_j$ which defines our Koszul complex. It is clear by definition that $\mathrm{IGrass}(j, F)$ is equal to the set of zeros of this cosection. Moreover, an easy calculation shows that locally these equations define a reduced subscheme of $\mathrm{Grass}(j, F)$. The dimension count shows that locally the equations give a regular sequence, so the Koszul complex is acyclic. ∎

The situation for the orthogonal group is very similar, but there is one difference. The special orthogonal group does not act transitively on the isotropic flags; only the orthogonal group does. This leads to some minor differences, which we highlight below. We just formulate the results we need, as the proofs are the same as in the case of a symplectic group.

We first consider the group of type B_n. Let F be a vector space of odd dimension $2n + 1$ with a nondegenerate symmetric form $\langle\ ,\ \rangle$. The simple roots of Φ are $\alpha_j = \epsilon_j - \epsilon_{j+1}$ for $j = 1, \ldots n - 1$ and $\alpha_n = \epsilon_n$. We take $\mathbf{G} = \mathrm{SO}(F)$. This group is not simply connected. Since we have in mind some applications to nilpotent orbits in the corresponding Lie algebra, we do not need to discuss spinor groups.

We start with a description of the homogeneous spaces $\mathbf{G}/\mathbf{B}$ in terms of flags. Let us recall that a subspace $R \subset F$ is *isotropic* if the restriction of $\langle \ , \ \rangle$ to R is zero.

We consider the set

$$\mathrm{IFlag}(F) = \{(R_1, \ldots, R_n) \in \mathrm{Flag}(1, 2, \ldots, n; F) \mid R_n \text{ is isotropic}\}.$$

The orthogonal group $\mathrm{O}(F)$ acts transitively on the set $\mathrm{IFlag}(F)$. Indeed, for each flag $(R_1 \ldots, R_n) \in \mathrm{IFlag}(F)$ we can choose a hyperbolic basis $e_1, \ldots, e_n, e, \bar{e}_n, \ldots, \bar{e}_1$ of F so $e_1, \ldots, e_i$ is a basis of R_i $(1 \le i \le n)$ and $e_1, \ldots, e_n, e$ is a basis of $R_n^\vee$. Now the orthogonal group $\mathrm{O}(F)$ operates transitively on hyperbolic bases. We will show that even $\mathrm{SO}(F)$ does. Indeed, let $g \in \mathrm{O}(F)$ be an element from $\mathrm{O}(F)$ which in given hyperbolic basis sends for each i e_i to e_i, $\bar{e}_i$ to $\bar{e}_i$ and e to $-e$. The element g fixes he flag $(R_1, \ldots, R_n)$ where R_i is spanned by $e_1, \ldots, e_i$ for $i = 1, \ldots, n$. This means that for every $h \in \mathrm{O}(F)$ we have $h(R_1, \ldots, R_n) = hg(R_1, \ldots, R_n)$. One of these elements has to lie in $\mathrm{SO}(F)$.

As for symplectic group we now identify $\mathbf{G}/\mathbf{B}$ with $\mathrm{IFlag}(F)$. Moreover, we can again give the relative version of the whole setup. Let $\mathcal{F}$ be an orthogonal vector bundle of dimension $2n + 1$ over a scheme X, i.e. a vector bundle $\mathcal{F}$ equipped with a map $\langle \ , \ \rangle : S_2\mathcal{F} \to \mathcal{O}_X$ for which the restriction to each fiber gives a nondegenerate symmetric form on it. We can construct a relative isotropic flag variety $\mathrm{IFlag}(\mathcal{F})$ with the structure map $p : \mathrm{IFlag}(\mathcal{F}) \to X$. The relative version of Bott's Theorem (4.3.1) is true if we replace cohomology groups with higher direct images. We leave the formulation of this result to the reader.

For each $j = 1, \ldots, n$ we consider the maximal parabolic subgoup $\mathbf{P}_j$ in $\mathbf{G} = \mathrm{SO}(2n + 1)$ which corresponds to the subset of all simple roots except α_j. The space $\mathbf{G}/\mathbf{P}_j$ can be identified with the isotropic Grassmannian $\mathrm{IGrass}(j, F)$ of isotropic subspaces of dimension j in F. This is a closed subset in $\mathrm{Grass}(F)$, so we can talk about the tautological subbundle $\mathcal{R}_j$ on $\mathrm{IGrass}(j, F)$. For any isotropic subspace from $\mathrm{IGrass}(j, F)$ we define the orthogonal complement

$$R^\vee = \{x \in F \mid \forall y \in R \ \langle x, y \rangle = 0\}.$$

The space $R^\vee$ contains R and has dimension $2n + 1 - j$. The correspondence $R \mapsto R^\vee$ defines a tautological bundle $\mathcal{R}_j^\vee$ of dimension $2n + 1 - j$

on IGrass(j, F). We have the inclusions of bundles on IGrass(j, F)

$$\mathcal{R}_j \subset \mathcal{R}_j^\vee \subset F \times \text{IGrass}(j, F).$$

The bundle $\mathcal{R}_j^\vee / \mathcal{R}_j$ is an orthogonal bundle of dimension $2(n - j) + 1$. Indeed, we have a map of vector bundles

$$S_2(\mathcal{R}_j^\vee / \mathcal{R}_j) \to \mathcal{O}_{\text{IGrass}(j,F)}$$

which on each fiber is induced by the form $\langle \, , \, \rangle$, so it is nondegenerate.

This means that for each dominant weight $\mu = (\mu_1, \ldots, \mu_{n-j})$ for the root system of type B_{n-j} we can talk about the bundle $V_\mu(\mathcal{R}_j^\vee / \mathcal{R}_j)$. Its fiber over a point corresponding to to the isotropic space R is $V_\mu(R^\vee / R)$.

(4.3.7) Corollary. *Let us consider the vector bundle* $\mathcal{V}_{\beta,\mu} = K_\beta \mathcal{R}_j \otimes V_\mu$ *($\mathcal{R}_j^\vee / \mathcal{R}_j$) over* IGrass($r$, F), *where* $\beta = (\beta_1, \ldots, \beta_j)$ *is a dominant integral weight for the root system of type* A_{j-1} *and* $\mu = (\mu_1, \ldots, \mu_{n-j})$ *is the integral dominant weight for the root system of type* B_{n-j}. *Let us consider the weight*

$$\gamma = (-\beta_j, \ldots, -\beta_1, \mu_1, \ldots, \mu_{n-j}).$$

Then one of the mutually exclusive possibilities occurs:

(1) *There exists* $\sigma \in W$, $\sigma \neq 1$ *such that* $\sigma(\gamma) = \gamma$. *Then all cohomology groups* $H^i(\text{IGrass}(j, F), \mathcal{V}_{\beta,\mu})$ *are 0 for* $i \geq 0$,
(2) *There exists unique* $\sigma \in W$ *such that* $\sigma \cdot (\gamma) := \alpha$ *is a dominant integral weight for the root system of type* B_n. *Then all cohomology groups* $H^i(\text{IGrass}(j, F), \mathcal{V}_{\beta,\mu})$ *are 0 for* $i \neq l(\sigma)$, *and*

$$H^{l(\sigma)}(\text{IGrass}(j, F), \mathcal{V}_{\beta,\mu}) = V_\alpha(F).$$

The proof of (4.3.7) is identical to that of (4.3.4).

Let us look more closely at the isotropic Grassmannian IGrass(j, F) as a subset of the Grassmannian Grass(j, F).

(4.3.8) Proposition. *The isotropic Grassmannian* IGrass(j, F) *is locally a complete intersection in* Grass(j, F). *The structure sheaf of* IGrass(j, F) *can be resolved by locally free sheaves over* Grass(j, F) *by means if the Koszul complex*

$$0 \to \bigwedge^{\binom{j+1}{2}} (S_2 \mathcal{R}_j) \to \ldots \to S_2 \mathcal{R}_j \overset{\psi}{\to} \mathcal{O}_{\text{Grass}(j,F)}.$$

The proof is identical to that of (4.3.6).

Finally we consider the group of type D_n. Let F be a vector space of even dimension $2n$ with a nondegenerate symmetric form $\langle \ , \ \rangle$. The simple roots of Φ are $\alpha_j = \epsilon_j - \epsilon_{j+1}$ for $j = 1, \ldots n - 1$ and $\alpha_n = \epsilon_{n-1} + \epsilon_n$. We take $\mathbf{G} = \mathrm{SO}(F)$. Again we are not interested in the spinor group.

The description of the homogeneous spaces $\mathbf{G}/\mathbf{B}$ in terms of flags is now different. We again start with the set $\mathrm{IFlag}(F)$ of isotropic flags:

$$\mathrm{IFlag}(F) = \{(R_1, \ldots, R_n) \in \mathrm{Flag}(1, 2, \ldots, n; F) \mid R_n \text{ is isotropic}\}.$$

As above, the orthogonal group $\mathrm{O}(F)$ acts transitively on the set $\mathrm{IFlag}(F)$. However, here $\mathrm{IFlag}(F)$ has two connected components and $\mathrm{SO}(F)$ operates transitively on each of them. In order to see this, let us fix a hyperbolic basis $e_1, \ldots, e_n, \bar{e}_n, \ldots, \bar{e}_1$. We associate to it a flag $(R_1^0, \ldots, R_n^0)$ where R_i^0 is spanned by $e_1, \ldots, e_i$. For a given flag $(R_1, \ldots, R_n)$ there exists $h \in \mathrm{O}(F)$ such that $hR_i^0 = R_i$ for $i = 1, \ldots, n$. Both flags are in the same component if $h \in \mathrm{SO}(F)$. Both components are homogeneous spaces for $\mathrm{SO}(F)$, so they are connected. The only thing to show is that they do not coincide. Let $(R_1, \ldots, R_n)$ be in both components. Then there exist elements $h_1 \in \mathrm{SO}(F)$ and $h_2 \in \mathrm{O}(F) \setminus \mathrm{SO}(F)$ such that $h_j R_i^0 = R_i$ for $i = 1, \ldots, n$, $j = 1, 2$. This means that $h_2^{-1} h_1$ fixes R_i^0 for $i = 1, \ldots, n$. Now simple linear algebra shows that $\det(h_2^{-1} h_1) = 1$, which is a contradiction.

We will denote two components of $\mathrm{IFlag}(F)$ by $\mathrm{IFlag}^+(F)$ and $\mathrm{IFlag}^-(F)$. We identify $\mathbf{G}/\mathbf{B}$ with $\mathrm{Flag}^+(F)$.

Moreover, we can again give the relative version of the whole setup. Let $\mathcal{F}$ be an orthogonal vector bundle of dimension $2n$ over a scheme X, i.e. a vector bundle $\mathcal{F}$ equipped with a map $\langle \ , \ \rangle : S_2\mathcal{F} \to \mathcal{O}_X$ for which the restriction to each fiber gives a nondegenerate symmetric form on it. We can construct a relative isotropic flag variety $\mathrm{IFlag}(\mathcal{F})$ with the structure map $p : \mathrm{IFlag}(\mathcal{F}) \to X$. The relative version of Bott's theorem (4.3.1) is true if we replace cohomology groups with higher direct images. We leave the formulation of this result to the reader.

For each $j = 1, \ldots, n$ we consider the maximal parabolic subgoup $\mathbf{P}_j$ in $\mathbf{G} = \mathrm{SO}(2n)$ which corresponds to the subset of all simple roots except α_j. Similar arguments to those for type B_n show that for $j = 1, \ldots, n - 2$ the space $\mathbf{G}/\mathbf{P}_j$ can be identified with the isotropic Grassmannian $\mathrm{IGrass}(j, F)$ of isotropic subspaces of dimension j in F. For $j = n$ the isotropic Grassmannian $\mathrm{IGrass}(n, F)$ has two connected components $\mathrm{IGrass}^+(F)$ and $\mathrm{IGrass}^-(F)$. They can be identified with the homogeneous spaces $\mathrm{SO}(F)/\mathbf{P}_j$ for $j = n - 1, n$.

For the remainder of this section we assume that $1 \le j \le n - 2$. Each $\mathrm{SO}(F)/\mathbf{P}_j$ is a closed subset in $\mathrm{Grass}(j, F)$, so we can talk about the tautological subbundle $\mathcal{R}_j$ on IGrass(j, F). For any isotropic subspace from IGrass(j, F) we define the orthogonal complement

$$R^\vee = \{x \in F \mid \forall y \in R \; \langle x, y \rangle = 0\}.$$

The space $R^\vee$ contains R and has dimension $2n - j$. The correspondence $R \mapsto R^\vee$ defines a tautological bundle $\mathcal{R}_j^\vee$ of dimension $2n - j$ on IGrass(j, F). We have the inclusions of bundles on IGrass(j, F)

$$\mathcal{R}_j \subset \mathcal{R}_j^\vee \subset F \times \text{IGrass}(j, F).$$

The bundle $\mathcal{R}_j^\vee / \mathcal{R}_j$ is an orthogonal bundle of dimension $2(n - j) + 1$. Indeed, we have a map of vector bundles

$$S_2(\mathcal{R}_j^\vee / \mathcal{R}_j) \to \mathcal{O}_{\text{IGrass}(j,F)}$$

which on each fiber is induced by the form $\langle \; , \; \rangle$, so it is nondegenerate.

This means that for each dominant weight $\mu = (\mu_1, \ldots, \mu_{n-j})$ for the root system of type D_{n-j} we can talk about the bundle $V_\mu(\mathcal{R}_j^\vee / \mathcal{R}_j)$. Its fiber over a point corresponding to to the isotropic space R is $V_\mu(R^\vee / R)$.

(4.3.9) Corollary. *Let us consider the vector bundle* $\mathcal{V}_{\beta,\mu} = K_\beta \mathcal{R}_j \otimes V_\mu(\mathcal{R}_j^\vee / \mathcal{R}_j)$ *over* IGrass(j, F), *where* $\beta = (\beta_1, \ldots, \beta_j)$ *is a dominant integral weight for the root system of type* A_{j-1} *and* $\mu = (\mu_1, \ldots, \mu_{n-j})$ *is the integral dominant weight for the root system of type* D_{n-j}. *Let us consider the weight*

$$\gamma = (-\beta_j, \ldots, -\beta_1, \mu_1, \ldots, \mu_{n-j}).$$

Then one of the mutually exclusive possibilities occurs:

(1) There exists $\sigma \in W$, $\sigma \neq 1$ such that $\sigma(\gamma) = \gamma$. Then all cohomology groups $H^i(\text{IGrass}(j, F), \mathcal{V}_{\beta,\mu})$ are 0 for $i \geq 0$.

(2) There exists unique $\sigma \in W$ such that $\sigma \cdot (\gamma) := \alpha$ is a dominant integral weght for the root system of type D_n. Then all cohomology groups $H^i(\text{IGrass}(j, F), \mathcal{V}_{\beta,\mu})$ are 0 for $i \neq l(\sigma)$, and

$$H^{l(\sigma)}(\text{IGrass}(j, F), \mathcal{V}_{\beta,\mu}) = V_\alpha(F).$$

The proof of (4.3.8) is identical to that of (4.3.4).

Let us look more closely at the isotropic Grassmannian IGrass(j, F) as a subset of the Grassmannian Grass(j, F).

(4.3.10) Proposition. *The isotropic Grassmannian* IGrass(j, F) *is locally a complete intersection in* Grass(j, F). *The structure sheaf of* IGrass(j, F) *can be resolved by locally free sheaves over* Grass(j, F) *by means if the Koszul complex*

$$0 \to \overset{\binom{j+1}{2}}{\bigwedge} (S_2 \mathcal{R}_j) \to \ldots \to S_2 \mathcal{R}_j \overset{\psi}{\to} \mathcal{O}_{\mathrm{Grass}(j,F)}.$$

The proof is identical to that of (4.3.5).

Exercises for Chapter 4

The General Linear Group

1. (a) Calculate the cohomology groups of bundles $\mathcal{L}(1, 4, 7, 5)$, $\mathcal{L}(3, 2, 1, 5)$ on $\mathbf{G/B}$ for $\mathbf{G} = \mathbf{GL}(4, \mathbf{C})$.
    (b) Calculate the cohomology of vector bundles $K_{(3,2,1)}\mathcal{Q} \otimes K_{(7,6,1)}\mathcal{R}$ and of $K_{(7,6,6)}\mathcal{R}$ on Grass(3, E) with dim $E = 6$.

2. Calculate the cohomology groups of bundles $K_\lambda \mathcal{Q}^*$ on the Grassmannian with tautological sequence

    $$0 \to \mathcal{R} \to F \to \mathcal{Q} \to 0,$$

    where dim $\mathcal{R} = r$, dim $\mathcal{Q} = q$.

3. Let E be an n-dimensional space. Consider the Grassmannian Grass(r; E) with the tautological sequence

    $$0 \to \mathcal{R} \to E \times \mathrm{Grass}(r; E) \to \mathcal{Q} \to 0.$$

    Let ξ be a subbundle of $S_2 E \times \mathrm{Grass}(r; E)$ fitting into the exact sequence

    $$0 \to \xi \to S_2 E \times \mathrm{Grass}(r; E) \to S_2 \mathcal{Q} \to 0.$$

    Notice that ξ also fits into an exact sequence

    $$0 \to S_2 \mathcal{R} \to \xi \to \mathcal{R} \otimes \mathcal{Q} \to 0.$$

    Calculate the cohomology groups of $\bigwedge^2 \xi$ and $\bigwedge^3 \xi$ using the information from the Schur complexes associated to the map $S_2 E \times \mathrm{Grass}(r; E) \to S_2 \mathcal{Q}$. Calculate this cohomology using the filtration induced by the second sequence on $\bigwedge^2 \xi$, $\bigwedge^3 \xi$.

4. We recall from Proposition (3.3.5) that the tangent bundle $T_{\mathrm{Grass}(r;E)}$ can be identified with $\mathcal{R}^* \otimes \mathcal{Q}$.

(a) Calculate the cohomology groups of the exterior algebra on the tangent bundle $T_{\text{Grass}(r;E)}$. Prove that the higher cohomology groups vanish.

(b) Calculate the cohomology groups of the exterior algebra on the cotangent bundle $T^*_{\text{Grass}(r;E)}$. Prove that only the group $H^i(\text{Grass}(r; E),$ $\bigwedge^j(T^*_{\text{Grass}(r;E)}))$ is nonzero if and only if when $i = j$. Prove that $H^i(\text{Grass}(r; E), \bigwedge^i(T^*_{\text{Grass}(r;E)}))$ consists of $P(i, r, n - r)$ copies of trivial representation, where $P(i, r, n - r)$ is the number of partitions of i contained in the $r \times (n - r)$ rectangle.

5. Some characteristic free cases of Bott's theorem:

(a) Let $\lambda = (\lambda_1, \ldots, \lambda_n)$ be such that for some $s < t$ we have $\lambda_1 \geq \ldots \geq$ $\lambda_s > \lambda_{s+1} = \ldots = \lambda_{t-1} \geq \lambda_{t+1} - 1 \geq \ldots \geq \lambda_n - 1$ $\lambda_t = \lambda_{s+1} +$ $t - s$. Then $H^i(\mathbf{G}/\mathbf{B}, \mathcal{L}(\lambda)) = 0$ for $i \neq t - s - 1$ and $H^{t-s}(\mathbf{G}/\mathbf{B},$ $\mathcal{L}(\lambda)) = L_{\nu'} E$ where $\nu = (\lambda_1, \ldots, \lambda_s, \lambda_t - (t - s) + 1, \lambda_{s+1} +$ $1, \ldots, \lambda_{t-1} + 1, \lambda_{t+1}, \ldots, \lambda_n)$.

(b) Let $\lambda = (\lambda_1, \ldots, \lambda_n)$ be such that for some $s < t$ we have $\lambda_1 \geq \ldots \geq$ $\lambda_s > \lambda_{s+1} = \ldots = \lambda_{t-1}, \lambda_{s+1} < \lambda_t < \lambda_{s+1} + t - s$. Then $H^i(\mathbf{G}/\mathbf{B},$ $\mathcal{L}(\lambda)) = 0$ for all i.

Other Classical Groups

6. Let F be a symplectic space of dimension $2n$. Consider the isotropic Grassmannian $\text{IGrass}(j, F)$. Let $\lambda = (\lambda_1, \ldots, \lambda_j)$ be a partition. Prove that if $\lambda_1 \leq 2n - j + 1$, then the cohomology of $K_\lambda \mathcal{R}_j$ can be zero or can contain only a trivial representation of $\text{Sp}(F)$. More precisely, the cohomology is nonzero precisely when λ is one of the partitions occurring in Proposition (6.4.3).

7. Let $\text{IGrass}(r; F)$ be the isotropic Grassmannian of r-dimensional isotropic subspaces in a symplectic space $(F, \langle -, - \rangle)$ of dimension $2n$. Calculate the cohomology groups of the exterior powers of the vector bundle $\mathcal{R}^\vee$.

8. Formulate and prove the analogues of exercises 6 and 7 for the even and odd orthogonal groups.

9. Let $(F, \langle \, , \, \rangle)$ be a symplectic space of dimension $2n$. For $1 \leq r \leq n$, let $\text{IGrass}(r; F)$ be the isotropic Grassmannian of r-dimensional isotropic spaces in F with tautological subbundle $\mathcal{R}$. Consider the filtration $0 \subset$ $\mathcal{R} \subset \mathcal{R}^\vee \subset F \times \text{IGrass}(r; F)$ of the vector bundles on $\text{IGrass}(r; F)$. The factor $(F \times \text{IGrass}(r; F))/\mathcal{R}^\vee$ can be identified with $\mathcal{R}^*$. Therefore we

have an epimorphism π of vector bundles on IGrass$(r; F)$ which is a composition

$$\pi : (T_{\text{Grass}(r;F)})|_{\text{IGrass}(r;F)} = \mathcal{R}^* \otimes ((F \times \text{IGrass}(r; F))/\mathcal{R})$$
$$\to \mathcal{R}^* \otimes \mathcal{R}^* \to \bigwedge^2 \mathcal{R}^*.$$

(a) Prove that the embedding IGrass$(r; F) \subset$ Grass$(r; F)$ allows one to identify the vector bundle $T_{\text{IGrass}(r;F)}$ with the kernel of π.

(b) Prove the exact sequence

$$0 \to \mathcal{R}^* \otimes (\mathcal{R}^\vee/\mathcal{R}) \to T_{\text{IGrass}(r;F)} \to D_2\mathcal{R}^* \to 0.$$

10. Let $(F, \langle -, - \rangle)$ be an orthogonal space of dimension n. For $1 \le r \le \frac{n}{2}$, let IGrass$(r; F)$ be the isotropic Grassmannian of r-dimensional isotropic spaces in F with tautological subbundle $\mathcal{R}$. Consider the filtration $0 \subset \mathcal{R} \subset \mathcal{R}^\vee \subset F \times$ IGrass$(r; F)$ of the vector bundles on IGrass$(r; F)$. The factor $(F \times \text{IGrass}(r; F))/\mathcal{R}^\vee$ can be identified with $\mathcal{R}^*$. Therefore we have an epimorphism π of vector bundles on IGrass$(r; F)$ which is a composition

$$\pi : (T_{\text{Grass}(r;F)})|_{\text{IGrass}(r;F)} = \mathcal{R}^* \otimes ((F \times \text{IGrass}(r; F))/\mathcal{R})$$
$$\to \mathcal{R}^* \otimes \mathcal{R}^* \to S_2\mathcal{R}^*.$$

(a) Prove that the embedding IGrass$(r; F) \subset$ Grass$(r; F)$ allows to identify the vector bundle $T_{\text{IGrass}(r;F)}$ with the kernel of π.

(b) Prove the exact sequences

$$0 \to \mathcal{R}^* \otimes (\mathcal{R}^\vee/\mathcal{R}) \to T_{\text{IGrass}(r;F)} \to \bigwedge^2 \mathcal{R}^* \to 0,$$
$$0 \to T_{\text{IGrass}(r;F)} \to \mathcal{R}^* \otimes F/\mathcal{R} \to S_2\mathcal{R}^* \to 0.$$

11. Use the results of exercises 9 and 10 to calculate cohomology groups of exterior powers of tangent and cotangent bundles on isotropic Grassmannians of isotropic subspaces of maximal dimension. Compare to the results of exercise 4.

Tensor Product Multiplicities

12. (Brauer and Klimyk's formula.) Let λ, μ be two dominant weights for a reductive group **G**. Prove that the decomposition of the tensor product

$V_\lambda \otimes V_\mu$ can be calculated as follows. Consider the set of weights $\Sigma(\mu) = \{v_1, \ldots, v_N\}$ occurring in V_μ (with $N = \dim V_\mu$). For each weight in

$$\lambda + \Sigma(\mu) = \{\lambda + v_1, \ldots, \lambda + v_N\}$$

calculate the Euler characteristic of the line bundle $\mathcal{L}(\lambda + v_i)$ on $\mathbf{G}/\mathbf{B}$. Then make cancellations if the same representation occurs with a positive and a negative sign. The remaining representations occur with positive sign and give the irreducible representations occurring in $V_\lambda \otimes V_\mu$.

5

The Geometric Technique

In this chapter we develop the basic technique for calculating syzygies. It applies to the subvarieties Y in an affine space X with a desingularization Z which is a total space of a vector bundle over some projective variety V, which is a subbundle of the trivial bundle $X \times V$ over V. In such situation the Koszul complex of sheaves on $X \times V$ resolving the structure sheaf of Z has terms that are pullbacks of vector bundles over V. Taking the direct image of this Koszul complex by the projection $p : X \times V \to V$, one gets the formula expressing terms on the free resolution of the coordinate ring of Y in terms of cohomology of bundles on V. One also gets interesting complexes by taking direct images of the Koszul complex twisted by a pullback of a vector bundle on V. In this chapter we discuss the general construction and properties of direct images of Koszul complexes. The examples will be given in following chapters.

The chapter is organized as follows. In section 5.1 we state the properties of the twisted direct images $F(\mathcal{V})_\bullet$ of Koszul complexes. In particular we give the expressions for their terms and homology. We also state the criteria for $F(\mathcal{V})_\bullet$ to be acyclic, the duality theorem for such complexes, and the result expressing the codimension and degree of Y in terms of the complex $F(\mathcal{V})_\bullet$.

In section 5.2 we give the actual construction of complexes $F(\mathcal{V})_\bullet$. It involves constructing certain double complexes of sheaves on $X \times V$ resolving the Koszul complex.

In section 5.3 we prove the other statements announced in section 5.1. In some of the proofs in sections 5.2 and 5.3 we rely on the machinery of derived categories. The necessary information is collected in section 1.2.5.

Section 5.4 contains the equivariant setup. We prove that if a reductive group $\mathbf{G}$ acts on X and the action stabilizes Y, the variety V is a homogeneous space $\mathbf{G}/\mathbf{P}$, and the bundle $\mathcal{V}$ is a homogeneous bundle, then the terms and homology of the complexes $F(\mathcal{V})_\bullet$ also carry an action of $\mathbf{G}$. We also discuss the results of Kempf on rational singularities of the subvarieties Y and on the geometry of the desingularization Z in the case when V is a homogeneous space.

136

In section 5.5 we give more explicit description of the differentials of $F(\mathcal{V})_\bullet$.

Section 5.6 describes the technique of degeneration sequences which allows us to compare complexes $F(\mathcal{V})_\bullet$ supported in different subvarieties.

5.1. The Formulation of the Basic Theorem

Throughout this chapter we work over the algebraically closed field **K** of arbitrary characteristic.

Let us consider the projective variety V of dimension m. Let $X = A_{\mathbf{K}}^N$ be the affine space. The space $X \times V$ can be viewed as a total space of trivial vector bundle $\mathcal{E}$ of dimension N over V. Let us consider the subvariety Z in $X \times V$ which is the total space of a subbundle $\mathcal{S}$ in $\mathcal{E}$. We denote by q the projection $q : X \times V \longrightarrow X$ and by q' the restriction of q to Z. Let $Y = q(Z)$. We get the basic diagram

$$
\begin{array}{ccc}
Z & \subset & X \times V \\
\downarrow q' & & \downarrow q \\
Y & \subset & X
\end{array}
$$

The projection from $X \times V$ onto V is denoted by p, and the quotient bundle $\mathcal{E}/\mathcal{S}$ by $\mathcal{T}$. Thus we have the exact sequence of vector bundles on V,

$$
0 \longrightarrow \mathcal{S} \longrightarrow \mathcal{E} \longrightarrow \mathcal{T} \longrightarrow 0.
$$

The dimensions of $\mathcal{S}$ and $\mathcal{T}$ will be denoted by s, t respectively. The coordinate ring of X will be denoted by A. It is a polynomial ring in N variables over **K**. We will identify the sheaves on X with A-modules.

(5.1.1) Proposition.

(a) *The locally free resolution of the sheaf* $\mathcal{O}_Z$ *as an* $\mathcal{O}_{X \times V}$-*module is given by the Koszul complex*

$$
\mathcal{K}(\xi)_\bullet : 0 \to \overset{t}{\bigwedge}(p^*\xi) \to \dots \to \overset{2}{\bigwedge}(p^*\xi) \to p^*(\xi) \to \mathcal{O}_{X \times V}
$$

where $\xi = \mathcal{T}^*$. *The differentials in this complex are homogeneous of degree 1 in the coordinate functions on X.*

(b) *The direct image* $p_*(\mathcal{O}_Z)$ *can be identified with the the sheaf of algebras* $\mathrm{Sym}(\eta)$ *where* $\eta = \mathcal{S}^*$.

Proof. Let us identify X with the vector space E of dimension N over **K**. The bundle $\mathcal{S}$ is the s-dimensional subbundle of the N-dimensional trivial bundle over V. By the universal property (3.3.2) of the Grassmannian, there exists a map $f : V \longrightarrow \text{Grass}(s, E)$ such that $S = f^*(\mathcal{R})$.

Let us consider the complex $\mathcal{K}_\bullet(\mathcal{Q}^*)$ from (3.3.3). We set

$$\mathcal{K}(\xi)_\bullet := f^*\mathcal{K}_\bullet(\mathcal{Q}^*).$$

The proposition follows by the same arguments as in the proof of (3.3.3). ∎

The idea of the geometric technique is to use the Koszul complex $\mathcal{K}(\xi)_\bullet$ to construct for each vector bundle $\mathcal{V}$ on V the free complex $F(\mathcal{V})_\bullet$ of A-modules with the homology supported in Y. These complexes are the main subject of this book. In many cases the complex $F(\mathcal{O}_V)_\bullet$ gives the free resolution of the defining ideal of Y.

In this section we state the theorems establishing the existence and basic properties of complexes $F(\mathcal{V})_\bullet$. The most important is the basic theorem (5.1.2) below, which gives the terms and the precise description of homology of complexes $F(\mathcal{V})_\bullet$. The next two sections will be devoted to the proofs of all the results that follow.

Before we state the basic theorem, let us introduce the twisted Koszul complex. For every vector bundle $\mathcal{V}$ on V we introduce the complex

$$\mathcal{K}(\xi, \mathcal{V})_\bullet := \mathcal{K}(\xi)_\bullet \otimes_{\mathcal{O}_{X \times V}} p^*\mathcal{V}.$$

This complex is a locally free resolution of the $\mathcal{O}_{X \times V}$-module $M(\mathcal{V}) := \mathcal{O}_Z \otimes p^*\mathcal{V}$.

Now we are ready to state the basic theorem.

(5.1.2) Basic Theorem. *For a vector bundle $\mathcal{V}$ on V we define free graded A-modules*

$$F(\mathcal{V})_i = \bigoplus_{j \geq 0} H^j \left(V, \overset{i+j}{\bigwedge} \xi \otimes \mathcal{V} \right) \otimes_k A(-i - j).$$

(a) There exist minimal differentials

$$d_i(\mathcal{V}) : F(\mathcal{V})_i \to F(\mathcal{V})_{i-1}$$

of degree 0 such that $F(\mathcal{V})_\bullet$ is a complex of free graded A-modules with

$$H_{-i}(F(\mathcal{V})_\bullet) = \mathcal{R}^i q_* M(\mathcal{V}).$$

In particular the complex $F(\mathcal{V})_\bullet$ is exact in positive degrees.

(b) *The sheaf $\mathcal{R}^i q_* M(\mathcal{V})$ is equal to $H^i(Z, M(\mathcal{V}))$ and it can be also identified with the graded A-module $H^i(V, \mathrm{Sym}(\eta) \otimes \mathcal{V})$.*

(c) *If $\phi : M(\mathcal{V}) \to M(\mathcal{V}')(n)$ is a morphism of graded sheaves then there exists a morphism of complexes*

$$f_\bullet(\phi) : F(\mathcal{V})_\bullet \to F(\mathcal{V}')_\bullet(n)$$

Its induced map $H_{-i}(f_\bullet(\phi))$ can be identified with the induced map

$$H^i(Z, M(\mathcal{V})) \to H^i(Z, M(\mathcal{V}'))(n).$$

This theorem will be proven in section 5.2.

If $\mathcal{V}$ is a trivial bundle of rank one on V, then the complex $F(\mathcal{V})_\bullet$ is denoted simply by $F_\bullet$.

The next theorem gives the criterion for the complex $F_\bullet$ to be the free resolution of the coordinate ring of Y.

(5.1.3) Theorem. *Let us assume that the map $q' : Z \longrightarrow Y$ is a birational isomorphism. Then the following properties hold:*

(a) *The module $q'_* \mathcal{O}_Z$ is the normalization of $\mathbf{K}[Y]$.*

(b) *If $\mathcal{R}^i q'_* \mathcal{O}_Z = 0$ for $i > 0$, then $F_\bullet$ is a finite free resolution of the normalization of $\mathbf{K}[Y]$ treated as an A-module.*

(c) *If $\mathcal{R}^i q'_* \mathcal{O}_Z = 0$ for $i > 0$ and $F_0 = H^0(V, \bigwedge^0 \xi) \otimes A = A$, then Y is normal and it has rational singularities.*

The complexes $F(\mathcal{V})_\bullet$ satisfy a Grothendieck type duality. Let ω_V denote the canonical divisor on V.

(5.1.4) Theorem. *Let $\mathcal{V}$ be a vector bundle on V. Let us introduce the dual bundle*

$$\mathcal{V}^\vee = \omega_V \otimes \bigwedge^t \xi^* \otimes \mathcal{V}^*.$$

Then

$$F(\mathcal{V}^\vee)_\bullet = F(\mathcal{V})_\bullet^*[m - t].$$

This result can be applied to give a criterion for the twisted module to be Cohen–Macaulay.

(5.1.5) Corollary. *Let us assume that $\dim Z = \dim Y$. Assume that for some vector bundle $\mathcal{V}$ on V we have $\mathcal{R}^i q'_* (\mathcal{O}_Z \otimes p^* \mathcal{V}) = 0$ for $i > 0$. Then the* *module $\mathcal{R}^0 q'_*(\mathcal{O}_Z \otimes p^*\mathcal{V})$ is a maximal Cohen–Macaulay module supported in Y if and only if $\mathcal{R}^i q'_*(\mathcal{O}_Z \otimes p^*\mathcal{V}^\vee) = 0$ for $i > 0$. In that case the module $\mathcal{R}^0 q'_*(\mathcal{O}_Z \otimes p^*\mathcal{V}^\vee)$ is also a maximal Cohen–Macaulay module, dual to $\mathcal{R}^0 q'_*(\mathcal{O}_Z \otimes p^*\mathcal{V})$ in the sense of (1.2.26).*

Proof. We apply (5.1.4) to the complexes $F(\mathcal{V})_\bullet$ and $F(\mathcal{V}^\vee)_\bullet$. Our assumption implies that

$$\text{codim } Y = \dim X - \dim Y = \dim X - \dim Z = \dim X$$
$$-(\dim X + m - t) = t - m.$$

Now (5.1.4) implies that the length of $F(\mathcal{V})_\bullet$ equals $t - m$ if and only if $F(\mathcal{V}^\vee)_\bullet$ has all the terms in nonnegative degrees. This establishes the first claim. The duality statement follows because the two complexes are dual to each other. ∎

If the complex $F(\mathcal{V})_\bullet$ satisfies the conditions of Corollary (5.1.5), we say that it has the Cohen–Macaulay property.

In particular, when codim $Y = 1$ the complexes with Cohen–Macaulay property have length one, so they are just matrices. The determinant of such a matrix equals $g^{\text{rank } \mathcal{V}}$, where g is an irreducible equation of Y. In that case the complex $F(\mathcal{V})_\bullet$ is called *a determinantal complex*. We will analyze such complexes for the case of discriminants and resultants in chapter 9.

We conclude this section by showing that the complex $F_\bullet$ contains the information about the codimension and the degree of Y. This fact will be useful in the cases when q is not necessarily a birational map.

(5.1.6) Theorem.
    (a) $\text{codim}_X Y = \max \{ i \mid F_i \neq 0 \}$.
    (b) *Let us assume that* $\dim Z = s + m < \dim X = N$. *Let* $r = N - m - s$. *Then we have*

$$\deg(q') \deg Y = \sum_{i,j} (-1)^{i+r} \frac{(i+j)^r}{r!} h^j \left(V, \bigwedge^{i+j} \xi \right)$$

where by definition $\deg(q')$ *is 0 when* $\dim Y < \dim Z$.

Theorems (5.1.3), (5.1.4), and (5.1.6) as well as Corollary (5.1.5) will be proved in section 5.3.

5.2. The Proof of the Basic Theorem

Before we prove Theorem (5.1.2) we recall several facts we will need. The first one is the result on an equivalence of categories of graded modules and sheaves.

Let S be a graded ring with $S_0 = A$ a finitely generated **K**-algebra and S_1 a finitely generated A-module. For a graded S-module M we denote by $M^{\sim}$ the corresponding sheaf on Proj S. For a sheaf $\mathcal{F}$ on X we define

$$\Gamma_*(\mathcal{F}) = \bigoplus_{n \in Z} \Gamma(\text{Proj } S, \mathcal{F}(n)).$$

We define an equivalence relation $\approx$ on graded S-modules by saying $M \approx M'$ if there exists an integer d such that $M_{\geq d} \simeq M'_{\geq d}$. Here $M_{\geq d} = \bigoplus_{n \geq d} M_n$. We say that a graded S-module M is quasifinitely generated if M is equivalent to a finitely generated module. In this setting we have

(5.2.1) Proposition. *The functors $^{\sim}$ and Γ_* induce an equivalence of categories between the category of quasifinitely generated graded S-modules modulo the equivalence $\approx$ and the category of coherent $\mathcal{O}_{\text{Proj } S}$-modules.*

Proof. This is exercise 5.9 in section II.5 of [H1]. ∎

We proceed with the discussion of some general facts concerning free complexes.

Let A be a graded ring with $A_0 = \mathbf{K}$.

(5.2.2) Proposition.

(a) *Let $G_{\bullet}$ be a complex of finitely generated graded free A-modules. The complex $G_{\bullet}$ decomposes into a direct sum*

$$G_{\bullet} = G'_{\bullet} \bigoplus G''_{\bullet}$$

where $G'_{\bullet}$ is a minimal complex and $G''_{\bullet}$ is exact. The terms of the complex $G_{\bullet}$ are

$$G'_i = H_i(G_{\bullet} \otimes_A \mathbf{K}) \otimes_k A.$$

(b) *Let $M_{\bullet}$ be a complex of graded A-modules with $M_i = 0$ for $i < i_0$. Then there exists a minimal complex $G_{\bullet}$ of free graded A-modules and a map $\phi : G_{\bullet} \to M_{\bullet}$ which is a quasiisomorphism.*

Proof. We will start with the proof of (a). Let $G_i = \bigoplus_{1 \leq j \leq g_i} A(-e_{j,i})$, where $g_i = \dim G - i$. Let us consider the differential $d_i : G_i \to G_{i-1}$. We can identify d_i with the matrix over A where the (k, j)th entry is homogeneous of degree $e_{j,i} - e_{k,i-1}$. We will prove that we can change the basis in G_i in a homogeneous way so the module G_i will decompose as follows:

$$G_i = B_i \oplus U_i \oplus B_i', \qquad (*)$$

so the differential d_i has a block decomposition

$$d_i = \begin{pmatrix} 0 & 0 & I \\ 0 & d_i' & 0 \\ 0 & 0 & 0 \end{pmatrix}.$$

If this is done, then obviously we can set $G_i' = U_i$ and $G_i'' = B_i \oplus B_i'$, and the proposition follows.

To get our decomposition we fix an index m, choose a basis in G_m in the appropriate way, and then spread this choice to the left and to the right by induction.

We start by changing the basis in G_m and G_{m-1} in a homogeneous way to bring d_m to the canonical form

$$d_m = \begin{pmatrix} d_m' & 0 \\ 0 & d_m'' \end{pmatrix}$$

where d_m' is a minimal matrix with homogeneous entries and d_m'' is an identity matrix. This means we can write $G_m = W_m \oplus V_m$, $G_{m-1} = W_{m-1} \oplus V_{m-1}$, with d_m' corresponding to the map $W_m \to W_{m-1}$ and d_m'' corresponding to the map $V_m \to V_{m-1}$. Now we notice that the rows of d_{m+1} corresponding to V_m are zero because $G_\bullet$ is a complex. Therefore it is really a map from G_{m+1} to U_m. Bringing this map to the canonical form as above, we see that we can decompose $W_m = U_m \oplus B_m$ and $G_{m+1} = W_{m+1} \oplus V_{m+1}$ in such way that d_{m+1} is a direct sum of the minimal map from W_{m+1} to U_m and the identity map from V_{m+1} to B_m. We get the required choice of basis in G_m by setting $B_m' = V_m$. In fact we have also chosen the direct summand $B_{m-1} = V_{m-1}$ in G_{m-1} and the direct summand $B_{m+1}' = V_{m+1}$ in G_{m+1}.

Next we show how to extend our choice of basis to the right. Assume that we have the block decomposition $(*)$ for $i > j$ together with the decomposition $G_j = B_j \oplus W_j$. We notice that $B_j \subset \operatorname{Ker} d_j$. Therefore we treat d_j as a map from W_j to G_{j-1}. We bring it to the canonical form, which means we can write $W_j = U_j \oplus B_j'$ and $G_{j-1} = B_{j-1} \oplus W_{j-1}$, so d_j is a direct sum of the minimal map from U_j to W_{j-1} and the identity from B_j' to B_{j-1}.

Similarly, let us assume that the decomposition $(*)$ is achieved for $i < j$ together with the decomposition $G_j = W_j \oplus B_j'$. We notice that the image of

d_{j+1} is contained in W_j, so we can treat is as a map from G_{j+1} to W_j. Reducing this map to the canonical form, we get the decompositions $W_j = B_j \oplus U_j$ and $G_{j+1} = W_{j+1} \oplus B'_{j+1}$ such that d_{j+1} is a direct sum of a minimal map from W_{j+1} to U_j and the identity map from B'_{j+1} to B_j. This completes the proof of (a).

Let us prove (b). It is enough to construct the quasiisomorphism $\phi : G_{\bullet} \to M_{\bullet}$ where $G_{\bullet}$ is a complex of free graded A-modules. We can achieve minimality by applying (a) and taking $G'_{\bullet}$ as our complex.

Let us denote the submodule $\operatorname{Ker} d$ in M_i by Z_i and the submodule $d(M_{i+1})$ by B_i. Then we have exact sequences

$$0 \to B_i \to Z_i \to H_i \to 0,$$

$$0 \to Z_i \to M_i \to B_{i-1} \to 0.$$

Let

$$0 \to \mathcal{B}_i \to \mathcal{Z}_i \to \mathcal{H}_i \to 0,$$

$$0 \to \mathcal{Z}_i \to \mathcal{M}_i \to \mathcal{B}_{i-1} \to 0$$

be the short exact sequences of free complexes covering the above maps. For each i we consider the map of complexes η_i given by the composition

$$\mathcal{M}_i \to \mathcal{B}_{i-1} \to \mathcal{Z}_{i-1} \to \mathcal{M}_{i-1}.$$

It is clear that $\eta_{i-1}\eta_i = 0$. We consider the double complex

$$\to \ldots \mathcal{M}_i \overset{\eta_i}{\longrightarrow} \mathcal{M}_{i-1} \overset{\eta_{i-1}}{\longrightarrow} \mathcal{M}_{i-2} \to \ldots.$$

We define $G_{\bullet}$ to be the total complex of this double complex. By construction it is equipped with the natural map ϕ to $M_{\bullet}$. It is clear from the spectral sequence associated to our double complex that ϕ is a quasiisomorphism. ∎

After these preparations we can proceed with the proof of (5.1.2).

Let us embed V in a projective space. Let $\mathcal{O}_V(1)$ be the ample line bundle corresponding to this embedding. We can assume that the higher cohomology of the sheaves $\mathcal{O}_V(n)$ vanishes for $n > 0$. This can be achieved by Serre's theorem ([H1, Proposition III.5.3]). Let us denote by R the homogeneous coordinate ring of V in this embedding.

The key step in the construction of complexes $F(V)_{\bullet}$ is the existence of a certain right resolution of the twisted Koszul complex $\mathcal{K}(\xi, V)_{\bullet}$.

(5.2.3) Lemma. *There exists a right resolution*

$$0 \to \mathcal{K}(\xi, \mathcal{V})_{\bullet} \to \mathcal{P}(\mathcal{V})_{\bullet\bullet}$$

such that the following properties hold:

(a) *Each module* $\mathcal{P}(\mathcal{V})_{ij}$ *is a direct sum of sheaves* $A(i) \otimes \mathcal{O}_V(n)$, *where* $n > 0$, *and therefore is* q_*-*acyclic.*

(b) *Each column*

$$0 \to \mathcal{K}(\xi, \mathcal{V})_j \to \mathcal{P}(\mathcal{V})_{\bullet j}$$

is a q_*-*acyclic resolution of* $\mathcal{K}(\xi, \mathcal{V})_j$ *by coherent* $\mathcal{O}_{X \times V}$-*modules which is the tensor product of* $A(-j)$ *with the* Γ-*acyclic resolution of* $\bigwedge^j(\xi) \otimes \mathcal{V}$.

Proof. Let us start with the dual complex $\mathcal{K}(\xi, \mathcal{V})^*_{\bullet}$. Its j-th term equals

$$\mathcal{K}(\xi, \mathcal{V})^*_j = \overset{-j}{\bigwedge}(p^*\xi^*) \otimes \mathcal{V}^*.$$

This is a complex of sheaves over $X \times V$ whose differential is homogeneous of degree 1 with respect to the generators of A. Let us apply to this complex the functor Γ_* from (5.2.1). We get a complex of bigraded $A \otimes R$-modules. The generators of the j-th term here have A-degree $-j$. Now we replace this complex by the complex of equivalent modules by cutting out in each module the components in nonpositive R-degrees. We get a complex $C(\mathcal{V})_{\bullet}$ of bigraded $A \otimes R$-modules where the j-th term has generators in A-degree $-j$ and in positive R-degree.

Next we consider the minimal free resolution $\widehat{C}(\mathcal{V})_{\bullet\bullet}$ of the complex $C(\mathcal{V})_{\bullet}$. Each module $\widehat{C}_{ij}$ has generators in positive R-degree.

Now we construct $\mathcal{P}(\mathcal{V})_{\bullet\bullet}$ by applying the functor $\widetilde{}$ to the complex $\widehat{C}(\mathcal{V})_{\bullet\bullet}$, and dualizing. By (5.2.1) it is the right resolution of the Koszul complex. Properties (a) and (b) are obviously satisfied. ∎

Consider the double complex $q_*(\mathcal{P}(\mathcal{V})_{\bullet\bullet})$. By Lemma (5.2.3) (a) it is a double complex of free graded A-modules.

Let us consider the total complex associated to $q_*(\mathcal{P}(\mathcal{V})_{\bullet\bullet})$,

$$G(\mathcal{V})_{\bullet} := Tot_{\bullet}(q_*(\mathcal{P}(\mathcal{V})_{\bullet\bullet})).$$

Since the resolution $\mathcal{P}(\mathcal{V})_{\bullet\bullet}$ is q_*-acyclic, we get

$$H_{-i}G(\mathcal{V})_{\bullet} = \mathcal{R}^i q_* M(\mathcal{V}).$$

Now we apply Proposition (5.2.2) to the complex $G(\mathcal{V})_\bullet$. We want to calculate the components of the minimal part of this complex. It is enough to calculate the homology of the complex $G(\mathcal{V})_\bullet \otimes \mathbf{K}$. We consider the double complex of vector spaces $q_*(\mathcal{P}_{\bullet\bullet}) \otimes \mathbf{K}$. The horizontal differentials in this double complex are 0 because the horizontal maps in $\mathcal{P}_{\bullet\bullet}$ are by Lemma (5.2.3) (a) the matrices with entries of degree 1 in A. By Lemma (5.2.3) (b) the homology of each column $q_*(\mathcal{P}_{\bullet j})$ consists of cohomology groups $H^\cdot(V, \bigwedge^j(\xi) \otimes \mathcal{V})$. Therefore

$$H^l(G(\mathcal{V})_\bullet \otimes \mathbf{K}) = \bigoplus_{j \geq 0} H^j\left(V, \overset{l+j}{\bigwedge} \xi \otimes \mathcal{V}\right).$$

Proposition (5.2.2) applied to $G(\mathcal{V})_\bullet$ gives

$$G(\mathcal{V})_\bullet = F(\mathcal{V})_\bullet \bigoplus L(\mathcal{V})_\bullet.$$

for some exact complex $L(\mathcal{V})_\bullet$. This proves part (a) of theorem (5.1.2).

Let us prove part (b). The first part of the statement follows from the fact that the module on X is determined by its global sections. The second part follows from the spectral sequence of the composition of maps $X \times V \to V \to *$, from the fact that p is affine, and from (5.1.1) (b).

Before we turn to part (c), let us state some facts about the complexes $F(\mathcal{V})_\bullet$ that follow easily from the proof of part (a) of Theorem (5.1.2).

(5.2.4) Proposition.

(a) The component

$$(d_i)^{(j,j')} : H^j\left(V, \overset{i+j}{\bigwedge} \xi \otimes \mathcal{V}\right) \otimes_k A(-i - j)$$

$$\to H^{j'}\left(V, \overset{i-1+j'}{\bigwedge} \xi \otimes \mathcal{V}\right) \otimes_k A(-i + 1 - j')$$

of the differential $d_i(\mathcal{V}) : F(\mathcal{V})_i \to F(\mathcal{V})_{i-1}$ is of homogeneous degree $j - j' + 1$.
(b) The component $(d_i)^{(j,j')}$ is zero if $j < j'$.

Proof. Part (a) follows from the fact that d_i is a homogeneous map of degree 0. Part (b) follows from minimality of the complex $F(\mathcal{V})_\bullet$. ∎

We need a result characterizing the complex $F(\mathcal{V})_\bullet$.

(5.2.5) Proposition. *The complex $F(V)_\bullet$ is the unique minimal free complex quasiisomorphic to $Rq_*(\mathcal{O}_Z \otimes p^*V)$. In particular, the complex $F(V)_\bullet$ does not depend on the choice of the sheaf $\mathcal{O}_V(1)$ and of the resolution $\mathcal{P}(V)_{\bullet\bullet}$.*

Proof. We constructed the complex $F(V)_\bullet$ as a minimal free complex quasiisomorphic to the total complex of the double complex $q_*(\mathcal{P}(V)_{\bullet\bullet})$. However, the total complex of $\mathcal{P}(V)_{\bullet\bullet}$ is a q_*-acyclic complex quasiisomorphic to $\mathcal{O}_Z \otimes p^*(V)$. The statement now follows from (5.2.2) (b) and the fact implicitly contained in (5.2.2) that every quasiisomorphism between minimal free complexes has to be an isomorphism. ■

Now we conclude the proof of part (c) of Theorem (5.1.2). A morphism $\phi : M(V) \to M(V')(n)$ induces the map

$$Rq_*(\phi) : Rq_*(\mathcal{O}_Z \otimes p^*V) \to Rq_*(\mathcal{O}_Z \otimes p^*V')(n)$$

in the derived category of bounded complexes of A-modules. However, every map between free complexes in this derived category is represented by a genuine map ψ of complexes ([H2], the dual version of Proposition I.4.7, or [GM], the dual version of Theorem 21, chapter III, section 5). The map ψ does not need to be homogeneous. However, since both complexes are graded, the homogeneous component ψ_0 of ψ of degree zero will also be a map of complexes. Moreover, since the induced map $H(\psi)_*$ has degree zero, the map $\psi - \psi_0$ is a map of free complexes inducing the trivial map on homology. Such a map is homotopic to zero, and thus ψ is homotopic to the map ψ_0 of degree zero. •

5.3. The Proof of Properties of Complexes $F(V)_\bullet$

In this section we prove Theorems (5.1.3), (5.1.4), and (5.1.6).

Proof of Theorem (5.1.3). First of all, we notice that parts (b) and (c) follow from Theorem (5.1.2) and part (a) of (5.1.3). Thus it is enough to prove part (a).

This statement follows from the following elementary lemma applied to the normalization of Y.

(5.3.1) Lemma. *Let $q : Z \to Y$ be a desingularization of Y. Let us assume that Y is normal. Then $q_*\mathcal{O}_Z = \mathcal{O}_Y$.*

Proof. The question is local on Y, so we can assume that $Y = \operatorname{Spec} A$ where A is a normal domain. The sheaf $q_* \mathcal{O}_Z$ is the sheaf associated to the ring $\Gamma(Z, \mathcal{O}_Z)$. Therefore it is enough to show that $\Gamma(Z, \mathcal{O}_Z) = A$. Since q is birational, it is clear that $\Gamma(Z, \mathcal{O}_Z)$ is contained in the field of fractions of A. It is also a finitely generated A-module, because q is proper. This proves the lemma. ∎

Proof of Theorem (5.1.4). We use the duality theorem for proper morphisms (Theorem (1.2.22)) for the map $f = q : X \times V \to X$, for $F^\bullet = \bigwedge^\bullet (p^* \xi) \otimes p^* V$, and for $G^\bullet = \mathcal{O}_X$. The complex $F(V)_\bullet$ is, by its construction, a free graded minimal representative of the object $\underline{R} f_*(F^\bullet)$. Therefore the right side of the theorem gives

$$\underline{R} \operatorname{Hom}_X^\bullet(\underline{R} f_*(F^\bullet), G^\bullet) = \underline{R} \operatorname{Hom}_X^\bullet(F(V)_\bullet, \mathcal{O}_X).$$

Now $\underline{R}$ can be dropped because $F(V)_\bullet$ is its own projective resolution (we calculate $\underline{R} \operatorname{Hom}$ as $R_{II} R_I$: compare Lemma 6.3, p. 66 in [H2]), and we are left with the complex $F(V)_\bullet^*$.

To identify the right side we notice that by (1.2.21) (c) we have

$$f^!(G^\bullet) = f^*(G^\bullet) \otimes \omega_{X \times V / V}[n] = \mathcal{O}_{X \times V} \otimes p^*(\omega_V)[n],$$

because f is smooth. Therefore the inside term on the left side in (1.2.22) can be written as

$$\underline{R} \operatorname{Hom}_{X \times V}^\bullet(F^\bullet, f^!(G^\bullet)) = \underline{R} \operatorname{Hom}_{X \times V}^\bullet \left(\bigwedge^\bullet (p^* \xi) \otimes p^* V, p^*(\omega_V)[n] \right).$$

By proposition 5.16 (p. 113) of [H2], with $L = \bigwedge^\bullet (p^* \xi) \otimes p^* V)$ we can identify the right hand side with

$$\underline{R} \operatorname{Hom}_{X \times V}^\bullet(\mathcal{O}_{X \times V}, \mathcal{O}_{X \times V}) \otimes \bigwedge^\bullet (p^* \xi)^* \otimes p^* V^* \otimes p^*(\omega_V)$$

which can be written as

$$\underline{R} \operatorname{Hom}_{X \times V}^\bullet(\mathcal{O}_{X \times V}, \mathcal{O}_{X \times V}) \otimes \bigwedge^\bullet (p^* \xi) \otimes \bigwedge^t (\xi^*) \otimes p^* V^* \otimes p^*(\omega_V)$$

where $t = \operatorname{rank} \xi$.

We can drop $\underline{R}$ in the above expression, because in the left place we have a locally free complex (again we calculate $\underline{R} \operatorname{Hom}$ as $R_{II} R_I$, using Lemma 6.3, p. 66 in [H2]). This means the complex above is quasiisomorphic to $\bigwedge^\bullet (p^* \xi) \otimes \bigwedge^t (p^* \xi)^* \otimes p^* V^* \otimes p^*(\omega_V)$. Therefore the left hand side in

(1.2.22) can be identified with

$$\underline{R}f_*\left(\overset{\bullet}{\bigwedge}(p^*\xi) \otimes \overset{t}{\bigwedge}(p^*\xi)^* \otimes p^*\mathcal{V}^* \otimes p^*(\omega_V) \right)$$

$$= F\left(\overset{t}{\bigwedge}(p^*\xi)^* \otimes p^*\mathcal{V}^* \otimes p^*(\omega_V) \right)_\bullet,$$

as claimed in (5.1.4). ■

Proof of Theorem (5.1.6). We start with part (a). Let us consider the canonical sheaf ω_Z. By the adjunction formula ([H1, Proposition II.8.20]), $\omega_Z = \omega_{X\times V}|_Z \otimes \bigwedge^t \xi^*$. Since X is just an affine space, $\omega_{X\times V} = p^*K$. Therefore $\omega_Z = \mathcal{O}_Z \otimes K \otimes \bigwedge^t \xi^*$. By the Grauert–Riemenschneider theorem (1.2.28) we know that $\mathcal{R}^i q_*(\omega_Z) = 0$ for $i > \dim Z - \dim Y$. Therefore the terms $F(K \otimes \bigwedge^t \xi^*)_i$ are zero for $i < \dim Y - \dim Z$. By the duality (5.1.4) this means that $F_i = 0$ for $i > \text{codim}_X Y$. It remains to show that for $i = \dim Z - \dim Y$ we have $\mathcal{R}^i q_*(\omega_Z) \neq 0$. After shrinking Y we can assume that q is smooth and projective. Then the last claim follows from the upper-semicontinuity theorem ([H1, III.12.11]) and the adjunction formula ([H1, II.8.20]), since each fiber Z_y is smooth of dimension i, so $H^i(Z_y, \omega_{Z_y})$ is one dimensional (hence nonzero) by Serre duality. This proves (a).

To prove (b) we consider the graded Hilbert function

$$P(F_\bullet, t) = \sum_{i,j\geq 0} (-1)^i t^{i+j} h^j\left(V, \overset{i+j}{\bigwedge}\xi \right)(1-t)^{-N}.$$

Writing $P(F_\bullet, t) = \sum_{a\geq 0} P(a)t^a$, we know that for big a the function $P(a)$ is polynomial in a. We also know that $P(F_\bullet, t)$ is the alternating sum of graded Hilbert functions of the homology modules $\mathcal{R}^i \mathcal{O}_Z$ of $F_\bullet$. The homology modules $\mathcal{R}^i \mathcal{O}_Z$ are supported in Y. Moreover, the modules $\mathcal{R}^i \mathcal{O}_Z$ for $i > 0$ are supported in the locus of points in Y where the fibers of q' have dimension at least 1, which is a proper subvariety of Y. The sheaf $q_* \mathcal{O}_Z$ is generically of rank $\deg q'$. This means that $P(a)$ is a polynomial of degree $\leq N - r$ and that the highest coefficient of $P(a)$ equals $(N - r)! \deg q' \deg Y$ in the case $\dim Y = \dim Z$ and is zero otherwise. Statement (b) of (5.1.6) now follows by standard calculation. ■

(5.3.2) Remarks. *The geometric method was first applied to determinantal varieties ([Ke 0],[L2], [JPW]). The general forms of statements (5.1.2), (5.1.3), (5.1.4), related to derived categories, were first used to deal with* *examples related to nilpotent orbit closures and discriminants ([W2], [W3], [Br5]). We follow the approach from [Br5] to prove the first part of (5.1.2) without derived categories.*

5.4. The G-Equivariant Setup

In this section we consider the special case of the construction from section 5.1 related to the situation when the variety V is a homogeneous space. This is the most important class of known examples where the geometric method applies. In fact, all examples considered in the following chapters are of this kind.

Let **G** be a linearly reductive group, and let **P** be a parabolic subgroup in **G**. We assume that the variety V is the homogeneous space **G**/**P**. We also assume that the group **G** acts linearly on the affine space X, so X can be identified with a representation of **G**. Let U be a **P**-submodule of X. We associate to U the vector bundle $Z = \mathbf{G} \times^{\mathbf{P}} U$, which is by definition the orbit space $\mathbf{G} \times U/\mathbf{P}$ with **P** acting by $p \cdot (g, y) = (gp^{-1}, py)$. The projection $\mathbf{G} \times U \to \mathbf{G}$ induces the **G**-equivariant morphism

$$p : Z = \mathbf{G} \times^{\mathbf{P}} U \to V = \mathbf{G}/\mathbf{P}.$$

Since U is a submodule of a **G**-module X, we have the embedding

$$Z = \mathbf{G} \times^{\mathbf{P}} U \to \mathbf{G} \times^{\mathbf{P}} X = \mathbf{G}/\mathbf{P} \times X$$

The identification on the right hand side is made by the morphism $(g, x) \mapsto (g\mathbf{P}, gx)$.

We denote by q the projection $\mathbf{G}/\mathbf{P} \times X \to X$, and by q' its restriction to Z. As before, we denote $Y = q'(Z)$.

This places us in the situation of section 5.1. Let W be another **P**-module. We associate to W the vector bundle $\mathcal{V}(W) := \mathbf{G} \times^{\mathbf{P}} W$. We can apply the construction from section 5.1 to get the complex $F(\mathcal{V}(W))_\bullet$ of free graded modules over the ring $A = \mathbf{K}[X] = \mathrm{Sym}(X^*)$. Let us recall that $F(\mathcal{V}(W))_i$ is given by the formula

$$F(\mathcal{V}(W))_i = \bigoplus_{j \geq 0} H^j\left(V, \bigwedge^{i+j} \xi \otimes \mathcal{V}(W)\right) \otimes_k A(-i - j).$$

Since the group **G** acts naturally on the bundles $\mathcal{V}(W)$ and ξ, it acts rationally on the cohomology groups $H^j(V, \bigwedge^{i+j} \xi \otimes \mathcal{V}(W))$. Therefore **G** acts rationally on free modules $F(\mathcal{V}(W))_\bullet$ via the diagonal action.

(5.4.1) Theorem. *Let* **G**, **P**, *V*, *X*, *Z*, *and* $\mathcal{V}(W)$ *be as above. Then the complex* $F(\mathcal{V}(W))_\bullet$ *can be constructed in such way that all the differentials*

$$d_i(\mathcal{V}(W)) : F(\mathcal{V}(W))_i \to F(\mathcal{V}(W))_{i-1}$$

are **G**-*equivariant.*

Proof. We just have to follow the proof of Theorem (5.1.2) to assure that each step can be made **G**-equivariant. Before we do that, we need a **G**-equivariant analogue of (5.2.2).

Let A be a graded ring over **K** with $A_0 = $ **K**. Let us assume that **G** acts rationally on A, i.e., **G** acts as a group of automorphisms of the graded ring A, so that each graded component A_i is a representation of **G** and that the multiplication maps are **G**-equivariant. Let $M = \sum_{i \geq i_0} M_i$ be a graded A-module. We say that the group **G** acts rationally on M (compatibly with the action on A) if each graded component M_i is a **G**-representation and the structure maps for the A-module M are **G**-equivariant. We will call a complex $M_\bullet$ of finitely generated modules over A **G**-*equivariant* if **G** acts rationally on each module G_i and the differentials are **G**-equivariant.

Let M be a graded A-module on which **G** acts rationally. Then we can choose a minimal set of generators for M which forms a **G** submodule in M. Indeed, the projection $M \to M/A^+M$ splits as a map of **G**-modules. This means that every projective graded finitely generated A-module P on which **G** acts rationally is of the form $P = \sum_j P_j \otimes A(-j)$ for some finite dimensional **G**-representations P_j. Therefore a complex $G_\bullet$ of finitely generated graded free A-modules is **G**-equivariant if each G_i is of the form $G_i = \sum_j G_{i,j} \otimes A(-j)$ for some finite dimensional representations $G_{i,j}$ of **G**, and the differentials $d_i : G_i \to G_{i-1}$ are **G**-equivariant.

We can now recover all the standard results on minimal free resolutions of graded modules and complexes in **G**-equivariant form. In particular we have

(5.4.2) Proposition.

    (a) Let $G_\bullet$ be a **G**-*equivariant complex of finitely generated graded free A-modules. The complex $G_\bullet$ decomposes into a direct sum*

$$G_\bullet = G'_\bullet \oplus G''_\bullet,$$

    where $G'_\bullet$ is a minimal complex, $G''_\bullet$ is exact, and both $G'_\bullet$ and $G''_\bullet$ are **G**-*equivariant. The terms of the complex $G_\bullet$ are*

$$G'_i = H_i(G_\bullet \otimes_A \mathbf{K}) \otimes_\mathbf{K} A.$$

*(b) Let $M_{\bullet}$ be a **G**-equivariant complex of graded A-modules with $M_i = 0$ for $i < i_0$. Then there exists a minimal **G**-equivariant complex $G_{\bullet}$ of free graded A-modules and a map $\phi : G_{\bullet} \to M_{\bullet}$ which is a quasiisomorphism.*

Proof. To prove both statements we just repeat the proof of (5.2.2). For (a) we notice that the decompositions

$$G_i = B_i \oplus W_i \oplus B_i' \qquad (*)$$

can be chosen in **G**-equivariant way. In the proof of (b) all modules Z_i, B_i, etc. are the modules with the rational **G** actions, so all the exact sequences are **G**-equivariant. Therefore the covering complexes of free modules can be also choosen in a **G**-equivariant way.

Now we go through all the steps of the proof of (5.1.2).

(1) The embedding of V in the projective space can be chosen in a **G**-equivariant way. Indeed, we can use any positive line bundle $\mathcal{L}(\alpha)$ (cf. section 4.3). Then the sheaf $\mathcal{O}_V(1)$ corresponds to the ample line bundle whose total space admits an action of **G**. Therefore the homogeneous coordinate ring R of V in this embedding admits a rational **G**-action.

The key step in the construction of complexes $F(\mathcal{V})_{\bullet}$ is the existence of a certain right resolution of the twisted Koszul complex $\mathcal{K}(\xi, V)_{\bullet}$.

(2) The twisted Koszul complex $\mathcal{K}(\xi, \mathcal{V}(W))_{\bullet}$ of sheaves on $X \times V$ consists of vector bundles admitting **G**-action. Therefore, after applying the functor Γ_* to its dual, we get a **G**-equivariant complex of bigraded $A \otimes R$-modules. Therefore its minimal resolution $\widehat{C}(\mathcal{V}(W))_{\bullet\bullet}$ is a **G**-equivariant complex, and thus the double complex $\mathcal{P}(\mathcal{V}(W))_{\bullet\bullet}$ is a **G**-equivariant complex of sheaves.

(3) It follows that the double complex $q_*(\mathcal{P}(\mathcal{V})_{\bullet\bullet})$ is a **G**-equivariant double complex of graded free A-modules. The rest of the proof follows by applying (5.4.2).

This concludes the proof of Theorem (5.4.1). ■

(5.4.3) Remark. *The construction we applied in this section is also possible when the group **G** is only assumed to be reductive. In such case we cannot claim that the complex $F(\mathcal{V}(W))_{\bullet}$ is **G**-equivariant. We get the **G**-action on the terms of $F(\mathcal{V}(W))_i$, and we can claim the **G**-equivariance of linear strands of $F(\mathcal{V}(W))_{\bullet}$. However, the higher degree maps come from the spectral sequence, so some lifting is required. Therefore the higher degree maps need not be **G**-equivariant.*

5.5. The Differentials in Complexes $F(V)_\bullet$.

In this section we discuss the description of differentials in the complexes of type $F(V)_\bullet$. The point is that if one follows through the proof of Proposition (5.2.2), one gets an inductive procedure for calculating the differential in $F(V)_\bullet$ which is not convenient to use.

The following result, due to Eisenbud and Schreyer, allows us to describe the differential in a closed form.

(5.5.1) Theorem ([ES]). *Let* $\mathbf{F}$ *be a double complex*

$$
\begin{array}{ccccccc}
 & & \uparrow & & \uparrow & & \\
\cdots & \to & F_j^{i+1} & \xrightarrow{d_h} & F_{j+1}^{i+1} & \to & \cdots \\
 & & d_v \uparrow & & \uparrow d_v & & \\
\cdots & \to & F_j^i & \xrightarrow{d_h} & F_{j+1}^i & \to & \cdots \\
 & & \uparrow & & \uparrow & &
\end{array}
$$

in some abelian category. Assume that $F_j^i = 0$ *for* $i \ll 0$. *Suppose that the vertical differential of* $\mathbf{F}$ *splits, so that for each* i, j *there is a decomposition* $F_j^i = G_j^i \oplus d_v(G_j^{i-1}) \oplus H_j^i$ *such that the kernel of* d_v *in* F_j^i *is* $H_j^i \oplus d_v(G_j^{i-1})$, *and such that* d_v *maps* G_j^{i-1} *isomorphically to* $d_v(G_j^i)$. *Let us write* $s : F_j^i \to H_j^i$ *for the projection corresponding to this decomposition and* $p : F_j^i \to d_v(G_j^{i-1}) \to G_j^{i-1}$ *for the composition of the projection with the inverse of* d_v *restricted to* G_j^{i-1}. *Then the total complex of* $\mathbf{F}$ *is homotopic to the complex*

$$
\cdots \to \bigoplus_{i+j=k} H_j^i \xrightarrow{d} \bigoplus_{i+j=k-1} H_j^i \to \cdots
$$

with differential

$$
d = \sum_{\ell \geq 0} s(d_h p)^\ell d_h.
$$

Proof. We write $d_t = d_v \pm d_h$ for the differential of the total complex. We note first that $s(d_h p)^j d_h$ takes H_j^i to $H_{j+\ell+1}^{i-\ell}$. Since $F_{j+\ell+1}^{i-\ell} = 0$ for $\ell \gg 0$, the sum in the definition of d is finite.

Let F denote $\mathbf{F}$ without a differential, i.e. viewed as a bigraded module. We will first show that F is the direct sum of three components

$$
G = \bigoplus_{i,j} G_j^i, \qquad d_t(G), \quad \text{and} \quad H = \bigoplus_{i,j} H_j^i
$$

and that d_t is a monomorphism on G.

The same statements with d_v replacing d_t are true by hypothesis. In particular, any element of F can be written in the form $g' + d_v(G) + h$ with $g' \in G^i_j$, $g \in G^{i-1}_j$, $h \in H^i_j$ for some i, j. Modulo $G + d_t(G) + H$, such an element can be written as $d_h(G) \in F^{i-1} + j + 1$. Since $F^s_t = 0$ for $s \ll 0$, we can use induction on i and assume $d_h(g) \in G + d_t(G) + H$, so we see that $F = G + d_t(G) + H$.

Suppose

$$g' \in G = \bigoplus_{i,j} G^i_j, \qquad g \in G = \bigoplus_{i,j} G^i_j, \quad \text{and} \quad h \in H = \bigoplus_{i,j} H^i_j$$

be such that $g' + d_t(g) + h = 0$. We need to show that $g' = g = h = 0$. Write $g = \sum_{k=a}^b g^{k-1}_{\ell-k}$ with $g^s_t \in G^s_t$. If $b - a = -1$, then $d_t = 0$ and the desired result is a special case of the hypothesis. In any case, there is no componnt of g in $G^b_{\ell-b-1}$, so the component of $d_t(g)$ in $G^b_{\ell-b}$ is equal to $d_v g^{b-1}_{\ell-b}$. From the hypothesis we see that $d_v g^{b-1}_{\ell-b} = 0$, so $g^{b-1}_{\ell-b} = 0$, and we are done by induction on $b - a$. This shows that $F = G \oplus d_t(G) \oplus H$ and that d_t is an isomorphism from G to $d_t(G)$.

The modules $G \oplus d_t(G)$ form a double complex contained in **F** that we will call **G**. Since $d_t : G \to d_t(G)$ is an isomorphism, the total complex of **G** is split exact. It follows that the total complex tot(**F**) is homotopic to **F**/tot(**G**), and the modules in the last complex are isomorphic to $\bigoplus_{i+j=k} H^i_j$. We will complete the proof by showing that the induced differential on tot(**F**)/tot(**G**) is the differential d defined in the statement of the theorem.

Choose $h \in H^i_j$. The image of h under the induced differential is the unique element $h' \in H$ such that $d_t(h) \equiv h' \pmod{G} + d_t(G)$. Now

$$d_t h = d_h h \equiv s d_h h + (d_v p) d_h h \pmod{G}.$$

However,

$$d_v p \equiv d_h p \equiv s(d_h p) + d_v p(d_h p) \pmod{G + d_t(G)}.$$

Continuing this way, and using again the fact that $F^i_j = 0$ for $i \ll 0$, we obtain

$$d_t h \equiv \sum_\ell s(d_h p)^\ell d_h h \pmod{G + d_t(G)},$$

as required. ∎

Let us specialize to the situation where our abelian category is the category of graded A-modules for some graded ring $A = \bigoplus_{d\geq 0} A_d$, with $A_0 = \mathbf{K}$. Assume that the modules F^i_j above are free A-modules.

(5.5.2) Corollary. *Let* **F** *be a double complex of graded free A-modules. Assume that the differential d_v is of degree 0 and that d_h is minimal. Then the assumptions of Theorem (5.5.1) are satisfied, and H is a minimal complex homotopically equivalent to* tot(**F**).

Proof. The only claim needing verification is that the vertical differential splits. Since $A_0 = \mathbf{K}$ and the modules F_j^i are free, each column of **F** is obtained from some complex of vector spaces over **K** by tensoring with A. The splitting can also be chosen over **K**. ∎

The most efficient general procedure to calculate the differential on the complexes of type $F_\bullet(\mathcal{V})$ consists in applying Corollary (5.5.2) to the complex $q_*(\mathcal{P}(\mathcal{V})_{\bullet\bullet})$ constructed in the course of the proof of Theorem (5.1.2) in section 5.2. Still, that procedure cannot be carried to its completion for large complexes. We will see in the following chapters that for equivariant complexes representation theory is the best tool for identifying the differentials.

5.6. Degeneration Sequences

So far we discussed the complexes $F(\mathcal{V})_\bullet$ and their properties. They often give the terms of the minimal resolution of the module $q_*(\mathcal{O}_Z \otimes p^*(\mathcal{V}))$. Sometimes it is useful to consider the exact sequences formed by such modules. This is especially useful in the "equivariant" situations, i.e. when our projective variety V is a homogeneous space. Such analysis allows sometimes to compare the resolutions of two orbit closures Y and Y_1 such that $Y_1 \subset Y$, i.e., Y_1 is a degeneration of Y.

Let us consider the basic diagram

$$
\begin{array}{ccc}
Z & \subset & X \times V \\
\downarrow q' & & \downarrow q \\
Y & \subset & X
\end{array}
$$

We assume that $V = \mathbf{G}/\mathbf{P}$ for some reductive algebraic group **G** and a parabolic subgroup **P**.

The variety Z is a total space of a vector subbundle of $X \times V$ which can be identified with η^*. Assume that η is a homogeneous bundle, i.e., it is of the form $\eta^* = \mathbf{G} \times^{\mathbf{P}} U$ for some rational **P**-module U. We denote $B := \mathrm{Sym}(U^*)$. This is a polynomial ring with a rational **P**-action.

Let $I \subset B$ be a **P**-equivariant ideal. We have a corresponding **G**-equivariant sheaf of ideals $\mathcal{I} \subset \mathcal{O}_Z$. The degeneration technique comes from trying to exploit the resolution of B/I as a B-module.

(5.6.1) Proposition. *Assume that we can find a **P**-equivariant resolution*

$$0 \to G_m \to G_{m-1} \to \ldots \to G_1 \to G_0$$

of B/I with $G_i = W_i \otimes B$. Then we have an induced exact sequence

$$0 \to \mathcal{G}_m \to \mathcal{G}_{m-1} \to \ldots \to \mathcal{G}_1 \to \mathcal{G}_0 \qquad (*)$$

*of vector bundles $\mathcal{G}_j = (\mathbf{G} \times^{\mathbf{P}} G_j) \otimes \mathcal{O}_Z$ which is a resolution of $\mathcal{O}_Z/\mathcal{I}$. Assume that higher cohomology groups $H^i(\mathbf{G}/\mathbf{P}, \mathcal{G}_j) = 0$ for $i \geq 1$, $0 \leq j \leq m$. Then we have a **G**-equivariant acyclic sequence of A-modules*

$$0 \to M_m \to M_{m-1} \to \ldots \to M_1 \to M_0$$

where $M_j = H^0(\mathbf{G}/\mathbf{P}, \mathcal{G}_j)$.

Proof. Decompose the exact sequence $(*)$ into short exact sequences, and use long cohomology sequences. ∎

   The existence of such a **P**-equivariant resolution is in general a rather subtle question. There are, however, two cases when such a resolution exists.
   The first case occurs when the unipotent radical **N** of **P** acts on B trivially. Denote by **L** a Levi factor of **P**.

(5.6.2) Proposition. *Assume that **N** acts trivially on B and that **L** is linearly reductive. Then the resolution $(*)$ of $\mathcal{O}_Z/\mathcal{I}$ exists.*

Proof. Since **N** acts on B trivially, B is really an **L**-module. Since **L** is linearly reductive, we can construct an **L**-equivariant resolution of B/I by the arguments used in section 5.4. ∎

   The other case occurs when the ideal I is a complete intersection defined by some **P**-semiinvariants.
   Let's recall that if an algebraic group acts rationally on a vector space U, then the ring of semiinvariants

$$SI(\mathbf{H}, U) = \bigoplus_{\chi \in \mathrm{char}(\mathbf{H})} SI(\mathbf{H}, U)_\chi,$$

where

$$SI(\mathbf{H}, U)_\chi = \{ f \in \mathrm{Sym}(U^*) \mid h \circ f = \chi(h)f \; \forall h \in \mathbf{H} \},$$

is the ring of functions that transform according to a certain character of **H**.

There is an important special case when one can predict which semiinvariants one should look at. It occurs when the rational **H**-module U has an open **H**-orbit. In such case one can classify the **H**-semiinvariants. This is due to the following result of Sato and Kimura.

(5.6.3) Lemma (Sato–Kimura [SK]). *Let* **H** *be a linear algebraic group acting rationally on a vector space* U. *Assume that this action has an open orbit. Then the ring* $SI(\mathbf{H}, U)$ *is a polynomial ring. Moreover, the characters of the generators are linearly independent. The generators of the ring of semiinvariants can be described as follows. Assume* Ox *is the open orbit of* **H** *in* U. *Let* $U \setminus Ox = D_1 \cup \ldots \cup D_t$ *be a decomposition into irreducible components. Assume that the first s components have codimension 1 in* U, *while the other components have codimension bigger than 1. Then the generators of* $SI(\mathbf{H}, U)$ *are the irreducible equations* $v_1, \ldots, v_s$ *of* $D_1, \ldots, D_s$.

For the proof of this lemma the reader should consult [Kr1, Theorem 2, section 3.6].

Coming back to our basic situation, i.e. $\mathbf{H} = \mathbf{P}$, $U = \eta^*$, and using the Koszul complex, we have

(5.6.4) Proposition. *Assume that the ideal* I *is a complete intersection defined by* **P**-*semiinvariants* $v_1, \ldots, v_s$ *of weights* $\lambda_1, \ldots, \lambda_s$. *Then the Koszul complex gives a resolution* (∗) *of length s, with* $\mathcal{G}_j = \bigoplus_{1 \le t_1 < \ldots < t_j \le s} \mathcal{O}_Z \otimes p^* \mathcal{L}(\lambda_{t_1} + \ldots + \lambda_{t_j})$.

The natural examples of the **P**-equivariant ideals arise as follows. Let $Y_1 \subset Y$ be a **G**-equivariant subset of Y. We consider the schematic preimage of Y_1 in the fiber of Z, i.e. the ideal of the schematic intersection $q'^{-1}(Y_1) \cap p^{-1}(e)$, where $e \in \mathbf{G}/\mathbf{P}$ is the coset of the identity. This is a **P**-equivariant ideal $I(Y_1)$ in the polynomial ring $B = \mathrm{Sym}(U^*)$. There is a corresponding sheaf of ideals $\mathcal{I}(Y_1) \subset \mathcal{O}_Z$. We call $I(Y_1)$ the *degeneration ideal* corresponding to a degeneration $Y_1 \subset Y$.

Exercises for Chapter 5

Cones over Nonsingular Curves

1. Let C be a nonsingular curve of genus g. Assume that L is a very ample line bundle on C. The bundle L defines the embedding i_L:

$C \to \mathbf{P}(H^0(C, L))$. Let $X = H^0(C, L)$ and let Y be the cone over $i_L(C)$. Consider

$$Z = \{(x, y) \in X \times C \mid x \in [i_L(y)]\},$$

where $[i_L(y)]$ is the line containing $i_L(y)$. Prove that we are in the situation of section 5.1 with $V = C, \eta = p^*(L)$.

2. Calculate the homology of complexes $F(M)_\bullet$ for any line bundle M on C, using (5.1.2) (b). Interpret the meaning of all theorems from section 5.1, in particular of the duality statement (5.1.4).

The Representations of SL(2). *Binary Forms*

We assume that E is a vector space over $\mathbf{K}$ of dimension 2. Let $X = (S_d E)^*$. We identify X with the set of homogeneous polynomials of degree d in variables x, y. We write the general element of X as

$$f = \sum_{i=0}^{d} \phi_i x^i y^{d-i}.$$

We have $A = \mathrm{Sym}(S_d E)$.

3. Let us fix d. We take $V = \mathrm{Grass}(1, E) = \mathbf{P}^1$. We write

$$0 \to \mathcal{R} \to E \times \mathbf{P}^1 \to \mathcal{Q} \to 0,$$

the tautological sequence on $\mathbf{P}^1$. We have identifications $\mathcal{R} = \mathcal{O}_{\mathbf{P}^1}(-1)$, $\mathcal{Q} = \mathcal{O}_{\mathbf{P}^1}(1)$. Prove that for each $0 \leq p \leq d$ we have an exact sequence

$$0 \to S_{d-p+1}\mathcal{R} \otimes S_{p-1}E \to S_d E \times \mathbf{P}^1 \to S_p \mathcal{Q} \otimes S_{d-p}E \to 0.$$

4. Consider the variety $Z_p \subset X \times \mathbf{P}^1$ given by

$$Z_p = \{(f, S) \in X \times \mathrm{Grass}(1, E^*) \mid f \text{ has a } p\text{-tuple root at } S\}.$$

Identifying S with the fiber of $\mathcal{Q}^*$, prove that Z_p comes from the construction of section 5.1 with $\xi = S_{d-p+1}\mathcal{R} \otimes S_{p-1}E, \eta = S_p \mathcal{Q} \otimes S_{d-p}E$. The variety $X_p := q'(Z_p)$ is the set of binary forms with a p-tuple root.

5. Let $p = d$. The terms of the complex $F_\bullet$ are given by

$$F_0 = A,$$

$$F_i = S_{i,1}E \otimes \overset{i+1}{\bigwedge}(S_{d-1}E) \otimes A(-i + 1)$$

for $i \geq 1$. Deduce that X_d is the cone over a rational normal curve in $\mathbf{P}^{d-1}$. Prove that the equations of X_d are 2×2 minors of the matrix

$$\begin{pmatrix} \phi_0 & \phi_1 & \cdots & \phi_{d-1} \\ \phi_1 & \phi_2 & \cdots & \phi_d \end{pmatrix}.$$

6. Let $p < d$. Prove that the terms of the complex $F_\bullet$ are

$$F_0 = A \oplus S_{d-p,1} E \otimes S_{p-1} E \otimes A(-1),$$

$$F_i = S_{(i+1)(d-p+1)-1,1} E \otimes \overset{i+1}{\bigwedge}(S_{p-1} E) \otimes A(-i-1)$$

for $i \geq 1$. Conclude that for $2 \leq p < d$ the variety X_p is not normal, but its normalization has rational singularities. Analyze the complex $F_\bullet$ in the case $p = 1$.

7. Prove that the normalization of X_p can be described as the geometric invariant theory quotient

$$\bar{X}_p = (E^* \times S_{d-p} E^*)//\mathbf{C}^*,$$

where the multiplicative group $\mathbf{C}^*$ acts on $E^* \times (S_{d-p} E)^*$ by the formula

$$t(l, g) = (t^{-1} l, t^p g)$$

for $t \in \mathbf{C}^*$, $l \in E^*$, $g \in (S_{d-p} E)^*$.

Highest Weight Vector Orbit Varieties

8. Let $V = \mathbf{G}/\mathbf{P}$. Let λ be a $\mathbf{P}$-regular weight, i.e. a weight for which the line bundle $\mathcal{L}(\lambda)$ is induced from $\mathbf{G}/\mathbf{P}$. Then $\mathcal{L}(\lambda)$ induces an embedding of $\mathbf{G}/\mathbf{P}$ into a projective space $\mathbf{P}(V_\lambda)$. Let $X = V_\lambda$, define

$$Z = \{(x, \bar{g}) \in X \times \mathbf{G}/\mathbf{P} \mid x = g(v_\lambda) \},$$

where v_λ is a highest weight vector in V_λ (i.e. a $\mathbf{U}$-invariant). Then $Y = q'(Z)$ can be identified with the closure of the orbit of v_λ in V_λ. Show that this variety is normal with rational singularities. Use (5.1.2) (b) to show that the coordinate ring of Y is $\bigoplus_{n \geq 0} V_{n\lambda}$.

6

The Determinantal Varieties

This is the first of a series of chapters where we apply the techniques of chapter 5 to concrete examples. We consider the determinantal varieties for the generic, generic symmetric, and generic skew symmetric matrices. We describe explicitly the terms of their minimal free resolutions over fields of characteristic 0. We also show that in characteristic $p > 0$ the resolution can be different than in characteristic 0.

In section 6.1 we deal with ideals of minors of generic matrices over a field of characteristic 0. We prove Lascoux's result providing the description of terms in minimal free resolutions of these ideals. We also treat in more detail the special cases of Eagon–Northcott and Gulliksen–Negard complexes.

Section 6.2 is devoted to determinantal ideals in positive characteristic. We prove Hashimoto's result that the resolution of 2×2 minors of a 5×5 generic matrix over a field of characteristic 3 is different than the corresponding resolution over a field of characteristic 0. Here we make use of theory of Schur complexes developed in section 2.4.

Section 6.3 deals with the ideals of minors of a generic symmetric matrix. Again we calculate the terms of a minimal free resolutions of such ideals over a field of characteristic 0. We also describe explicitly the special cases of resolutions of length 3 and 6.

In section 6.4 we consider the ideals of Pfaffians of generic skew symmetric matrices. Again we give the terms of minimal free resolutions over fields of characteristic 0 and describe the resolutions explicitly for the cases when they have lengths 3 and 6.

In section 6.5 we discuss the equivariant modules supported in determinantal varieties. We use the desingularizations of the determinantal varieties to construct several families of such modules. We prove some results about the Grothendieck group of equivariant modules.

Finally, in sections 6.6 and 6.7 we discuss the analogues of the results from section 6.5 for the modules supported in symmetric and skew symmetric determinantal varieties.

6.1. The Lascoux Resolution

We work over a fixed field $\mathbf{K}$. Consider the space X of $m \times n$ matrices over $\mathbf{K}$. We denote by $\phi_{i,j}$ the (i, j)th coordinate function on X. The coordinate ring $A = \mathbf{K}[X]$ is isomorphic to the polynomial ring $\mathbf{K}[\phi_{i,j}]_{1 \leq i \leq m, \ 1 \leq j \leq n}$.

We identify X with the space $\mathrm{Hom}_{\mathbf{K}}(F, G)$, where F and G are two vector spaces over $\mathbf{K}$ of dimensions m, n respectively. Since $\mathrm{Hom}_{\mathbf{K}}(F, G)$ is naturally isomorphic to $F^* \otimes G$, we can identify the coordinate ring A with the symmetric algebra $\mathrm{Sym}(F \otimes G^*)$. Let $\{f_i\}_{1 \leq i \leq m}$ and $\{g_j\}_{1 \leq j \leq n}$ be the bases in F, G respectively. Then the coordinate function $\phi_{i,j}$ is identified with the element $f_i \otimes g_j^*$ from $\mathrm{Sym}(F \otimes G^*)$. We denote by

$$\Phi : F \otimes_{\mathbf{K}} A \to G \otimes_{\mathbf{K}} A$$

the generic map $\Phi(f_i \otimes 1) = \sum_{j=1}^{n} \phi_{i,j}(g_j \otimes 1)$.

The group $\mathrm{GL}(F) \times \mathrm{GL}(G)$ acts on X, and therefore on A by row and column operations.

For each r satisfying $0 \leq r < \min(m, n)$ we consider the subvariety Y_r of X given by

$$Y_r = \{\phi \in X \mid \mathrm{rank}\,\phi \leq r\},$$

called the *determinantal variety*. The variety Y_r can be identified with the set of matrices Φ whose minors of size $r + 1$ vanish. We denote the ideal generated by minors of Φ of size $r + 1$ in A by I_{r+1}. We call the ideal I_{r+1} the *determinantal ideal of* $(r + 1) \times (r + 1)$ *minors of the generic matrix*.

In this section we will apply the technique developed in chapter 5 to determinantal varieties. We will use the complexes $F_\bullet(\mathcal{V})$ to describe the minimal free resolution of the quotient ring A/I_{r+1} and to establish the following properties:

(1) The coordinate ring $\mathbf{K}[Y_r]$ is normal and Cohen–Macaulay.
(2) I_{r+1} is the defining ideal of Y_r.
(3) The Hilbert function of $\mathbf{K}[Y_r]$ is independent of the characteristic of $\mathbf{K}$.

Property (3) suggests that the minimal resolution of $\mathbf{K}[Y_r]$ could be independent of characteristic of the base field $\mathbf{K}$. In fact, we will show in the next section that this is not true.

Let us fix m, n, and r. We take $Y = Y_r$ and $V = \mathrm{Grass}(r, G)$, with the tautological sequence

$$0 \to \mathcal{R} \to G \times V \to \mathcal{Q} \to 0.$$

Here dim $\mathcal{R} = r$ and dim $\mathcal{Q} = q = n - r$. We consider the variety

$$Z = \{(\phi, R) \in X \times V \mid \mathrm{Im}\, \phi \subset R\}.$$

In fact the variety Z is the total space of the bundle $F \otimes \mathcal{R}$. This means that we are in the setting of the section 5.1 with $\mathcal{E} = F^* \otimes G$. Therefore $\mathcal{S} = F^* \otimes \mathcal{R}$ and $\mathcal{T} = F^* \otimes \mathcal{Q}$. We consider the basic diagram

$$
\begin{array}{ccc}
Z & \subset & X \times V \\
\downarrow q' & & \downarrow q \\
Y & \subset & X
\end{array}
$$

Applying Proposition (5.1.1), we see that the resolution of $\mathcal{O}_Z$ as an $\mathcal{O}_{X \times V}$-module is given by the Koszul complex

$$\mathcal{K}(F \otimes \mathcal{Q}^*)_{\bullet} : 0 \to \overset{mq}{\bigwedge}(p^*(F \otimes \mathcal{Q}^*)) \to \ldots \to p^*(F \otimes \mathcal{Q}^*) \to \mathcal{O}_{X \times V}.$$

Let us notice that the following special properties hold in our case.

(6.1.1) Proposition.
   (a) The variety Z is a desingularization of Y.
   (b) The higher direct images $R^i q'_*(\mathcal{O}_Z)$ are zero for $i > 0$.
   (c) The Hilbert function of the normalization of $\mathbf{K}[Y]$ does not depend on the characteristic of $\mathbf{K}$.

Proof. To prove (a) we consider the open subset

$$U = \{(\phi, R) \in Z \mid \mathrm{rank}\, \phi = r\}.$$

By the definition of Z we see that if $(\phi, R) \in U$ then $R = \mathrm{Im}\, \phi$. Therefore the algebraic map $\phi \mapsto (\phi, \mathrm{Im}\, \phi)$ is the inverse of the map $q'|U$. This means $q'|U$ is an isomorphism, so q' is a birational isomorphism.

In order to establish (b) we apply Proposition (5.1.1)(b). We see that $q_* \mathcal{O}_Z = \mathrm{Sym}(F \otimes \mathcal{R}^*)$. Using Theorem (5.1.2)(b) (applied with $\mathcal{V} = \mathcal{O}_V$), we see that it is enough to show that

$$H^i(V, \mathrm{Sym}(F \otimes \mathcal{R}^*)) = 0 \qquad \text{for} \quad i > 0.$$

By the Cauchy formula (3.2.5), every symmetric power $S_t(F \otimes \mathcal{R}^*)$ has a filtration with associated graded object $\sum_{|\lambda|=t} L_\lambda F \otimes L_\lambda \mathcal{R}^*$. By the version (4.1.12) of Kempf's theorem for Grassmannians, the higher cohomology of each $L_\lambda F \otimes L_\lambda \mathcal{R}^*$ vanishes. This proves part (b). Notice that if the characteristic of $\mathbf{K}$ is zero, then we can use the version (4.1.9) of the Bott's theorem for Grassmannians to establish the vanishing of higher cohomology groups of $\mathrm{Sym}(F \otimes \mathcal{R}^*)$. Since the dimension of $H^0(V, L_\lambda \mathcal{R}^*)$ does not depend on the characteristic of $\mathbf{K}$ by Kempf's theorem, (c) follows from (5.1.2)(b) and part (b). ∎

It follows from (6.1.1) and (5.1.3) that the complex $F_\bullet$ gives a minimal resolution of the normalization of the coordinate ring $\mathbf{K}[Y]$ as an A-module. We want to calculate the terms of the complex $F_\bullet$. Since the cohomology depends on the characteristic of $\mathbf{K}$, we have to treat the cases of characteristic 0 and p separately. Thus for the remainder of this section we assume that char $\mathbf{K} = 0$.

By Theorem (5.1.2) the terms of the complex $F_\bullet$ are given by the cohomology groups $H^i(V, \bigwedge^{i+j}(F \otimes \mathcal{Q}^*))$. Applying the Cauchy formula (2.3.3), we know that

$$\bigwedge^t (F \otimes \mathcal{Q}^*) = \bigoplus_{|\lambda|=t} L_\lambda F \otimes K_\lambda \mathcal{Q}^*.$$

Therefore we have

$$H^i\left(V, \bigwedge^t(F \otimes \mathcal{Q}^*)\right) = \bigoplus_{|\lambda|=t} L_\lambda F \otimes H^i(V, K_\lambda \mathcal{Q}^*).$$

We calculate the groups $H^i(V, K_\lambda \mathcal{Q}^*)$. Using the version of Bott's theorem for Grassmannians (Corollary (4.1.9)), we see that we have to apply Bott's algorithm (4.1.5) to the sequence

$$(0, \lambda) + \rho = (n - 1, \ldots, n - r, \lambda_1 + n - r - 1, \ldots, \lambda_q).$$

The first r numbers in this sequence are consecutive, and the last q numbers form a decreasing sequence. Let us assume that $(0, \lambda)$ is regular. Let s be the biggest number such that $\lambda_s + n - r - s > n - 1$. Then $\lambda_{s+1} + n - r - s - 1 < n - r$. In terms of λ this means that

$$\lambda_s \geq r + s, \qquad \lambda_{s+1} < s + 1.$$

We denote the set of partitions satisfying the above inequalities by $P(r, s)$. Reordering $(0, \lambda) + \rho$ means that the numbers $\lambda_1 + n - r - 1, \ldots, \lambda_s + n - r - s$ go in front of $n - 1, \ldots, n - r$. Let us denote the corresponding permutation by $w(s)$. Clearly $\ell(w(s)) = rs$. The conditions for $\lambda \in P(r, s)$ mean that if the partition λ has the Durfee square size s (cf. section 1.1.2), then the sequence $(0, \lambda)$ is regular if and only if λ contains an additional $s \times r$ rectangle to the right of the Durfee square. In that case the partition $w(s) \cdot (0, \lambda) = (\lambda_1 - r, \ldots, \lambda_s - r, s^r, \lambda_{s+1}, \ldots, \lambda_q)$. This means we have proved

(6.1.2) Proposition. *The i-th term of the complex $F_\bullet$ is given by*

$$F_i = \bigoplus_{s \geq 0} \bigoplus_{\lambda \in P(r,s),\ |\lambda|-rs=i} L_\lambda F \otimes K_{w(s)\cdot(0,\lambda)} G^* \otimes_K A.$$

Let us try to rewrite this result in a more symmetric way. Every partition λ from $P(r, s)$ can be written as

$$\lambda = (r + s + \alpha_1, \ldots, r + s + \alpha_s, \beta_1, \ldots, \beta_{n-s})$$

where α and β are two partitions. In this setup we have

$$w(s) \cdot (0, \lambda) = (s + \alpha_1, \ldots, s + \alpha_s, s^r, \beta_1, \ldots, \beta_{q-s}).$$

The dual partition $(w(s) \cdot (0, \lambda))'$ can in this notation be expressed as

$$(w(s) \cdot (0, \lambda))' = (r + s + \beta_1', \ldots r + s + \beta_s', \alpha_1', \ldots, \alpha_{m-s}').$$

The term corresponding to λ appears in F_i with $i = s^2 + |\alpha| + |\beta|$.

We can rewrite (6.1.2) in terms of partitions α and β. For any s we consider the set

$$Q(s) = \{(\alpha, \beta) | \alpha \subset (m - r - s)^s, \beta \subset (s)^{(n-r-s)}\}.$$

For $(\alpha, \beta) \in Q(s)$ we denote

$$P_1(\alpha, \beta) = (r + s + \alpha_1, \ldots, r + s + \alpha_s, \beta_1, \ldots, \beta_{n-s}),$$
$$P_2(\alpha, \beta) = (r + s + \beta_1', \ldots, r + s + \beta_s', \alpha_1', \ldots, \alpha_{m-s}').$$

Then (6.1.2) can be rewritten in following way:

(6.1.3) Proposition.

$$F_i = \bigoplus_{s \geq 0} \bigoplus_{(\alpha,\beta) \in Q(s),\ i=s^2+|\alpha|+|\beta|} L_{P_1(\alpha,\beta)} F \otimes L_{P_2(\alpha,\beta)} G^* \otimes_K A.$$

This way of writing the term is symmetric in F and G^*.

Now we can state the basic result on the syzygies of determinantal varieties.

(6.1.4) Theorem (Lascoux [L2]). *Assume that char $K = 0$. The complex $F_\bullet$ is a minimal resolution of the coordinate ring $K[Y_r]$. Therefore the i-th module in this minimal resolution is given by the formula (6.1.3).*

Proof. The only thing that we have to prove is that $K[Y_r]$ is normal. However, it is clear from (6.1.3) that $F_0 = A$. By (5.1.3)(c) the normality follows. ∎

Let us state some other consequences of Theorem (6.1.4).

(6.1.5) Corollary. *Assume that* char $\mathbf{K} = 0$.

(a) *The determinantal ideal I_{r+1} is a prime ideal. The quotient ring A/I_{r+1} is a normal domain.*

(b) *The quotient ring A/I_{r+1} is a Cohen–Macaulay ring with rational singularities.*

(c) *Let us assume that $m \geq n$. Then the type of the module A/I_{r+1} is equal to the dimension of the module $L_{(n-r)^{(m-n)}}G^*$. Therefore A/I_{r+1} is a Gorenstein ring if and only if $m = n$.*

(d) *The t-th homogeneous component of the ring A/I_{r+1} decomposes as a $\mathrm{GL}(F) \times \mathrm{GL}(G)$-module as follows:*

$$(A/I_{r+1})_t = \bigoplus_{|\lambda|=t, \ \lambda_1 \leq r} L_\lambda F \otimes L_\lambda G^*$$

Proof. To prove (a) let us notice that from (6.1.3) it follows at once that $F_1 = \bigwedge^{r+1} F \otimes \bigwedge^{r+1} G^* \otimes_{\mathbf{K}} A(-r-1)$. The map from F_1 to F_0 has to be $\mathrm{GL}(F) \times \mathrm{GL}(G)$-invariant. By the Cauchy formula (3.2.5) it has to be given by $(r+1) \times (r+1)$ minors of the matrix Φ. Since the variety Y is irreducible, its defining ideal has to be prime.

Part (b) follows from (5.1.3)(c).

To see that (c) is true, let us assume that $m \geq n$. We look at the last module in the resolution. It is clear from (6.1.3) that it is

$$F_{(m-r)(n-r)} = L_{(m)^{(n-r)}} F \otimes L_{(n)^{(n-r)},(n-r)^{(m-n)}} G^* \otimes_{\mathbf{K}} A(-m(n-r)).$$

The result follows, since the codimension of Y_r equals $(m-r)(n-r)$. This is also a way to see that A/I_{r+1} is Cohen–Macaulay without using rational singularities.

Finally, (d) follows from (5.1.2)(b) and from Corollary (4.1.9). ■

The remainder of this section is devoted to a more explicit description of the resolution of determinantal ideals in some important special cases. Historically these cases preceded Lascoux's paper. We preserve all our notation, assuming again that $m \geq n$.

(6.1.6) The Eagon–Northcott Complex. *Let us consider the case $r = n - 1$. The complex $F_\bullet$ is the resolution of the ideal I_n of maximal minors of the matrix Φ. We notice that all modules corresponding to elements of $Q(s)$ are zero for $s \geq 2$. Obviously the only contribution from the set $Q(0)$ is*

$$F_0 = L_{(0)} F \otimes L_{(0)} G^* \otimes_{\mathbf{K}} A = A.$$

*The elements from the set $Q(1)$ give the contribution to the terms F_i for $i > 0$.
In fact $F_i = L_{(n+i-1)}F \otimes L_{(n,1^{i-1})}G^* \otimes_K A(-n-i+1)$ for $i \geq 1$, and identifying these Schur functors, we get*

$$F_i = \bigwedge^{n+i-1} F \otimes \bigwedge^n G^* \otimes D_{i-1}G^* \otimes_K A(-n-i+1).$$

*We used the divided power because in such form the description of our complex
will be characteristic free.*

*Since our complex is the minimal resolution of I_n and the differentials
are $GL(m) \times GL(n)$-equivariant (section 5.4), we can use the representation
theory to identify completely the differentials in $F_\bullet$. The differential*

$$d_1 : \bigwedge^n F \otimes \bigwedge^n G^* \otimes_K A(-n) \longrightarrow A$$

has to be the composition

$$\bigwedge^n F \otimes \bigwedge^n G^* \otimes_K A(-n) \to S_n(F \otimes G^*) \otimes_K A(-n) \to A,$$

*where the left map is the embedding via $n \times n$ minors (cf. (3.2.5)) and the
right map is the multiplication in A. Similarly, for $i \geq 1$ the map*

$$d_{i+1} : \bigwedge^{n+i} F \otimes \bigwedge^n G^* \otimes D_i G^* \otimes_K A(-n-i)$$
$$\to \bigwedge^{n+i-1} F \otimes \bigwedge^n G^* \otimes D_{i-1}G^* \otimes_K A(-n-i+1)$$

is determined by its homogeneous component

$$\bigwedge^{n+i} F \otimes \bigwedge^n G^* \otimes D_i G^* \to \bigwedge^{n+i-1} F \otimes \bigwedge^n G^* \otimes D_{i-1}G^* \otimes A_1$$
$$= \bigwedge^{n+i-1} F \otimes \bigwedge^n G^* \otimes D_{i-1}G^* \otimes F \otimes G^*$$

*and thus has to be (up to a scalar we choose to be equal to 1) the tensor
product of diagonalizations*

$$\bigwedge^{n+i} F \to \bigwedge^{n+i-1} F \otimes F, \quad D_i G^* \to D_{i-1}G^* \otimes G^*$$

tensored with $\bigwedge^n G^*$.

*One can describe d_{i+1} by an explicit formula. If $\{f_1, \ldots, f_m\}$ is a basis of
F, $\{g_1, \ldots, g_n\}$ is a basis of G, and $i_1, \ldots, i_n$ are nonnegative integers such*

that $i_1 + \ldots + i_n = i$, then the image

$$d_{i+1}(f_{u_1} \wedge \ldots \wedge f_{u_{n+i}} \otimes g_1^{*(i_1)} \ldots g_n^{*(i_n)})$$

equals

$$\sum_{s,t}(-1)^{s+1}\phi_{u_s,t}\ f_{u_1} \wedge \ldots \wedge \hat{f}_{u_s} \wedge \ldots \wedge f_{u_{n+i}} \otimes g_1^{*(i_1)} \ldots g_t^{*(i_t-1)} \ldots g_n^{*(i_n)}$$

One can check easily that for $i \geq 1$ one has $d_i d_{i+1} = 0$.
Our result generalizes to arbitrary characteristic.

(6.1.7) Proposition. *The complex $F_\bullet$ defined above is a minimal free resolution of the A-module A/I_n over a field $\mathbf{K}$ of arbitrary characteristic, and thus when $\mathbf{K}$ is replaced by an arbitrary commutative ring.*

Proof. The proof is a repetition of proof of Lascoux's theorem in a characteristic free setting. We notice that in the case under consideration the bundle $\mathcal{Q}^*$ used in our Koszul complex is a line bundle. Therefore $\bigwedge^i(F \otimes \mathcal{Q}^*) = \bigwedge^i F \otimes S_i\mathcal{Q}^*$. The cohomology of bundles $S_i\mathcal{Q}^*$ has a characteristic free description by Serre's theorem ([H1, chapter 3, section 5]). This proves that the terms of the complex $F_\bullet$ are given by the formulas above. Then the reasoning we have just given in characteristic 0 case allows us to identify the differentials. Of course there exists an alternative proof of the general version based on Buchsbaum–Eisenbud acyclicity criterion (1.2.12) (see for example [BV, section 2C]). ∎

(6.1.8) The Gulliksen–Negard Complex ([GN]). *This is the case $m = n$ and $r = n - 2$. The nonzero terms of $F_\bullet$ are*

$$F_0 = L_{(0)}F \otimes L_{(0)}G^* \otimes_\mathbf{K} A(0) = A(0),$$
$$F_1 = L_{(n-1)}F \otimes L_{(n-1)}G^* \otimes_\mathbf{K} A(-n+1),$$
$$F_2 = (L_{(n)}F \otimes L_{(n-1,1)}G^* \otimes_\mathbf{K} A(-n)) \oplus (L_{(n-1,1)}F \otimes L_{(n)}G^* \otimes_\mathbf{K} A(-n)),$$
$$F_3 = L_{(n,1)}F \otimes L_{(n,1)}G^* \otimes_\mathbf{K} A(-n-1),$$
$$F_4 = L_{(n,n)}F \otimes L_{(n,n)}G^* \otimes_\mathbf{K} A(-2n).$$

We give an explicit description of the differentials. The idea is to construct the middle strand of $F_\bullet$ (consisting of F_3, F_2, and F_1) as a complex first. We treat the map

$$\Phi : F \otimes_\mathbf{K} A(-1) \to G \otimes_\mathbf{K} A$$

as a complex with nonzero terms appearing in homological degrees 1 *and* 0.
The complex

$$\Phi^*[-1] : G^* \otimes_K A \to F^* \otimes_K A(1)$$

has nonzero terms in degrees 0 *and* -1. *Let* $f_1, \ldots, f_m$ *and* $g_1, \ldots g_n$ *be the bases of* F *and* G *respectively. We have two equivariant maps of complexes*

$$\mathrm{Ev} : \Phi \otimes \Phi^* \to A,$$

$$\mathrm{Tr} : A \to \Phi \otimes \Phi^*$$

given by formulas

$$\mathrm{Ev}(f_i \otimes f_j^*) = \delta_{i,j}, \qquad \mathrm{Ev}(g_i \otimes g_j^*) = \delta_{i,j},$$

$$\mathrm{Ev}(f_i \otimes g_j^*) = 0, \qquad \mathrm{Ev}(g_i \otimes f_j^*) = 0,$$

and

$$\mathrm{Tr}(1) = \sum_{i=0}^{n} g_i \otimes g_i^* - \sum_{j=0}^{m} f_j \otimes f_j^*.$$

This implies that the composition $\mathrm{Ev}\,\mathrm{Tr} = 0$. *We consider the complex of complexes*

$$A[-1] \to \Phi \otimes \Phi^*[-1] \to A[-1],$$

and let $H_\bullet$ *be the complex which is the homology in the middle term. The nonzero terms of the complex* $H_\bullet$ *are*

$$F \otimes G^* \otimes_K A(-1) \to U \otimes_K A \to G \otimes F^* \otimes_K A(1),$$

where U *is the homology of the complex of vector spaces*

$$K \to (F \otimes F^*) \oplus (G \otimes G^*) \to K$$

with the maps coming from trace and evaluations according to the formulas given.

Now we can see that after tensoring $H_\bullet$ *by* $\bigwedge^n F \otimes \bigwedge^n G^*$, *shifting the grading by* $-n$, *and shifting the homological degree, we get the complex with the terms* F_3, F_2, F_1. *Reasoning as with the Eagon–Northcott complex, we see that* $H_\bullet \otimes \bigwedge^n F \otimes \bigwedge^n G^*$ *has to be a middle strand of* $F_\bullet$. *We* *augment our complex from both sides. We set $F_0' = F_0$, $F_4' = F_4$. We define the maps*

$$d_1 : F_1 \to F_0$$

by sending the generator $f_1 \wedge f_2 \wedge \ldots \wedge \hat{f_i} \wedge \ldots \wedge f_m \otimes g_1^* \wedge \ldots \wedge \hat{g_j^*} \wedge \ldots \wedge g_n^* \otimes 1$ to the determinant $M(i, j)$ of the matrix Φ with the i-th row and j-th column deleted. We also define the map

$$d_4 : F_4 \to F_3$$

by sending the generator to $\sum_{i,j=1}^n (-1)^{i+j} M(i, j) f_i \otimes g_j^* \otimes 1$. Here we identify $L_{(n,1)}F$ with F, and $L_{(n,1)}G^*$ with G^*. Reasoning as with the Eagon–Northcott complex, we see that $H_\bullet \otimes \bigwedge^n F \otimes \bigwedge^n G^*$ has to be a middle strand of $F_\bullet$.

(6.1.9) Remarks.

(a) *The fact that determinantal ideals are perfect was first proved by Eagon and Hochster in [HE]. The proof using the straightening law was given by DeConcini, Eisenbud, and Procesi in [DEP1].*

(b) *The idea of using higher direct images to calculate syzygies is due to Kempf. In his thesis ([Ke1]) he constructed the Eagon–Northcott complex in the way described above, using Serre's theorem. Lascoux in his groundbreaking paper [L2] constructed the syzygies in the general case (in characteristic 0) using this method. The precise description of the differentials in Lascoux's resolution was given by P. Roberts in his unpublished preprint [R1.].*

(c) *Several special cases of the resolutions of determinantal ideals were known before Lascoux's proof. In addition to the Eagon–Northcott and Gulliksen–Negard complexes mentioned above, Poon ([Pn]) treated the case $m = r + 3$, $n = r + 2$. In all these papers the approach was algebraic. Various criteria for the exactness of the complex and localization were used. The resolutions in these cases were proven to be characteristic free.*

6.2. The Resolutions of Determinantal Ideals in Positive Characteristic

Throughout this section we assume that **K** is a field of characteristic p. We retain the rest of the notation from the previous section. Let us recall that we constructed the free complex $F_\bullet$ of A-modules with the i-th term

$$F_i = \bigoplus_{j \geq 0} H^j \left(V, \bigwedge^{i+j} (F \otimes \mathcal{Q}^*) \right) \otimes_K A,$$

which is a minimal resolution of the normalization of the coordinate ring $K[Y_r]$. We will use the Schur complexes (section 2.3) to obtain some information about the terms of the complex $F_\bullet$ in characteristic p. Our information is far from being complete, because we do not have the analogue of Bott's theorem. Still, we can prove that $K[Y_r]$ is normal and that the determinantal ideal I_{r+1} is radical. This will establish properties (1)–(3) from the beginning of the previous section in characteristic p.

We will also show that even though the Hilbert function of $K[Y_r]$ does not depend on the characteristic of K, the resolution $F_\bullet$ does. More precisely, for $m = n = 5$ and $r = 1$ the complex $F_\bullet$ is different in characteristic 0 and 3. This is the example given by Hashimoto.

Consider the natural epimorphism $\pi : G^* \times V \to \mathcal{R}^*$ defined over $V = $ Grass(r, G). Applying (2.4.12)(c), we see that the complex $L_\lambda \pi$ gives a right resolution of $K_\lambda \mathcal{Q}^*$. By (2.4.12)(a), the terms of $L_\lambda \pi$ have a filtration whose associated graded object is $\bigoplus_\mu K_{\lambda/\mu} G * \otimes L_\mu \mathcal{R}^*$. By Corollary (4.1.12) we see that each term of this filtration is Γ-acyclic. This means that $L_\lambda \pi$ is a Γ-acyclic resolution of $K_\lambda \mathcal{Q}^*$, so the complex $\Gamma(L_\lambda \pi)$ can be used to calculate the cohomology of $K_\lambda \mathcal{Q}^*$. We want to identify $\Gamma(L_\lambda \pi)$ more precisely.

Let us consider the Schur complex $L_\lambda(\mathrm{id})$, where $\mathrm{id} : G^* \to G^*$ is the identity map. Applying (2.4.12)(a), we see that there exists a natural subcomplex $(L_\lambda(\mathrm{id}))(r + 1)$ whose terms have a filtration with the associated graded object

$$\bigoplus_{\mu, \mu_1 > r} K_{\lambda/\mu} G^* \otimes L_\mu G^*.$$

(6.2.1) Proposition. *The complex $\Gamma(L_\lambda \pi)$ is naturally isomorphic to $L_\lambda(\mathrm{id})/ L_\lambda(\mathrm{id})(r + 1)$.*

Proof. We have a commutative diagram

$$\begin{array}{ccc} G^* & \xrightarrow{\mathrm{id}} & G^* \\ \downarrow \mathrm{id} & & \downarrow \pi \\ G^* & \xrightarrow{\pi} & \mathcal{R}^* \end{array}$$

This induces a map of complexes $\bar{\theta} : L_\lambda(\mathrm{id}) \to L_\lambda(\pi)$. Applying the functor Γ to this map, we get a map of complexes $\Gamma(\bar{\theta}) : L_\lambda(\mathrm{id}) \to \Gamma(L_\lambda \pi)$. We want to show that $\Gamma(\bar{\theta})$ is an epimorphism with the kernel $L_\lambda(\mathrm{id})(r+1)$. First of all it is clear that the image of $L_\lambda(\mathrm{id})(r+1)$ is zero for dimension reasons. This means we get the induced map

$$\theta : L_\lambda(\mathrm{id})/L_\lambda(\mathrm{id})(r+1) \to \Gamma(L_\lambda\pi).$$

Let us define the subcomplexes $\mathcal{X}_{<\nu}$ and $\mathcal{X}_{\leq\nu}$ of $L_\lambda(\mathrm{id})/L_\lambda(\mathrm{id})(r+1)$ to be the subcomplexes with the associated graded objects

$$\bigoplus_{\mu,\mu<\nu} K_{\lambda/\mu}G^* \otimes L_\mu G^* \quad \text{and} \quad \bigoplus_{\mu,\mu\leq\nu} K_{\lambda/\mu}G^* \otimes L_\mu G^*$$

respectively (cf. exercise 21 of chapter 2). Let $\mathcal{Y}_{<\nu}$ and $\mathcal{Y}_{\leq\nu}$ be the analogous subcomplexes of $L_\lambda\pi$. Let $\mathcal{Z}_{<\nu} = \Gamma(\mathcal{Y}_{<\nu})$, $\mathcal{Z}_{\leq\nu} = \Gamma(\mathcal{Y}_{\leq\nu})$. Then the map θ induces a commutative diagram of complexes

$$
\begin{array}{ccccccc}
0 & \to & \mathcal{X}_{<\nu} & \to & \mathcal{X}_{\leq\nu} & \to & K_{\lambda/\nu}G^* \otimes L_\nu G^* \\
  &   & \downarrow   &   & \downarrow   &   & \downarrow \\
0 & \to & \mathcal{Z}_{<\nu} & \to & \mathcal{Z}_{\leq\nu} & \to & K_{\lambda/\nu}G^* \otimes L_\nu G^*
\end{array}
$$

The vertical map on the right hand side is induced by the map $L_\nu G^* \to \Gamma(L_\nu\mathcal{R}^*) = L_\nu G^*$, and the identifications are such that this map becomes an identity. This means that the right hand side map in the second row is onto. Thus we see by induction on ν that all vertical maps are isomorphisms. This proves (6.2.1). ∎

(6.2.2) Proposition. $H^i(V, K_\lambda \mathcal{Q}^*) = H_{|\lambda|-i-1}(L_\lambda(\mathrm{id})(r+1))$.

Proof. The statement follows immediately if we observe that $L_\lambda(\mathrm{id})$ is exact, so

$$H^i(K_\lambda \mathcal{Q}^*) = H^{i+1}(L_\lambda(\mathrm{id})(r+1)) = H_{|\lambda|-i-1}(L_\lambda(\mathrm{id})(r+1))$$

because of our convention for writing degrees. ∎

With these preparations out of the way, we are ready to analyze the low terms of the complex $F_\bullet$.

(6.2.3) Proposition.
  (a) We have $F_i = 0$ for $i < 0$, $F_0 = A$, $F_1 = \bigwedge^{r+1} F \otimes \bigwedge^{r+1} G^* \otimes_{\mathbf{K}} A$.
  (b) The coordinate ring $\mathbf{K}[Y_r]$ is normal, and the ideal I_{r+1} is prime. The resolution $q : Z \to Y_r$ is rational, so the ideal I_{r+1} is perfect.
  (c) The dimension of the graded components of A/I_{r+1} does not depend on the field $\mathbf{K}$.

Proof. First we prove (a). The first claim is equivalent to saying that $H^i(V, K_\lambda Q^*) = 0$ for $i > |\lambda|$. This is clear, since in (6.2.1) we constructed the Γ-acyclic resolution of $K_\lambda Q^*$ of length $|\lambda|$. The second claim is equivalent to the claim that $H^{|\lambda|}(V, K_\lambda Q^*) = 0$. This again is clear, since the resolution constructed in (6.2.1) either has length smaller than $|\lambda|$ or is equal to $L_\lambda(\mathrm{id})$ and therefore exact. This establishes the normality of $\mathbf{K}[Y_r]$, so $F_\bullet$ is the resolution of $\mathbf{K}[Y_r]$. The third claim is equivalent to saying that the ideal I_{r+1} is the defining ideal of Y_r. Indeed, F_1 is given by the generators of the defining ideal, and our claim says they occur in degree $r + 1$, and their number is equal to the number of minors of order $r + 1$ of the generic matrix.

On the other hand we see from (6.1.1)(c) that the coordinate ring $\mathbf{K}[Y_r]$ has the same Hilbert function as in characteristic 0, so the dimension of the i-th graded component of $\mathbf{K}[Y_r]$ equals to $\sum_{|\lambda|=i,\ \lambda_1 \le r} \dim L_\lambda F \otimes L_\lambda G$. On the other hand, using the straightening formula (3.2.5), we see that the dimension of the i-th graded component of A/I_{r+1} is at most $\sum_{|\lambda|=i,\ \lambda_1 \le r} \dim L_\lambda F \otimes L_\lambda G$. This proves the third part of (a).

Let us prove (b). The first two claims were already established in the course of proving (a). The third claim of (b) follows from (1.2.26) when we observe that by the argument in the proof of theorem (5.1.6)(a) (section 5.3), $\omega_Z = \mathcal{O}_Z \otimes p^*(\mathcal{O}_V(m - n))$. We can assume that $m \ge n$, so by Kempf's theorem and (5.1.3)(b) we have $\mathcal{R}^i q_*(\omega_Z) = 0$ for $i > 0$.

Part (c) of the proposition follows from (6.1.1)(c), as observed above. ∎

Next we describe an example of a resolution of a determinantal ideal not being characteristic free. First of all let us notice that from (6.2.3)(c) it follows that the determinantal variety is defined over $\mathbf{Z}$. Let us denote by B the polynomial ring $\mathbf{Z}[X_{i,j}]$ ($1 \le i \le m,\ 1 \le j \le n$). Let us denote by I_{r+1} the ideal in B generated by the minors of rank $r + 1$ of the generic matrix.

It follows from (6.2.3)(c) that B/I_{r+1} is a free $\mathbf{Z}$-module. Therefore, by (1.2.11), calculating the functors $\mathrm{Tor}_i^B(B/I_{r+1}, \mathbf{Z})$ from the Koszul complex on the variables $X_{i,j}$ over B, we see that the ranks of the modules in the resolution of the specialization A/I_{r+1} are calculated as the ranks of homology groups of a complex of free $\mathbf{Z}$-modules tensored with $\mathbf{K}$. It follows that these ranks are the same in all but finitely many characteristics (including characteristic 0) and they can increase in finitely many characteristics.

If the B-module B/I_{r+1} had a minimal B-free resolution $G_\bullet$ (in the sense that its differentials would all have positive degree), then the minimal resolution of the A-module A/I_{r+1} would be just $G_\bullet \otimes \mathbf{K}$. The ranks of $\mathrm{Tor}_i^A(A/I_{r+1}, \mathbf{K})$ would be equal to the ranks of the graded pieces of the modules in $G_\bullet$, as tensoring with $\mathbf{K}$ would make all the differentials zero. Thus if such a resolution $G_\bullet$ existed, the ranks of terms of the complex $F_\bullet$ would not depend on the characteristic of $\mathbf{K}$.

In fact we will show that in the case of 2×2 minors of the 5×5 matrix the resolutions of A/I_2 in characteristic 3 and in characteristic 0 are different.

The resolutions in characteristic 0 were described in the previous section. Let us assume that characteristic of $\mathbf{K}$ equals 3 and that $\dim F = \dim G = 5$. The result we will prove is

(6.2.4) Proposition. *Let us assume* char $\mathbf{K} = 3$ *and* $\dim F = \dim G = 5$. *Then cohomology groups* $H^1(V, \bigwedge^5(F \otimes Q^*))$ *and* $H^2(V, \bigwedge^5(F \otimes Q^*))$ *contain a composition factor* $\bigwedge^5 F \otimes \bigwedge^5 G^*$.

Proof. We first analyze the cohomology of $K_{2,2,1}Q^*$. By (6.2.1) it is enough to calculate the cohomology of $L_{2,2,1}(\mathrm{id})(2)$. This complex looks like this:

$$\ldots \to 0 \to K_{2,1}G^* \otimes \overset{2}{\bigwedge} G^* \overset{d}{\to} G^* \otimes G^* \otimes L_{2,1}G^* \to \ldots \qquad (**)$$

with the term $K_{2,1}G^* \otimes \bigwedge^2 G^*$ occurring in degree 2. Let us recall the exact sequence

$$0 \to K_{2,1}G^* \to \overset{2}{\bigwedge} G^* \otimes G^* \to \overset{3}{\bigwedge} G^* \to 0$$

(cf. exercise 2, chapter 2).

The diagonal map $\Delta : \bigwedge^3 G^* \to \bigwedge^2 G^* \otimes G^*$ induces a map $\partial : \bigwedge^3 G^* \to K_{2,1}G^*$ over a field of characteristic 3, because over $\mathbf{Z}$ we have $m\Delta = 3\,\mathrm{id}_{\bigwedge^3 G^*}$.

Let us consider the composition

$$\alpha : \overset{5}{\bigwedge} G^* \overset{\Delta}{\to} \overset{3}{\bigwedge} G^* \otimes \overset{2}{\bigwedge} G^* \overset{\partial \otimes 1}{\to} K_{2,1}G^* \otimes \overset{2}{\bigwedge} G^*.$$

We will show that the composition $d\alpha$ of α with the map d from $(**)$ is zero. We notice that the differential d is a composition

$$K_{2,1}G^* \otimes \overset{2}{\bigwedge} G^* \overset{d' \otimes 1}{\to} G^* \otimes G^* \otimes G^* \otimes \overset{2}{\bigwedge} G^* \overset{1 \otimes 1 \otimes v}{\to} G^* \otimes G^* \otimes L_{2,1}G^*,$$

where v is the canonical epimorphism $\bigwedge^2 G^* \otimes G^* \to L_{2,1}G^*$.

Let $g_1, g_2, \ldots, g_5$ denote the basis of G^*. Let x, y, z be three of the basis elements. We denote

$$d'\partial(x \otimes y \otimes z) = a(y \otimes z \otimes x) + b(z \otimes y \otimes x) + \ldots$$

for some $a, b \in \mathbf{K}$. Notice that a, b do not depend on the choice of x, y, z. To prove that the composition $d\alpha$ is zero it is enough to prove that sum of tensors in $d\alpha(g_1 \wedge g_2 \wedge \ldots \wedge g_5)$ having g_1, g_2, g_3 in a tableau from $L_{2,1}G^*$ is zero. To see this we notice that such tableaux can come only from the summands $g_u \wedge g_4 \wedge g_5 \otimes g_v \wedge g_w$ for $1 \leq u \leq 3$, $\{u, v, w\} = \{1, 2, 3\}$, with appropriate signs. Using the formula above for those summands, we see that our sum equals

$$(ag_4 \wedge g_5 + bg_5 \wedge g_4) \otimes [(g_1 \wedge g_2)(g_3) - (g_1 \wedge g_3)(g_2) + (g_2 \wedge g_3)(g_1)],$$

so it is zero, since $(g_1 \wedge g_2)(g_3) - (g_1 \wedge g_3)(g_2) + (g_2 \wedge g_3)(g_1)$ is 0 in $L_{2,1}G^*$.

This consideration shows that the representation $\bigwedge^5 G^*$ occurs in $H^1(V, K_{2,2,1}Q^*)$. Since the Euler characteristic of $K_{2,2,1}Q^*$ is independent of the characteristic of $\mathbf{K}$, we see that $\bigwedge^5 G^*$ has to occur in some other cohomology group of $K_{2,2,1}Q^*$, which is easily seen (by analyzing the last map in $L_{2,2,1}(\mathrm{id})(2)$) to be $H^2(V, K_{2,2,1}Q^*)$. This means that $L_{2,2,1}F \otimes \bigwedge^5 G^*$ occurs in $H^1(V, L_{2,2,1}F \otimes K_{2,2,1}Q^*)$ and in $H^2(V, L_{2,2,1}F \otimes K_{2,2,1}Q^*)$.

We need a lemma on the composition factors of the Schur functors in characteristic 3. ∎

(6.2.5) Lemma. *Let us assume that* char $\mathbf{K} = 3$. *The representation* $\bigwedge^5 F$ *occurs as a composition factor in* $L_\lambda F$ *only when* $\lambda = (2, 2, 1)$ *or* $\lambda = (5)$.

Proof. This follows from exercise 9 in chapter 2, or from the tables in [Jm]. ∎

Lemma (6.2.2) shows that $\bigwedge^5 F \otimes \bigwedge^5 G^*$ occurs as a composition factor in cohomology groups $H^1(V, L_{2,2,1}F \otimes K_{2,2,1}Q^*)$ and in $H^2(V, L_{2,2,1}F \otimes K_{2,2,1}Q^*)$. Now we use the straightening formula (3.2.7). The exterior power $\bigwedge^5(F \otimes Q^*)$ has a natural filtration whose associated graded object is $\sum_{|\lambda|=5} L_\lambda F \otimes K_\lambda Q^*$. This means that we can linearly order the partitions of 5 by setting $\lambda(1) = (5)$, $\lambda(2) = (4, 1)$, $\lambda(3) = (3, 2)$, $\lambda(4) = (3, 1, 1)$, $\lambda(5) = (2, 2.1)$, $\lambda(6) = (2, 1, 1, 1)$, and $\lambda(7) = (1^5)$. Thus there exist natural subbundles $\mathcal{Z}_i$ of $\bigwedge^5(F \otimes Q^*)$ (which include $\mathcal{Z}_0 = 0$ and $\mathcal{Z}_7 = \bigwedge^5(F \otimes Q^*)$) are such that the sequences

$$0 \to \mathcal{Z}_{i-1} \to \mathcal{Z}_i \to L_{\lambda(i)}F \otimes K_{\lambda(i)}Q^* \to 0$$

are exact.

The subbundles $\mathcal{Z}_i$ are homogeneous, so their cohomology groups have the structure of $GL(F) \times GL(G)$-modules. The long exact sequences of cohomology associated to sequences above are the equivariant sequences of $GL(F) \times GL(G)$-modules. Therefore, to prove (6.2.4) it is enough to show that the representation $\bigwedge^5 F \otimes \bigwedge^5 G^*$ cannot occur as a composition factor in the cohomology groups of $L_\lambda F \otimes K_\lambda \mathcal{Q}^*$ for $\lambda \neq (2, 2, 1)$. For all λ other than $\lambda = (5)$, this claim is clear by Lemma (6.2.5). In order to prove it for $\lambda = (5)$ we notice that the complex $L_{(5)}(\text{id})(2)$ equals

$$0 \to D_5 G^* \to D_4 G^* \otimes G^* \to 0.$$

This means that the only nonzero cohomology group equals $K_{(4,1)}G^*$. However, $K_{(4,1)}G^*$ has the same composition series as $L_{(2,1,1,1)}F$, and therefore by Lemma (6.2.5) it has no composition factor $\bigwedge^5 G^*$. This proves Proposition (6.2.4).•

Proposition (6.2.4) is the main ingredient in the proof of Hashimoto's theorem

(6.2.6) Theorem (Hashimoto, [Ha1]). *Let $m = n = 5, r = 1$. Then the terms of the minimal free resolution of the coordinate ring of the determinantal variety $\mathbf{K}[Y_r] = A / I_{r+1}$ over A are different in characteristic 0 and in characteristic 3.*

Proof. By (6.1.4) and (6.2.3) it is enough to show that the terms of $F_\bullet$ depend on the characteristic. It follows from (6.1.4) that in characteristic 0 we have $H^2(V, \bigwedge^5(F \otimes \mathcal{Q}^*)) = 0$. However, (6.2.4) states that in characteristic 3 we have $H^2(V, \bigwedge^5(F \otimes \mathcal{Q}^*)) \neq 0$. Therefore the complex $F_\bullet$ depends on the characteristic of $\mathbf{K}$. ∎

(6.2.7) Remarks. *Theorem (6.2.6) was proven by Hashimoto in [Ha1]. In fact, he proved that for every r the resolution of the ideal I_{r+1} for $m = n = r + 4$ is different in characteristic 3 and in characteristic 0. This result was re-proved by Roberts and Weyman in [RW]. The proof given above follows their approach. One should also note that it is known that if $m \leq r + 3$, then the resolution of A / I_{r+1} is independent of the characteristic of $\mathbf{K}$ (Eagon and Northcott ([EN1], [EN2]) for $m = r + 1$, and Akin, Buchsbaum, and Weyman ([ABW1], for $m = r + 2$). Subsequently Hashimoto constructed more examples of jumps in syzygies of determinantal varieties in positive characteristics (cf. [Ha4]).*

6.3. The Determinantal Ideals for Symmetric Matrices

In this section we apply the methods of chapter 5 to the ideals of minors of symmetric matrices. It turns out that the application of our methods is not so straightforward and one needs to mix them with other ways of calculating the minimal resolutions to get final results.

We work over a field $\mathbf{K}$ of characteristic 0.

Let us consider the space X^s of $n \times n$ symmetric matrices over $\mathbf{K}$. We denote by $\phi_{i,j}$ the (i, j)th coordinate function on X^s. The coordinate ring $A^s = \mathbf{K}[X]$ is isomorphic to the polynomial ring $\mathbf{K}[\phi_{i,j}]_{1 \le i \le j \le n}$, because $\phi_{i,j} = \phi_{j,i}$.

The general linear group $GL(n)$ acts on X^s by the simultaneous row and column operations. As in section 6.1, we introduce the equivariant setting in order to exploit this action. We identify X^s with the space $S_2 E^*$, where E is the vector space over $\mathbf{K}$ of dimension n. We can identify the coordinate ring A^s of X^s with the symmetric algebra $\mathrm{Sym}(S_2 E)$. Let $\{e_i\}_{1 \le i \le n}$ be the basis in E. Then the coordinate function ϕ_{ij} is identified with the element $e_i e_j$ from $\mathrm{Sym}(S_2 E)$. We denote by

$$\Phi : E \otimes_{\mathbf{K}} A^s \to E^* \otimes_{\mathbf{K}} A^s$$

the generic map $\Phi(e_i \otimes 1) = \sum_{j=1}^{n} \phi_{i,j}(e_j^* \otimes 1)$.

For each r satisfying $0 \le r \le n$ we consider the subvariety Y_r^s of X given by

$$Y_r^s = \{ \phi \in X \mid \mathrm{rank}\, \phi \le r \}$$

and called *the symmetric determinantal variety*. The variety Y_r^s can be identified with the set of symmetric matrices Φ whose minors of size $r + 1$ vanish. We denote the ideal generated by minors of Φ of size $r + 1$ in A^s by I_{r+1}^s. We call the ideal I_{r+1}^s the *symmetric determinantal ideal of* $(r + 1) \times (r + 1)$ *minors of the generic symmetric matrix*.

Let us denote by $Q_s(m)$ the set of partitions λ that in the hook notation (compare section 1A.2) can be written $\lambda = (a_1, \ldots, a_t | b_1, \ldots, b_t)$, where for each j we have $a_j = b_j + s$.

(6.3.1) Theorem.

(a) *The coordinate ring* $\mathbf{K}[Y_r^s]$ *is normal and Cohen–Macaulay. Its t-th homogeneous component decomposes to* $GL(E)$*-representations as follows:*

$$\mathbf{K}[Y_r^s]_t = \bigoplus_{\lambda, \ |\lambda|=2t, \ \lambda_i' \text{ even for all } i, \lambda_1 \le r} L_\lambda E.$$

(b) I^s_{r+1} is the defining ideal of Y^s_r.

(c) The i-th term G_i of the minimal free resolution of A^s/I^s_{r+1} as an A^s-module is given by the formula

$$G_i = \bigoplus_{\lambda\in Q_{r-1}(2t),\ \text{rank}\,\lambda\ \text{even},\ i=t-r\frac{1}{2}\,\text{rank}\,\lambda} L_\lambda E \otimes_K A^s.$$

Before we go further, let us explain statement (c) in more detail. Every partition λ from $Q_{r-1}(2t)$ with even rank can be written in the form

$$\lambda = \lambda(\alpha, u) = (\alpha_1 + 2u + r - 1, \ldots, \alpha_{2u} + 2u + r - 1, \alpha'_1, \ldots, \alpha'_v),$$

where $u = \frac{1}{2}\,\text{rank}\,\lambda$ and α is a partition satisfying $\alpha'_1 \le 2u$.

Example. *Let $r = 2, u = 2, \alpha = (2, 1, 1)$. The partition $\lambda(\alpha, u)$ is given pictorially by*

where the boxes corresponding to α are filled by $\bullet$, the boxes corresponding to α' are filled by $\circ$, and the boxes providing additional $r - 1$ elements for diagonal hook lengths are filled by X.

The representations occurring in the resolution of A^s/I^s_{r+1} are the functors $L_{\lambda(\alpha,u)}E$ for all choices of α and u. The term $L_{\lambda(\alpha,u)}E$ occurs in the i-th term in the resolution, where $i = \frac{1}{2}|\lambda| - r\frac{1}{2}\,\text{rank}\,\lambda$. However in our case $\frac{1}{2}|\lambda| = |\alpha| + 2u^2 + u(r - 1)$ and $r\frac{1}{2}\,\text{rank}\,\lambda = ru$. Therefore our term occurs in G_i with $i = |\alpha| + 2u^2 - u$.

Proof of Theorem (6.3.1). We will use two incidence varieties. The first one will allow us to establish the first two statements of the theorem. The second one will allow us to describe the minimal resolution.

1. The first incidence variety. Let us fix n and r. We take $Y = Y^s_r$ and $V = \text{Grass}(n - r, E)$ with the tautological sequence

$$0 \to \mathcal{R} \to E \times V \to \mathcal{Q} \to 0.$$

Here dim $\mathcal{R} = n - r$ and dim $\mathcal{Q} = q = r$. For a subspace R in E we denote by i the embedding of R into E. We consider the variety

$$Z_1 = \{(\phi, R) \in X \times V \mid \phi\, i = 0\}.$$

The variety Z_1 is the total space of the bundle $\mathcal{S}_1 = S_2 \mathcal{Q}^*$. The bundle $\xi^* = \mathcal{T}_1$ defined by the exact sequence

$$0 \to S_2 \mathcal{Q}^* \to S_2 E^* \times V \to \mathcal{T}_1 \to 0.$$

This means that we are in the setting of section 5.1 with $\mathcal{E} = S_2 E^*$.

We consider the basic diagram

$$
\begin{array}{ccc}
Z_1 & \subset & X \times V \\
\downarrow q' & & \downarrow q \\
Y & \subset & X
\end{array}
$$

(6.3.2) Proposition. *The map $q' : Z_1 \to Y_r^s$ is a birational isomorphism.*

Proof. Let us consider the open subset U in Y_r^s consisting of all symmetric maps ϕ of rank equal to r. Then it is clear from the construction that if $\phi \in U$ and $(\phi, R) \in Z_1$ then $R = \operatorname{Ker} \phi$. Therefore the map $\phi \mapsto (\phi, \operatorname{Ker} \phi)$ is the inverse of q' on the open subset U. ∎

We apply Proposition (5.1.1)(b). It follows that the direct image $p_*(\mathcal{O}_{Z_1})$ can be identified with the symmetric algebra $\operatorname{Sym}(S_2 \mathcal{Q})$. Therefore by Proposition (2.3.8)(a) and Corollary (4.1.9) it has no higher cohomology. Since the morphism p is affine, we see by [H1, chapter III, exercise 8.2] that $H^i(X \times V, \mathcal{O}_{Z_1}) = 0$ for $i > 0$. This means that $\mathcal{R}^i q'_*(\mathcal{O}_{Z_1}) = 0$ for $i > 0$, because Y is an affine variety. Using (5.1.3)(a), we see that $q'_*(\mathcal{O}_{Z_1})$ is the normalization of $\mathbf{K}[Y_r^s]$ and this normalization has rational singularities.

Finally let us show that $q'_*(\mathcal{O}_{Z_1}) = \mathbf{K}[Y_r^s]$. We just saw that $q'_*(\mathcal{O}_{Z_1})$ can be identified with the A^s-module $H^0(V, \operatorname{Sym}(S_2 \mathcal{Q}))$. Applying (4.1.9) and the formula (2.3.8)(a), we see that this module decomposes into the representations of $\operatorname{GL}(E)$ in the following way:

$$H^0(V, \operatorname{Sym}(S_2 \mathcal{Q})) = \bigoplus_{\lambda,\, \lambda_1 \le r,\ \lambda'_i \text{ even for all } i} L_\lambda E.$$

To prove that the ring $\mathbf{K}[Y_r^s]$ is normal, it is therefore enough to show that each representation on the right hand side of the above formula occurs in

$\mathbf{K}[Y_r^s]$. We use the explicit description of $\mathbf{U}$-invariants in $\mathrm{Sym}(S_2 E)$ exhibited in the proof of (2.3.8)(a). It is obvious that these functions do not vanish on Y_r^s and therefore the corresponding representation occurs in $\mathbf{K}[Y_r^s]$.

The description of $\mathbf{U}$-invariants in $\mathrm{Sym}(S_2 E)$ given in the proof of (2.3.8)(a) has another application. It is clear from what we just proved that the defining ideal of Y_r^s consists of all representations $L_\lambda E$ occurring in (2.3.8)(a) (for arbitrary t) for which $\lambda_1 > r$. However, from (2.3.8)(a) we see that the $\mathbf{U}$-invariant corresponding to such a representation is contained in I_{r+1}^s. Since I_{r+1}^s is $GL(E)$-invariant, the whole representation $L_\lambda E$ is contained in I_{r+1}^s. This shows that the defining ideal of Y_r^s is equal to I_{r+1}^s.

This completes the proof of first two parts of Theorem (6.3.1).

2. The second incidence variety. Let us fix n and r. We assume that r is even, say $r = 2u$. We take $V = \mathrm{Grass}(n - u, E)$ with the tautological sequence

$$0 \to \mathcal{R} \to E \times V \to \mathcal{Q} \to 0,$$

where $\dim \mathcal{R} = n - u$ and $\dim \mathcal{Q} = q = u$. For a subspace R in E we denote by i the embedding of R into E. We consider the variety

$$Z_2 = \{(\phi, R) \in X \times V \mid i^* \phi\, i = 0\}.$$

The variety Z_2 is the total space of the bundle $\mathcal{S}_2$ defined by the following sequence:

$$0 \to \mathcal{S}_2 \to S_2 E^* \times V \to S_2 \mathcal{R}^* \to 0.$$

This means that we are in the setting of the section 5.1 with $\mathcal{E} = S_2 E^* \times V$. It follows that $\mathcal{T}_2 = S_2 \mathcal{R}^*$.

Note that the variety Y, which is by definition $q(Z_2)$, is not equal to Y_u^s. We consider the basic diagram

$$
\begin{array}{ccc}
Z_2 & \subset & X \times V \\
\downarrow q' & & \downarrow q \\
Y & \subset & X
\end{array}
$$

Applying Proposition (5.1.1), we see that the resolution of $\mathcal{O}_Z$ as an $\mathcal{O}_{X \times V}$-module is given by the Koszul complex

$$\mathcal{K}_\bullet(S_2 \mathcal{R}) : 0 \to \overset{\frac{1}{2}(n-u)(n-u+1)}{\bigwedge} (p^*(S_2 \mathcal{R})) \to \ldots \to p^*(S_2 \mathcal{R}) \to \mathcal{O}_{X \times V}.$$

We calculate the terms of the complex $F_\bullet$. By the formula (2.3.9)(a) we see that

$$\bigwedge^t (S_2\mathcal{R}) = \bigoplus_{\lambda \in Q_1(2t)} K_\lambda \mathcal{R}.$$

Therefore we have to calculate the cohomology groups $H^i(V, K_\lambda\mathcal{R})$ for $\lambda \in Q_1$. Using Corollary (4.1.9), we see that we have to consider a sequence

$$(0, \lambda) + \rho = (n - 1, \ldots, n - u, \lambda_1 + n - u - 1, \ldots, \lambda_{n-u}).$$

The first u numbers in this sequence are consecutive, and the last $n - u$ numbers form a decreasing sequence. Let us assume that $(0, \lambda)$ is regular. Let s be the biggest number such that $\lambda_s + n - u - s > n - 1$. Then $\lambda_{s+1} + n - u - s - 1 < n - u$. In terms of λ this means that

$$\lambda_s \geq u + s, \qquad \lambda_{s+1} < s + 1.$$

The only nonzero cohomology group of $K_\lambda\mathcal{R}$ will be

$$H^{us}(V, K_\lambda\mathcal{R}) = K_\mu E,$$

where $\mu = (\lambda_1 - u, \ldots, \lambda_s - u, s^u, \lambda_{s+1}, \ldots, \lambda_{n-u})$

Now we use the fact that $\lambda \in Q_1$. This means that there exists a partition α (with $\alpha_1 \leq n - 2u - s + 1$) such that

$$\lambda = \lambda(\alpha, s) = (\alpha_1 + s + u, \ldots, \alpha_s + s + u, s^{u-1}, \alpha_1', \ldots, \alpha_{n-2u-s+1}').$$

In terms of α the partition μ equals

$$\mu = \mu(\alpha, s) = (\alpha_1 + s, \ldots, \alpha_s + s, s^{2u-1}, \alpha_1', \ldots, \alpha_{n-2u-s+1}').$$

The term corresponding to $\mu(\alpha, s)$ occurs in $F_\bullet$ in the place $|\alpha| + \frac{1}{2}s(s - 1)$. We have proved the following proposition.

(6.3.3) Proposition. *The terms of the complex $F_\bullet$ are given by the formula*

$$F_i = \bigoplus_{(s,\alpha),\ \alpha_1 \leq n-2u-s+1,\ i=|\alpha|+\frac{1}{2}s(s-1)} K_{\mu(\alpha,s)} E \otimes_{\mathbf{K}} A^s.$$

We proceed with the analysis of the complex $F_\bullet$. First of all, it is clear that $F_i = 0$ for $i < 0$. Therefore $F_\bullet$ is acyclic. Let us analyze the terms F_0 and F_1. Clearly $F_0 \otimes_A \mathbf{K} = K_0 E \oplus K_{(1^{2u})} E$ and $F_1 \otimes_A \mathbf{K} = K_{(2,1^{2u})} E \oplus$

$K_{(2^{2u+1})}E \oplus K_{(3^{2u+2})}E$. Using the Littlewood–Richardson rule and the plethysm formula (2.3.8)(a), we see that the only nonzero equivariant maps go from $K_{(2^{2u+1})}E$ to $K_0 E$ and from $K_{(2,1^{2u})}E \oplus K_{(3^{2u+2})}E$ to $K_{(1^{2u})}E$. We conclude that the complex $F_\bullet$ is a direct sum of the resolutions of two modules M_1 and M_2 that are presented as follows:

$$K_{(2^{2u+1})}E \otimes_{\mathbf{K}} A^s(-2u - 1) \to K_0 E \otimes_{\mathbf{K}} A^s \to M_1 \to 0,$$

$$(K_{(3^{2u+2})}E \otimes_{\mathbf{K}} A^s(-3u - 3)) \oplus (K_{(2,1^{2u})}E \otimes_{\mathbf{K}} A^s(-u - 1))$$
$$\to K_{(1^{2u})}E \otimes_{\mathbf{K}} A^s(-r) \to M_2 \to 0.$$

We observe that the map presenting M_1 in has to be nonzero. Indeed, if not, then the module M_1 would be a free module, but we know that $H_0(F_\bullet)$ has to be supported in Y_r^s. Therefore we conclude that the map presenting M_1 has to be given by the minors of order $2u + 1$ of the generic matrix Φ. This allows us to identify the module M_1 with $\mathbf{K}[Y_{2u}^s]$. We will use the complex $F_\bullet$ to establish the third part of Theorem (6.3.1) in the case of Y_{2u}^s.

In order to do that it remains to show which representations in (6.3.3) belong to the resolution of M_1 and which belong to the resolution of M_2.

(6.3.4) Proposition. *The representation $K_{\mu(\alpha,s)}E$ from (6.3.3) occurs in the resolution of M_1 if and only if s is even.*

Proof. Let us recall that by the graded analogue of (1.2.11) the terms in the minimal free resolution of A^s / I_{2u+1}^s, tensored with $A^s/(A^s)^+ = \mathbf{K}$ are isomorphic to the functors $\mathrm{Tor}_i^{A^s}(A^s / I_{2u+1}^s, \mathbf{K})$. By Theorem (5.1.2) and (1.2.11) this means that $F_i \otimes_{A^s} \mathbf{K}$ is isomorphic as a $\mathrm{GL}(E)$-module to $\mathrm{Tor}_i^{A^s}(A^s / I_{2u+1}^s, \mathbf{K})$. This isomorphism preserves homogeneous degrees.

On the other hand, $\mathrm{Tor}_i^{A^s}(A^s / I_{2u+1}^s, \mathbf{K})$ can be calculated using the Koszul complex on the homogeneous ideal $(A^s)^+$. Let us denote this complex by $K_\bullet(\{\phi_{(i,j)}\})$. Then

$$\mathrm{Tor}_i^{A^s}(A^s / I_{2u+1}^s, \mathbf{K}) = H_i(K_\bullet(\{\phi_{i,j}\}) \otimes_{A^s} A^s / I_{2u+1}^s).$$

Let us write down explicitly the homogeneous strand of the tensor product on the right hand side. It is the following complex of $\mathrm{GL}(E)$-modules:

$$K(i + t)_\bullet : \ldots \to K(i + t)_{i+1} \to K(i + t)_i \to K(i + t)_{i-1} \to \ldots,$$

where $K(i + t)_i = \bigwedge^i (S_2 E) \otimes \mathbf{K}[Y_{2u}^s]_t$.

Proposition (6.3.4) is an immediate consequence of the following lemma.

(6.3.5) Lemma. *Let* $\mu(\alpha, s)$ *be a partition described in (6.3.3). Let* $i = |\alpha| + \frac{1}{2}s(s - 1)$ *and* $t = su$. *Then the representation* $K_{\mu(\alpha,s)}E$ *occurs in* $H_i(K(i + t))$ *if and only if* s *is even.*

Proof. We will show the following facts

1. If s is even then $K_{\mu(\alpha,s)}E$ occurs with multiplicity 1 in $K(i + t)_i$ and it does not occur in neither $K(i + t)_{i-1}$ nor $K(i + t)_{i+1}$.
2. If s is odd then $K_{\mu(\alpha,s)}E$ does not occur in $K(i + t)_i$.

Let us observe that by the first part of Theorem (6.3.1)

$$\mathbf{K}[Y^s_{2u}]_t = \bigoplus_{\lambda \; \lambda'_i \text{ even for all } i, \lambda_1 \leq 2u} L_\lambda E$$

Therefore we can write

$$K(i + t)_i = \bigoplus_{\beta, \gamma} K_\beta E \otimes K_\gamma E$$

where we sum over all partitions β from $Q_1(2i)$ (compare the formula (2.3.9) a)) and all partitions γ such that $|\gamma| = 2t$, γ_i is even for all i, $\gamma'_1 \leq 2u$.
Since we assume that $\mathbf{K}$ has characteristic 0, $K_{\mu(\alpha,s)}E = L_{\mu(\alpha,s)'}E$
Let us first assume that s is even, say $s = 2p$.
First we treat a case when $\alpha'_1 = 2p$. Let us denote

$$\beta_0 = (2p - 1 + \alpha_1, \ldots, 2p - 1 + \alpha_{2p}, \alpha'_2, \ldots, \alpha'_{n-2u-2p+1})$$
$$\gamma_0 = (2p^{2v})$$

Then it follows from the Littlewood–Richardson rule that $K_{\mu(\alpha,s)}E$ occurs with multiplicity 1 in $K_{\beta_0}E \otimes K_{\gamma_0}E$ and therefore it occurs in $K(i + t)_i$. Let us assume that $K_{\mu(\alpha,s)}E$ occurs in $K_\beta E \otimes K_\gamma E$ occurring in $K(i + t)_{i-1}$ or $K(i + t)_{i+1}$. We observe that the length of the leg of any diagonal box in β cannot be bigger than the length of the leg of the corresponding square in $\mu(\alpha, s)$. Therefore $\beta \subset \beta_0$. This means that $\mu(\alpha, s)$ cannot occur in $K(i + t)_{i+1}$. Let us consider the case of $K(i + t)_{i-1}$. Now $\beta \subset \beta_0$ and β differs from β_0 by two boxes. At most one of those boxes lies below the $2p$-th row of β.

We also see that to obtain $\mu(\alpha, s)$ from β and γ by Littlewood–Richardson rule we have to add a column of $2u$ boxes to each of the first $2p$ columns of β. We conclude that $\gamma = (2p + 2, 2p^{2u-1})$. Since one of the columns of $\mu(\alpha, s)/\beta_0$ has length bigger than $2u$, the Littlewood–Richardson rule gives a contradiction.

If $\alpha_1' < 2p$ the argument is similar. We take

$$\beta_0 = (2p + \alpha_1, \ldots, 2p + \alpha_{2p-1}, \alpha_1', \ldots, \alpha_{n-2u-2p+1}')$$
$$\gamma_0 = (2p^{2u})$$

Again we see that $K_{\mu(\alpha,s)}E$ occurs in $K_{\beta_0}E \otimes K_{\gamma_0}E$ but it does not occur in neither $K(i+t)_{i-1}$ nor in $K(i+t)_{i+1}$. The reason for not occurring in $K(i+t)_{i+1}$ is the condition $\beta \subset \beta_0$, and for not occurring in $K(i+t)_{i-1}$ is the too long column in $\mu(\alpha, s)/\beta_0$.

Let us suppose that the rank s of $\mu(\alpha, s)$ is odd and that $K_{\mu(\alpha,s)}E$ appears in $K(i+t)_i$. Let us assume that $K_{\mu(\alpha,s)}E$ occurs in $K_\beta E \otimes K_\gamma E$ as above. We observe that for each diagonal square of $\mu(\alpha, s)$ the difference between the length of its leg and its arm is $2u - 1$ while for each diagonal square of β this difference is -1. This means that the only possibility for β is

$$\beta = (s + \alpha_1, \ldots, s + \alpha_s, \alpha_2', \ldots, \alpha_{n-2u-2p+1}')$$

We can see now by applying Littlewood–Richardson rule that the only way to get $K_{\mu(\alpha,s)}E$ in the tensor product $K_\beta E \otimes K_\gamma E$ is when $\gamma = (s^{2u})$ and when $\mu(\alpha, s)$ is obtained from β by adding u boxes in each of the first s columns. The resulting partition however has rank $s + 1$ which is even. This gives a contradiction and the Lemma (6.3.5) is proven. ■

This completes the proof of Theorem (6.3.1) for even r.

3. The proof of the third part of Theorem (6.3.1) for odd r. Let us assume that r is odd. We consider the affine space X'^s of all symmetric $n + 1$ by $n + 1$ matrices. The coordinate ring $A'^s = \mathbf{K}[X'^s]$ is a polynomial ring $\mathbf{K}[\phi_{i,j}']$, $1 \le i, j \le n + 1$. As before, we assume that $\phi_{i,j}' = \phi_{j,i}'$. We identify X'^s with the space $S_2 E'^*$, where E' is a vector space of dimension $n + 1$ over k. This allows us to identify A'^s with the symmetric algebra $\mathrm{Sym}(S_2 E')$.

Let us denote by Φ' the generic map

$$\Phi' : E' \otimes_{\mathbf{K}} A'^s \to E'^* \otimes_{\mathbf{K}} A'^s$$

defined by the formula

$$\Phi(e_j') = \sum_{1 \le i \le n+1} \phi_{i,j}' e_i'^*,$$

where $\{e_j'\}_{1 \le j \le n+1}$ is the basis of E'.

Let us denote by $I_{r+2}'^s$ the determinantal ideal of $(r + 2) \times (r + 2)$ minors of the matrix Φ'. By the previous considerations we know the terms of the minimal free resolution $G_\bullet'$ of $A'^s / I_{r+2}'^s$ over A'^s.

Let B be a ring A'^s localized at the powers of the element $T = \phi_{n+1,n+1}$. We denote by $H'_\bullet$ the complex $G'_\bullet \otimes_{A'^s} B$. Over the ring B we can change the basis of $E' \otimes_K B$ in such way that the map $\Phi' \otimes_{A'^s} B$ has the form

$$\begin{pmatrix} \Phi & 0 \\ 0 & 1 \end{pmatrix}.$$

We consider the ring homomorphism $f : B \to A^s$, where we set

$$f(\phi'_{i,j}) = \begin{cases} \phi_{i,j} & \text{for } 1 \le i, j \le n, \\ 0 & \text{for } j = n+1, 1 \le i \le n, \\ 1 & \text{for } i = j = n+1. \end{cases}$$

Observe that $f(I'^s_{r+2}) = I^s_{r+1}$ and that the length of the complex $H'_\bullet$ is equal to depth(I^s_{r+1}) by the first part of Theorem (6.3.1). Therefore, by the Eagon–Northcott theorem (1.2.14), we see that the complex $H_\bullet := H'_\bullet \otimes_B A^s$ is a nonminimal resolution of I^s_{r+1}.

We will use the complex $H_\bullet$ to get the information about the minimal resolution $G_\bullet$. First of all we observe that the definition of f allows us to identify E' with $E \oplus K$. Therefore the terms of $H_\bullet$ are given by

$$H_i = \bigoplus_{\lambda \in Q_r(2t),\ \mathrm{rank}\,\lambda\ \mathrm{even},\ i = t - (r-1)\frac{1}{2}\,\mathrm{rank}\,\lambda} L_\lambda E' \otimes_K A^s. \tag{$*$}$$

Using the direct sum decomposition (2.3.1), we know that

$$L_\lambda E' = \bigoplus_{\{\mu \mid \mu \in \lambda,\ \mu_i \ge \lambda_i - 1\ \text{for all } i\}} L_\mu E. \tag{$**$}$$

For each λ occurring in the decomposition $(*)$ we denote by $\lambda(0)$ the partition obtained from λ by removing a box from each of the first $2t$ rows. We notice that each $\lambda(0)$ has the form described in (6.3.1)(c) and that in this way we obtain all the representations of that form. Therefore the following lemma finishes the proof.

(6.3.6) Lemma. *Let us fix $L_\lambda E'$ occurring in $(*)$. Then $L_{\lambda(0)} E$ is the only representation among the terms of $(**)$ occurring in $\mathrm{Tor}^{A^s}_\bullet(A^s/I^s_{r+1}, K)$.*

Proof. We again use the complexes

$$K(i+t)_\bullet : \ldots \to K(i+t)_{i+1} \to K(i+t)_i \to K(i+t)_{i-1} \to \ldots,$$

where $K(i+t)_i = \bigwedge^i (S_2 E) \otimes K[Y^s_{2u}]_t$. We know that $\mathrm{Tor}^{A^s}_i(A^s/I^s_{r+1}, K)_{i+t} = H_i(K(i+t)_\bullet)$. First we will show that for a given $L_\lambda E'$ from $(*)$, the only μ from the corresponding decomposition $(**)$ that appears in $K(i + s)_i$ for some s is $\lambda(0)$. This follows from two simple observations. By the Littlewood–Richardson rule (2.3.4), for any partition ν such that $L_\nu E$ occurs in $K(i + s)_i$ for some s and for any diagonal box in ν, the difference in length of its arm and leg is at most $r - 1$. On the other hand, for every partition μ such that $L_\mu E$ occurs in $(**)$ and μ is not $\lambda(0)$, there exists a diagonal box for which the difference in length of its arm and leg equals r. Therefore the only representations that have the chance to occur in $\operatorname{Tor}_i^{A^s}(A^s/I_{r+1}^s, \mathbf{K})_{i+t}$ are $L_{\lambda(0)} E$. However, we know by Lemma (6.3.5) that all of the representations $L_{\lambda(0)} E$ occur in $\operatorname{Tor}_i^{A^s}(A^s/I_{r+1}^s, \mathbf{K})_{i+t}$. This completes the proof of Theorem (6.3.1). ∎

(6.3.7) Corollary.
(a) The coordinate ring A^s/I_{r+1}^s is a normal domain.
(b) The ideal I_{r+1}^s is perfect.
(c) The ideal I_{r+1}^s is Gorenstein if and only if $n - r$ is odd.

Proof. Only the last part of the corollary requires a proof. We see by the third part of (6.3.1) that the last module in the minimal resolution of A^s/I_{r+1} is $L_{(n^{n-r+1})}E$ for $n - r$ odd and $L_{(n^{n-r},n-r)}E$ for $n - r$ even. The dimension of this representation is 1 precisely when $n - r$ is odd. ∎

We conclude this section with several examples of resolutions of minors of symmetric matrices of small length.

(6.3.8) The Goto–Józefiak–Tachibana Complex. *Let us consider the case $r = n - 2$. This means we consider the resolution of A^s/I_{n-1}^s. Using Theorem (6.3.1)(c), we get the following description of the terms in the resolution:*

$$G_0 = A^s(0),$$
$$G_1 = L_{(n-1,n-1)}E \otimes_{\mathbf{K}} A^s(-n + 1),$$
$$G_2 = L_{(n,n-1,1)}E \otimes_{\mathbf{K}} A^s(-n),$$
$$G_3 = L_{(n,n,2)}E \otimes_{\mathbf{K}} A^s(-n - 1).$$

As in the case of the Eagon–Northcott complex (6.1.6), the equivariance of differentials determines each component uniquely up to a scalar.

To give the description of differentials we identify part of the complex containing modules G_1, G_2, G_3 in terms of Schur complexes. Let us consider the Schur complex $S_2(\Phi)$, where

$$\Phi : E \otimes_{\mathbf{K}} A^s(-1) \to E^* \otimes_{\mathbf{K}} A^s$$

is the complex given by the generic map. Its nonzero terms are

$$\bigwedge^2 E \otimes_{\mathbf{K}} A^s(-2) \to E \otimes E^* \otimes_{\mathbf{K}} A^s(-1) \to S_2 E^* \otimes_{\mathbf{K}} A^s.$$

The evaluation map $\alpha : E \otimes E^* \otimes A^s(-1) \to A^s(-1)$ gives rise to an epimorphism of complexes EV: $S_2(\Phi) \to A^s[-1]$. Define $B_2(\Phi) := \operatorname{Ker} \mathrm{EV}$. This is a complex of free A^s-modules. Now $B_2(\Phi) \otimes \bigwedge^n E^{\otimes 2}(-n+1)$ has terms isomorphic to G_3, G_2, G_1 and therefore has to be isomorphic to the linear strand of $G_{\bullet}$. We next define the differential

$$d_1 : G_1 \to G_0 = A^s.$$

Let $e_1, \ldots, e_n$ be a basis of E. We send a typical generator $e_i^* e_j^* \otimes 1$ to $M(i, j)$, the determinant of the matrix we get from Φ when we delete the i-th row and the j-th column. It is easy to check that in this way we obtain a complex with the same terms as $G_{\bullet}$. The equivariance implies this complex is isomorphic to $G_{\bullet}$ and therefore a resolution of A^s/I_{n-1}^s.

One can define a characteristic free version of this complex. For this the reader should consult the original paper of Józefiak ([J]). (Goto and Tachibana defined this complex in [GT] and proved it gives a resolution in all characteristics except 2.)

(6.3.9) The Complex of Length 6. *This is the case of $r = n - 3$. Our complex will be the minimal resolution of the A^s-module A^s/I_{n-2}^s. The calculation of the terms given by (6.3.1)(c) gives*

$G_0 = A^s(0)$,

$G_1 = L_{(n-2,n-2)} E \otimes_{\mathbf{K}} A^s(-n+2)$,

$G_2 = L_{(n-1,n-2,1)} E \otimes_{\mathbf{K}} A^s(-n+1)$,

$G_3 = (L_{(n-1,n-1,2)} E \otimes_{\mathbf{K}} A^s(-n)) \oplus (L_{(n,n-2,1,1)} E \otimes_{\mathbf{K}} A^s(-n))$,

$G_4 = L_{(n,n-1,2,1)} E \otimes_{\mathbf{K}} A^s(-n-1)$,

$G_5 = L_{(n,n,2,2)} E \otimes_{\mathbf{K}} A^s(-n-2)$,

$G_6 = L_{(n,n,n,n)} E \otimes_{\mathbf{K}} A^s(-2n)$.

We construct the middle strand with the linear differential in terms of Schur complexes. We start with the evaluation map EV: $S_2(\Phi) \to A^s[-1]$. Then we define the trace map to be its dual TR: $A^s[-1] \to \bigwedge^2(\Phi)$. Let us consider

the Schur complex $L_{(2,2)}(\Phi)$. *Trace and evaluation maps induce the following diagram:*

$$\bigwedge^2(\Phi) \xrightarrow{\alpha} L_{(2,2)}(\Phi) \xrightarrow{\beta} S_2(\Phi).$$

The map α is defined as the composition

$$\bigwedge^2(\Phi) \xrightarrow{1 \otimes TR} \bigwedge^2(\Phi) \otimes \bigwedge^2(\Phi) \longrightarrow L_{(2,2)}(\Phi),$$

where the second map is the canonical projection. The map β is the composition

$$L_{(2,2)}(\Phi) \longrightarrow S_2(\Phi) \otimes S_2(\Phi) \xrightarrow{EV \otimes 1} S_2(\Phi),$$

where the first map is the canonical injection.

It is easy to check that $\beta\alpha = 0$. We define

$$B_{(2,2)}(\Phi) = \operatorname{Ker} \beta / \operatorname{Im} \alpha.$$

The complex $B_{(2,2)}(\Phi) \otimes \bigwedge^n E^{\otimes 2}$ has the same terms as the middle linear strand of $G_\bullet$ (up to the shifts in degree). Since its differential is easily checked to be nonzero on every summand $L_\lambda E \otimes A^s$, the equivariance implies this complex is isomorphic to the middle strand of $G_\bullet$. The remaining maps are easy to define.

The generator corresponding to the tableau

$$T = \begin{array}{|c|c|c|c|c|}\hline i_1 & i_2 & \cdots & \cdots & i_{n-2} \\ \hline j_1 & j_2 & \cdots & \cdots & j_{n-2} \\ \hline \end{array}$$

with $T(1, w) = i_w$, $T(2, w) = j_w$ for $w = 1, \dots, n - 2$ is sent by d_1 to the minor $M(i_1, \dots, i_{n-2}; j_1, \dots, j_{n-2})$ of the submatrix of Φ consisiting of rows $i_1, \dots, i_{n-2}$ and columns $j_1, \dots, j_{n-2}$. The element $d_6(1)$ is defined as

$$\sum_{1 \le i < j \le n} \sum_{1 \le k < l \le n} (-1)^{i+j+k+l}$$
$$M(1, 2, \dots, \hat{i}, \dots, \hat{j}, \dots, n; 1, 2, \dots, \hat{k}, \dots, \hat{l}, \dots, n) T(i, j; k, l),$$

where $T(i, j; k, l)$ denotes the tableau of the shape $(2, 2)$ with $T(1, 1) = i$, $T(1, 2) = j$, $T(2, 1) = k$, $T(2, 2) = l$. In this formula we do not write the two rows of length n in $L_{(n,n,2,2)}E$ because they have to be filled by all numbers from 1 to n.

(6.3.10) Remarks. *The determinantal ideals for symmetric matrices were first shown to be perfect by Kutz [Ku]. The proof was based on the method* *of principal radical systems. The terms of the minimal resolution were first calculated in the paper of Lascoux [L2]. The special case of the resolution of length 3 was treated before by algebraic methods in [J] and [GT]. This section follows the paper [JPW].*

The resolutions of ideals of minors of symmetric matrices are not characteristic free. The relevant example is given in [Ha3]. In [Ha6] Hashimoto shows that for $n \geq 11$, $r = 2$ the third syzygy of A^s / I^s_{r+1} is different from the char $\mathbf{K} = 0$ case when char $\mathbf{K} = 3$. J. Andersen [An] showed in her University of Minnesota thesis that in the case $r = 2$, $n \geq 7$ the fifth syzygy of A^s / I^s_{r+1} is different from the char $\mathbf{K} = 0$ case when char $\mathbf{K} = 2$.

6.4. The Determinantal Ideals for Skew Symmetric Matrices

We work over a field $\mathbf{K}$ of characteristic 0.

Let us consider the space X^a of $n \times n$ skew symmetric matrices over $\mathbf{K}$. We denote by $\phi_{i,j}$ the (i, j)th coordinate function on X^s. The coordinate ring $A = \mathbf{K}[X]$ is isomorphic to the polynomial ring $\mathbf{K}[\phi_{i,j}]_{1 \leq i < j \leq n}$, because $\phi_{i,j} = -\phi_{j,i}$.

The general linear group $\mathrm{GL}(n)$ acts on X^a by the simultaneous row and column operations. As in the previous sections, we introduce the equivariant setting in order to exploit this action. We identify X^a with the space $\bigwedge^2 E^*$, where E is the vector space over $\mathbf{K}$ of dimension n. We can identify the coordinate ring A^a of X^a with the symmetric algebra $\mathrm{Sym}(\bigwedge^2 E)$. Let $\{e_i\}_{1 \leq i \leq n}$ be the basis in E. Then the coordinate function $\phi_{i,j}$ is identified with the element $e_i \wedge e_j$ from $\mathrm{Sym}(\bigwedge^2 E)$. We denote by

$$\Phi : E \otimes_{\mathbf{K}} A^a \to E^* \otimes_{\mathbf{K}} A^a$$

the generic map $\Phi(e_i \otimes 1) = \sum_{j=1}^n \phi_{ij}(e_j^* \otimes 1)$.

For each even $r = 2u$ satisfying $0 \leq r \leq n$ we consider the subvariety Y_r^a of X,

$$Y_r^a = \{ \phi \in X^a \mid \mathrm{rank}\,\phi \leq r \},$$

called *the skew symmetric determinantal variety*. We assume that r is even because it is well known from linear algebra that a skew symmetric matrix has even rank.

The variety Y_r^a can be identified with the set of skew symmetric matrices Φ whose Pfaffians of size $2u + 2$ vanish. We denote the ideal generated by Pfaffians of Φ of size $2u + 2$ in A^a by I_{2u+2}^a. We call the ideal I_{2u+2}^a *the skew symmetric determinantal ideal of $(2u + 2) \times (2u + 2)$ Pfaffians of the generic skew symmetric matrix.*

Let us denote by $Q_s(m)$ the set of partitions λ that in the hook notation (compare section 1.1.2) can be written $\lambda = (a_1, \ldots, a_t | b_1, \ldots, b_t)$, where for each j we have $a_j = b_j + s$.

We will use the geometric technique developed in chapter 5 to prove the following properties.

(6.4.1) Theorem.

(a) *The coordinate ring* $\mathbf{K}[Y_{2u}^a]$ *is normal and Cohen–Macaulay. Its t-th homogeneous component decomposes to* $GL(E)$-*representations as follows:*

$$\mathbf{K}[Y_{2u}^a]_t = \bigoplus_{\lambda, \; |\lambda|=2t, \; \lambda_i \text{ even for all } i, \; \lambda_1 \leq 2u} L_\lambda E.$$

(b) I_{2u+2}^a *is the defining ideal of* Y_{2u}^a.

(c) *The i-th term G_i of the minimal free resolution of A^a/I_{2u+2}^a as an A^a-module is given by the formula*

$$G_i = \bigoplus_{\lambda \in Q_{2u+1}(2t), \; i = t - u \text{ rank } \lambda} L_\lambda E \otimes_{\mathbf{K}} A^a.$$

Let us look at statement (c) of the theorem more closely. Every partition λ from $Q_{2u+1}(2t)$ can be written

$$\lambda = \lambda(\alpha, v) = (\alpha_1 + 2u + 1 + v, \ldots, \alpha_v + 2u + 1 + v, \alpha_1', \ldots, \alpha_w'),$$

where $v = \text{rank } \lambda$ and α is a partition satisfying $\alpha_1' \leq v$.

Example. *Let $u = 1$, $v = 2$, $\alpha = (3, 1)$. The partition $\lambda(\alpha, v)$ is given pictorially by*

where the boxes corresponding to α are filled by •, the boxes corresponding to α' are filled by ○, and the boxes providing an additional $2u + 1$ elements for diagonal hook lengths are filled by X.

The representations occurring in the resolution of A^a/I_{2u+2}^a are the functors $L_{\lambda(\alpha, v)} E$ for all choices of α and v. The term $L_{\lambda(\alpha, v)} E$ occurs in the i-th term in the resolution, where $i = \frac{1}{2}|\lambda| - (u-1)\,\text{rank}\,\lambda$. However, in our case $\frac{1}{2}|\lambda| = |\alpha| + v^2 + v(2u-1)$ and $(u-1)\,\text{rank}\,\lambda = (u-1)v$. Therefore our term occurs in G_i with $i = |\alpha| + \frac{1}{2}v(v+1)$.

Proof of Theorem (6.4.1). As in the previous section, we use two incidence varieties. One of them will allow us to prove the first two parts of the theorem, and the other will allow us to describe the minimal resolution.

1. The first incidence variety. Let us fix n and $r = 2u$. We take $Y = Y_{2u}^a$ and $V = \text{Grass}(n - 2u, E)$ with the tautological sequence

$$0 \to \mathcal{R} \to E \times V \to \mathcal{Q} \to 0.$$

Here $\dim \mathcal{R} = n - 2u$ and $\dim \mathcal{Q} = q = 2u$. For a subspace R in E we denote by i the embedding of R into E. We consider the variety

$$Z_1 = \{(\phi, R) \in X \times V \,|\, \phi i = 0\}.$$

The variety Z_1 is the total space of the bundle $\mathcal{S}_1 = \bigwedge^2 \mathcal{Q}^*$. The bundle $\mathcal{T}_1$ is defined by the exact sequence

$$0 \to \overset{2}{\bigwedge} \mathcal{Q}^* \to \overset{2}{\bigwedge} E^* \times V \to \mathcal{T}_1 \to 0.$$

This means that we are in the setting of section 5.1 with $\mathcal{E} = \bigwedge^2 E^* \times V$.

We consider the basic diagram

$$
\begin{array}{ccc}
Z_1 & \subset & X \times V \\
\downarrow q' & & \downarrow q \\
Y & \subset & X
\end{array}
$$

(6.4.2) Proposition. *The map $q' : Z_1 \to Y_{2u}^a$ is a birational isomorphism.*

Proof. Let us consider the open U subset in Y_{2u}^a consisting of all symmetric maps ϕ of rank equal to $2u$. Then it is clear from the construction that if $\phi \in U$ and $(\phi, R) \in Z_1$, then $R = \text{Ker}\,\phi$. Therefore the map $\phi \mapsto (\phi, \text{Ker}\,\phi)$ is the inverse of q' on the open subset U. ∎

We apply Proposition (5.1.1)(b). It follows that the direct image $p_*(\mathcal{O}_{Z_1})$ can be identified with the symmetric algebra $\text{Sym}(\bigwedge^2 \mathcal{Q})$. Therefore by the formula (2.3.8)(b) and by Corollary (4.1.9) it has no higher cohomology. Since the morphism p is affine, we see by [H1, chapter III, exercise 8.2] that $H^i(X \times V, \mathcal{O}_{Z_1}) = 0$ for $i > 0$. This means that $\mathcal{R}^i q'_*(\mathcal{O}_{Z_1}) = 0$ for $i > 0$, because Y is an affine variety. Using Theorem (5.1.3)(a), we see that $q'_*(\mathcal{O}_{Z_1})$ is the normalization of $K[Y_r^a]$ and this normalization has rational singularities.

Finally let us show that $q'_*(\mathcal{O}_{Z_1}) = K[Y_{2u}^a]$. We just saw that $q'_*(\mathcal{O}_{Z_1})$ can be identified with the A^a-module $H^0(V, \mathrm{Sym}(\bigwedge^2 \mathcal{Q}))$. Using (2.3.8)(b) and Corollary (4.1.9), we see that this module decomposes into the representations of $GL(E)$ in the following way:

$$H^0\left(V, \mathrm{Sym}(\overset{2}{\bigwedge} \mathcal{Q})\right) = \bigoplus_{\lambda, \lambda_1 \leq 2u, \ \lambda_i \text{ even for all } i} L_\lambda E.$$

To prove that the ring $K[Y_{2u}^a]$ is normal it is therefore enough to show that each representation on the right hand side of the above formula occurs in $K[Y_{2u}^a]$.

We use the explicit description of U-invariants in $\mathrm{Sym}(\bigwedge^2 E)$ given in the proof of (2.3.8)(b). It is obvious that these functions do not vanish on $Y_2^a u$ and therefore the corresponding representations occur in $K[Y_{2u}^a]$.

The description of U-invariants in $\mathrm{Sym}(\bigwedge^2 E)$ has another application. It is clear from what we just proved that the defining ideal of Y_{2u}^a consists of all representations $L_\lambda E$ occurring in (2.3.8)(b) (for arbitrary t) for which $\lambda_1 > 2u$. Also, by (2.3.8)(b) we see that the U-invariant corresponding to such representation is contained in I_{2u+2}^a. Since I_{2u+2}^a is $GL(E)$-equivariant, the whole representation $L_\lambda E$ is contained in I_{2u+2}^a. This shows that the defining ideal of Y_{2u}^a is equal to I_{2u+2}^a.

This completes the proof of first two parts of Theorem (6.4.1).

2. The second incidence variety. Let us fix n and $r = 2u$. We consider $V = \mathrm{Grass}(n - u, E)$ with the tautological sequence

$$0 \to \mathcal{R} \to E \times V \to \mathcal{Q} \to 0,$$

where $\dim \mathcal{R} = n - u$ and $\dim \mathcal{Q} = q = u$. For a subspace R in E we denote by i the embedding of R into E. We consider the variety

$$Z_2 = \{(\phi, R) \in X \times V \mid i^* \phi i = 0\}.$$

The variety Z_2 is the total space of the bundle $\mathcal{S}_2$ defined by the following sequence:

$$0 \to \mathcal{S}_2 \to \overset{2}{\bigwedge} E^* \times V \to \overset{2}{\bigwedge} \mathcal{R}^* \to 0.$$

This means that we are in the setting of the section 5.1 with $\mathcal{E} = \bigwedge^2 E^* \times V$. It follows that $\mathcal{T}_2 = \bigwedge^2 \mathcal{R}^*$.

Note that the variety Y, which is by definition $q(Z_2)$, is not a priori equal to Y_{2u}^a.

We consider the basic diagram

$$
\begin{array}{ccc}
Z_2 & \subset & X \times V \\
\downarrow q' & & \downarrow q \\
Y & \subset & X
\end{array}
$$

Applying Proposition (5.1.1), we see that the resolution of $\mathcal{O}_Z$ as an $\mathcal{O}_{X \times V}$-module is given by the Koszul complex

$$
\mathcal{K}\left(\bigwedge^2 \mathcal{R} \right)_{\bullet} : 0 \to
$$

$$
\bigwedge^{\frac{1}{2}(n-u)(n-u-1)} \left(p^* \left(\bigwedge^2 \mathcal{R} \right) \right) \to \cdots \to p^* \left(\bigwedge^2 \mathcal{R} \right) \to \mathcal{O}_{X \times V}.
$$

We calculate the terms of the complex $F_{\bullet}$. By the formula (2.3.9)(b) we see that

$$
\bigwedge^t \left(\bigwedge^2 \mathcal{R} \right) = \bigoplus_{\lambda \in Q_{-1}(2t)} K_\lambda \mathcal{R}.
$$

Therefore we have to calculate the cohomology groups $H^i(V, K_\lambda \mathcal{R})$ for $\lambda \in Q_{-1}$. Using Corollary (4.1.9), we see that we have to consider a sequence

$$
(0, \lambda) + \rho = (n - 1, \ldots, n - u, \lambda_1 + n - u - 1, \ldots, \lambda_{n-u}).
$$

The first u numbers in this sequence are consecutive, and the last $n - u$ numbers form a decreasing sequence. Let us assume that $(0, \lambda)$ is regular. Let s be the biggest number such that $\lambda_s + n - u - s > n - 1$. Then $\lambda_{s+1} + n - u - s - 1 < n - u$. In terms of λ this means that

$$
\lambda_s \geq u + s, \qquad \lambda_{s+1} < s + 1.
$$

The only nonzero cohomology group of $K_\lambda \mathcal{R}$ will be

$$
H^{us}(V, K_\lambda \mathcal{R}) = K_\mu E,
$$

where $\mu = (\lambda_1 - u, \ldots, \lambda_s - u, s^u, \lambda_{s+1}, \ldots, \lambda_{n-u})$.

Now we use the fact that $\lambda \in Q_{-1}$. This means that there exists a partition α (with $\alpha_1 \leq n - 2u - s + 1$) such that

$$
\lambda = \lambda(\alpha, s) = (\alpha_1 + s + u, \ldots, \alpha_s + s + u, s^{u+1}, \alpha'_1, \ldots, \alpha'_{n-2u-s+1}).
$$

In terms of α the partition μ equals

$$
\mu = \mu(\alpha, s) = (\alpha_1 + s, \ldots, \alpha_s + s, s^{2u+1}, \alpha'_1, \ldots, \alpha'_{n-2u-s+1}).
$$

The term corresponding to $\mu(\alpha, s)$ occurs in $F_{\bullet}$ in the place $|\alpha| + \frac{1}{2}s(s + 1)$. We have proved the following proposition.

(6.4.3) Proposition. *The terms of the complex $F_{\bullet}$ are given by the formula*

$$F_i = \bigoplus_{(s,\alpha),\ \alpha_1 \leq n-2u-s-1,\ i=|\alpha|+\frac{1}{2}s(s+1)} K_{\mu(\alpha,s)}E \otimes_{\mathbf{K}} A^a.$$

We proceed with the analysis of the complex $F_{\bullet}$. First of all, it is clear that $F_i = 0$ for $i < 0$. Therefore $F_{\bullet}$ is acyclic. Let us analyze the terms F_0 and F_1. Clearly $F_0 = K_0 E$ and $F_1 = K_{(1^{2u+2})}E$. This means that the complex $F_{\bullet}$ is the resolution of $\mathbf{K}[Y_{2u}^a]$, and the last part of Theorem (6.4.1) follows.●

We finish this section by giving two examples in low codimension.

(6.4.4) The Buchsbaum–Eisenbud complex ([BE3]). *This is the case $n = 2t + 1, r = 2t - 2$. The complex $G_{\bullet}$ gives the resolution of the module A^a / I_{2t}^a. The calculation of the terms of the complex $G_{\bullet}$ from Theorem (6.4.1) gives*

$$G_3 = L_{(2t+1,2t+1)}E \otimes A^a(-2t - 1),$$
$$G_2 = L_{(2t+1,1)}E \otimes A^a(-t - 1),$$
$$G_1 = L_{(2t)}E \otimes A^a(-t),$$
$$G_0 = A^a(0).$$

The middle strand $G_2 \longrightarrow G_1$ is just $\Phi \otimes \bigwedge^n E$. where

$$\Phi : E \otimes_{\mathbf{K}} A^a(-1) \to E^* \otimes_{\mathbf{K}} A^a$$

is the generic map.

We augment the complex $G_2' \longrightarrow G_1'$ by two maps $d_1 : G_1 \longrightarrow G_0$ and $d_3 : G_3 \longrightarrow G_2$ given by the formulas

$$d_1(e_1 \wedge \ldots \wedge \hat{e}_j \wedge \ldots \wedge e_{2t+1}) = \mathrm{Pf}(\hat{j})$$

$$d_3(1) = \sum_{j=1}^{2t+1} (-1)^j \, \mathrm{Pf}(\hat{j}) e_j$$

where $\mathrm{Pf}(\hat{j})$ is the Pfaffian of the $2t \times 2t$ skew symmetric matrix obtained by deleting in Φ the j-th row and the j-th column.

In the second formula we identify $L_{(2t+1,1)}E$ with E. One can check directly that these formulas define the $\mathrm{GL}(E)$-equivariant complex with the same terms as $G_{\bullet}$. Then the argument we used in (6.1.6) for the Eagon–Northcott complex applies, and we see that $G_{\bullet}'$ is isomorphic to $G_{\bullet}$ and thus it is a resolution of A^a / I_{2t}^a. One can use the Buchsbaum–Eisenbud acyclicity criterion (1.2.12) to show that the complex defined in this way gives a resolution of A^a/I_{2t}^a when **K** is replaced by an arbitrary commutative ring.

(6.4.5) Remarks. *Buchsbaum and Eisenbud proved in [BE3] that the complex $G_\bullet$ defined above is the universal resolution of Gorenstein ideals of codimension 3. This means that if S is a commutative ring and J is a Gorenstein ideal of codimension 3 in S, then there exists a homomorphism of rings $\psi : A^a \to S$ such that $J = \psi(I_{2t}^a)$, i.e., J is the ideal of $2t \times 2t$ Pfaffians of the $(2t+1) \times (2t+1)$ matrix $\psi(\Phi)$. The resolution of S/J as an S-module is given by $G_\bullet \otimes_{A^a} S$.*

(6.4.6) The Józefiak–Pragacz complex ([JP2]). *This is the complex of length 6 which arises when $n = 2t + 2$, $r = 2t - 2$. The complex $G_\bullet$ gives the resolution of A^a/I_{2t}^a. The nonzero terms are*

$G_0 = A^a(0),$

$G_1 = L_{(2t)}E \otimes_K A^a(-t),$

$G_2 = L_{(2t+1,1)}E \otimes_K A^a(-t-1),$

$G_3 = (L_{(2t+2,1,1)}E \otimes_K A^a(-t-2)) \oplus (L_{(2t+1,2t+1)}E \otimes_K A^a(-2t-1)),$

$G_4 = L_{(2t+2,2t+1,1)}E \otimes_K A^a(-2t-2),$

$G_5 = L_{(2t+2,2t+2,2)}E \otimes_K A^a(-2t-3),$

$G_6 = L_{(2t+2,2t+2,2t+2)}E \otimes_K A^a(-3t-3).$

As before, we identify the linear strands of $G_\bullet$ using Schur complexes and trace and evaluation maps.

The complex $G_\bullet$ has two linear strands (apart from the trivial ones occurring at both ends). The first one is the complex $H_\bullet^1$ with the terms $L_{(2t+2,1,1)}E \otimes_K A^a(-t-2)$, $L_{(2t+1,1)}E \otimes_K A^a(-t-1)$, $L_{(2t)}E \otimes_K A^a(-t)$. Since Φ is a skew symmetric matrix, we can identify Φ and $\Phi^*[-1]$. We will denote the identifying isomorphism by τ. We have a map EV of complexes

$$\bigwedge^2(\Phi) \longrightarrow \Phi \otimes \Phi \xrightarrow{1\otimes\tau} \Phi \otimes \Phi^*[-1] \xrightarrow{EV} A^a[-1].$$

Define

$$B_{(1,1)}(\Phi) := \operatorname{Ker} EV.$$

We define

$$H_\bullet^1 = B_{(1,1)}(\Phi) \otimes \bigwedge^{2t+2} E(-t)[-1].$$

This is the first linear strand of $G_\bullet$.

The second linear strand, $H_\bullet^2$, is the complex with the terms $L_{(2t+2,2t+2,2)}E \otimes_K A^a(-2t-3)$, $L_{(2t+2,2t+1,1)}E \otimes_K A^a(-2t-2)$, $L_{(2t+1,2t+1)}E \otimes_K A^a(-2t-1)$.

We have a map TR *of complexes*

$$A^a[-1] \xrightarrow{\mathrm{TR}} \Phi \otimes \Phi^*[-1] \xrightarrow{1 \otimes \tau} \Phi \otimes \Phi \longrightarrow S_2(\Phi).$$

Define

$$A_2(\Phi) = \mathrm{Coker\,TR}.$$

Finally,

$$H_\bullet^2 = A_2(\Phi) \otimes \bigwedge^{2t+2} E^{\otimes 2}(-2t-1)[-3].$$

Of course we can define the 0*th linear strand to be* $H_\bullet^0 A^a(0)$, *and the strand* $H_\bullet^3 = L_{(2t+2,2t+2,2t+2)}E \otimes A^a(-2t-3)$. *Notice that up to grading and homological shifts,* $H_\bullet^i$ *is dual to* $H_\bullet^{3-i}$.

It remains to define the maps d_i from $H_\bullet^i$ to $H_\bullet^{i-1}$ The map d_1 is nonzero on the term $L_{(2t)}E \otimes A^a(-t)$, and it sends the generator $e_1 \wedge \ldots \wedge \hat{e}_i \wedge \ldots \wedge \hat{e}_j \wedge \ldots \wedge e_n$ to $\mathrm{Pf}(\hat{i}, \hat{j})$, where the last symbol denotes the Pfaffian of the $2t \times 2t$ skew symmetric matrix we get from Φ by deleting the i-th and j-th rows and the i-th and j-th columns. The map d_3 can be defined as the dual of d_1 (up to shifts in grading and homological degree).

The most difficult part is the definition of d_2. We just sketch its construction. We start with the map of complexes:

$$S_2(\Phi)[-1] \xrightarrow{\Delta \otimes \mathrm{TR}} \Phi \otimes \Phi \otimes S_2(\Phi) \xrightarrow{1 \otimes 1 \otimes \Delta} \Phi \otimes \Phi \otimes \Phi \otimes \Phi$$
$$\xrightarrow{m_{1,3} \otimes m_{2,4}} \bigwedge^2(\Phi) \otimes \bigwedge^2(\Phi).$$

Here $m_{1,3} \otimes m_{2,4}$ denotes the map which multiplies the first factor by the third one, and the second factor by the fourth one. We can check easily that this map induces the map of complexes

$$A_2(\Phi)[-1] \longrightarrow B_{(1,1)}(\Phi) \otimes B_{(1,1)}(\Phi)$$

However, up to grading and homological shifts, $B_{(1,1)}(\Phi)$ is the same as $H_\bullet^1$. Therefore we have a map $d_1' : B_{(1,1)}(\Phi) \longrightarrow A^a$. We define d_2 as a composition

$$A_2(\Phi)[-1] \longrightarrow B_{(1,1)}(\Phi) \otimes B_{(1,1)}(\Phi) \xrightarrow{1 \otimes d_1'} B_{(1,1)}(\Phi).$$

Now it is easy to check that the maps d_i' define a GL(E)-equivariant double complex

$$H_\bullet^3 \longrightarrow H_\bullet^2 \longrightarrow H_\bullet^1 \longrightarrow H_\bullet^0.$$

In order to check that this complex is isomorphic to $G_\bullet$ it is enough to show that the components of the differentials are nonzero when restricted to every summand $L_\lambda E \otimes A^a$. Indeed, inducting on the homological and the homogeneous degree, we see by exactness of $G_\bullet$ that in given homological and homogeneous degrees the module of cycles in $G_\bullet$ has only one irreducible representation of the required kind, so it has to he the one covered by the differential we just constructed.

Checking that the differentials defined above have the required properties can be carried out using the $\mathbf{U}$-invariant elements in each representation.

(6.4.7) Remarks. *The minimal resolutions of ideals of Pfaffians are not, in general, characteristic free. In fact, Kurano showed in [K3], [K4] that the relations between 4×4 Pfaffians of an 8×8 skew symmetric matrix are not spanned by linear relations when char $\mathbf{K} = 2$. The more general family of similar examples is provided in [Ha5]. In the case of complex (6.4.6) Pragacz showed in [P_i] that the ranks of syzygies do not depend on characteristic of K.*

6.5. Modules Supported in Determinantal Varieties

We preserve the notation from section 6.1. We will study the GL(F) $\times$ GL(G)-equivariant modules supported in determinantal varieties. We are interested in the structure of such modules. An interesting family of modules supported in determinantal varieties are the direct images of equivariant sheaves on the desingularization we studied in section 6.1. Our constructions allow us to calculate the minimal resolutions of such modules, given by the appropriate complexes $F(\mathcal{V})_\bullet$.

It turns out that in addition to the desingularization used in section 6.1, we have two other resolutions of singularities of the determinantal variety Y_r, leading to three such families. The choice of desingularization does not make a difference when investigating the coordinate rings $A_r := \mathbf{K}[Y_r]$, but different desingularizations lead to different families of equivariant modules.

It turns out that each of these families generates the Grothendieck group of graded GL(F) $\times$ GL(G)-equivariant modules supported in Y_r. This means that, at least in principle, the resolution of an equivariant GL(F) $\times$ GL(G)-module supported in Y_r can be obtained from complexes $F(\mathcal{V})_\bullet$.

We show that some of the terms of the complexes $F(\mathcal{V})_\bullet$ depend on the characteristic of the field $\mathbf{K}$.

When studying determinantal varieties Y_r in section 6.1, we used a desingularization

$$Z_r^{(2)} = \{(\phi, \bar{R}) \in X \times \text{Grass}(r, G) \mid \text{Im } \phi \subset R\}.$$

In fact we could have made another choice:

$$Z_r^{(1)} = \{(\phi, R) \in X \times \text{Grass}(m - r, F) \mid \phi \mid_{\bar{R}} = 0\}.$$

The choice of desingularization was irrelevant when studying the coordinate rings of determinantal varieties, but it makes a difference when looking at twisted complexes $F(\mathcal{V})_\bullet$. In order to make the situation symmetric it is also necessary to study the fibered product

$$Z_r = Z_r^{(1)} \times_{X_r} Z_r^{(2)} = \{(\phi, R, \bar{R}) \in X \times \text{Grass}(r, G)$$
$$\times \text{Grass}(m - r, F) \mid \text{Im } \phi \subset R, \ \phi \mid_{\bar{R}} = 0\}.$$

Throughout this section we denote by $0 \to \mathcal{R} \to F \times \text{Grass}(m - r, F) \to Q \to 0$ the tautological sequence on $\text{Grass}(m - r, F)$, and by $0 \to \bar{\mathcal{R}} \to G \times \text{Grass}(r, G) \to \bar{Q} \to 0$ the tautological sequence on $\text{Grass}(r, G)$.

Let $\mathcal{C}_r(F, G)$ be the category of graded A_r-modules with rational $\text{GL}(F) \times \text{GL}(G)$-action compatible with the module structure, and equivariant degree 0 maps. We denote by $K_0'(A_r)$ the Grothendieck group of the category $\mathcal{C}_r(F, G)$.

For an equivariant graded module $M \in \text{Ob}(\mathcal{C}_r(F, G))$ and for $q \in \mathbf{Z}$, we denote by $M(q)$ the module M with gradation shifted by q, i.e. $M(q)_n = M_{q+n}$. For $M \in \text{Ob}(\mathcal{C}_r(F, G))$ we define the *graded character of M*,

$$\text{char}(M) = \sum_{n \in \mathbf{Z}} \text{char}(M_n) q^n \in \text{Rep}(\text{GL}(F) \times \text{GL}(G^*))[[q]][q^{-1}].$$

where $\text{Rep}(\text{GL}(F) \times \text{GL}(G^*))$ denotes the representation ring of $\text{GL}(F) \times \text{GL}(G^*)$ and char is the product of character maps described in (2.2.10).

We recall from section 2.2, that an integral weight for $\text{GL}(m)$ is just an m-tuple $\alpha = (\alpha_1, \ldots, \alpha_m)$ of integers. The weight α is *dominant* if $\alpha_1 \geq \ldots \geq \alpha_m$.

Let $\alpha = (\alpha_1, \ldots, \alpha_m)$ be an integral weight for $\text{GL}(F)$. We set $\alpha^{(1)} = (\alpha_1, \ldots, \alpha_r)$, $\alpha^{(2)} = (\alpha_{r+1}, \ldots, \alpha_m)$. Let $\beta = (\beta_1, \ldots, \beta_n)$ be an integral weight for $\text{GL}(G^*)$. We define $\beta^{(1)} = (\beta_1, \ldots, \beta_r)$, $\beta^{(2)} = (\beta_{r+1}, \ldots, \beta_n)$.

Let $\alpha = (\alpha^{(1)}, \alpha^{(2)})$. Assume that both $\alpha^{(1)}$ and $\alpha^{(2)}$ are dominant. Let β be a dominant weight for $\text{GL}(G^*)$. For each such pair (α, β) we define a sheaf

$$\mathcal{M}(\alpha, \beta) = p^{(1)*}(L_{\alpha^{(1)'}} Q \otimes L_{\alpha^{(2)'}} \mathcal{R}) \otimes L_{\beta'} G^* \otimes \mathcal{O}_{Z^{(1)}}$$

of graded modules on $Z^{(1)}$. Symmetrically, let α be a dominant weight for $GL(F)$, and let β be a weight for $GL(G^*)$ such that $\beta^{(1)}$ and $\beta^{(2)}$ are both dominant. We define a sheaf

$$\mathcal{N}(\alpha, \beta) = L_{\alpha'} F \otimes p^{(2)*}(L_{\beta^{(1)\nu}} \bar{\mathcal{Q}} \otimes L_{\beta^{(2)\nu}} \bar{\mathcal{R}}) \otimes \mathcal{O}_{Z^{(2)}}$$

of graded modules on $Z^{(2)}$.

Finally, let α, β be weights such that $\alpha^{(1)}$, $\alpha^{(2)}$, $\beta^{(1)}$, $\beta^{(2)}$ are dominant. We define a sheaf

$$\mathcal{P}(\alpha, \beta) = p^{(1)*}(L_{\alpha^{(1)\nu}} \mathcal{Q} \otimes L_{\alpha^{(2)\nu}} \mathcal{R}) \otimes p^{(2)*}(L_{\beta^{(1)\nu}} \bar{\mathcal{Q}} \otimes L_{\beta^{(2)\nu}} \bar{\mathcal{R}}) \otimes \mathcal{O}_Z$$

of graded modules on Z. We define the equivariant graded modules

$$M(\alpha, \beta) = H^0(Z^{(1)}, \mathcal{M}(\alpha, \beta)),$$

$$N(\alpha, \beta) = H^0(Z^{(2)}, \mathcal{N}(\alpha, \beta)),$$

$$P(\alpha, \beta) = H^0(Z, \mathcal{P}(\alpha, \beta)).$$

Let us start with providing some examples of modules from three families. We look first at the family $M(\alpha, \beta)$. Since in this case $M(\alpha, \beta) = M(\alpha, (0)) \otimes_K L_{\beta'} G^*$, we can assume that $\beta = (0)$.

(6.5.1) Example. *Let $\alpha^{(2)} = (0)$. The sheaf $\mathcal{M}(\alpha, \beta)$ equals $L_{\alpha^{(1)\nu}} \mathcal{Q} \otimes \mathcal{O}_{Z^{(1)}}$. Therefore the direct image $p_*^{(1)} \mathcal{M}(\alpha, \beta)$ equals $L_{\alpha^{(1)\nu}} \mathcal{Q} \otimes \mathrm{Sym}(\mathcal{Q} \otimes G^*)$. Using the straightening formula (2.3.2) and the characteristic free version [Bo2] of the Littlewood–Richardson rule, we see that the higher cohomology of $\mathcal{M}(\alpha, \beta)$ vanishes. The j-th homogeneous component $M_j(\alpha, (0))$ has a filtration with the associated graded object*

$$\bigoplus_{|\mu|=j} (L_{\alpha^{(1)\nu}} F \otimes L_{\mu'} F)_{\leq r} \otimes L_{\mu'} G^*,$$

where $(L_{\alpha^{(1)\nu}} F \otimes L_{\mu'} F)_{\leq r}$ denotes a factor of $(L_{\alpha^{(1)\nu}} F \otimes L_{\mu'} F)$ consisting of all $L_\nu F$ in the tensor product with $\nu_1' \leq r$. In characteristic zero we can write the direct sum instead of filtration.

(6.5.2) Example. *The simplest examples of modules $M(\alpha, (0))$ are provided by the ones corresponding to line bundles on $\mathrm{Grass}(m - r, F)$. The line bundles on $\mathrm{Grass}(m - r, F)$ correspond to tensor powers $\mathcal{O}_{\mathrm{Grass}(m-r,F)}(t) = (\bigwedge^r \mathcal{Q})^{\otimes t}$ for $t \in \mathbf{Z}$. For $t \geq 0$ this is a special case of the example (6.5.1). Let us describe the minimal free resolution of the module $M((n^r), (0))$. Assume that the characteristic of $\mathbf{K}$ is zero. The terms come from the cohomology of the vector bundles $(\bigwedge^r \mathcal{Q})^{\otimes t} \otimes \bigwedge^\bullet (\mathcal{R} \otimes G^*)$. Decomposing by Cauchy's*

formula we see that one needs to apply Bott's theorem (4.1.9) to the weights
$(t^r, \lambda_1, \ldots, \lambda_{m-r})$. *The description is similar to the description of syzygies of
determinantal varieties. The surviving terms decompose to families depend-
ing on a parameter s saying how many parts* $\lambda_1, \ldots, \lambda_s$ *have to move in front
of r parts t. Notice that if* $\lambda_1 \le t$, *there are no exchanges and we have* $s = 0$.
The condition for s terms being exchanged is $\lambda_s \ge s + t + r$, $\lambda_{t+1} \le s + t$.
In such case we write $\lambda := \lambda(t, s, \alpha, \beta) = (\alpha_1 s + t + r, \ldots, \alpha_t + s + t + r, \beta_1, \ldots, \beta_{m-r-t})$. *The corresponding term is* $K_\mu F \otimes L_\lambda G^*$ *where* $\mu :=$
$\mu(t, s, \alpha, \beta) = ((\alpha_1 + s + t, \ldots, \alpha_t + s + t, (s + t)^r \beta_1, \ldots, \beta_{m-r-t})$. *The
term corresponding to the set of data* t, s, α, β *occurs in homological de-
gree* $|\alpha| + |\beta| + s^2 + st$.

In particular, take $r = 3, t = 1, s = 2$. *Take* $\alpha = (3, 1)$, $\beta = (3, 2)$. *The
partitions* λ, μ *are*

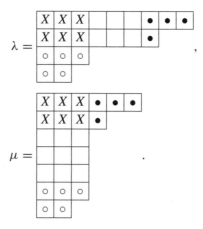

Here the boxes corresponding to α *are filled by* •, *the boxes corresponding
to* β *are filled by* ○, *and the boxes corresponding to the* $(s + t) \times s$ *rectangle
both partitions have to contain are filled by* X.

(6.5.3) Theorem. *The group* $K_0'(A_r)$ *is generated by the classes of the mod-
ules of each of the families* $M(\alpha, \beta)(q)$, $N(\alpha, \beta)(q)$, $P(\alpha, \beta)(q)$ *where* α *and
* β *are both dominant weights and* $q \in \mathbf{Z}$.

(6.5.4) Theorem.
  (a) *The group* $K_0'(A_r)$ *is isomorphic to the additive subgroup of the ring*
     $\mathrm{Rep}(GL(F) \times GL(G^*))[[q]][q^{-1}]$ *generated by shifted graded charac-
     ters of modules* $M(\alpha, \beta)$ *(or* $N(\alpha, \beta)$, *or* $P(\alpha, \beta)(q))$ *for* α, β *dominant,*
  (b) *The group* $K_0'(A_r)$ *is isomorphic to the additive group of the ring*
     $\mathrm{Rep}(GL(F) \times GL(G^*))[q][q^{-1}]$.

We provide the proofs only for the family $M(\alpha, \beta)$. The proofs for the family $N(\alpha, \beta)$ are symmetric. The proofs for the family $P(\alpha, \beta)$ and the transition formulas are given in [W6].

We start with basic observations about the cohomology groups of the sheaves $\mathcal{M}(\alpha, \beta)$.

Since $p^{(1)}$ is an affine map, $R^i p_*^{(1)} \mathcal{O}_{Z^{(1)}} = 0$ for $i > 0$. We also have $p_*^{(1)} \mathcal{O}_{Z^{(1)}} = \mathrm{Sym}(\mathcal{Q} \otimes G^*)$. Using the Leray spectral sequence and a projection formula (assuming α arbitrary, β dominant), we have

$$H^i(Z^{(1)}, \mathcal{M}(\alpha, \beta)) = H^i(\mathrm{Grass}(m - r, F), K_{\alpha^{(1)\prime}}\mathcal{Q}$$
$$\otimes K_{\alpha^{(2)\prime}}\mathcal{R} \otimes \mathrm{Sym}(\mathcal{Q} \otimes G^*)).$$

(6.5.5) Proposition. *Let α, β be dominant weights for $\mathrm{GL}(F)$, $\mathrm{GL}(G^*)$ respectively. Then $H^i(Z^{(1)}, \mathcal{M}(\alpha, \beta)) = 0$ for $i > 0$.*

Proof. By definition

$$\mathcal{M}(\alpha, \beta) = p^{(1)*}(L_{\alpha^{(1)\prime}}\mathcal{Q} \otimes L_{\alpha^{(2)\prime}}\mathcal{R}) \otimes L_{\beta\prime}G^* \otimes \mathcal{O}_{Z^{(1)}}.$$

Using this, we see that

$$p_*^{(1)}\mathcal{M}(\alpha, \beta) = L_{\alpha^{(1)\prime}}\mathcal{Q} \otimes L_{\alpha^{(2)\prime}}\mathcal{R} \otimes L_{\beta\prime}G^* \otimes \mathrm{Sym}(\mathcal{Q} \otimes G^*).$$

Using the straightening law (3.2.5) and the fact that the tensor product of two Schur functors has a filtration whose associated graded object is a direct sum of Schur functors ([Bo2]), we are reduced to proving that if α is dominant then

$$H^i(\mathrm{Grass}(m - r, F), L_{\alpha^{(1)\prime}}\mathcal{Q} \otimes L_{\alpha^{(2)\prime}}\mathcal{R}) = 0$$

for $i > 0$. In characteristic 0 this follows at once from (4.1.9). The characteristic free version follows from (4.1.12). ∎

We can get additional information about the cohomology of $\mathcal{M}(\alpha, \beta)$ when **K** is a field of characteristic zero. Let α be a weight with $\alpha^{(1)}$, $\alpha^{(2)}$ dominant. We define the number $l(\alpha)$ as follows. Consider the weight $\alpha + \rho$ where $\rho = (m - 1, m - 2, \ldots, 1, 0) = (u_1, \ldots, u_m)$. Define, by reverse induction on s (from $s = r$ to $s = 1$) the numbers

$$\delta_s = \min\{t \mid t \geq \delta_{s+1},$$
$$t + m - s \notin \{\delta_{s+1} + m - s - 1, \ldots, \delta_r + m - r, \alpha_{r+1} + m - r - 1, \ldots, \alpha_m\}\}.$$

By construction the weight $(\delta_1, \ldots, \delta_r, \alpha_{r+1}, \ldots, \alpha_m) + \rho$ is not orthogonal to any root. By (4.1.9) there exists a unique l such that $H^l(\text{Grass}(m - r, F), K_{\delta'}\mathcal{Q} \otimes K_{\alpha^{(2)'}}\mathcal{R}) \neq 0$. We define $l(\alpha) := l$.

(6.5.6) Proposition. *Let* $\mathbf{K}$ *be a field of characteristic zero. Let* $\alpha = (\alpha^{(1)}, \alpha^{(2)})$ *be a weight for* $\text{GL}(F)$ *with* $\alpha^{(1)}, \alpha^{(2)}$ *dominant. Let* $l(\alpha)$ *be defined as above. Assume that* β *is a dominant weight for* $\text{GL}(G^*)$.

(a)

$$H^i(Z^{(1)}, \mathcal{M}(\alpha, \beta)) = 0 \qquad \text{for} \quad i > l(\alpha).$$

(b)

$$H^{l(\alpha)}(Z^{(1)}, \mathcal{M}(\alpha, \beta)) \neq 0.$$

Proof. We can assume that $\beta = (0)$ because, by the projection formula, tensoring with $K_{\beta'}G^*$ commutes with taking cohomology. This means we are reduced to calculating the cohomology

$$H^*(\text{Grass}(m - r, F), L_{\alpha^{(1)'}}\mathcal{Q} \otimes L_{\alpha^{(2)'}}\mathcal{R} \otimes \text{Sym}(\mathcal{Q} \otimes G^*)).$$

This can be rewritten as

$$\bigoplus_{\delta \in \alpha^{(1)} \otimes \gamma} H^*(\text{Grass}(m - r, F), L_{\delta'}\mathcal{Q} \otimes L_{\alpha^{(2)'}}\mathcal{R} \otimes L_{\gamma'}G^*).$$

By the Littlewood–Richardson rule (2.3.4), every weight occurring in the tensor product $\alpha^{(1)} \otimes \gamma$ is bigger than or equal to $\alpha^{(1)}$ termwise. Also, since $\dim \mathcal{Q} = r \leq \dim G^*$, all such weights δ will occur in $\alpha' \otimes \gamma$ for some γ.

Consider $\alpha = (\alpha^{(1)}, \alpha^{(2)})$ satisfying the assumptions of the proposition. Let δ_0 be the weight constructed in defining $l(\alpha)$. This, by definition, is the termwise minimal weight for which $L_{\delta_0'}\mathcal{Q} \otimes L_{\alpha^{(2)'}}\mathcal{R}$ has nonzero cohomology. This cohomology occurs in degree $l(\alpha)$. Also it is clear from (4.1.9) that for δ that is bigger termwise than δ_0 the cohomology of $L_{\delta'}\mathcal{Q} \otimes L_{\alpha^{(2)'}}\mathcal{R}$, if nonzero, occurs in degree $\leq l(\alpha)$. This proves both parts of the proposition. ∎

We also get information on the support of cohomology modules $H^i(Z^{(1)}, \mathcal{M}(\alpha, \beta))$.

In oder to state the result we need to introduce one more notion. The permutation $\sigma \in \Sigma_m$ is an *r-Grassmannian permutation* if $\sigma(1) > \ldots > \sigma(r)$, $\sigma(r + 1) > \ldots > \sigma(m)$. For each *r*-Grassmannian permutation σ we denote by C_σ the *Weyl chamber* of all weights $(\gamma_1, \ldots, \gamma_m)$ such that the entries in

$\gamma + \rho$ are pairwise different and ordered in the same way as the sequence $\sigma(1), \ldots, \sigma(m)$.

Let σ be an r-Grassmannian permutation of length i. For α arbitrary and β dominant we define the A_r-module $H^i(\text{Grass}(m - r, F), \mathcal{M}(\alpha, \beta))_\sigma$ to be the part of $H^i(\text{Grass}(m - r, F), \mathcal{M}(\alpha, \beta))$ consisting of all cohomology modules of sheaves $K_{\delta'}\mathcal{Q} \otimes K_{\alpha^{(2)'}}\mathcal{R} \otimes K_{\gamma'}G^*$ for which the weight $(\delta, \alpha^{(2)}) \in C_\sigma$. It is clear that this is a direct summand of the A_r-module $H^i(Z^{(1)}, \mathcal{M}(\alpha, \beta))$.

(6.5.7) Proposition. *Let* **K** *be a field of characteristic* 0. *Asume that* α *is arbitrary and* β *is dominant.*

(a) *The module* $H^i(Z^{(1)}, \mathcal{M}(\alpha, \beta))_\sigma$ *is nonzero if and only if there exists* $\delta = (\delta_1, \ldots, \delta_r)$ *such that* $\delta \geq \alpha^{(1)}$ *(termwise) and* $(\delta, \alpha^{(2)}) \in C_\sigma$.
(b) *The support of* $H^i(Z^{(1)}, \mathcal{M}(\alpha, \beta))_\sigma$ *is the determinantal variety* Y_{s-1} *for* $s = \sigma(r + 1)$.

Proof. As in the proof of (6.5.6), we can assume that $\beta = (0)$. The first part of the proposition follows as in the proof of (6.5.6). Let us choose an r-Grassmannian permutation σ of length i. We are interested in the support of the cohomology modules of

$$\mathcal{M}(\alpha, \beta)_\sigma = \bigoplus_\gamma \bigoplus_{\delta \in \alpha^{(1)} \otimes \gamma, \ \delta \in C_\sigma} K_{\delta'}\mathcal{Q} \otimes K_{\alpha^{(2)'}}\mathcal{R} \otimes K_{\gamma'}G^*.$$

Let $\sigma(r + 1) = s$. Then we can increase $\delta_1, \ldots, \delta_{s-1}$ as we please to still get the weights $(\delta, \alpha^{(2)})$ from C_σ. On the other hand, the indices $\delta_s, \ldots, \delta_r$ can increase only by a limited number if we are to get a weight from C_σ. Now the use of Littlewood–Richardson rule and (6.1.5)(d) shows that the support of the module $H^i(Z^{(1)}, \mathcal{M}(\alpha, \beta))_\sigma$ equals Y_{s-1}. ∎

We proceed to prove Theorems (6.5.3) and (6.5.4) for the family $M(\alpha, \beta)$. Each equivariant sheaf $\mathcal{M}$ on $Z^{(1)}$ has its Euler characteristic class

$$\chi(\mathcal{M}) = \sum_{i \geq 0} (-1)^i [H^i(Z^{(1)}, \mathcal{M})] \in K_0'(A_r).$$

(6.5.8) Proposition. *The group* $K_0'(A_r)$ *is generated by the Euler characteristic classes* $\chi(\mathcal{M}(\alpha, \beta))$ *for* $(\alpha^{(1)}, \alpha^{(2)}, \beta$ *dominant).*

Proof. Let M be a graded A_r-module with a rational $\text{GL}(F) \times \text{GL}(G^*)$ action. Then the natural morphism

$$M \to (q_*^{(1)})(q^{(1)*}M)$$

has a kernel and cokernel supported in Y_{r-1}. It is therefore enough to show.

(1) If M is a module supported in Y_{r-1}, then its class in $K'_0(A_r)$ is in the span of classes $\chi(\mathcal{M}(\alpha, \beta))$ for $(\alpha^{(1)}, \alpha^{(2)})$, β dominant.
(2) The class of $q^{(1)*} M$ is contained in the span of sheaves $\mathcal{M}(\alpha, \beta)$ in the Grothendieck goup of equivariant sheaves on $Z^{(1)}$.

We start with the proof of (1). Since in this argument all constructions will commute with tensoring by $L_{\beta'} G^*$, we will drop it from our notation, dealing with sheaves $\mathcal{M}(\alpha^{(1)}, \alpha^{(2)}) := \mathcal{M}(\alpha^{(1)}, \alpha^{(2)}, (0))$.

It is enough to show that the Euler characteristic of each sheaf of type $\mathcal{M}(\alpha)$, for Y_{r-1} denoted by $\hat{\mathcal{M}}(\alpha)$, is in the subgroup of $K'_0(A_r)$ generated by Euler characteristics of sheaves $\mathcal{M}(\alpha)$ for Y_r.

Consider the Grassmannian $\mathrm{Grass}(m - r + 1, F)$ with the tautological sequence

$$0 \to \hat{\mathcal{R}} \to F \times \mathrm{Grass}(m - r + 1, F) \to \hat{\mathcal{Q}} \to 0.$$

Consider the flag variety $\mathrm{Flag}(m - r, m - r + 1; F)$ with universal flag $0 \subset \mathcal{R} \subset \hat{\mathcal{R}} \subset F$. Let

$$v_1 : \mathrm{Flag}(m - r, m - r + 1; F) \to \mathrm{Grass}(m - r, F),$$

$$v_2 : \mathrm{Flag}(m - r, m - r + 1; F) \to \mathrm{Grass}(m - r + 1, F)$$

denote the natural projections.

We have by definition

$$\hat{\mathcal{M}}(\alpha_1^{(1)}, \ldots, \alpha_{r-1}^{(1)}, \alpha_1^{(2)}, \ldots, \alpha_{m-r+1}^{(2)})$$
$$= v_{2*}(L_{\alpha^{(1)\prime}} \hat{\mathcal{Q}} \otimes L_{\alpha_1^{(2)\prime}}(\hat{\mathcal{R}}/\mathcal{R}) \otimes L_{(\alpha_2^{(2)\prime}, \ldots, \alpha_{m-r+1}^{(2)\prime})} \mathcal{R} \otimes \mathrm{Sym}(\hat{\mathcal{Q}} \otimes G^*)).$$

■

The higher direct images of the tensor product sheaf on the right hand side vanish. This sheaf has a Koszul type resolution of locally free modules over $\mathrm{Sym}(\mathcal{Q} \otimes G^*)$ on $\mathrm{Flag}(m - r, m - r + 1; F)$ with terms

$$L_{\alpha^{(1)\prime}} \hat{\mathcal{Q}} \otimes L_{\alpha_1^{(2)\prime}}(\hat{\mathcal{R}}/\mathcal{R}) \otimes L_{(\alpha_2^{(2)\prime}, \ldots, \alpha_{m-r+1}^{(2)\prime})} \mathcal{R}$$

$$\otimes \bigwedge^{\bullet}(\mathrm{Ker}(\mathcal{Q} \to \hat{\mathcal{Q}}) \otimes G^*) \otimes \mathrm{Sym}(\hat{\mathcal{Q}} \otimes G^*),$$

which can be rewritten as

$$L_{\alpha^{(1)\prime}} \hat{\mathcal{Q}} \otimes L_{\alpha_1^{(2)\prime} + \bullet}(\hat{\mathcal{R}}/\mathcal{R}) \otimes L_{(\alpha_2^{(2)\prime}, \ldots, \alpha_{m-r+1}^{(2)\prime})} \mathcal{R} \otimes \bigwedge^{\bullet}(G^*) \otimes \mathrm{Sym}(\hat{\mathcal{Q}} \otimes G^*),$$

because $\mathrm{Ker}(\mathcal{Q} \to \hat{\mathcal{Q}})$ is isomorphic to $\hat{\mathcal{R}}/\mathcal{R}$, as can be seen from the commutative diagram

$$
\begin{array}{ccccccccc}
0 & \to & \mathcal{R} & \to & F \times \mathrm{Flag}(m-r, m-r+1; F) & \to & \mathcal{Q} & \to & 0 \\
 & & \downarrow & & \downarrow & & \downarrow & & \\
0 & \to & \hat{\mathcal{R}} & \to & F \times \mathrm{Flag}(m-r, m-r+1; F) & \to & \hat{\mathcal{Q}} & \to & 0
\end{array}
$$

of vector bundles over $\mathrm{Flag}(m-r, m-r+1; F)$.

We push down the terms of the resolution by v_{1*}. The bundle $\mathcal{Q}$ is induced from $\mathrm{Grass}(m-r, F)$. Thus each term

$$
L_{\alpha^{(1)}} \hat{\mathcal{Q}} \otimes L_{\alpha_1^{(2)'}+t}(\hat{\mathcal{R}}/\mathcal{R}) \otimes L_{(\alpha_2^{(2)'}, \ldots, \alpha_{m-r+1}^{(2)'})} \mathcal{R} \otimes \overset{t}{\bigwedge}(G^*) \otimes \mathrm{Sym}(\hat{\mathcal{Q}} \otimes G^*)
$$

gives us a sheaf $\mathcal{M}(\gamma^{(1)}, \gamma^{(2)})$, possibly with sign. Taking Euler characteristics gives an expression of $\chi(\hat{\mathcal{M}}(\alpha^{(1)}, \alpha^{(2)}))$ as a combination of terms of the type $\chi(\mathcal{M}(\gamma^{(1)}, \gamma^{(2)}))$.

To prove statement (2) we notice that q_{1*} is a sheaf of graded $\mathrm{Sym}(\mathcal{Q} \otimes G^*)$-modules. We can take its free $\mathrm{GL}(\mathcal{Q}) \times \mathrm{GL}(\mathcal{R}) \times \mathrm{GL}(G^*)$-equivariant resolution. Its terms are, up to filtration, direct sums of sheaves of type $\mathcal{M}(\alpha^{(1)}, \alpha^{(2)}, \beta)$. This completes the proof of Proposition (6.5.4). ■

(6.5.9) Remarks. *Notice that the proof of Proposition (6.5.8) shows that if $(\alpha^{(1)}, \alpha^{(2)})$ is dominant, then the Euler characteristic of $\hat{\mathcal{M}}(\alpha^{(1)}, \alpha^{(2)}, \beta)$ is in the subgroup of $K_0'(A_r)$ spanned by the Euler characteristics of sheaves $\mathcal{M}(\alpha, \beta)$ with α, β dominant.*

Theorem (6.5.3) for modules of type $M(\alpha, \beta)$ follows from the following statement:

(6.5.10) Proposition. *The classes $\chi(\mathcal{M}(\alpha^{(1)}, \alpha^{(2)}, \beta))$ such that $(\alpha^{(1)}, \alpha^{(2)})$ is dominant generate the group $K_0'(A_r)$.*

Proof. We first reduce the proof to the case when the characteristic of **K** equals zero. First of all we notice that the representation ring of the general linear group $\mathrm{GL}(F)$ is spanned by the classes of Schur modules $L_\alpha F$. This follows from the description of irreducible representations of $\mathrm{GL}(F)$ given in Theorem (2.2.9). Moreover, the Littlewood–Richardson rule holds in the representation ring of $\mathrm{GL}(F)$ in a characteristic free way, by the remark following Theorem (2.3.4). Finally, even though Bott's theorem is true only over a field of characteristic zero, the Euler characteristic of any line bundle over a flag variety is characteristic free. Our statement involves only Euler characteristics and the characters of modules $M(\alpha, \beta)$ for α, β dominant. Both these notions are independent of the characteristic and can be described using only Schur modules and the Littlewood–Richardson rule. We see that the proof in characteristic zero implies that the same statement holds over a field of arbitrary characteristic.

Let H be the subgroup of $K'_0(A_r)$ spanned by the Euler characteristic classes of sheaves $\mathcal{M}(\alpha, \beta)$ with α and β dominant. Consider an arbitrary sheaf $\mathcal{M}(\alpha^{(1)}, \alpha^{(2)}, \beta)$. We use induction on $s := s(\alpha^{(1)}, \alpha^{(2)}) = \alpha_1^{(2)} - \alpha_r^{(1)}$. If $s \leq 0$, then $(\alpha^{(1)}, \alpha^{(2)})$ is dominant and there is nothing to prove. Suppose that for $(\gamma^{(1)}, \gamma^{(2)})$ with smaller s the corresponding sheaves are in H.

We identify $\mathcal{M}(\alpha^{(1)}, \alpha^{(2)}, \beta)$ with its direct image

$$p_{1*}\mathcal{M}(\alpha^{(1)}, \alpha^{(2)}, \beta) = L_{\alpha^{(1)'}}\hat{\mathcal{Q}} \otimes L_{\alpha^{(2)'}}(\mathcal{R}) \otimes \mathrm{Sym}(\mathcal{Q} \otimes G^*).$$

Consider the subsheaf $\mathcal{M}_{<s}(\alpha^{(1)}, \alpha^{(2)}, \beta)$ of $\mathcal{M}(\alpha^{(1)}, \alpha^{(2)}, \beta)$ consisting of all summands $L_{\gamma^{(1)'}}\mathcal{Q} \otimes L_{\gamma^{(2)'}}\mathcal{R} \otimes S_\delta G^*$ such that $\gamma_1^{(2)} - \gamma_r^{(1)} < s$. It follows from the Littlewood–Richardson rule (2.3.4) that it is a $\mathrm{Sym}(\mathcal{Q} \otimes G^*)$ submodule. Denote by $\mathcal{M}_s(\alpha^{(1)}, \alpha^{(2)}, \beta)$ the factor $\mathcal{M}(\alpha^{(1)}, \alpha^{(2)}, \beta)/\mathcal{M}_{<s}(\alpha^{(1)}, \alpha^{(2)}, \beta)$. We have an exact sequence

$$0 \to \mathcal{M}_{<s}(\alpha^{(1)}, \alpha^{(2)}, \beta) \to \mathcal{M}(\alpha^{(1)}, \alpha^{(2)}, \beta) \to \mathcal{M}_s(\alpha^{(1)}, \alpha^{(2)}, \beta) \to 0.$$

The support of all cohomology groups of the sheaf $\mathcal{M}_s(\alpha^{(1)}, \alpha^{(2)}, \beta)$ is contained in Y_{r-1}. Indeed, if we multiply the summand $L_{\gamma^{(1)'}}\mathcal{Q} \otimes L_{\gamma^{(2)'}}\mathcal{R} \otimes S_\delta G^*$ by $\bigwedge^r \mathcal{Q} \otimes \bigwedge^r G^*$ corresponding to $r \times r$ minors, we add one to each entry of $\gamma^{(1)'}$, so s has to decrease. This means that the ideal of $r \times r$ minors annihilates all cohomology groups of $\mathcal{M}_s(\alpha^{(1)}, \alpha^{(2)}, \beta)$. Now Remark (6.5.9) and an induction on r imply that the class $\chi(\mathcal{M}_s(\alpha^{(1)}, \alpha^{(2)}, \beta))$ is in H.

Consider the relative version of the $\mathrm{GL}(\mathcal{Q}) \times \mathrm{GL}(\mathcal{R}) \times \mathrm{GL}(G^*)$-equivariant free resolution of the sheaf $\mathcal{M}_{<s}(\alpha^{(1)}, \alpha^{(2)}, \beta)$. Its i-th term $\mathcal{F}_i$ is a direct sum of sheaves $\mathcal{M}(\gamma^{(1)}, \gamma^{(2)}, \beta)$ and each term occurring in $\mathcal{F}_i$ has smaller s than $(\alpha^{(1)}, \alpha^{(2)})$. Indeed, by induction on i and by the Littlewood–Richardson rule (2.3.4) it follows that on multiplying $\mathcal{M}(\gamma^{(1)}, \gamma^{(2)}, \beta)$ occurring in $\mathcal{F}_i$ by summands of $\mathrm{Sym}(\mathcal{Q} \otimes G^*)$, the invariant s in all resulting summands can only decrease.

We conclude that, by induction on s, the Euler characteristic $\chi(\mathcal{M}_{<s}(\alpha^{(1)}, \alpha^{(2)}, \beta))$ is in H and Proposition (6.5.10) is proven. ∎

Proof of Theorem (6.5.4). Part (a) is a consequence of the fact that the characters of $M(\alpha, \beta)(-i)$ are linearly independent in $\mathrm{Rep}(\mathrm{GL}(F) \times \mathrm{GL}(G^*))$

$[[q]][q^{-1}]$. To prove (b) we define the homomorphism of abelian groups

$$\Psi : \mathrm{Rep}(\mathrm{GL}(F) \times \mathrm{GL}(G^*))[q][q^{-1}] \to K_0'(A_r)$$

by sending the class $[(L_{\alpha'} F \otimes L_{\beta'} G^*)q^i]$ to $[M(\alpha, \beta)(-i)]$. By Theorem (6.5.3) the homomorphism Ψ is an epimorphism. It is also a monomorphism, because the characters of modules $M(\alpha, \beta)(-i)$ are linearly independent in $\mathrm{Rep}(\mathrm{GL}(F) \times \mathrm{GL}(G^*))[[q]][q^{-1}]$ and therefore in $K_0'(A_r)$. ∎

Theorem (6.5.3) implies that for any graded $\mathrm{GL}(F) \times \mathrm{GL}(G^*)$-equivariant A_r-module M the class $[M]$ can be expressed as a linear combination of classes $[M(\alpha, \beta)(-i)]$. In particular we can describe the class of A_{r-1}.

(6.5.11) Proposition. *The class of A_{r-1} in $K_0'(A_r)$ is given by the formula*

$$[A_{r-1}] = [A_r] - \sum_{i=0}^{n-r}[M((i+1, 1^{r-1}), (1^{r+i}, 0^{n-r-i}))].$$

Proof. Let $\mathcal{A}$ denote the sheaf of algebras $\mathrm{Sym}(\mathcal{Q} \otimes G^*)$ over $\mathrm{Grass}(m - r, F)$. Consider the relative Eagon–Northcott complex over $\mathcal{A}$,

$$0 \to \mathcal{E}_{n-r} \to \mathcal{E}_{n-r-1} \to \ldots \to \mathcal{E}_1 \to \mathcal{E}_0 \to \mathcal{A},$$

where

$$\mathcal{E}_i = D_i \mathcal{Q} \otimes \bigwedge^r \mathcal{Q} \otimes \bigwedge^n G^* \otimes \mathcal{A}(-i-r).$$

This is a sheaf resolution of a sheaf $\mathcal{B}$ of algebras. Using the relative version of the straightening law (3.2.5), we deduce that the sheaf $\mathcal{B}$ has a filtration whose associated graded object is

$$[\mathcal{B}] = \bigoplus_{\lambda=(\lambda_1,\ldots,\lambda_{r-1})} L_{\lambda'} \mathcal{Q} \otimes L_{\lambda'} G^*.$$

Kempf's theorem (4.1.10) implies that the higher cohomology of $\mathcal{B}$ vanishes and that the sections of $\mathcal{B}$ are isomorphic to A_{r-1}. We also have the identification $\mathcal{E}_i = \mathcal{M}((i + 1), 1^{r-1}), (1^{r+i}, 0^{n-r-i}))$. This concludes the proof. ∎

(6.5.12) Remarks. *The relative Eagon–Northcott complex used in the proof of (6.5.11) and its sections is the simplest example of a degeneration sequence discussed in section 5.6.*

For the remainder of this section we assume that $\mathbf{K}$ is a field of characteristic 0. We calculate the depth of modules $M(\alpha, \beta)$. Of course the depth of $M(\alpha, \beta)$ is independent of β, so we assume that β is zero and consider $M(\alpha) := M(\alpha, (0))$.

We start with arbitrary $(\alpha^{(1)}, \alpha^{(2)})$, not necessarily dominant. We assume only that the sheaf $\mathcal{M}(\alpha^{(1)}, \alpha^{(2)})$ has no higher cohomology.

(6.5.13) Proposition. *The sheaf $\mathcal{M}(\alpha^{(1)}, \alpha^{(2)})$ has no higher cohomology if and only if $\alpha_r^{(1)} \geq \alpha_1^{(2)} - t$ where t is such that $\alpha_1^{(2)} = \ldots = \alpha_t^{(2)} > \alpha_{t+1}^{(2)}$.*

Proof. Let $\delta_1, \ldots, \delta_r$ be the numbers defined before the statement of Proposition (6.5.6). The condition of Proposition (6.5.13) means that $\delta_r \geq \alpha_1^{(2)}$, and that is equivalent to $l(\alpha) = 0$. ∎

We want to find the projective dimension of $M(\alpha^{(1)}, \alpha^{(2)})$, because by the Auslander–Buchsbaum formula (1.2.7)′ we have

$$\text{depth}(A, M) + \text{pd}_A(M) = mn.$$

The resolution of $M(\alpha^{(1)}, \alpha^{(2)})$ is given by the complex $F(L_{\alpha^{(1)}} \mathcal{Q} \otimes L_{\alpha^{(2)}} \mathcal{R})_\bullet$, whose i-th term is given by

$$\bigoplus_{j \geq 0} H^j \left(\text{Grass}(m - r, F), L_{\alpha^{(1)}} \mathcal{Q} \otimes L_{\alpha^{(2)}} \mathcal{R} \otimes \bigwedge^{i+j} (\mathcal{R} \otimes G^*) \right) \otimes A(-i - j).$$

In order to study the top of the resolution, we look first at the last term of the Koszul complex. The corresponding $\text{GL}(F)$-weight is

$$t(\alpha) = (\alpha_1^{(1)}, \ldots, \alpha_r^{(1)}, \alpha_1^{(2)} + n, \ldots, \alpha_{m-r}^{(2)} + n).$$

Let's look at the weight

$$t(\alpha) + \rho = (a_1, \ldots, a_r, b_1, \ldots, b_{m-r}).$$

For $j = 1, \ldots, m - r$ we define the sequences $t_j(\alpha)$ inductively by setting

$$t_j(\alpha) = (a_1, \ldots, a_r, d_1, \ldots, d_j, b_{j+1}, \ldots, b_{m-r}),$$

where d_{j+1} is defined as

$$d_{j+1} = \max\{t \mid t \leq b_{j+1},\ b < d_j,\ t \notin \{a_1, \ldots, a_r\}\}.$$

Notice that in this definition the condition $t < d_j$ could be skipped, because each b_j is essentially lowered to the first possible number that is not one of the a_i's or previous d_K's. Let us also define numbers

$$q_j = n - (b_j - d_n), \ p_j = q_j - \#\{i \mid a_i < d_j\}$$

for $j = 1, \ldots, m - r$.

(6.5.14) Lemma. *For each* $j = 1, \ldots, m - r$ *we have* $p_j \geq n - r$.

Proof. Let us imagine that we construct the sequences $t_j(\alpha)$ by the following process. We look at b_j and start lowering it by 1 until we reach a number that is not equal to any a_i or d_K for $k < j$. Then every step of lowering by one accounts for some a_i satisfying $d_j + 1 \leq a_i \leq b_j$ for some d_K (which comes from a previous $a_K > b_j$). This means that we have a set of $b_j - d_j$ a_K's which is disjoint from the set $\{i \mid a_i < d_j\}$. Therefore

$$p_j = n - (b_j - d_j) - \#\{i \mid a_i < d_j\} \geq n - r.$$

This concludes the proof of the lemma. ∎

(6.5.15) Theorem. *Let* **K** *be a field of characteristic zero. Let* $(\alpha^{(1)}, \alpha^{(2)})$ *satisfy the condition of Proposition (6.5.13). Then the projective dimension of* $M(\alpha^{(1)}, \alpha^{(2)})$ *over* A *equals* $\sum_{j=1}^{m-r} p_j$.

Proof. Let us decompose the terms of the complex $F(L_{\alpha^{(1)\prime}} Q \otimes L_{\alpha^{(2)\prime}} R)_\bullet$ using the Cauchy formula (2.3.3) and the Littlewood–Richardson rule (2.3.4). The GL(F) weights of all terms have the form $(\alpha^{(1)}, \delta^{(2)})$ with $\delta^{(2)}$ containing $\alpha^{(2)}$ such that the difference between corresponding terms of $\delta^{(2)\prime}$ and $\alpha^{(2)\prime}$ does not exceed n. Also, all such weights occur. For all weights of this form we need to find the supremum of the numbers $|\delta^{(2)}| - |\alpha^{(2)}| - l(w)$ where w is a permutation ordering the weight $(\alpha^{(1)\prime}, \delta^{(2)\prime}) + \rho$. It is clear tht the top number is obtained from the sequence $t_{m-r}(\alpha)$. We need therefore to find in which term $F(L_{\alpha^{(1)\prime}} Q \otimes L_{\alpha^{(2)\prime}} R)_i$ the corresponding term occurs. The homogeneous degree is $\sum_{j=1}^{m-r} (n - (b_j - d_j))$, and the length of the permutation w is $\sum_{j=1}^{m-r} \#\{i \mid a_i < d_j\}$. The statement of the theorem follows. ∎

(6.5.16) Corollary. *Let* **K** *be a field of characteristic zero. Assume that the weight* $(\alpha^{(1)\prime}, \alpha^{(2)\prime})$ *satisfies the condition of Proposition (6.5.13). The depth of* $M(\alpha^{(1)}, \alpha^{(2)})$ *equals* $mn - \sum_{j=1}^{m-r} p_j$. *The module* $M(\alpha^{(1)}, \alpha^{(2)})$ *is a maximal Cohen–Macaulay module over* A_r *if and only if every number* p_j *equals* $n - r$.

Let us look more closely at the weights giving maximal Cohen–Macaulay modules over A_r. Let us look at the process of getting sequences $t_j(\alpha)$ by lowering the numbers b_j to d_j. At each stage we modify the sequence $(a_1, \ldots, a_r)$ as follows. We define inductively the sequences $(a_1^{(j)}, \ldots, a_r^{(j)})$ by setting

$$(a_1^{(0)}, \ldots, a_r^{(0)}) = (a_1, \ldots, a_r),$$

$$(a_1^{(j)}, \ldots, a_r^{(j)}) = \begin{cases} (a_1^{(j-1)}, \ldots, \hat{a}_i^{(j-1)}, \ldots a_r^{(j)}, d_j) & \text{if } b_j = a_i^{(j-1)}, \\ (a_1^{(j-1)}, \ldots a_r^{(j-1)}) & \text{otherwise.} \end{cases}$$

Now $p_j = n - r$ for $j = 1, \ldots, m - r$ if and only if we have $b_j \geq \max\{a_1^{(j-1)}, \ldots a_r^{(j-1)}\}$ for each $j = 1, \ldots, m - r$. Indeed, each a_i either induces a number between b_j and $d_j + 1$ or is smaller than d_j, so the overall number by which the projective dimension decreases at the j-th stage is $n - r$. Stating the last result, we have

(6.5.17) Corollary. *Let $(\alpha^{(1)}, \alpha^{(2)})$ satisfy the condition of Proposition (6.5.13). Define the sets $(a_1^{(j)}, \ldots a_r^{(j)})$ as above. Then $M(\alpha^{(1)}, \alpha^{(2)})$ is a maximal Cohen–Macaulay A_r-module if and only if for every $j = 1, \ldots, m - r$ we have*

$$b_j \geq \max\{a_1^{(j-1)}, \ldots a_r^{(j-1)}\}.$$

We conclude the discussion of modules $M(\alpha)$ by showing an example of a module of type $M(\alpha)$ whose Cohen–Macaulay property depends on the characteristic of the field $\mathbf{K}$.

(6.5.18) Example. *Let us set $r = 2$, $m = 3$, $n = 4$. Consider the module $M((2, 0, 0))$. It consists of sections of the sheaf*

$$\mathcal{M} = S_2 \mathcal{Q} \otimes \operatorname{Sym}(\mathcal{Q} \otimes G^*)$$

over Grass$(1, F)$. *The module $M((2, 0, 0))$ is maximal Cohen–Macaulay over fields of characteristic $\neq 2$, but in characteristic 2 it is not maximal Cohen–Macaulay.*

Proof. The higher cohomology groups of $\mathcal{M}(2, 0, 0)$ vanish. This follows from the Cauchy formula (3.2.5), the Kempf vanishing theorem (4.1.10), and the fact that the tensor product of Schur functors has a filtration with factors isomorphic to Schur modules. (In characteristic 0 one could just use Bott's theorem (4.1.9).)

A minimal free resolution of $M(2, 0, 0)$ is given by the complex

$$F((2, 0, 0))_\bullet = S_2 Q \otimes \bigwedge^\bullet (R \otimes G^*).$$

In this case dim $R = 1$, so for each $i, 0 \le i \le 4$, we deal with the cohomology of $S_2 Q \otimes S_i R \otimes \bigwedge^i G^*$. In characteristic 0 we get the resolution

$$0 \to L_{3,2} F \otimes \bigwedge^3 G^* \otimes A(-3) \to L_{3,1} F \otimes \bigwedge^2 G^* \otimes A(-2) \to S_2 F \otimes A.$$

In fact one can see that this complex is acyclic over all fields $\mathbf{K}$ with char $\mathbf{K} \ne 2$.

If char $\mathbf{K} = 2$, then the bundle $S_2 Q \otimes S_4 R$ has nonzero cohomology. In fact

$$H^1(\mathrm{Grass}(1, F), S_2 Q \otimes S_4 R) = H^2(\mathrm{Grass}(1, F), S_2 Q \otimes S_4 R) = L_{3,3} F.$$

Therefore the resolution of $M((2, 0, 0))$ over a field of characteristic 2 is

$$0 \to L_{3,3} F \otimes \bigwedge^4 G^* \otimes A(-4) \to L_{3,3} F \otimes \bigwedge^4 G^* \otimes A(-4) \oplus L_{3,2} F$$

$$\otimes \bigwedge^3 G^* \otimes A(-3) \to L_{3,1} F \otimes \bigwedge^2 G^* \otimes A(-2) \to S_2 F \otimes A.$$

We see that $M((2, 0, 0))$ is a maximal Cohen–Macaulay module over A_r over fields of characteristic $\ne 2$, but in characteristic 2 its depth drops by one. ∎

6.6. Modules Supported in Symmetric Determinantal Varieties

We preserve the notation from section 6.3. To construct a family of modules supported in Y_r^s we use the first incidence variety from section 6.3.

Let us fix n and r. We take $Y = Y_r^s$ and $V = \mathrm{Grass}(n - r, E)$ with the tautological sequence

$$0 \to R \to E \times V \to Q \to 0.$$

Here dim $R = n - r$ and dim $Q = q = r$. For a subspace R in E, we denote by i the embedding of R into E. We consider the variety

$$Z_1 = \{(\phi, R) \in X \times V \mid \phi\, i = 0\}.$$

As usual, we have the diagram

$$
\begin{array}{ccc}
Z_1 & \subset & X \times V \\
\downarrow q' & & \downarrow q \\
Y & \subset & X
\end{array}
$$

The variety Z_1 is the total space of the bundle $\mathcal{S}_1 = S_2 \mathcal{Q}^*$. The bundle $\mathcal{T}_1$ is defined by the exact sequence

$$0 \to S_2 \mathcal{Q}^* \to S_2 E^* \times V \to \mathcal{T}_1 \to 0.$$

Let $\mathcal{C}_r^s(E)$ be the category of graded $A_r^s := \mathbf{K}[Y_r^s]$-modules with rational $GL(E)$-action compatible with the module structure, and equivariant degree 0 maps. We denote $K_0'(A_r^s)$ the Grothendieck group of the category $\mathcal{C}_r^s(E)$.

For an equivariant graded module $M \in \mathrm{Ob}(\mathcal{C}_r^s(E))$ and for $q \in \mathbf{Z}$, we denote by $M(q)$ the module M with gradation shifted by q, i.e. $M(q)_n = M_{q+n}$. For $M \in \mathrm{Ob}(\mathcal{C}_r^s(E))$ we define the *graded character of M*,

$$\mathrm{char}(M) = \sum_{n \in \mathbf{Z}} \mathrm{char}(M_n)\, q^n \in \mathrm{Rep}(GL(E))[[q]][q^{-1}],$$

where $\mathrm{Rep}(GL(E))$ denotes the representation ring of $GL(E)$, and char denotes the character map described in (2.2.10).

Let $\alpha = (\alpha_1, \ldots, \alpha_n)$ be an integral weight for $GL(E)$. We set $\alpha^{(1)} = (\alpha_1, \ldots, \alpha_r)$, $\alpha^{(2)} = (\alpha_{r+1}, \ldots, \alpha_n)$.

Let $\alpha = (\alpha^{(1)}, \alpha^{(2)})$. Assume that both $\alpha^{(1)}$ and $\alpha^{(2)}$ are dominant. We define a sheaf

$$\mathcal{M}^s(\alpha) = p^*(L_{\alpha^{(1)}} \mathcal{Q} \otimes L_{\alpha^{(2)}} \mathcal{R}) \otimes \mathcal{O}_{Z_1}$$

of graded modules on Z_1.

We define the equivariant graded A_r^s-modules

$$M^s(\alpha) = H^0(Z_1, \mathcal{M}^s(\alpha)).$$

(6.6.1) Example. *The simplest examples of modules $M^s(\alpha)$ are provided by the pushdowns of line bundles on $\mathrm{Grass}(n - r, E)$. The line bundles on $\mathrm{Grass}(n - r, E)$ correspond to tensor powers $\mathcal{O}_{\mathrm{Grass}(n-r,E)}(m) = (\bigwedge^r \mathcal{Q})^{\otimes m}$ for $m \in \mathbf{Z}$. Notice that the formula (5.1.4) for the dual bundle implies that the canonical module $K_{A_r^s} = M((r + 1)^r, 0^{n-r}) \otimes \bigwedge^n F^{\otimes -r-1}$.*

For $m \geq 0$ the corresponding modules occur as a subset of a family from the next example.

(6.6.2) Example. *Let $\alpha^{(2)} = (0)$. The sheaf $\mathcal{M}^s(\alpha)$ equals $L_{\alpha^{(1)}} \mathcal{Q} \otimes \mathcal{O}_{Z_1}$. Therefore the direct image $p_* \mathcal{M}^s(\alpha)$ equals $L_{\alpha^{(1)}} \mathcal{Q} \otimes \mathrm{Sym}(S_2 \mathcal{Q})$. The higher cohomology of $\mathcal{M}^s(\alpha)$ vanishes by Bott's theorem (4.1.9). The j-th* *homogeneous component $M_j^s(\alpha)$ decomposes as follows:*

$$\bigoplus_{|\mu|=j,\ \mu_i'\ \text{even for all } i, \mu_1 \leq r} (L_{\alpha^{(1)'}} E \otimes L_{\mu'} E)_{\leq r},$$

where $(L_{\alpha^{(1)'}} E \otimes L_{\mu'} E)_{\leq r}$ denotes a factor of $(L_{\alpha^{(1)'}} E \otimes L_{\mu'} E)$ consisting of all $L_\nu E$ in the tensor product with $\nu_1' \leq r$.

The main results of this section are the analogues of the results from the previous section.

(6.6.3) Theorem. *The Grothendieck group $K_0'(A_r^s)$ is generated by the classes of the modules $M^s(\alpha)(q)$, where α is a dominant weight and $q \in \mathbf{Z}$.*

(6.6.4) Theorem.
  (a) *The Grothendieck group $K_0'(A_r^s)$ is isomorphic to the additive subgroup of the ring $\mathrm{Rep}(\mathrm{GL}(E))[[q]][q^{-1}]$ generated by shifted graded characters of modules $M^s(\alpha)$ for α dominant.*
  (b) *The group $K_0'(A_r^s)$ is isomorphic to the additive group of the ring $\mathrm{Rep}(\mathrm{GL}(E))[q][q^{-1}]$.*

The proof of Theorems (6.6.3) and (6.6.4) follows the same steps as the proof of Theorems (6.5.3) and (6.5.4). We only give the statements, leaving the detais as an exercise to the reader.

(6.6.5) Proposition. *Let α be a dominant weight for $\mathrm{GL}(E)$. Then $H^i(Z_1, \mathcal{M}^s(\alpha)) = 0$ for $i > 0$.*

Proof. This is an analogue of (6.5.5). ∎

Let α be a weight with $\alpha^{(1)}$, $\alpha^{(2)}$ dominant. We define the number $l(\alpha)$ as follows. Consider the weight $\alpha + \rho$, where $\rho = (n-1, n-2, \ldots, 1, 0) = (u_1, \ldots, u_n)$. Define, by reverse induction on s (from $s = r$ to $s = 1$), the numbers

$$\delta_s = \min\{t \mid t \geq \delta_{s+1},$$
$$t+n-s \notin \{\delta_{s+1}+n-s-1, \ldots, \delta_r+n-r, \alpha_{r+1}+n-r-1, \ldots, \alpha_n \}\}.$$

By construction the weight $(\delta_1, \ldots, \delta_r, \alpha_{r+1}, \ldots, \alpha_n) + \rho$ is not orthogonal to any root. By (4.1.9) there exists a unique l such that $H^l(\mathrm{Grass}\ (n-r, E), K_{\delta'}\mathcal{Q} \otimes K_{\alpha^{(2)'}}\mathcal{R}) \neq 0$. We define $l(\alpha) := l$.

(6.6.6) Proposition. *Let* **K** *be a field of characteristic zero. Let* $\alpha = (\alpha^{(1)}, \alpha^{(2)})$ *be a weight for* GL(F) *with* $\alpha^{(1)}, \alpha^{(2)}$ *dominant. Let* $l(\alpha)$ *be defined as above. Then* $H^i(Z_1, \mathcal{M}^s(\alpha)) = 0$ *for* $i > l(\alpha)$.

Proof. This is an analogue of the first statement of (6.5.6). Notice that the analogue of the second statement need not be true. ∎

Each equivariant sheaf $\mathcal{M}^s$ on Z_1 has its Euler characteristic class

$$\chi(\mathcal{M}^s) = \sum_{i \geq 0}(-1)^i[H^i(Z_1, \mathcal{M}^s)] \in K_0'(A_r^s).$$

(6.6.7) Proposition. *The group* $K_0'(A_r^s)$ *is generated by the Euler characteristic classes* $\chi(\mathcal{M}^s(\alpha))$ *for* $(\alpha^{(1)}, \alpha^{(2)})$ *dominant.*

Proof. This is an analogue of (6.5.8). For the proof one has to use the same type of induction, using the flag variety Flag($n - r, n - r + 1; E$). ∎

(6.6.8) Remarks. *Notice that, as in the previous section, we can also show that if* $(\alpha^{(1)}, \alpha^{(2)})$ *is dominant, then the Euler characteristic of* $\hat{\mathcal{M}}^s(\alpha^{(1)}, \alpha^{(2)})$ *is in the subgroup of* $K_0'(A_r^s)$ *spanned by the Euler characteristics of sheaves* $\mathcal{M}^s(\alpha, \beta)$ *with* α *dominant.*

Theorem (6.6.3) follows from the following statement:

(6.6.9) Proposition. *The classes* $\chi(\mathcal{M}^s(\alpha^{(1)}, \alpha^{(2)})$ *such that* $(\alpha^{(1)}, \alpha^{(2)})$ *is dominant generate the group* $K_0'(A_r^s)$.

Proof. The proof is analogous to (6.5.10). We proceed by induction on $s := s(\alpha^{(1)}, \alpha^{(2)}) = \alpha_1^{(2)} - \alpha_r^{(1)}$. ∎

Proof of Theorem (6.6.4). Part (a) is a consequence of the fact that the characters of $M^s(\alpha)(-i)$ are linearly independent in Rep(GL(E))[[q]][q^{-1}]. To prove (b) we define the homomorphism of abelian groups

$$\Psi : \text{Rep(GL}(E))[q][q^{-1}] \to K_0'(A_r^s)$$

by sending the class $[(L_{\alpha'}E)q^i]$ to $[M^s(\alpha)(-i)]$. By Theorem (6.6.3) the homomorphism Ψ is an epimorphism. It is also a monomorphism, because the characters of modules $M^s(\alpha)(-i)$ are linearly independent in Rep(GL(E))[[q]][q^{-1}] and therefore in $K_0'(A_r^s)$. ∎

We finish this section with a criterion for a module $M^s(\alpha)$ to be maximal Cohen–Macaulay.

We use Corollary (5.1.5). Before we start, let us work out the duality statement for the variety Z_1. To this end we need to calculate the bundle $\bigwedge^m \xi^* \otimes K_V$, where $m = \text{rank}\,\xi$. Using (3.3.5), we get $K_V = (\bigwedge^r \mathcal{Q})^{-n+r} \otimes (\bigwedge^{n-r} \mathcal{R})^r$. Similarly, $\bigwedge^m \xi^* = \bigwedge^{r(r+1)/2} S_2 \mathcal{Q} \otimes \bigwedge^{n(n+1)/2} S_2 F^*$. Putting these facts together, we conclude that the dualizing bundle is

$$\omega_{Z_1} = K_{(-n-1+r)^n} F^* \otimes \left(\bigwedge^r \mathcal{Q} \right)^{-n+r+1}.$$

Therefore, for the weight

$$\alpha = (\alpha_1^{(1)}, \ldots, \alpha_r^{(1)}, \alpha_1^{(2)}, \alpha_{n-r}^{(2)}),$$

the dual weight is given by the formula

$$\alpha^\vee = (-n+r+1-\alpha_r^{(1)}, \ldots -n+r+1-\alpha_1^{(1)}, -\alpha_{n-r}^{(2)}, \ldots, -\alpha_1^{(2)}).$$

(6.6.10) Proposition. *The module $M^s(\alpha)$ is maximal Cohen–Macaulay provided that $l(\alpha) = l(\alpha^\vee) = 0$.*

Proof. Let's denote $\mathcal{V}(\alpha) = K_{\alpha^{(1)}} \mathcal{Q} \otimes K_{\alpha^{(2)}} \mathcal{R}$. One applies Corrolary (5.1.5) to the complexes $F(\mathcal{V}(\alpha))_\bullet$ and $F(\mathcal{V}(\alpha^\vee))_\bullet$. The vanishing of the higher cohomology is assured by Proposition (6.6.6). ■

(6.6.11) Example. *Let us assume that $\alpha^{(2)} = 0$. Then $M^s(\alpha)$ has nonvanishing higher cohomology by Proposition (6.6.6). The condition $l(\alpha^\vee) = 0$ is true when the inequality $-n+r+1-\alpha_1^{(1)} \geq -n+r$ is satisfied, i.e. when $\alpha_1^{(1)} \leq 1$. This proves that the modules*

$$M^s(1^u) = H^0\left(\text{Grass}(n-r, F), \bigwedge^u \mathcal{Q} \otimes \text{Sym}(S_2 \mathcal{Q}) \right)$$

are maximal Cohen–Macaulay for $0 \leq u \leq r$.

6.7. Modules Supported in Skew Symmetric Determinantal Varieties

We preserve the notation of section 6.4.

To construct a family of modules supported in Y_r^a we use the first incidence variety from section 6.4.

Let us fix n and r, with r even. We take $Y = Y_r^a$ and $V = \text{Grass}(n - r, E)$ with the tautological sequence

$$0 \to \mathcal{R} \to E \times V \to \mathcal{Q} \to 0.$$

Here $\dim \mathcal{R} = n - r$ and $\dim \mathcal{Q} = q = r$. For a subspace R in E we denote by i the embedding of R into E. We consider the variety

$$Z_1 = \{ (\phi, R) \in X \times V \mid \phi \, i = 0 \}.$$

Let $\mathcal{C}_r^a(E)$ be the category of graded $A_r^a := \mathbf{K}[Y_r^a]$-modules with rational $\text{GL}(E)$-action compatible with the module structure, and equivariant degree 0 maps. We denote by $K_0'(A_r^a)$ the Grothendieck group of the category $\mathcal{C}_r^a(E)$.

For an equivariant graded module $M \in \text{Ob}(\mathcal{C}_r^a(E))$ and for $q \in \mathbf{Z}$ we denote by $M(q)$ the module M with gradation shifted by q, i.e. $M(q)_n = M_{q+n}$. For $M \in \text{Ob}(\mathcal{C}_r^a(E))$ we define the *graded character* of M,

$$\text{char}(M) = \sum_{n \in \mathbf{Z}} \text{char}(M_n) \, q^n \in \text{Rep}(\text{GL}(E))[[q]][q^{-1}],$$

where $\text{Rep}(\text{GL}(E))$ denotes the representation ring of $\text{GL}(E)$ and char denotes the character map defined in (2.2.10).

Let $\alpha = (\alpha_1, \ldots, \alpha_n)$ be an integral weight for $\text{GL}(E)$. We set $\alpha^{(1)} = (\alpha_1, \ldots, \alpha_r)$, $\alpha^{(2)} = (\alpha_{r+1}, \ldots, \alpha_n)$.

Let $\alpha = (\alpha^{(1)}, \alpha^{(2)})$. Assume that both $\alpha^{(1)}$ and $\alpha^{(2)}$ are dominant. We define a sheaf

$$\mathcal{M}^a(\alpha) = p^*(L_{\alpha^{(1)}} \mathcal{Q} \otimes L_{\alpha^{(2)}} \mathcal{R}) \otimes \mathcal{O}_{Z_1}$$

of graded modules on Z_1.

We define the equivariant graded A_r^a-modules

$$M^a(\alpha) = H^0(Z_1, \mathcal{M}^a(\alpha)).$$

(6.7.1) Example. *The simplest examples of modules $M^a(\alpha)$ are provided by the pushdowns of line bundles on $\text{Grass}(n - r, E)$. The line bundles on $\text{Grass}(n - r, E)$ correspond to tensor powers $\mathcal{O}_{\text{Grass}(n-r,E)}(m) = (\bigwedge^r \mathcal{Q})^{\otimes m}$ for $m \in \mathbf{Z}$. For $m \geq 0$ the corresponding modules occur as a subset of a family from the next example.*

(6.7.2) Example. *Let $\alpha^{(2)} = (0)$. The sheaf $\mathcal{M}^a(\alpha)$ equals $L_{\alpha^{(1)}} \mathcal{Q} \otimes \mathcal{O}_{Z_1}$. Therefore the direct image $p_* \mathcal{M}^a(\alpha)$ equals $L_{\alpha^{(1)}} \mathcal{Q} \otimes \text{Sym}(S_2 \mathcal{Q})$. The higher cohomology of $\mathcal{M}^a(\alpha)$ vanishes by Bott's theorem (4.1.9). The j-th* *homogeneous component $M_j^a(\alpha)$ decomposes as follows:*

$$\bigoplus_{|\mu|=j,\ \mu_i'\ \text{even for all } i,\ \mu_1 \le r} (L_{\alpha^{(1)\prime}} E \otimes L_{\mu'} E)_{\le r},$$

where $(L_{\alpha^{(1)\prime}} E \otimes L_{\mu'} E)_{\le r}$ denotes a factor of $(L_{\alpha^{(1)\prime}} E \otimes L_{\mu'} E)$ consisting of all $L_\nu E$ in the tensor product with $\nu_1' \le r$.

(6.7.3) Theorem. *The Grothendieck group $K_0'(A_r^a)$ is generated by the classes of the modules $M^a(\alpha)(q)$, where α is a dominant weight and $q \in \mathbb{Z}$.*

(6.7.4) Theorem.
    (a) *The Grothendieck group $K_0'(A_r^a)$ is isomorphic to the additive subgroup of the ring $\mathrm{Rep}(GL(E))[[q]][q^{-1}]$ generated by shifted graded characters of modules $M^a(\alpha)$ for α dominant.*
    (b) *The group $K_0'(A_r^a)$ is isomorphic to the additive group of the ring $\mathrm{Rep}(GL(E))[q][q^{-1}]$.*

Again the proofs of (6.7.3) and (6.7.4) follow the same steps as the proofs of (6.5.3) and (6.5.4). We just give the statements, leaving the proofs to the reader.

(6.7.5) Proposition. *Let α be a dominant weight for $GL(E)$. Then $H^i(Z_1, \mathcal{M}^a(\alpha)) = 0$ for $i > 0$.*

Proof. This is the analogue of (6.5.5). ∎

Let α be a weight with $\alpha^{(1)}$, $\alpha^{(2)}$ dominant. We define the number $l(\alpha)$ as follows. Consider the weight $\alpha + \rho$, where $\rho = (n-1, n-2, \ldots, 1, 0) = (u_1, \ldots, u_n)$. Define, by reverse induction on s (from $s = r$ to $s = 1$) the numbers

$$\delta_s = \min\{t \mid t \ge \delta_{s+1},$$
$$t + n - s \notin \{\delta_{s+1} + n - s - 1, \ldots, \delta_r + n - r, \alpha_{r+1} + n - r - 1, \ldots, \alpha_n\}\}.$$

By construction the weight $(\delta_1, \ldots, \delta_r, \alpha_{r+1}, \ldots, \alpha_n) + \rho$ is not orthogonal to any root. By (4.1.9) there exists a unique l such that $H^l(\mathrm{Grass}(n-r, E), K_{\delta'} \mathcal{Q} \otimes K_{\alpha^{(2)\prime}} \mathcal{R}) \ne 0$. We define $l(\alpha) := l$.

(6.7.6) Proposition. *Let $\mathbf{K}$ be a field of characteristic zero. Let $\alpha = (\alpha^{(1)}, \alpha^{(2)})$ be a weight for $GL(F)$ with $\alpha^{(1)}$, $\alpha^{(2)}$ dominant. Let $l(\alpha)$ be defined as above.*

Then

$$H^i(Z_1, \mathcal{M}^a(\alpha)) = 0 \qquad \text{for} \quad i > l(\alpha).$$

Proof. This is the analogue of the first statement of (6.5.6). Notice that the analogue of the second statement need not be true. ∎

Each equivariant sheaf $\mathcal{M}^a$ on Z_1 has its Euler characteristic class

$$\chi(\mathcal{M}^a) = \sum_{i \geq 0} (-1)^i [H^i(Z_1, \mathcal{M}^a)] \in K_0'(A_r^a).$$

(6.7.7) Proposition. *The group $K_0'(A_r^a)$ is generated by the Euler characteristic classes $\chi(\mathcal{M}^a(\alpha))$ for $(\alpha^{(1)}, \alpha^{(2)}$ dominant).*

Proof. This is the analogue of (6.5.8). We use a similar induction, using the flag variety $\text{Flag}(n - r, n - r + 2; E)$. ∎

(6.7.8) Remarks. *Notice that, as in section 6.5, we can also show that if $(\alpha^{(1)}, \alpha^{(2)})$ is dominant, then the Euler characteristic of $\hat{\mathcal{M}}^a(\alpha^{(1)}, \alpha^{(2)})$ is in the subgroup of $K_0'(A_r^a)$ spanned by the Euler characteristics of sheaves $\mathcal{M}^a(\alpha, \beta)$ with α dominant.*

Theorem (6.7.3) follows from the following statement

(6.7.9) Proposition. *The classes $\chi(\mathcal{M}^a(\alpha^{(1)}, \alpha^{(2)})$ such that $(\alpha^{(1)}, \alpha^{(2)})$ is dominant generate the group $K_0'(A_r^a)$.*

Proof. This is the analogue of (6.5.10). Again we induct on $s := s(\alpha^{(1)}, \alpha^{(2)}) = \alpha_1^{(2)} - \alpha_r^{(1)}$. ∎

Proof of Theorem (6.7.4). Part (a) is a consequence of the fact that the characters of $M^a(\alpha)(-i)$ are linearly independent in $\text{Rep}(GL(E))[[q]][q^{-1}]$. To prove (b) we define the homomorphism of abelian groups

$$\Psi : \text{Rep}(GL(E))[q][q^{-1}] \to K_0'(A_r^a)$$

by sending the class $[(L_{\alpha'}E)q^i]$ to $[M^a(\alpha)(-i)]$. By Theorem (6.7.3) the homomorphism Ψ is an epimorphism. It is also a monomorphism, because the characters of modules $M^a(\alpha)(-i)$ are linearly independent in $\mathrm{Rep}(\mathrm{GL}(E))$ $[[q]][q^{-1}]$ and therefore in $K_0'(A_r^a)$. $\blacksquare$

Next we give a criterion for a module $M^a(\alpha)$ to be maximal Cohen–Macaulay.

We use Corollary (5.1.5). Before we start, let us work out the duality statement for the variety Z_1. To this end we need to calculate the bundle $\bigwedge^m \xi^* \otimes K_V$ where $m = \mathrm{rank}\,\xi$. Using (3.3.5), we get $K_V = (\bigwedge^r \mathcal{Q})^{-n+r} \otimes (\bigwedge^{n-r} \mathcal{R})^r$. Similarly, $\bigwedge^m \xi^* = \bigwedge^{r(r-1)/2} S \bigwedge^2 \mathcal{Q} \otimes \bigwedge^{n(n-1)/2} \bigwedge^2 F^*$. Putting these facts together, we conclude that the dualizing bundle is

$$\omega_{Z_1} = K_{(-n+1+r)^n} F^* \otimes \left(\bigwedge^r \mathcal{Q} \right)^{-n+r-1}.$$

Therefore for the weight

$$\alpha = (\alpha_1^{(1)}, \ldots, \alpha_r^{(1)}, \alpha_1^{(2)}, \alpha_{n-r}^{(2)}),$$

the dual weight is given by the formula

$$\alpha^\vee = (-n+r-1-\alpha_r^{(1)}, \ldots -n+r-1-\alpha_1^{(1)}, -\alpha_{n-r}^{(2)}, \ldots, -\alpha_1^{(2)}).$$

(6.7.10) Proposition. *The module $M^a(\alpha)$ is maximal Cohen–Macaulay provided that $l(\alpha) = l(\alpha^\vee) = 0$.*

Proof. Let's denote $\mathcal{V}(\alpha) = K_{\alpha^{(1)}} \mathcal{Q} \otimes K_{\alpha^{(2)}} \mathcal{R}$. One applies Corrolary (5.1.5) to the complexes $F_\bullet(\mathcal{V}(\alpha))$ and $F_\bullet(\mathcal{V}(\alpha^\vee))$. The vanishing of the higher cohomology is assured by Proposition (6.7.6). $\blacksquare$

(6.7.11) Example. *Let us assume that $\alpha^{(2)} = 0$. Then $M^a(\alpha)$ has nonvanishing higher cohomology by Proposition (6.6.6). Let us also assume $\alpha_1^{(1)} \le 1$. This means we consider the modules*

$$M^a(1^v) = H^0\left(\mathrm{Grass}(n-r, F), \bigwedge^v \mathcal{Q} \otimes \mathrm{Sym}\left(\bigwedge^2 \mathcal{Q} \right) \right).$$

Then condition $l(\alpha^\vee) = 0$ is equivalent to $v \ge 3$ (which implies $r = 2u \ge 4$). This proves that the modules $M^a(1^v)$ are maximal Cohen–Macaulay for $3 \le v \le r$.

Exercises for Chapter 6

Analogues of Determinantal Varieties for Other Classical Groups

The Symplectic Group

Let F be a symplectic space of dimension $2n$, let G be a vector space, $\dim G = m$. Consider $X = \mathrm{Hom}_K(F, G)$. This space can be identified with the set of m-tuples of vectors from F.

1. For $1 \le r \le n$ define

    $$Y_r = \{\phi \in X \mid \exists R \in \mathrm{IGrass}(r, F), \ \phi(R) = 0 \}.$$

    Prove that Y_r has a resolution of singularities which is a total space of a vector bundle over $\mathrm{IGrass}(r, F)$

2. Prove, by using Theorem (5.1.2)(b), that in that case Y_n is normal and has rational singularities.

3. Let $m \le n$. Calculate the complex $F_\bullet$, and use it to show that the variety Y_n is a complete intersection given by the vanishing of $\binom{m}{2}$ $\mathrm{Sp}(F)$-invariants of degree 2, given by the representation $\bigwedge^2 G^*$.

4. Use exercises 2 and 3 to show that there exists an acyclic complex $K(n, \lambda; F)_\bullet$ resolving the irreducible representation $V_\lambda F$ of the group $\mathrm{Sp}(F)$ in terms of Schur functors, with the i-th term

    $$K(n, \lambda; F)_i := \bigoplus_{\mu \in Q_{-1}(2i)} K_{\lambda/\mu} F.$$

5. Let $r = n$, $\dim G = m > n$. Calculate the terms of the complex $F_\bullet$. Prove that the variety Y_n is a complete intersection in a determinantal variety of matrices of rank $\le n$, given by $\binom{m}{2}$ $\mathrm{Sp}(F)$-invariants.

6. Let $r < n$. Prove that Y_r is normal and has rational singularities. Assume that $m \le 2n - r + 1$. Then the terms of $F_\bullet$ contain only trivial $\mathrm{Sp}(F)$-representations. More precisely, $F_\bullet$ is a specialization of the resolution of $2(n - r + 1)$ Pfaffians of the generic skew symmetric $m \times m$ matrix (described in section 6.3), where our $m \times m$ skew symmetric matrix is a matrix of $\mathrm{Sp}(F)$-invariants in $A = \mathrm{Sym}(F \otimes G^*)$, given by the representation $\bigwedge^2 G^* \subset A_2$.

7. For the example $2n = 6$, $r = 2$, $m = 6$ calculate the terms of the complex $F_\bullet$ and prove that Y_r is not a complete intersection in a determinantal variety.

8. Prove that the defining ideal of Y_r is generated by $2(n - r + 1)$ Pfaffians of the skew symmetric $m \times m$ matrix of $Sp(F)$-invariants in $A = Sym(F \otimes G^*)$, given by the representation $\bigwedge^2 G^* \subset A_2$.

9. Let $1 \leq r < n$. Define the variety
$$Y_{2n-r} = \{\phi \in X \mid \exists R \in IGrass(r, F), \ \phi(R^\vee) = 0\}.$$

   Prove that Y_{2n-r} has a resolution of singularities which is a total space of a vector bundle over $IGrass(r, F)$. Prove that Y_{2n-r} is normal , with rational singularities.

10. Prove that the defining ideal of Y_{2n-r} is generated by $Sp(F)$-invariants in A_2 (given by the representation $\bigwedge^2 G^*$) and the $(r + 1) \times (r + 1)$ minors of the matrix ϕ.

The Orthogonal Group

We will formulate the exercises for the even orthogonal group. The formulations for the odd orthogonal group are left to the reader.

Let F be an orthogonal space of dimension $2n$; let G be a vector space, $\dim G = m$. Consider $X = Hom_K(F, G)$. This space can be identified with the set of m-tuples of vectors from F.

11. For $1 \leq r \leq n$ let
$$Y_r = \{\phi \in X \mid \exists R \in IGrass(r, F), \ \phi(R) = 0 \}.$$

   Prove that Y_r has a resolution of singularities which is a total space of a vector bundle over $IGrass(r, F)$. (Hint: $\xi = \mathcal{R} \otimes G^*$.)

12. Prove, by using Theorem $(5.1.2)(b)$, that in that case Y_n is normal and has rational singularities.

13. Let $m \leq n$. Calculate the complex $F_\bullet$ to show that the variety Y_n is a complete intersection given by the vanishing of $\binom{m+1}{2}$ $O(F)$-invariants of degree 2, given by the representation $S_2 G^*$.

14. Let λ be a dominant weight for $SO(F)$ with integer coordinates. Use exercises 12 and 13 to show that there exists an acyclic complex $K(n, \lambda; F)_\bullet$ resolving the irreducible representation $V_\lambda F$ of the group $SO(F)$ in terms of Schur functors, with the i-th term
$$K(n, \lambda; F)_i := \bigoplus_{\mu \in Q_1(2i)} K_{\lambda/\mu} F.$$

15. Let $r = n$, $\dim G = m > n$. Calculate the terms of the complex $F_\bullet$. Prove that the variety Y_n is a complete intersection in a determinantal variety of matrices of rank $\leq n$, given by $\binom{n+1}{2}$ $SO(F)$-invariants.

16. Let $r < n$. Prove that Y_r is not normal, but that its normalization has rational singularities. Assume that $m \le 2n - r$. Then the terms of $F_\bullet$ contain only trivial $SO(F)$-representations.

17. For the example $2n = 8, r = 2, m = 6$ calculate the terms of the complex $F_\bullet$.

18. Prove that the defining ideal of Y_r is generated by $(n - r + 1) \times (n - r + 1)$ minors of the symmetric $m \times m$ matrix of $SO(F)$-invariants in $A = \text{Sym}(F \otimes G^*)$, given by the representation $S_2 G^* \subset A_2$.

19. Let $1 \le r \le n$. Define the variety

$$Y_{2n-r} = \{\, \phi \in X \mid \exists R \in \text{IGrass}(r, F), \ \phi(R^\vee) = 0 \,\}.$$

Prove that Y_{2n-r} has a resolution of singularities which is a total space of a vector bundle over $\text{IGrass}(r, F)$. (Hint: $\xi = \mathcal{R}^\vee \otimes G^*$.) Prove that Y_{2n-r} is normal, with rational singularities.

20. Prove that the defining ideal of Y_{2n-r} is generated by $O(F)$-invariants in A_2 (given by the representation $S_2 G^*$) and the $(r + 1) \times (r + 1)$ minors of the matrix ϕ.

The First Fundamental Theorem for the General Linear Group

Let E be a vector space of dimension n. Let $X = E^{\otimes m} \oplus E^{*\otimes p}$. We identify X with $\text{Hom}_K(G, E) \oplus \text{Hom}_K(E, H)$ where G, H are two vector spaces, $\dim G = m$, $\dim H = p$. We have $A = \text{Sym}(G \otimes E^* \oplus E \otimes H^*)$. We can identify X with the set of m-tuples of vectors from E and p-tuples of covectors from E^*.

21. For each pair $(r, s), r + s = n, r \le m, s \le p$, consider the variety

$$Y_{r,s} = \{(\phi, \psi) \in X \mid \psi\phi = 0, \ \text{rank}\,\phi \le r, \ \text{rank}\,\psi \le s \,\}.$$

Prove that $Y_{r,s}$ has a desingularization $Z_{r,s}$ with $V = \text{Grass}(r, E), \xi = G \otimes Q^* \oplus \mathcal{R} \otimes H^*$. Prove that $Y_{r,s}$ is normal and has rational singularities.

22. Choose $m + p = n, r = m, s = p$. Prove that in this case the complex $F_\bullet$ is a Koszul complex on the $GL(E)$-invariants in A of bidegree $(1, 1)$ which correspond to the representation $G \otimes H^*$.

23. Let $\lambda = (\lambda_1, \ldots, \lambda_r), \mu = (\mu_1, \ldots, \mu_s)$ be two partitions. Take the isotypic component of type $K_\lambda G \otimes K_\mu H^*$ to obtain the complex $K(r, s, \lambda, \mu; E)_\bullet$ with the following properties:
    (a) $K(r, s, \lambda, \mu; E)_\bullet$ is acyclic.

(b) The i-th term of $K(r, s, \lambda, \mu; E)_\bullet$ is

$$K(r, s, \lambda, \mu; E)_i = \bigoplus_{|v|=i} K_{\lambda/v} E^* \otimes K_{\mu/v'} E.$$

(c) The complex $K(r, s, \lambda, \mu; E)_\bullet$ resolves the representation $K_{(\mu, -\lambda)} E$, where $(\mu, -\lambda) = (\mu_1, \ldots, \mu_s, -\lambda_r, \ldots, -\lambda_1)$.

Isotropic Grassmannians Revisited

24. Prove that the equations of isotropic Grassmannians defined in exercises 1, 2, 3 in chapter 4 generate (together with the Plücker relations) the defining ideals of the cones over isotropic Grassmannians.

Differentials in the Resolutions of Ideals of Minors of a Generic Matrix

25. We work with the notation of section 6.1. Recall that by (6.1.3) the i-th term in the Lascoux complex equals

$$F_i = \bigoplus_{s \geq 0} \bigoplus_{(\alpha, \beta) \in Q(s), \ i = s^2 + |\alpha| + |\beta|} L_{P_1(\alpha, \beta)} F \otimes L_{P_2(\alpha, \beta)} G^* \otimes_K A.$$

Denote

$$F_i^{(s)} = \bigoplus_{(\alpha, \beta) \in Q(s), \ i = s^2 + |\alpha| + |\beta|} L_{P_1(\alpha, \beta)} F \otimes L_{P_2(\alpha, \beta)} G^* \otimes_K A,$$

so $F_i = \bigoplus_{s \geq 0} F_i^{(s)}$. Prove that the differential $d_i : F_i \to F_{i-1}$ has only components of degree 1 taking $F_i^{(s)}$ to $F_{i-1}^{(s)}$ and components of degree $r + 1$ taking $F_i^{(s)}$ to $F_{i-1}^{(s-1)}$.

26. Prove that the only possible nonzero components of the degree 1 part of the differential $d_i^{1,s} : F_i^{(s)} \to F_{i-1}^{(s)}$ restricted to $L_{P_1(\alpha, \beta)} F \otimes L_{P_2(\alpha, \beta)} G^* \otimes_K A$ go to the terms $L_{P_1(\gamma, \delta)} F \otimes L_{P_2(\gamma, \delta)} G^* \otimes_K A$ with $\gamma \subset \alpha, \delta \subset \beta$, $|\alpha/\gamma| = 1, |\beta/\delta| = 1$.

27. Prove that the only possible nonzero components of the degree $r + 1$ part of the differential $d_i^{r+1,s} : F_i^{(s)} \to F_{i-1}^{(s-1)}$ are as follows. The map $d_i^{r+1,s}$ restricted to the term $L_{P_1(\alpha, \beta)} F \otimes L_{P_2(\alpha, \beta)} G^* \otimes_K A$ is zero unless $\alpha_s = 0$, $\alpha_1' \leq s - 1, \beta_s = 0, \beta_1' \leq s - 1$. If these conditions are satisfied, the only nonzero component of $d^{r+1,s}$ restricted to $L_{P_1(\alpha, \beta)} F \otimes L_{P_2(\alpha, \beta)} G^* \otimes_K A$ goes to $L_{P_1(\gamma, \delta)} F \otimes L_{P_2(\gamma, \delta)} G^* \otimes_K A$ from $F_{i-1}^{(s-1)}$, where $\gamma = (\alpha_1 + 1, \ldots, \alpha_{s-1} + 1)$, $\delta = (\beta_1 + 1, \ldots, \beta_{s-1} + 1)$. The coefficients of that component are the linear combinations of $(r + 1) \times (r + 1)$ minors of Φ.

Differentials in the Resolutions of Ideals of Minors
of a Generic Symmetric Matrix

28. We work with the notation of section 6.3. By Theorem (6.3.1) the i-th term G_i of the resolution of the ideal of $(r+1) \times (r+1)$ minors of a generic symmetric matrix is given by

$$G_i = \bigoplus_{u \geq 0} \bigoplus_{\lambda(\alpha,u); \ i=|\alpha|+2u^2-u} L_{\lambda(\alpha,u)} E \otimes_K A^s.$$

Denote

$$G_i^{(u)} = \bigoplus_{\lambda(\alpha,u); \ i=|\alpha|+2u^2-u} L_{\lambda(\alpha,u)} E \otimes_K A^s,$$

so $G_i = \bigoplus_{u \geq 0} G_i^{(u)}$. Prove that the differential $d_i : G_i \to G_{i-1}$ has only components of degree 1 taking $G_i^{(u)}$ to $G_{i-1}^{(u)}$ and components of degree $r+1$ taking $G_i^{(u)}$ to $G_{i-1}^{(u-1)}$.

29. Prove that the only possible nonzero components of the degree 1 part of the differential $d_i^{1,u} : G_i^{(u)} \to G_{i-1}^{(u)}$ restricted to $L_{\lambda(\alpha,u)} \otimes_K A^s$ go to the terms $L_{\lambda(\beta,u)} \otimes_K A^s$ with $\beta \subset \alpha$, $|\alpha/\beta| = 1$.

30. Prove that the only possible nonzero components of the degree $r+1$ part of the differential $d_i^{r+1,u} : G_i^{(u)} \to G_{i-1}^{(u-1)}$ are as follows. The map $d_i^{r+1,u}$ restricted to the term $L_{\lambda(\alpha,u)} \otimes_K A^s$ is zero unless $\alpha_{2u-1} = \alpha_{2u} = 0$, $\alpha_1' \leq 2u - 1$. If these conditions are satisfied, the only nonzero component of $d^{r+1,u}$ restricted to $L_{\lambda(\alpha,u)} \otimes_K A^s$ goes to $L_{\lambda(\beta,u-1)} \otimes_K A^s$, where $\beta = (\alpha_1 + 2, \ldots, \alpha_{2u-2} + 2)$. The coefficients of that component are the linear combinations of $(r+1) \times (r+1)$ minors of Φ.

Differentials in the Resolutions of Ideals of Pfaffians
of a Generic Skew Symmetric Matrix

31. We work with the notation of section 6.4. By Theoerm (6.4.1) the i-th term G_i of the resolution of the ideal of $(2r+2) \times (2r+2)$ Pfaffians of a generic skew symmetric matrix is given by

$$G_i = \bigoplus_{v \geq 0} \bigoplus_{\lambda(\alpha,v); \ i=|\alpha|+\frac{1}{2}v(v+1)} L_{\lambda(\alpha,v)} E \otimes_K A^a.$$

Denote

$$G_i^{(v)} = \bigoplus_{\lambda(\alpha,v); \ i=|\alpha|+\frac{1}{2}v(v+1)} L_{\lambda(\alpha,v)} E \otimes_K A^a,$$

so $G_i = \bigoplus_{v \geq 0} G_i^{(v)}$. Prove that the differential $d_i : G_i \to G_{i-1}$ has only components of degree 1 taking $G_i^{(v)}$ to $G_{i-1}^{(v)}$ and components of degree $r + 1$ taking $G_i^{(v)}$ to $G_{i-1}^{(v-1)}$.

32. Prove that the only possible nonzero components of the degree 1 part of the differential $d_i^{1,v} : G_i^{(v)} \to G_{i-1}^{(v)}$ restricted to $L_{\lambda(\alpha,v)} \otimes_{\mathbf{K}} A^a$ go to the terms $L_{\lambda(\beta,v)} \otimes_{\mathbf{K}} A^a$ with $\beta \subset \alpha$, $|\alpha/\beta| = 1$.

33. Prove that the only possible nonzero components of the degree $r + 1$ part of the differential $d_i^{r+1,v} : G_i^{(v)} \to G_{i-1}^{(v-1)}$ are as follows. The map $d_i^{r+1,v}$ restricted to the term $L_{\lambda(\alpha,v)} \otimes_{\mathbf{K}} A^a$ is zero unless $\alpha_v = 0$, $\alpha_1' \leq v - 1$. If these conditions are satisfied, the only nonzero component of $d^{r+1,v}$ restricted to $L_{\lambda(\alpha,v)} \otimes_{\mathbf{K}} A^a$ goes to $L_{\lambda(\beta,v-1)} \otimes_{\mathbf{K}} A^a$, where $\beta = (\alpha_1 + 1, \ldots, \alpha_{v-1} + 1)$. The coefficients of that component are the linear combinations of $(2r + 2) \times (2r + 2)$ Pfaffians of Φ.

Maximal Cohen–Macaulay Modules with Linear Resolutions

34. Consider the twisted sheaf $\mathcal{M}((n - r)^r, (0)) = K_{(n-r)^r} \mathcal{Q} \otimes \mathcal{O}_{Z_r^{(1)}}$ defined in section 6.5.

    (a) Prove that the sheaf $\mathcal{M}((n - r)^r, (0))$ has no higher cohomology, so the twisted complex $F(K_{(n-r)^r} \mathcal{Q})_\bullet$ provides a minimal resolution of $M((n - r)^r, (0))$.

    (b) Show that the complex $F(K_{(n-r)^r} \mathcal{Q})_\bullet$ has length $(m - r)(n - r)$ and that it has a linear differential. More precisely, the only nonvanishing terms of the complex $F(K_{(n-r)^r} \mathcal{Q})_\bullet$ are

    $$F(K_{(n-r)^r} \mathcal{Q})_i = H^0 \left(\text{Grass}(m - r, F), K_{(n-r)^r} \mathcal{Q} \otimes \bigwedge^i \xi \right)$$

    for $0 \leq i \leq (m - r)(n - r)$.

    (c) Use the duality from exercise 18, chapter 2, changing the Schur functors on F to the Schur functors on F^*, to identify the complex $F(K_{(n-r)^r} \mathcal{Q})_\bullet$ with the Schur complex $L_{(m-r)^{(n-r)}}(\Phi^*)$.

    (d) Use Buchsbaum–Eisenbud acyclicity criterion (1.2.12) to prove that over a field $\mathbf{K}$ of arbitrary characteristic the Schur complex $L_{(m-r)^{(n-r)}}(\Phi^*)$ is acyclic. Note that the Euler characteristic of this complex in the relative situation (i.e. when F, G are replaced by vector bundles over some scheme) gives the class occurring in the Porteous formula for the cohomology class of the degeneracy locus X_r. The module $M((n - r)^r, (0))$ is a maximal Cohen–Macaulay module supported in X_r with a linear resolution.

35. Consider the twisted sheaf $\mathcal{M}((n-r)^r, n-r-1, n-r-2, \ldots, 1, 0)$ defined in section 6.6. It is supported in the symmetric determinantal variety X_r^s of symmetric $n \times n$ matrices of rank $\leq r$.

   (a) Prove that the sheaf $\mathcal{M}((n-r)^r, n-r-1, n-r-2, \ldots, 1, 0)$ has no higher cohomology, so the twisted complex $F(K_{(n-r)^r} \mathcal{Q} \otimes K_{(n-r-1, n-r-2, \ldots, 1, 0)} \mathcal{R})_\bullet$ provides a minimal resolution of $\mathcal{M}((n-r)^r, n-r-1, n-r-2, \ldots, 1, 0)$.

   (b) Show that the complex $F(K_{(n-r)^r} \mathcal{Q} \otimes K_{(n-r-1, n-r-2, \ldots, 1, 0)} \mathcal{R})_\bullet$ has length $\binom{n-r+1}{2}$ and that it has a linear differential. More precisely, the only nonvanishing terms of the complex $F(K_{(n-r)^r} \mathcal{Q} \otimes K_{(n-r-1, n-r-2, \ldots, 1, 0)} \mathcal{R})_\bullet$ are

$$H^0\left(\mathrm{Grass}(n-r, F), K_{(n-r)^r} \mathcal{Q} \otimes K_{(n-r-1, n-r-2, \ldots, 1, 0)} \mathcal{R} \otimes \overset{i}{\bigwedge} \xi\right).$$

   for $0 \leq i \leq \binom{n-r+1}{2}$. Conclude that $\mathcal{M}((n-r)^r, n-r-1, n-r-2, \ldots, 1, 0)$ is a maximal Cohen–Macaulay module supported in X_r^s with a linear resolution.

36. Consider the twisted sheaf $\mathcal{M}((n-r-1)^r, n-r-1, n-r-2, \ldots, 1, 0)$ defined in section 6.7. It is supported in the skew symmetric determinantal variety X_r^a of skew symmetric $n \times n$ matrices of rank $\leq r$. Here we assume that $r = 2u$ is even.

   (a) Prove that the sheaf $\mathcal{M}((n-r-1)^r, n-r-1, n-r-2, \ldots, 1, 0)$ has no higher cohomology, so the twisted complex $F(K_{(n-r-1)^r} \mathcal{Q} \otimes K_{(n-r-1, n-r-2, \ldots, 1, 0)} \mathcal{R})_\bullet$ provides a minimal resolution of $\mathcal{M}((n-r-1)^r, n-r-1, n-r-2, \ldots, 1, 0)$.

   (b) Show that the complex $F(K_{(n-r-1)^r} \mathcal{Q} \otimes K_{(n-r-1, n-r-2, \ldots, 1, 0)} \mathcal{R})_\bullet$ has length $\binom{n-r}{2}$ and that it has a linear differential. More precisely, the only nonvanishing terms of the complex $F(K_{(n-r-1)^r} \mathcal{Q} \otimes K_{(n-r-1, n-r-2, \ldots, 1, 0)} \mathcal{R})_\bullet$ are

$$H^0\left(\mathrm{Grass}(n-r, F), K_{(n-r-1)^r} \mathcal{Q} \otimes K_{(n-r-1, n-r-2, \ldots, 1, 0)} \mathcal{R}) \otimes \overset{i}{\bigwedge} \xi\right)$$

   for $0 \leq i \leq \binom{n-r}{2}$. Conclude that $\mathcal{M}((n-r-1)^r, n-r-1, n-r-2, \ldots, 1, 0)$ is a maximal Cohen–Macaulay module supported in X_r^a with a linear resolution.

Resolutions of $K_\lambda(\Phi)$

37. Consider the Grassmannian $\mathrm{Grass}(m-n, F)$. Consider the incidence variety

$$Z_m = \{(\phi, R) \in X \times \mathrm{Grass}(m-r, F) \mid \phi \mid R = 0\}.$$

Denote by

$$0 \to \hat{\mathcal{R}} \to F \times \mathrm{Grass}(m - n, F) \to \hat{\mathcal{Q}} \to 0$$

the tautological sequence on $\mathrm{Grass}(m - n, F)$. Define

$$\mathcal{X}(\lambda) := K_\lambda \hat{\mathcal{Q}} \otimes \mathcal{O}_{Z_m}$$

The corresponding modules of sections are

$$X(\lambda) := H^0(Z_m, K_\lambda \hat{\mathcal{Q}} \otimes \mathcal{O}_{Z_m}).$$

(a) Prove that $H^i(Z_m, \mathcal{X}(\lambda)) = 0$ for $i > 0$,
(b) Prove that $p_* \mathcal{X}(\lambda) := K_\lambda \hat{\mathcal{Q}} \otimes \mathrm{Sym}(\hat{\mathcal{Q}} \otimes G^*)$ with $\mathcal{R}^i p_* \mathcal{X}(\lambda) = 0$ for $i > 0$.
(c) Conclude that $F(K_\lambda(\hat{\mathcal{Q}}))_\bullet$ gives a minimal free resolution of the module $X(\lambda)$.
(d) Assume that the partition λ has i boxes on the diagonal, i.e. $\lambda_{i+1} \leq i \leq \lambda_i$. Let us also assume that λ has exactly s nonzero parts. Show that the module $X(\lambda)$ has a minimal presentation

$$K_{(\lambda, 1^{n+1-s})}F \otimes \overset{n+1-s}{\bigwedge} G^* \otimes_{\mathbf{K}} A(-n - 1 + s)$$
$$\to K_\lambda F \otimes_{\mathbf{K}} A \to X(\lambda) \to 0.$$

(e) Show that the projective dimension of $X(\lambda)$ is equal to $(m - n)i$.
(f) Assume that $i = 1$. Show that the complex $F(K_\lambda \hat{\mathcal{Q}})_\bullet$ can be augmented by one more map $K_\lambda F \otimes_{\mathbf{K}} A \to K_\lambda G^* \otimes_{\mathbf{K}} A(|\lambda|)$ to get a longer minimal free resolution. In other words, the module $X(\lambda)$ turns out to be the first syzygy of the module $C(\lambda) := \mathrm{Coker}(K_\lambda(\Phi))$. The modules $C(\lambda)$ are therefore perfect modules supported in $I_n(\phi)$.

38. Let $i = 1$. Write $\lambda = (q, 1^{p-1})$.
   (a) Use the formula (2.32)(b) and exercise 5 of chapter 4 to prove that the argument from exercise 37 can be made characteristic free. Writing $\lambda = (q, 1^{p-1})$, the terms at the right end of the characteristic free version of the resolution of $C(\lambda)$ (defined as the cokernel of $L_{(p,1^{q-1})}(\phi)$) are as follows:

$$F_0 = L_{(p,1^{q-1})}G \otimes_{\mathbf{K}} A(p + q - 1),$$
$$F_1 = L_{(p,1^{q-1})}F \otimes_{\mathbf{K}} A,$$
$$F_i = L_{(i+n-1,1^{q-1})}F \otimes (L_{(n-p+1,1^{i-2})}G)^* \otimes_{\mathbf{K}} A(-n + p + i - 1)$$

for $2 \leq i \leq m - n + 1$,

(b) Use exercise 17 of chapter 2 to describe the differential in the resolution of $C(p, 1^{q-1})$ explicitly. Prove that $C(p, 1^{q-1})$ is a perfect module. This resolution was first described in [BE2]. The free resolutions of the modules $C(\lambda)$ for partitions with more boxes on the diagonal were analyzed in [Ar1] and [Ar2]. None of these modules are perfect.

39. Let $\lambda = (k^n)$.

(a) Prove that the module $X(\lambda) \otimes_K (\bigwedge^n G^*)^{\otimes k}$ is isomorphic (as an equivariant $GL(F) \times GL(G^*)$-module) to the k-th power of the ideal $I_n(\phi)$ of maximal minors of ϕ. The projective dimension of $I_n(\phi)^k$ equals $(m - n) \min(k, n)$.

(b) Prove that the complexes $F(L_{\lambda'}\hat{Q})_\bullet$ are characteristic free and linear when there are no terms coming from higher cohomology, i.e. when $\lambda_{i+1} = \ldots = \lambda_n = i$.

Resolutions of Powers of the Ideal of $2t \times 2t$ Pfaffians of a $(2t + 1) \times (2t + 1)$ Skew Symmetric Matrix

40. Take $\dim E = 2t + 1$, $V = \text{Grass}(1, E)$ with the tautological sequence

$$0 \to \hat{R} \to E \times V \to \hat{Q} \to 0.$$

Define

$$Z_{2n} = \{(\phi, R) \in X \times V \mid \phi \mid_R = 0\},$$

where i denotes the embedding of $\hat{R}$ into $E \times V$. Denote by p, q, q' the usual projections. Then $\mathcal{X}^a_\lambda = K_\lambda \hat{Q} \otimes \mathcal{O}_{Z_{2n}}$ on Z_{2n}, where $\lambda = (\lambda_1, \ldots, \lambda_{2t})$ is a partition into at most $2t$ parts. Denote

$$X^a(\lambda) = H^0(Z_{2t}, \mathcal{X}^a(\lambda)).$$

(a) $H^i(Z_{2t}, \mathcal{X}^a(\lambda)) = 0$ for $i > 0$.

(b) $p_* \mathcal{X}^a(\lambda) := K_\lambda \hat{Q} \otimes \text{Sym}(\bigwedge^2 \hat{Q})$ with $\mathcal{R}^i p_* \mathcal{X}^a(\lambda) = 0$ for $i > 0$.

(c) The complex $F(K_\lambda(\hat{Q}))_\bullet$ gives a minimal free resolution of the module $X^a(\lambda)$.

(d) Let $\lambda = (k^{2t})$. Show that the module $X^a(\lambda) = I^a_{2t}(\phi)^k$ has a linear free resolution for k even and has one dimensional representation outside the linear strand for k odd, $k < 2t + 1$.

(e) Show

$$\text{pd}_A(X(k^{2t})) = \begin{cases} k & \text{for } k \text{ even}, \\ k + 1 & \text{for } k \text{ odd}. \end{cases}$$

More precisely, the terms in the resolution are

$$F_i = L_{(k^{2t},i)} F \otimes_{\mathbf{K}} A(-i)$$

for $0 \le i \le \min(k, 2t)$, with the additional term

$$F_{k+1} = L_{((k+1)^{2t+1})} F \otimes_{\mathbf{K}} A\left(-\frac{k+1+2t}{2} \right)$$

occurring for k odd, $k < 2t$.

(f) Show, using Kempf's vanishing theorem and exercise 5 of chapter 4, that the Betti numbers of powers of $I_{2t}^a(\phi)^k$ do not depend on the characteristic of $\mathbf{K}$. The differentials in these resolutions were described explicitly in [BS] and [KU].

7

Higher Rank Varieties

In this chapter we investigate the higher rank varieties. They are the analogues of determinantal varieties for more complicated representations $L_\lambda E$. They were first considered in the paper [Po] of Porras.

In section 7.1 we look at the general case. We prove that higher determinantal varieties have rational singularities, and we find equations defining them set-theoretically. We also classify the rank varieties whose defining ideals are Gorenstein.

In section 7.2 we investigate the rank varieties for symmetric tensors of degree bigger than two. We prove that in this case the defining equations described in section 7.1 generate the radical ideal. We also analyze the cases of tensors of rank one, which correspond to the cones over multiple embeddings of projective spaces.

In section 7.3 we look at rank varieties for skew symmetric tensors of degree bigger than two. An interesting feature is that the normality of these rank varieties depends on the characteristic of the base field. We pay particular attention to the special case of syzygies of Plücker ideals defining the cones over Grassmannians embedded into projective space by Plücker embeddings.

7.1. Basic Properties

Let λ be a partition. Let E be a vector space of dimension n over $\mathbf{K}$. Consider the representation $X = K_{\lambda'} E^*$ as an affine space over $\mathbf{K}$. Its coordinate ring can be identified with $A^\lambda = \mathbf{K}[X] = \mathrm{Sym}_{\mathbf{K}}(L_\lambda E)$. For $\lambda_1 \le r < n$ we define the rank variety $Y_r^\lambda \subset X$ of tensors of rank $\le r$,

$$Y_r^\lambda = \{\phi \in K_{\lambda'} E \mid \exists S \subset E, \ \dim S = r, \ \phi \in K_{\lambda'} S \subset K_{\lambda'} E \}.$$

This means the tensor ϕ has a rank $\le r$ if there exists a basis $\{e_1, \ldots, e_n\}$ of E such that ϕ can be written using the tensors involving $e_1, \ldots, e_r$. The condition $r \ge \lambda_1$ assures that $X_r \ne \emptyset$.

228

(7.1.1) Examples.

(a) *Let $\lambda = (2)$. Then $A^\lambda = \mathrm{Sym}(S_2 E)$, and Y_r^λ is the rank variety Y_r^s for symmetric matrices analyzed in section 6.3.*

(b) *Let $\lambda = (1^2)$. Then $A^\lambda = \mathrm{Sym}(\bigwedge^2 E)$. Assume that r is even. Then Y_r^λ is the rank variety Y_r^a for skew symmetric matrices considered in section 6.4. If r is odd, we get Y_r^λ to be Y_{r-1}^a.*

In order to analyze the variety Y_r^λ we use the obvious incidence variety

$$Z_r^\lambda = \{(\phi, S) \in K_{\lambda'} E \times \mathrm{Grass}(r, E^*) \mid \phi \in K_{\lambda'} S \subset K_{\lambda'} E^*\}.$$

We can identify the Grassmannian $\mathrm{Grass}(r, E^*)$ with $\mathrm{Grass}(n - r, E)$. We write the tautological sequence

$$0 \to \mathcal{R} \to E \times \mathrm{Grass}(n - r, E) \to \mathcal{Q} \to 0$$

with $\dim \mathcal{R} = n - r$, $\dim \mathcal{Q} = r$. The subspace S becomes a fiber of $\mathcal{Q}^*$.

The variety Z_r^λ is an analogue of varieties Z_1 from sections 6.3, 6.4. We can use our incidence variety in the same way as in chapter 6.

(7.1.2) Proposition. *Let $\mathbf{K}$ be a field of characteristic 0.*

(a) *The coordinate ring $\mathbf{K}[Y_r^\lambda]$ is normal and has rational singularities. In particular, $\mathbf{K}[Y_r^\lambda]$ is Cohen–Macaulay.*

(b) *The ideal I_r^λ of functions vanishing on Y_r^λ is a span of all representations $L_\mu E$ with $\mu_1 > r$ inside of $\mathrm{Sym}(L_\lambda E)$.*

Proof. Let us use the notation from section 5.1, denoting by $p : Z_r^\lambda \to \mathrm{Grass}$ $(n - r, E), q : X \times \mathrm{Grass}(n - r, E) \to X$, and $q' : Z_r^\lambda \to Y_r^\lambda$ the projections. We will write ξ^λ for ξ and η^λ for η. The bundle η^λ can be identified with $L_\lambda \mathcal{Q}$, and ξ^λ fits into an exact sequence

$$0 \to \xi^\lambda \to L_\lambda E \to L_\lambda \mathcal{Q} \to 0.$$

Using Theorem (5.1.2) (b) we see that it is enough to show that $H^i(\mathrm{Grass}$ $(n - r, E), \mathrm{Sym}(\eta^\lambda)) = 0$ for $i > 0$. But since the field $\mathbf{K}$ has characteristic zero, it is clear that for each j, $\mathrm{Sym}_j(\eta^\lambda) = \mathrm{Sym}_j(L_\lambda \mathcal{Q})$ decomposes to a direct sum (with multiplicities) of Schur functors $L_\mu \mathcal{Q}$. By Corollary (4.1.9) we get the vanishing.

It is also clear that the ring $H^0(\mathrm{Grass}(n - r, E), \mathrm{Sym}(\eta^\lambda))$ is a factor of $\mathrm{Sym}(L_\lambda E)$ obtained by factoring out all representations $L_\mu E$ with $\mu_1 > r$. This proves the normality of $H^0(\mathrm{Grass}(n - r, E), \mathrm{Sym}(\eta^\lambda))$ and part (b). ∎

It is an interesting but difficult problem to determine the defining equations of varieties Y_r^λ. It turns out that in general case we have an easy set of equations defining Y_r^λ set theoretically. Let us look at the map

$$L_{\lambda/(1)}E \overset{1\otimes\mathrm{Tr}}{\to} L_{\lambda/(1)}E \otimes E \otimes E^* \overset{m\otimes 1}{\to} L_\lambda E \otimes E^*,$$

where $1 \otimes \mathrm{Tr}$ is the multiplication by the trace element $\sum_{i=1}^n e_i \otimes e_i^*$, and m is an epimorphism $m : L_{\lambda/(1)}E \otimes E \to L_\lambda E$ sending the element $T \otimes u$ where T is a tableau of shape $\lambda/(1)$ to a tableau T with u inserted in the upper left corner. The presentation of $L_{\lambda/(1)}E$ by generators and relations from section 2.1 implies the existence of such an epimorphism. The map above induces the map of free A^λ-modules

$$\Phi_\lambda : L_{\lambda/(1)}E \otimes_{\mathbf{K}} A^\lambda(-1) \to E^* \otimes_{\mathbf{K}} A^\lambda.$$

(7.1.3) Proposition. *The variety Y_r^λ is defined set-theoretically by $(r + 1) \times (r + 1)$ minors of Φ_λ.*

Proof. Let $\phi \in K_{\lambda'}E^*$. Denote by $\Phi_\lambda(\phi)$ the linear map $L_{\lambda/(1)}E \to E^*$ obtained from Φ_λ by substituting for each linear function on $K_{\lambda'}E^*$ (identified with an element of degree one in A^λ) its value on a tensor ϕ.

If a tensor $\phi \in K_{\lambda'}S \subset K_{\lambda'}E$ for some subspace S of dimension r in E^*, then the image $\mathrm{Im}\,\Phi_\lambda(\phi)$ is clearly contained in S. Indeed, choose a basis $\{e_1, \ldots, e_{n-r}\}$ of $\mathrm{Ker}(E \to S^*)$, and complement it to the basis $\{e_1, \ldots, e_n\}$ of E, so $e_{n-r+1}^*, \ldots, e_n^*$ are a basis of S. Consider the basis of $L_\lambda E$ consisting of standard tableaux with respect to this basis. If a tableau contains a number $\leq n - r$, it is zero when evaluated on ϕ. Therefore the image of $\Phi_\lambda(\phi)$ is contained in the span of $e_{n-r+1}^*, \ldots, e_n^*$ which is S.

The other implication follows similarly. If rank $\Phi_\lambda(\phi) \leq r$, then there exists a subspace S of dimension r in E^* containing the image of $\Phi_\lambda(\phi)$. Choosing a basis $\{e_1, \ldots, e_n\}$ as above, we can conclude that if a standard tableau T contains a number $\leq n - r$, then the number in the upper left corner is $\leq n - r$, so T was gotten from inserting that number in an upper left corner of a tableau T' of shape $\lambda/(i)$. If T evaluated on ϕ were not zero, then the image $\Phi_\lambda(\phi)(T')$ would not be contained in S. ∎

Next we give a criterion for the projection $q' : Z_r^\lambda \to Y_r^\lambda$ to be a birational isomorphism.

(7.1.4) Proposition. *Let λ be a partition, r a number such that $\lambda_1 \leq r < n$. Assume that λ is not one of the partitions $(2), (r - 2), (1), (r - 1)$. Then the projection $q' : Z_r^\lambda \to Y_r^\lambda$ is a birational isomorphism.*

Proof. Let λ be a partition with $\lambda_1 \leq r < n$. First we notice that if there is a tensor $\phi \in K'_\lambda E$ such that rank $\Phi_\lambda(\phi) = r$, then the projection $q' : Z^\lambda_r \to Y^\lambda_r$ is a birational isomorphism. Indeed, the map sending ϕ to $(\phi, \operatorname{Im} \Phi_\lambda(\phi))$ is the inverse map to q' on an open subset of X_r consisting of tensors ϕ for which rank $\Phi_\lambda(\phi) = r$. This subset is nonempty by our assumption. The proposition follows now from the next statement.

(7.1.5) Lemma. *Let λ be a partition with $\lambda_1 \leq n$. Assume λ is not one of the partitions* (1), (2), $(n - 2)$, $(n - 1)$. *Then there exists in $K_{\lambda'} E$ a tensor ϕ of rank n.*

Proof. Let X_{n-1} be the subset of $K_{\lambda'} E$ of tensors of rank $\leq n - 1$. It is enough to show that if λ is not one of the partitions listed in the proposition, then $X_{n-1} \neq K_{\lambda'} E$.

If $\lambda' = (d)$, then the tensor $e_1^d + \ldots + e_n^d$ is easily seen to be of rank n. If $\lambda_1 = n$, it is clear that every tableau of shape λ in order to be nonzero has to have all n numbers from $[1, n]$ in the first row, so every nonzero tensor $\phi \in K_{\lambda'} E$ has rank n.

This means that when writing $\lambda = (\lambda_1, \ldots, \lambda_s)$ we can assume $2 \leq \lambda_1 \leq n - 1$.

Consider the modification Z^λ_{n-1}. It is enough to show that $\dim Z^\lambda_{n-1} < \dim K_{\lambda'} E$. The dimension of Z^λ_{n-1} equals $\dim Z^\lambda_{n-1} = n - 1 + \dim K_{\lambda'} E'$, where E' is a vector space of dimension $n - 1$. It is therefore enogh to show that if λ is not one of the partitions listed in the proposition then $\dim K_{\lambda'} E - \dim K_{\lambda'} E' > n - 1$.

First consider the case $\lambda = (t)$. Since

$$\dim \bigwedge^t E - \dim \bigwedge^t E' = \binom{n}{t} - \binom{n-1}{t} = \binom{n-1}{t-1},$$

we see that we are done in the case $2 < t < n - 1$. Notice that for $t = 2$ and $t = n - 2$ we have the quality above.

Let $\lambda = (\lambda_1, \ldots, \lambda_s)$ be an arbitrary partition with $s \geq 2$, $2 \leq t = \lambda_1 \leq n - 1$. It is enough to show that

$$\dim K_{\lambda'} E - \dim K_{\lambda'} E' > \dim \bigwedge^t E - \dim \bigwedge^t E'. \tag{$*$}$$

In order to show this fact, we recall that the dimension of $K_{\lambda'} E$ is the number of standard tableaux of shape λ with entries from $[1, n]$. For every tableau S of shape (t) with number n occurring in S we construct the standard tableau $T(S)$ by setting $T(S)(i, j) = S(1, j)$ for all $(i, j) \in D(\lambda)$. This proves that the weak inequality (∗) holds. In order to prove that the inequality is sharp, we produce one more standard tableau of shape λ with entries $[1, n]$, containing n. This will be the tableau U given by setting $U(i, \lambda_i + 1 - u) = n + 1 - u$ in the case when the partition λ is not rectangular. If λ is rectangular, then it is easy to produce a standard tableau U containing n which is not constant in columns by taking $U(i, \lambda_i + 1 - u) = n + 1 - u$ for $i \geq 2$ and $U(1, j) = j$. Lemma 7.1.5 is proved. ■

(7.1.6) Remark. *Propositions (7.1.2) and (7.1.3) generalize to several tensors. If $X = K_{\lambda^{(1)}} E^* \oplus \ldots \oplus K_{\lambda^{(t)}} E^*$, we could define the subvariety Y_r to be the set of t-tuples of tensors $(\phi_1, \ldots, \phi_t) \in X$ which can be simultaneously expressed using tableaux involving r basis vectors. The role of the map Φ_λ is played by the map*

$$\phi_{\lambda^{(1)}, \ldots, \lambda^{(t)}} : L_{\lambda^{(1)}} E \otimes_K A \oplus \ldots \oplus L_{\lambda^{(t)}} E \otimes_K A \to E^* \otimes_K A$$

defined on the j-th component using the tensor ϕ_j.

Finally we address the question when the defining ideal of Y_r^λ is Gorenstein.

(7.1.7) Theorem. *The variety Y_r^λ is defined by Gorenstein ideals in the following cases:*

(a) $n = \dfrac{|\lambda| \dim L_\lambda Q}{r}$.

In the remaining cases $n > (|\lambda| \dim L_\lambda Q)/r$:

(b1) $\lambda = (r^k, 1^2)$, $r > 1$, and $n - (|\lambda| \dim L_\lambda Q)/r$ is positive, divisible by $|\lambda|/2$,

(b2) $\lambda = (r^k, 2)$, $r > 2$ is even, and $n - (|\lambda| \dim L_\lambda Q)/r$ is positive, divisible by $|\lambda|$,

(b3) $\lambda = (r^k, (r-1)^2)$, $r \geq 1$, and $n - (|\lambda| \dim L_\lambda Q)/r$ is divisible by $|\lambda|/2$,

(b4) $\lambda = (r^k, r - 2)$, $r > 2$ is even, and $n - (|\lambda| \dim L_\lambda Q)/r$ is divisible by $|\lambda|$,

(b5) $\lambda = (r^k)$, $n > k$, and n is divisible by k.

Proof. We use the duality statement (5.1.4). Denote $t = \dim \xi$. We calculate the bundle $\bigwedge^t \xi^* \otimes \omega_V$. By (3.3.5) $\omega_V = \mathcal{O}_V(-n)$. Also we have $\bigwedge^t \xi^* \cong \mathcal{O}_V((|\lambda| \dim L_\lambda Q)/r)$. Therefore the dualizing bundle is given by $\mathcal{O}_V((|\lambda| \dim L_\lambda Q)/r - n)$. This means that in the case $n = (|\lambda| \dim L_\lambda Q)/r$ the defining ideal is Gorenstein.

Also in the case $n < (|\lambda| \dim L_\lambda \mathcal{Q})/r$ it cannot happen that the defining ideal is Gorenstein, because the module of sections of the sheaf $\mathcal{O}_V$ $((|\lambda| \dim L_\lambda \mathcal{Q})/r - n) \otimes \text{Sym}(L_\lambda \mathcal{Q})$ has a representation of dimension > 1 in degree 0, so it cannot be isomorphic to $K[Y_r^\lambda]$.

However, for $n > (|\lambda| \dim L_\lambda \mathcal{Q})/r$ it can still happen that the defining ideal is Gorenstein. It happens when the module of sections of $\mathcal{O}_V$ $((|\lambda| \dim L_\lambda \mathcal{Q})/r - n) \otimes \text{Sym}(L_\lambda \mathcal{Q})$ is isomorphic to $K[Y_r^\lambda]$.

The sections of $\mathcal{O}_V((|\lambda| \dim L_\lambda \mathcal{Q})/r - n) \otimes \text{Sym}(L_\lambda \mathcal{Q})$ are given by all representations $L_\mu \mathcal{Q}$ from $\text{Sym}(L_\lambda \mathcal{Q})$ such that for the conjugate partition $\mu' = (\mu'_1, \dots, \mu'_r)$ we have

$$\mu'_r \geq n - \frac{|\lambda| \dim L_\lambda \mathcal{Q}}{r}.$$

Such situation can occur only when the ideal consisting of representations $L_\mu \mathcal{Q}$ satisfying the above condition is generated by the one dimensional bundle $L_\mu \mathcal{Q}$ with $\mu' = (r^x)$, where $x = n - (|\lambda| \dim L_\lambda \mathcal{Q})/r$. This can happen only when for $L_\lambda F$, with $\dim F = r$, the variety of tensors of rank $< r$ has codimension 1.

Let us classify such cases.

(7.1.8) Lemma. *Let* $\dim F = r$. *Let* λ *be a partition with* $\lambda_1 < r$. *Then the subvariety* Y_{r-1}^λ *has codimension 1 in the following cases:*

(b1) $\lambda = (1^2)$,
(b2) $\lambda = (2)$, r *even*,
(b3) $\lambda = ((r-1)^2)$,
(b4) $\lambda = (r-2)$, r *even*.

Proof. Let us assume that codim $Y_{r-1}^\lambda = 1$. Then we have the inequality

$$\dim L_\lambda F \leq 1 + r - 1 + \dim L_\lambda F'$$

or, equivalently,

$$\dim L_\lambda F - \dim L_\lambda F' \leq r,$$

where F' is a vector space of dimension $r - 1$. But setting $F = F' \oplus \mathbf{K}$ we get that

$$\sum_{\mu, \lambda/\mu \in \text{VS}, \ \lambda/\mu \neq \emptyset} \dim L_\mu F' = r.$$

The case $r = 1$ is trivial, as $\lambda = \emptyset$. In the case $r = 2$ we have $\lambda = (1^d)$ and the set of tensors of rank ≤ 1 is a cone over a rational normal curve of degree d by exercise 5 of chapter 5. Its codimension is equal to $d - 1$. The case $d = 2$ is covered under (b1).

Let us assume $r \geq 3$. If λ has at least two rows of different lengths, then there are at least three partitions μ on the left hand side of the last formula. Also, observe that if a partition μ has a row of length $< r - 1$, then $\dim L_\mu F' \geq r - 1$. Thus our equality cannot happen. Thus all rows of λ have the same length. However if there are at least three of them, we still have at least three possible partitions on the left hand side, so the equality cannot happen. Thus λ has two rows or one row. Now we can analyze the situation directly to see that if there are two rows, they have to have length $r - 1$ or 1, giving cases (b1) and (b3).

If there is one row, we can see that the only possibilities are that the length is $1, 2, r - 2, r - 1$. But the cases $r - 1$ and 1 are eliminated because there Y_{r-1}^λ has codimension zero. This completes the proof of the lemma. ∎

Now we can conclude the proof of Theorem (7.1.7). Indeed, we have four possible cases, but we have to take into account that the partition λ can have some rows of length r. Moreover, we have an additional case $\lambda = (r^k)$ where also the codimension of tensors of rank $< r$ is one. This leads to the cases (b1)–(b5) of Theorem (7.1.7). The divisibility condition comes from the fact that we need x to be divisible by the degree of the generating invariant of the subring of $SL(F)$-invariants in $\text{Sym}(L_\lambda F)$. This completes the proof of Theorem (7.1.7).●

7.2. Rank Varieties for Symmetric Tensors

In this section we consider rank varieties for symmetric tensors. Some of the results of the previous section can be strengthened in this case. Again E denotes a vector space over $\mathbf{K}$ of dimension n. We take $\lambda = (1^d)$ and so $X = D_d E^*$, $A^{(1^d)} = \text{Sym}(S_d E)$. We assume that $d \geq 3$, because in the case $d = 2$ we get the determinantal varieties for symmetric matrices which were discussed in section 6.3.

First we consider the case $r = n - 1$. Following the paper [Po] of Porras, we describe the whole minimal free resolution of the ideal $I_{n-1}^{(1^d)}$. The incidence variety defined in section 7.1 becomes

$$Z_{n-1}^{(1^d)} = \{(\phi, S) \in X \times \text{Grass}(n - 1, E^*) \mid \phi \in D_d S \subset D_d E^*\}.$$

The tautological sequence we use is

$$0 \to \mathcal{R} \to E \times \text{Grass}(1, E) \to \mathcal{Q} \to 0.$$

We recall that $\eta = S_d \mathcal{Q}$.

(7.2.1) Proposition. *In the case of symmetric tensors we have* $\xi^{(1^d)} = \mathcal{R} \otimes S_{d-1}E$.

Proof. The bundle $\xi^{(1^d)}$ fits into the exact sequence

$$0 \to \xi^{(1^d)} \to S_d E \to S_d \mathcal{Q} \to 0.$$

This means we need to show the exactness of the sequence

$$0 \to \mathcal{R} \otimes S_{d-1}E \to S_d E \to S_d \mathcal{Q} \to 0.$$

The maps are easily defined. The composition of both maps is zero. To check the exactness we need to do it locally. There the sequence is exact because of the direct sum decomposition (2.3.1) for the symmetric power. ∎

This means the calculation of the cohomology for the exterior powers of $\xi^{(1^d)}$ reduces to the calculation of cohomology of line bundles on the projective space. This is provided by Serre's theorem [H1, chapter III, Theorem 5.1]. Let us summarize.

(7.2.2) Proposition. *The complex $F_\bullet$ has terms given by*

$$F_0 = A^{(1^d)}, \quad F_i = K_{i,1^{n-1}}E \otimes \bigwedge^{n-1+i}(S_{d-1}E) \otimes_{\mathbf{K}} A^{(1^d)}(-i - n + 1)$$

for $i \geq 1$. The length of the complex is $\dim S_{d-1}E - n + 1 = \binom{n+d-2}{d-1} - n + 1$. *In fact, $F_\bullet$ is just an Eagon–Northcott complex associated to the maximal minors of the map*

$$\Phi_{(1^d)} : S_{d-1}E \otimes_{\mathbf{K}} A^{(1^d)}(-1) \to E^* \otimes_{\mathbf{K}} A^{(1^d)}.$$

(7.2.3) Corollary. *The ideal $I_r^{(1^d)}$ is generated by $(r+1) \times (r+1)$ minors of the matrix $\Phi_{(1^d)}$.*

Proof. For $r = n - 1$ it follows from Proposition (7.2.2). To prove the general case we use descending induction on r. Let $I_r'^{(1^d)}$ be the ideal generated by $(r + 1) \times (r + 1)$ minors of the matrix $\Phi_{(1^d)}$. Assume that $I_{r+1}^{(1^d)} = I_{r+1}'^{(1^d)}$. Let us also assume that $A/I_r'^{(1^d)}$ contains nilpotents. The set of nilpotents in this ring is a $GL(E)$-stable subspace. Therefore it has to contain a $\mathbf{U}^+$-invariant. But the weight of this $\mathbf{U}^+$-invariant contains $\leq r + 1$ basis vectors, otherwise it would be in $I_{r+1}^{(1^d)} = I_{r+1}'^{(1^d)}$. This means such $\mathbf{U}^+$-invariant (which is nilpotent modulo $I_r'^{(1^d)}$) exists already for $\dim E = r + 1$. This is a contradiction with Proposition (7.2.2). ∎

(7.2.4) Remark. *Proposition (7.2.2) and Corollary (7.2.3) are true for several symmetric tensors of degrees $d_1, \ldots, d_t$. The role of the map $\Phi_{(1^d)}$ is played by the map*

$$\phi_{1^{d_1}, \ldots, 1^{d_t}} : S_{d_1-1}E \otimes_K A^{(1^d)} \oplus \ldots \oplus S_{d_t-1}E \otimes_K A^{(1^d)} \to E^* \otimes_K A^{(1^d)},$$

where the j-th component is defined using the j-th tensor.

We also state the criterion for rank varieties of symmetric tensors to be Gorenstein.

(7.2.5) Theorem. *The variety $Y_r^{(1^d)}$ is Gorenstein in the following cases:*

(a) $n = \binom{r+d-1}{d-1}$,

(b) $n > \binom{r+d-1}{d-1}$, $r = 1$, n divisible by d.

Proof. This follows from Theorem (7.1.7). Analyzing all cases, we see that $\lambda = (1^d)$ can occur only in cases (a) and (b3) (with $r = 1$). This last case gives case (b) of our statement. ∎

The most interesting varieties are those of tensors of rank ≤ 1. The variety $Y_1^{(d)}$ is the cone over the d-tuple embedding of $\mathbf{P}^{n-1} = \mathbf{P}(E)$ into the projective space $\mathbf{P}(S_d E)$. There is a lot of interest in the resolutions of these varieties, especially trying to determine for which p they satisfy the property N_p of Green and Lazarsfeld introduced in [GL]. This is equivalent to asking about the smallest i for which $H^2(\bigwedge^i \xi) \ne 0$. The reader may consult [B1], for some recent results. In general the description of the resolution seems to be rather difficult, as it is connected to the problem of inner plethysm. The composition series of the bundle $\xi^{(1^d)}$ induced by the tautological exact sequence will always involve the term $S_d \mathcal{R}$, so the exterior powers are related to higher plethysm. One might hope, however, to determine all pairs (i, j) for which the cohomology group $H^i(\mathrm{Grass}(n-r, E^*), \bigwedge^{i+j} \xi^{(1^d)}) \ne 0$. One might hope that for every such pair (i, j) there will be a representation that will occur only few times in the spectral sequence and thus will allow one to determine that $H^i(\mathrm{Grass}(n-r, E^*), \bigwedge^{i+j} \xi^{(1^d)}) \ne 0$. Below we do it for the embeddings of $\mathbf{P}^2$. We use the approach of Ottaviani and Paoletti [OP].

Let us look at the resolutions of d-tuple embeddings of projective spaces in more detail.

In section 6.3 we exhibited the resolutions of such embeddings for $d = 2$. Proposition (7.2.2) provides the answer for the embeddings of $\mathbf{P}^1$.

It is not difficult to locate the top part of the resolution of $A^{(1^d)}/I_r^{\prime(1^d)}$.

(7.2.6) Proposition ([OP]). *Let us fix n, d. We choose the number j to be the minimal number such that $(j + 1)d \geq n$. Then the top part of the complex $F_\bullet$ is*

$$H^{n-j}\left(\mathrm{Grass}(1, E), \bigwedge^{\binom{d+n-1}{n-1}-1-j} \xi^{(1^d)}\right) = S_{(j+1)d-n}E^* \otimes \left(\bigwedge^n E\right)^{\otimes\binom{n+d-1}{n}}.$$

Proof. Using the duality statement (5.1.4) and taking into account that the canonical bundle on $\mathbf{P}^{n-1}$ is $\mathcal{O}(-n)$, we see that it is enough to locate the rightmost part of the complex $F(S_{d-n-1}\mathcal{Q})_\bullet$. But this is equivalent by (5.1.2) (b) to finding the cohomology of $\bigoplus_{j\geq 0} S_{d-n+jd}\mathcal{Q}$. It is now clear by (4.1.9) that the higher cohomology has to vanish and that the module $H^0(\mathrm{Grass}(1, E), \bigoplus_{j\geq 0} S_{d-n+jd}\mathcal{Q})$ is generated by the representation $S_{(j+1)d-n}V$ in degree j described in our statement. ∎

For the remainder of this section we assume that char $\mathbf{K} = 0$.

(7.2.7) Example. *Let us take $d = n = 3$. Then $j = 0$, and Proposition (7.2.6) says that the top of the resolution is one dimensional. This means that the resolution in question is self-dual and the ideal $I_1^{(1^3)}$ is Gorenstein. But this means that H^0 and H^2 strands of the resolution consist of one copy of one dimensional representation each. This allows to describe the terms of the resolution by calculating Euler characteristics of exterior powers of $\xi^{(1^d)}$. The terms of the complex $F_\bullet$ are*

$F_0 = (0, 0, 0)$,

$F_1 = (4, 2, 0)$,

$F_2 = (4, 3, 2) \oplus (5, 3, 1) \oplus (5, 4, 0) \oplus (6, 2, 1)$,

$F_3 = (5, 4, 3) \oplus (5, 5, 2) \oplus (6, 3, 3) \oplus (6, 4, 2) \oplus (6, 5, 1) \oplus (7, 3, 2)$
$\oplus (7, 4, 1)$,

$F_4 = (6, 5, 4) \oplus (6, 6, 3) \oplus (7, 4, 4) \oplus (7, 5, 3) \oplus (7, 6, 2) \oplus (8, 4, 3)$
$\oplus (8, 5, 2)$,

$F_5 = (7, 6, 5) \oplus (8, 6, 4) \oplus (8, 7, 3) \oplus (9, 5, 4)$,

$F_6 = (9, 7, 5)$,

$F_7 = (9, 9, 9)$,

where we write (a, b, c) instead of $K_{a,b,c}V \otimes A^{(1^d)}$. The homogeneous degree is easily seen from the size of each partition.

Let us fix $n = 3$. We want to determine the minimal j for which the cohomology modules $H^2(\text{Grass}(1, E) \bigwedge^{j+2} \xi^{(1^d)}) \neq 0$.

(7.2.8) Proposition (Ottaviani–Paoletti, [OP, Theorem 2.1]). *Let $n = 3$. We have $H^2(\text{Grass}(1, E), \bigwedge^{j+2} \xi^{(1^d)}) \neq 0$ for $j \geq 3d - 2$.*

Proof. It is enough to show that $H^2(\text{Grass}(1, E), \bigwedge^{3d} \xi^{(1^d)}) \neq 0$. Indeed, applying the duality (5.1.4) takes the H^2 strand to the dual of the H^0 strand, and since the H^0 strand has to be linearly exact we know that if the j-th term in this strand is nonzero, then all terms in degrees $\leq j$ also have to be nonzero.

By Serre duality it is enough to show that $H^0(\text{Grass}(1, E), S_{d-3}\mathcal{Q} \otimes \bigwedge^{d(d-3)/2} \xi^{(1^d)}) \neq 0$. Now everything follows from the following

(7.2.9) Lemma. *The sheaf $\bigwedge^j \xi^{(1^d)} \otimes S_t \mathcal{Q}$ has nonzero sections for $1 \leq j \leq \binom{n+d-1}{d} - 1$, $j + 1 \leq \binom{n+t-1}{n-1}$ and $t \geq 1$.*

But we have an exact sequence

$$0 \to \bigwedge^j \xi^{(1^d)} \to \bigwedge^j (S_d V) \times \text{Grass}(1, E) \to S_d \mathcal{Q} \otimes \bigwedge^{j-1} \xi^{(1^d)} \to 0.$$

This means that the sections of $\bigwedge^j \xi^{(1^d)} \otimes S_t \mathcal{Q}$ can be identified with the kernel

$$\text{Ker}\left(\bigwedge^j (S_d V) \otimes S_t V \overset{a_t}{\to} \bigwedge^{j-1} (S_d V) \otimes S_{t+d} V\right).$$

We use a Koszul complex

$$\ldots \to \bigwedge^{j+1} (S_d V) \otimes S_{t-d}\mathcal{Q} \to \bigwedge^j (S_d V) \otimes S_t \mathcal{Q} \to \bigwedge^{j-1} (S_d V) \otimes S_{t+d}\mathcal{Q} \to \ldots$$

with the differential being a composition

$$\bigwedge^{j+1} (S_d V) \otimes S_{t-d}\mathcal{Q} \to \bigwedge^j (S_d V) \otimes S_d V \otimes S_{t-d}\mathcal{Q}$$

$$\to \bigwedge^j (S_d V) \otimes S_d \mathcal{Q} \otimes S_{t-d}\mathcal{Q} \to \bigwedge^j (S_d V) \otimes S_t \mathcal{Q}.$$

Notice that if $t = pd + q$, this is just a twisted symmetric power $S_q \mathcal{Q} \otimes S_{j+p}$ $(S_d V \to S_d \mathcal{Q})$. The existence of this complex means that for $t \geq d$ the sections of $\bigwedge^{j+1}(S_d V) \otimes S_{t-d}\mathcal{Q}$ give the sections of $\bigwedge^j \xi^{(1^d)} \otimes S_t \mathcal{Q}$. In particular, for $d = t$ we get for each family of polynomials $s_0, \ldots, s_j$ an element

$$\sum_{i=0}^{j}(-1)^i s_0 \otimes \ldots \otimes \hat{s}_i \otimes \ldots \otimes s_j \otimes s_i$$

in Ker a_t. Let $1 \leq t < d$. If we can factor $s_i = u w_i$ with deg $u = d - t$, we see that the element

$$\sum_{i=0}^{j}(-1)^i s_0 \otimes \ldots \otimes \hat{s}_i \otimes \ldots \otimes s_j \otimes w_i$$

gives a nonzero section of $\bigwedge^j \xi^{(1^d)} \otimes S_t \mathcal{Q}$. This construction is possible as soon as we can find $j + 1$ linearly independent polynomials of degree $\leq t$, i.e. if $j + 1 \leq \binom{n+t-1}{n-1}$. •

7.3. Rank Varieties for Skew Symmetric Tensors

In this section we consider rank varieties for skew symmetric tensors. The results on the equations are less precise than for symmetric tensors because the ideal of minors of the map Φ_λ is not radical in this case. We follow the paper [Po] of Porras in extracting information about the resolutions and equations of the defining ideals of tensors of rank $\leq n - 1$.

One can describe the generators of the defining ideals quite precisely in the case of skew symmetric tensors of degree 3.

We also pay special attention to the skew symmetric tensors of degree d of minimal possible rank d. The variety $Y_d^{(d)}$ in this case is the cone over a Grassmannian Grass(r, E) embedded via Plücker embedding. The problem of finding higher syzygies of these ideals was posed by Study over 100 years ago.

As before E denotes a vector space over $\mathbf{K}$ of dimension n. We take $\lambda = (d)$ and so $X = \bigwedge^d E^*$, $A^{(d)} = \text{Sym}(\bigwedge^d E)$. We assume that $n - 3 \geq d \geq 3$, because in the remaining cases the only possible varieties we can get are the determinantal varieties for skew symmetric matrices which were considered in section 6.4.

For $d \leq r < n$ we defined in 7.1 the rank varieties $Y_r^{(d)}$.

First we consider the case $r = n - 1$. The incidence variety defined in section 7.1 becomes

$$Z_{n-1}^{(d)} = \{(\phi, S) \in X \times \text{Grass}(n - 1, E^*) \mid \phi \in \bigwedge^d S \subset \bigwedge^d E^*\}.$$

The tautological sequence we use is

$$0 \to \mathcal{R} \to E \times \mathrm{Grass}(1, E) \to \mathcal{Q} \to 0.$$

We recall that $\eta^{(d)} = \bigwedge^d \mathcal{Q}$.

(7.3.1) Proposition. *In the case of skew symmetric tensors of rank $\leq n - 1$ we have $\xi^{(d)} = \mathcal{R} \otimes \bigwedge^{d-1} \mathcal{Q}$.*

Proof. The bundle $\xi^{(d)}$ fits into the exact sequence

$$0 \to \xi^{(d)} \to \overset{d}{\bigwedge} E \to \overset{d}{\bigwedge} \mathcal{Q} \to 0.$$

This means we need to show the exactness of the sequence

$$0 \to \mathcal{R} \otimes \overset{d-1}{\bigwedge} \mathcal{Q} \to \overset{d}{\bigwedge} E \to \overset{d}{\bigwedge} \mathcal{Q} \to 0.$$

The maps are easily defined. The composition of both maps is zero. To check the exactness we need to do it locally. There the sequence is exact because of the direct sum decomposition (2.3.1) for the exterior power. ∎

We can also easily determine the top term of the resolution.

(7.3.2) Proposition ([Po]). *Let us assume $3 \leq d \leq n - 3$. Let $r = n - 1$. Denote $m = \dim \xi = \binom{n-1}{d-1}$. The top term of the complex $F_\bullet$ is $H^{n-1}(\mathrm{Grass}(n - 1, F), \bigwedge^m \xi) = K_{(m-n+1,q+1,\dots,q+1)} E \otimes A^{(d)}(-m)$, where $q = \binom{n-2}{d-2}$. In particular the only case in which I_{n-1} is a Gorenstein ideal is when $d = 3, n = 6$.*

Proof. The term in question comes from the cohomology of the top exterior power of ξ. It occurs in F_{m-n+1}. It is clear by (5.1.6) (a) and by (7.1.3) that $F_i = 0$ for $i > m - n + 1$. It remains to show that the term listed above is the only term in F_{m-n+1}. The only possible cohomology groups are

$$H^{n-1-j}\left(\mathrm{Grass}(n - 1, F), \overset{m-j}{\bigwedge} \xi \right)$$

for $j > 0$. But by (7.3.1), using the isomorphism $\bigwedge^{m-j} \xi = \bigwedge^m \xi \otimes \bigwedge^j \xi^*$, we see that we look at the weights

$$(q - \mu_{n-1}, \dots, q - \mu_1, m - j),$$

where μ is a partition such that $K_\mu \mathcal{Q}^*$ occurs in $\bigwedge^j (\bigwedge^{d-1} \mathcal{Q}^*)$. We also assume that the weight in question does not have a nonzero $(n - 1)$st cohomology group. This means

$$q - \mu_{n-1} \geq m - j - n + 2.$$

This implies

$$j \geq m - q - n + 2 = \binom{n-2}{d-1} - n + 2.$$

If we can show that the inequality above implies $j \geq n - 1$, we are done, as we have eliminated all possibilities for j. However the inequality

$$\binom{n-2}{d-1} - n + 2 \geq n - 1$$

fails only for $n = 6$, $d = 3$ and for $n = 7$, $d = 3, 4$. These three cases can be handled explicitly. ∎

Proposition (7.3.1) means that the calculation of the cohomology for the exterior powers of $\xi^{(d)}$ is not as easy as in the symmetric case. In order to perform the explicit calculation we would need to know the decomposition of $\bigwedge^i (\bigwedge^{d-1} Q)$ into Schur functors. This, as explained in section 2.3, is a very difficult problem. Still, we have an explicit formula for plethysm when $d - 1 = 2$. It also allows us to describe in some cases, as for symmetric tensors, the pairs (i, j) for which $H^i(\text{Grass}(1, E), \bigwedge^{i+j} \xi^{(d)}) \neq 0$.

Let us assume $d = 3$. Then the plethysm formula (2.3.9) (b) makes the problem of calculating a complex $F_\bullet$ a combinatorial exercise. Still, the analysis of the whole complex $F_\bullet$ is quite complicated. Here we just analyze the defining equations of the ideal of $Y_{n-1}^{(d)}$. Interested readers should consult [Po].

(7.3.3) Proposition ([Po]). *Let us assume $d = 3$. The ideal $I_{n-1}^{(d)}$ of the variety $Y_{n-1}^{(d)}$ has generators in all degrees i satisfying $[n/2] \leq i \leq (1 + 2n - (1 + 8n)^{1/2})/2$.*

Proof. We need to show that the terms occurring in the term F_1 of the complex $F_\bullet$ appear in homogeneous degrees i satisfying the above inequalities and that for each such i we get a nonzero contribution. We know that

$$F_1 = \bigoplus_{i>0} H^{i-1}\left(\text{Grass}(n-1, E), \bigwedge^i \xi\right) \otimes A^{(d)}(-i).$$

We also have by (2.3.9) (b)

$$\bigwedge^i \xi = \bigoplus_{\mu \in Q_{-1}(2i)} K_\mu Q \otimes S_i R.$$

This means we need to apply (4.1.9) to the weights

$$(\mu_1, \ldots, \mu_{n-1}, i)$$

with $\mu \in Q_{-1}(2i)$. Notice that such term gives a contribution to H^{i-1} if and only if $\mu_{n-i} \neq 0$ and $\mu_{n-i+1} = 0$. Therefore we need to estimate for which i we have a partition in $Q_{-1}(2i)$ with exactly $n - i$ parts. Let us look for a lower bound for i. For $n = 2t$ even the smallest partition of this kind is clearly $\mu = (t - 1, 2^2, 1^{t-3}) \in Q_{-1}(2t)$. For $n = 2t + 1$ the smallest such partition is clearly $\mu = (t, 1^t) \in Q_{-1}(2t)$. This proves the lower bound of the proposition.

Let us seek the biggest possible partition in $Q_{-1}(2i)$ with $n - i$ parts. Any such partition has to be contained in the rectangle $((n - i - 1)^{n-i})$, so we must have the inequality $2i \leq (n - i)(n - i + 1)$, which gives the upper bound in the proposition. Of course, for any i satisfying the inequalities of the proposition we can find the appropriate $\mu \in Q_{-1}(2i)$ by choosing any partition from $Q_{-1}(2i)$ containing the partition giving the lower bound and contained in the rectangle $((n - i + 1)^{n-i})$. ∎

(7.3.4) Corollary ([Po]). *Let d, n be as above, with $3 \leq d \leq n - 3$. For any r satisfying $3 \leq r \leq n - 1$ the ideal of $(r + 1) \times (r + 1)$ minors of $\Phi^{(d)}$ is not radical.*

Proof. It is enough to show that $I_r^{(d)}$ has some nonzero elements in degrees $\leq r$. This is clear, since the representations generating $I_r^{(d)}$ for $r + 1$ dimensional space will give the representations from $I_r^{(d)}$ for n dimensional space. By (7.3.3) they occur in degrees

$$\frac{1 + 2(r + 1) - (1 + 8(r + 1))^{1/2}}{2} \leq \frac{1 + 2(r + 1) - 3}{2} = r.$$

The corollary follows. ∎

Let us also state when the ideal $I_r^{(d)}$ is Gorenstein.

(7.3.5) Theorem. *The ideal $I_r^{(d)}$ is Gorenstein in the following cases:*

(a) $n = \binom{r-1}{d-1}$,
(b) $d = 2$, *and r is even,*
(c) $n > \binom{r-1}{d-1}$, $d = r - 2$, *and $n - \binom{r-1}{d-1}$ is divisible by $\frac{r-2}{2}$,*
(d) $d = r$.

Proof. This is a special case of Theorem (7.1.7). Cases (b2), (b4), and (b5) of (7.1.7) lead to cases (b), (c), and (d) of our statement. ∎

For the remainder of this section we investigate the case $r = d$. In this case $\eta = \bigwedge^d Q$ is one dimensional and therefore

$$A^{(d)}/I_d^{(d)} = \bigoplus_{n \geq 0} K_{(n^d)} E.$$

Let us describe the top term of the resolution.

(7.3.6) Proposition. *The top term in the resolution $F_\bullet$ is*

$$H^{d(n-d)-n+1}(\mathrm{Grass}(n-d, E), \bigwedge^{\binom{n}{d}-n} \xi) = K_{(\binom{n-1}{d-1})-1)^n)} E.$$

Proof. Let us use the duality statement (5.1.4). The canonical sheaf $K_{\mathrm{Grass}(n-d,E)}$ is equal to $\mathcal{O}_{\mathrm{Grass}(n-d,E)}(-n)$. The top exterior power $\bigwedge^{\binom{n}{d}-1} \xi$ is isomorphic (up to the twist by a power of determinant of E) to $\mathcal{O}_{\mathrm{Grass}(n-d,E)}(1)$. Therefore the dualizing bundle will be $\mathcal{O}_{\mathrm{Grass}(n-d,E)}(-n+1)$ or, in terms of tautological bundles $\bigwedge^d Q^{\otimes(-n+1)}$. Calculating the cohomology of the sheaf $\bigwedge^d Q^{\otimes(-n+1)} \otimes \mathrm{Sym}(\eta)$, we see that the generator occurs in degree $n-1$ and is a trivial representation. This means that by (5.1.4) the top term of the resolution occurs in homogeneous degree $\dim \xi - n + 1$, and it is a one dimensional representation. Since the variety $Y_d^{(d)}$ is normal with rational singularities by (5.1.2) (b) and (5.1.3), we see that this term has to occur in the term F_i with $i = \dim X - \dim Y_d^{(d)} = \binom{n}{d} - d(n-d) - 1$. Calculating the powers of the determinants involved, one gets the proposition. ∎

(7.3.7) Remark. *Note that for $d = 2$ we constructed the resolution of the Plücker ideal in section 6.4. It is the resolution of 4×4 Pfaffians of a generic skew symmetric $n \times n$ matrix. Indeed, the cone over the Grassmannian* Grass $(2, E)$ *can be identified with the set of skew-symmetric matrices of rank ≤ 1.*

We finish this section with the analysis of the case $d = 3, n = 6$. We will analyze the rank variety $Y_5^{(3)}$. The main goal is to show that this variety is not normal over a field **K** of characteristic 2, so the conclusion of Proposition (7.1.2) fails in positive characteristic.

Thus E is a vector space of dimension 6, and we work over the Grassmannian Grass$(1, E)$ with tautological sequence

$$0 \to \mathcal{R} \to E \times \mathrm{Grass}(1, E) \to Q \to 0.$$

We have $\eta^{(3)} = \bigwedge^3 Q, \xi^{(3)} = \mathcal{R} \otimes \bigwedge^2 Q.$

(7.3.8) Proposition. *Let* char $\mathbf{K} = 0$. *The complex* $F_\bullet$ *has the following nonzero terms:*

$$F_0 = A^{(d)}, \qquad F_1 = L_{(6,3)}E \otimes A^{(d)}(-3), \qquad F_2 = L_{(6,5,1)}E \otimes A^{(d)}(-4),$$
$$F_3 = L_{(6^2,5,1)}E \otimes A^{(d)}(-6), \qquad F_4 = L_{(6^3,3)}E \otimes A^{(d)}(-7),$$
$$F_5 = L_{(6^5)}E \otimes A^{(d)}(-10).$$

(7.3.9) Proposition. *Let* $\mathbf{K}$ *be a field of characteristic* 2. *Then the only nonzero cohomology groups of* $S_2\mathcal{R} \otimes \bigwedge^2 \mathcal{Q}$ *are*

$$H^1\left(\mathrm{Grass}(1, E), S_2\mathcal{R} \otimes \overset{2}{\bigwedge}\left(\overset{2}{\bigwedge}\mathcal{Q}\right)\right)$$

$$= H^2\left(\mathrm{Grass}(1, E), S_2\mathcal{R} \otimes \overset{2}{\bigwedge}\left(\overset{2}{\bigwedge}\mathcal{Q}\right)\right) = \overset{6}{\bigwedge}E.$$

Proof. We notice that in characteristic 2, $L_{(3,1)}\mathcal{Q}$ is not irreducible. In fact we have the natural exact sequence

$$0 \to M_{(2,1,1)}\mathcal{Q} \to L_{(3,1)}\mathcal{Q} \to \overset{4}{\bigwedge}\mathcal{Q} \to 0.$$

Also, we have a natural map

$$\psi : L_{3,1}\mathcal{Q} \to \overset{2}{\bigwedge}\left(\overset{2}{\bigwedge}\mathcal{Q}\right)$$

induced by the composition map

$$\overset{3}{\bigwedge}\mathcal{Q} \otimes \mathcal{Q} \overset{\Delta \otimes 1}{\to} \overset{2}{\bigwedge}\mathcal{Q} \otimes \mathcal{Q} \otimes \mathcal{Q} \overset{1 \otimes m}{\to} \overset{2}{\bigwedge}\mathcal{Q} \otimes \overset{2}{\bigwedge}\mathcal{Q}.$$

This map, however, is not an isomorphism in characteristic zero. Its image is isomorphic to $\bigwedge^4 \mathcal{Q}$; its kernel, to $M_{(2,1,1)}\mathcal{Q}$.

Next we notice that

$$H^s(\mathrm{Grass}(1, E), S_2\mathcal{R} \otimes L_{3,1}\mathcal{Q}) = 0$$

for all $s \geq 0$, by exercise 5 of chapter 4. Also, the only nonzero cohomology group of $S_2\mathcal{R} \otimes \bigwedge^4 \mathcal{Q}$ is $H^1(\mathrm{Grass}(1, E), S_2\mathcal{R} \otimes \bigwedge^4 \mathcal{Q}) = \bigwedge^6 E$. This we can deduce from identifying $S_2\mathcal{R}$ with $D_2\mathcal{R}$ and using the Γ-acyclic resolution $\bigwedge^2(E \to \mathcal{Q})$ of $D_2\mathcal{R}$, tensored with $\bigwedge^4 \mathcal{Q}$. The sections of this resolution give a complex

$$0 \to D_2 E \otimes \bigwedge^4 E \to E \otimes E \otimes \bigwedge^4 E \to \left(\bigwedge^2 E \otimes \bigwedge^4 E \right) \Big/ \bigwedge^6 E,$$

which has only one homology $- \bigwedge^6 E$ in degree 1.

Now, using the short exact sequences

$$0 \to M_{(2,1,1)} \mathcal{Q} \to L_{(3,1)} \mathcal{Q} \to \bigwedge^4 \mathcal{Q} \to 0$$

and

$$0 \to \bigwedge^4 \mathcal{Q} \to \bigwedge^2 \left(\bigwedge^2 \mathcal{Q} \right) \to M_{(2,1,1)} \mathcal{Q} \to 0$$

tensored with $S_2 \mathcal{R}$ and the long exact sequences of cohomology they induce, we deduce the proposition. ∎

(7.3.10) Proposition. *Let* char $\mathbf{K} = 2$. *Then the ring* $A^{(3)}/J_5$ *is not normal.*

Proof. Recall that $H^0(\text{Grass}(1, E), \text{Sym}(\bigwedge^3 \mathcal{Q}))$ can be identified with the normalization of A/J_5. By Proposition (7.3.9) the complex $F_\bullet$ has a nontrivial term in homological degree 0 and in homogeneous degree 2. It follows that the natural map

$$S_2(\bigwedge^3 E) \to H^0 \left(\text{Grass}(1, E), \text{Sym}_2 \left(\bigwedge^3 \mathcal{Q} \right) \right)$$

is not onto, and therefore the normalization of $A^{(3)}/J_5$ is not generated as an $A^{(3)}$-module by a unit in degree 0. ∎

Exercises for Chapter 7

Minimal Resolutions of the Ideal $I_3^{(3)}$ for $n = 6, 7$

1. Let $X = \bigwedge^d E^*$ be the set of skew symmetric tensors. Denote by $Y \subset X$ the set of 1-decomposable tensors, i.e. the set of tensors ϕ in X such that $\phi = \psi \wedge l$ where $l \in E^*$ is a linear form and $\psi \in \bigwedge^{d-1} E^*$. Prove that Y has a desingularization which is a total space of a vector bundle over the Grassmannian $\text{Grass}(n - 1, F)$. Identify $\xi = \bigwedge^d \mathcal{R}$ and $\eta = \mathcal{Q} \otimes \bigwedge^{d-1} \mathcal{R}$. Prove that Y is normal and has rational singularities.

2. In the situation of exercise 1, consider the twisted complex $F(S_d Q^*)_\bullet$. Prove that its homology modules are

$$H_{-i}(F(S_d Q^*)_\bullet) = \oplus_{j \geq 0} H^i \left(\mathrm{Grass}(n-1, E), \right.$$

$$\left. S_d Q^* \otimes \mathrm{Sym}_j \left(Q \otimes \bigwedge^{d-1} R \right) \right).$$

Prove that the nonzero homology occurs for $i = -d+1, \ldots, 0$. Prove that

$$H_{-d+1}(F(S_d Q^*)_\bullet) = A/I_d^{(d)}(-1).$$

3. In the situation of exercise 2, specialize to $d = 3$. Prove that the only nonzero homology modules of $F(S_3 Q^*)_\bullet$ are H_{-2} and H_0. Writing $N = H_0(F(S_3 Q^*)_\bullet)$, prove that the j-th graded component of N is

$$N_j = \bigoplus_{a+b+\ldots=j, \, j-3 \geq a} K_{(j-3,a,a,b,b,\ldots)} E.$$

4. In the situation of exercise 3, specialize to $n = 6$. We have

$$N_j = \bigoplus_{a+b+\ldots=j, \, j-3 \geq a \geq b} K_{(j-3,a,a,b,b,0)} E.$$

Prove that the complex $F(S_3 Q^*)_\bullet$ has the following nonzero terms

(0^6)

$\nwarrow$

$(2, 1^4, 0)$

$\uparrow$

$(3, 2, 1^4) \oplus (2^4, 1, 0)$

$\uparrow \qquad \nwarrow$

$(3^2, 2^2, 1^2) \qquad\qquad (5, 2^5)$

$\nwarrow \qquad\qquad \uparrow$

$(5, 3^3, 2^2) \oplus (4^3, 2^3)$

$\uparrow$

$(5, 4^2, 3^2, 2)$

$\uparrow$

$(5^2, 4^3, 2)$

$\uparrow$

$(5^5, 2)$

where we just write the partitions of the occurring Weyl functors. The vertical arrows denote the maps of degree 1 and skew arrows denote the maps of degree 2.

5. We still work with $X = \bigwedge^3 E^*$, $\dim E = 6$. Consider the variety $Y_5^{(3)}$ of tensors of rank ≤ 5 in X. Let $Z_5^{(3)}$ be its desingularization considered in section 7.3:

$$Z_5^{(3)} = \left\{ (\phi, S) \in X \times \mathrm{Grass}(5, E^*) \mid \phi \in \bigwedge^3 S \subset \bigwedge^3 E^* \right\}.$$

We write

$$0 \to \mathcal{R}' \to \bigwedge^3 E \times \mathrm{Grass}(5, E) \to \mathcal{Q}' \to 0$$

for the tautological sequence on $\mathrm{Grass}(5, E)$. We have $\eta' = \bigwedge^3 \mathcal{Q}'$, $\xi' = \mathcal{R}' \otimes \bigwedge^2 \mathcal{Q}'$. Consider the twisted complex $F'(\bigwedge^5 \mathcal{Q}'^{\otimes 3})_\bullet$. Prove that it is acyclic, and show that the A-module $H_0(F'(\bigwedge^5 \mathcal{Q}'^{\otimes 3})_\bullet)$ is isomorphic to N twisted by $\bigwedge^6 E^{\otimes -3}$. Prove that the complex $F'(\bigwedge^5 \mathcal{Q}'^{\otimes 3})_\bullet$ has the following nonzero terms (after twisting back by $\bigwedge^6 E^{\otimes 3}$):

$$(3^5, 0)$$
$$\uparrow$$
$$(4^2, 3^3, 1)$$
$$\uparrow$$
$$(5, 4^2, 3^2, 2)$$
$$\uparrow$$
$$(6, 4^3, 3^2) \oplus (5^3, 3^3)$$
$$\uparrow \qquad\qquad \nwarrow$$
$$(7, 4^5) \qquad\qquad\qquad\qquad (6^2, 5^2, 4^2)$$
$$\nwarrow \qquad\qquad \uparrow$$
$$(6^4, 5, 4) \oplus (7, 6, 5^4)$$
$$\uparrow$$
$$(7, 6^4, 5)$$
$$\nwarrow$$
$$(7^6)$$

The vertical arrows denote the maps of degree 1, and the skew arrows denote the maps of degree 2.

6. Prove that there exists a map Ψ of the complex constructed in exercise 5 to the complex constructed in exercise 4 induced by the isomorphism $H_0(F(S_3 \mathcal{Q}^*)_\bullet) = N$. Prove that the cone of Ψ gives a minimal resolution of the $A^{(3)}$-module $A^{(3)}/I_3^{(3)}$.

7. Let us specialize the situation from exercise 3 to $n = 7$. We have

$$N_j = \bigoplus_{a+b+c\ldots=j,\,j-3\geq a\geq b\geq c} K_{(j-3,a,a,b,b,c,c)}E.$$

The components N_j are nonzero for $j \geq 5$. Let N' be the span of all summands with $c \geq 1$. Prove that N' is an $A^{(3)}$-submodule of N. Denote $N'' = N/N'$. Prove that the minimal resolution of N' equals $F(Q^*)_\bullet \otimes \bigwedge^7 F$, in the notation of exercise 1.

8. Prove that the minimal resolution of the $A^{(3)}$-module N'' from exercise 7 can be obtained as follows. Take $V = \mathrm{Grass}(,E)$ with tautological sequence

$$0 \to \mathcal{R}'' \to E \times \mathrm{Grass}(2, E) \to \mathcal{Q}'' \to 0.$$

Here $\dim \mathcal{R}'' = 2$, $\dim \mathcal{Q}'' = 5$. Take the bundle ξ'' to be the kernel

$$0 \to \xi'' \to \bigwedge^3 E \times \mathrm{Grass}(2, E) \to \bigwedge^3 \mathcal{Q}'' \to 0.$$

The bundle ξ'' also fits into the exact seuence

$$0 \to \bigwedge^2 \mathcal{R}'' \otimes \mathcal{Q}'' \to \xi'' \to \mathcal{R}'' \otimes \bigwedge^2 \mathcal{Q}'' \to 0.$$

Prove that the minimal resolution of N'' is $F''((\bigwedge^5 \mathcal{Q}'')^{\otimes 3})_\bullet$.

9. Use the information obtained in exercise 8 to calculate the terms in the linear strand of the minimal resolution of $A^{(3)}/I_3^{(3)}$ for $n = 7$.

10. Consider the twisted sheaf

$$\mathcal{M} = K_{((r-1)(n-r-1))^r} \mathcal{Q} \otimes K_{((r-1)(n-r-1),(r-1)(n-r-2),\ldots,(r-1),0)} \mathcal{R} \otimes \mathcal{O}_{Z_r^{(r)}}$$

supported in the rank variety $Y_r^{(r)}$. Prove that $\mathcal{M}$ has no higher cohomology. Show that $M := H^0(Z_r^{(r)}, \mathcal{M})$ is a maximal Cohen–Macaulay module with a linear resolution supported in $Y_r^{(r)}$.

11. Finish the proof of Proposition (7.3.2). Calculate explicitly the resolution in the case $n = 6$, $d = 3$ and in the cases $n = 7$, $d = 3, 4$.

12. Let $X = S_d E^*$ be the set of symmetric tensors. Denote by $Y \subset X$ the set of 1-decomposable tensors, i.e. the set of tensors ϕ in X such that $\phi = \psi \circ l$ where $l \in E^*$ is a linear form. Prove that Y has a desingularition which is a total space of a vector bundle over the Grassmannian $\mathrm{Grass}(n - 1, F)$. Identify $\xi = S_d \mathcal{R}$ and $\eta = \mathcal{Q} \otimes S_{d-1}E$. Prove that Y is not normal but its normalization has rational singularities. Identify the normalization of Y.

Higher Rank Varieties for Orthogonal and Symplectic Groups

13. Let F be an orthogonal space of dimension $m = 2n + 1$ or $m = 2n$. We denote the nondegenerate symmetric form on F by $\langle \ , \ \rangle$. We choose a standard hyperbolic basis of F, denoted $\{e_1, \ldots, e_n, e, \bar{e}_n, \ldots, \bar{e}_1\}$ in the odd case and $\{e_1, \ldots, e_n, \bar{e}_n, \ldots, \bar{e}_1\}$ in the even case. The only nonzero values of the form $\langle \ , \ \rangle$ are $\langle e_i, \bar{e}_i \rangle = 1$, $\langle e, e \rangle = 1$. Let λ be a partition, and let $V_\lambda F$ be a representation of $SO(F)$ of highest weight λ. This is a space of tensors from $L_{\lambda'} F$ modulo the tensors containing a trace element (compare exercise 14 of chapter 6). For $\lambda'_1 \leq n$ we define the rank variety Y_r^λ as the set of tensors that (after the change of basis) can be written only in terms of $e_1, \ldots, e_r$. Construct the desingularization Z_r^λ of the variety Y_r^λ using the isotropic Grassmannian IGrass(r, F). Prove that the variety Y_r^λ is normal and has rational singularities.

14. Let F be an symplectic space of dimension $2n$. We denote the skew symetric nondegenerate form on F by $\langle \ , \ \rangle$. We choose a standard symplectic basis of F, denoted $\{e_1, \ldots, e_n, \bar{e}_n, \ldots, \bar{e}_1\}$. The only nonzero values of the form $\langle \ , \ \rangle$ are $\langle e_i, \bar{e}_i \rangle = 1$, $\langle e, e \rangle = 1$. Let λ be a partition, and let $V_\lambda F$ be a representation of $Sp(F)$ of highest weight λ. These are tensors from $L_{\lambda'} F$ modulo the tensors containing a trace element (compare exercise 4 of chapter 6). For $\lambda'_1 \leq n$ we define the rank variety Y_r^λ as the set of tensors that (after the change of basis) can be written only in terms of $e_1, \ldots, e_r$. Construct the desingularization Z_r^λ of the variety Y_r^λ using the isotropic Grassmannian IGrass(r, F). Prove that the variety Y_r^λ is normal and has rational singularities.

The Isotropic Grassmannian IGrass$(3, 6)$

15. Let F be a symplectic space of dimension 6. Consider the representation $V_{1^3} F$. It fits into the exact sequence

$$0 \to F \overset{t}{\to} \bigwedge^3 F \to V_{1^3} F \to 0,$$

where the map t is the multiplication by the element $t = \sum_{1 \leq i \leq n} e_i \wedge \bar{e}_1$ from $\bigwedge^2 F$ given by the form. Prove that the representation $V_{1^3} F$ has four $Sp(F)$-orbits (except the zero orbit). There is a general orbit, a hypersurface given by the vanishing of (the unique up to scalar) invariant Δ of degree 4, the orbit X given by the tensors where the partial derivatives of Δ (forming a representation $V_{1^3} F$ in degree 3) vanish, and the orbit Y which is the cone over IGrass$(3, F)$. Prove that codim $X = 4$, codim $Y = 7$.

16. Find the desingularization of X which is a homogeneous bundle over some isotropic flag variety. Conclude that the minimal rsolution of $K[X]$ has the following terms:

$$F_0 = A, \qquad F_1 = V_{1^3} F \otimes A(-3), \qquad F_2 = V_2 F \otimes A(-4),$$
$$F_3 = V_{1^2} F \otimes A(-6), \qquad F_4 = V_1 F \otimes A(-7).$$

17. Calculate the minimal resolution of $K[Y]$. Show that its terms are

$$F_0 = A, \qquad F_1 = V_2 F \otimes A(-2), \qquad F_2 = V_{2,1} F \otimes A(-3),$$
$$F_3 = V_{2,1,1} F \otimes A(-4),$$

$$F_4 = V_{2,1,1} F \otimes A(-6), \qquad F_5 = V_{2,1} F \otimes A(-7),$$
$$F_6 = V_2 F \otimes A(-8), \qquad F_7 = A(-10).$$

This representation is one of the so-called subexceptional series, corresponding to the entries in the row of the Freudenthal magic square. These representations are (using Bourbaki's notation for fundamental weights)
(a) $G = SL(2)$, $V = V(3\omega_1)$,
(b) $G = SL(2) \times SL(2) \times SL(2) \times \Sigma_3$, $V = V(\omega_1) \otimes V(\omega_1) \otimes V(\omega_1)$, where Σ_3 denotes a permutation group,
(c) $G = SP(6)$, $V = V(\omega_3)$,
(d) $G = SL(6)$, $V = V(\omega_3)$,
(e) $G = $ (spinor group.)(12), $V = V(\omega_5)$ (highest weight $(\frac{1}{2}, \frac{1}{2}, \frac{1}{2}, \frac{1}{2}, \frac{1}{2}, \frac{1}{2})$),
(f) $G = E_7$, $V = V(\omega_6)$.

The uniformity, described by Landsberg and Manivel in [LM], is with respect to the parameter m, which in the above cases takes values $-\frac{2}{3}, 0, 1, 2, 4, 8$. The dimension of the representation V is $6m + 8$. Again there are four orbits: the general one, the hypersurface defined by the vanishing of a (unique up to scalar) invariant of degree 4, X, and Y. The codimension of X is $m + 3$; the codimension of Y is $3m + 4$. Notice that item (d) on the list is the representation $\bigwedge^3 F$ for dim $F = 3$. The resolution of X in that case is described in (7.3.8). The resolution of Y is calculated in exercises 4, 5, 6.

8

The Nilpotent Orbit Closures

In this chapter we deal with another important class of varieties – the nilpotent orbit closures of the adjoint action of a simple algebraic group on its Lie algebra. These varieties play an important role in representation theory. All such orbit closures have desingularizations which are total spaces of vector bundles over homogeneous spaces. We describe the applications of the geometric method. The vector bundles involved in the construction of these desingularizations are more complicated than in the case of determinantal varieties. The explicit formula for the terms of complexes $F(\mathcal{L})_\bullet$ is not known in general. Still, one can prove some interesting results.

The first two sections of the chapter are devoted to the nilpotent orbit closures for the general linear group.

In section 8.1 we describe the desingularizations of these orbit closures explicitly. We apply theorems from chapter 5 to prove that all orbit closures are normal, are Gorenstein, and have rational singularities. We also describe the combinatorial way of estimating the terms of the complexes $F_\bullet$ in this case.

This method is then used in section 8.2 to describe the generators of the defining ideals of nilpotent orbit closures.

In section 8.3 we treat the case of general simple groups. We prove a theorem of Hinich and Panyushev saying that the normalization of every nilpotent orbit closure is Gorenstein and has rational singularities.

Finally, in sections 8.4 and 8.5 we look at the case of classical groups. We mainly work with examples showing how geometric method can be applied in special cases. We give examples of nonnormal orbit closures and discuss some special cases. In the case of the symplectic groups, for orbits corresponding to partitions with even parts, we prove the estimate on the weights of representations generating the defining ideals. We give conjectures for such estimates for all nilpotent orbits for classical groups.

8.1. The Closures of Conjugacy Classes of Nilpotent Matrices

Let E be a vector space of dimension n over a field $\mathbf{K}$. We consider the affine space $X = \operatorname{Hom}_{\mathbf{K}}(E, E)$ of $n \times n$ matrices. We identify the space X with $E^* \otimes E$. The general linear group $\operatorname{GL}(E)$ acts on X by conjugation; the element $g \in \operatorname{GL}(E)$ sends the matrix $\phi \in \operatorname{Hom}(E, E)$ to $g^{-1}\phi g$. The coordinate ring of X can be identified with the symmetric algebra $A = \operatorname{Sym}(E \otimes E^*)$.

We start with the brief analysis of the ring of invariants $A^{\operatorname{GL}(E)}$.

Let us denote by v_i $(i = 1, \ldots, n$) the unique (up to scalar) $\operatorname{GL}(E)$-invariant in $\bigwedge^i E \otimes \bigwedge^i E^*$. To fix the scalar we can choose

$$v_i = \sum_{I \subset [1,n],\ |I|=i} \phi_{I,I}$$

to be the sum of principal minors of the generic matrix ϕ. Notice that v_i is the coefficient of s^{n-i} in the characteristic polynomial $\chi(\phi, s) := \det(\phi - s\,\mathrm{Id})$ of ϕ.

The following proposition is a special case of Chevalley's theorem.

(8.1.1) Proposition. *The ring $A^{\operatorname{GL}(E)}$ is a polynomial ring in $v_1, \ldots, v_n$.*

Proof. We notice that by Cauchy decomposition

$$A = \bigoplus_{\alpha} L_\alpha E \otimes L_\alpha E^*.$$

Moreover, by the Littlewood–Richardson rule (2.3.4) we see that for each α the tensor product $L_\alpha E \otimes L_\alpha E^*$ contains precisely one trivial representation. This means that the dimension of the d-th graded component $A_d^{\operatorname{GL}(E)}$ is equal to the number of partitions of d with at most n parts. This in turn means that the Poincaré series of $A^{\operatorname{GL}(E)}$ is given by the formula

$$P_{A^{\operatorname{GL}(E)}}(t) := \sum_{d \geq 0} (\dim A_d^{\operatorname{GL}(E)}) t^d = \frac{1}{(1-t)\ldots(1-t^n)}.$$

This suggests that $A^{\operatorname{GL}(E)}$ is a polynomial ring in n variables, with generators of degree $1, \ldots, n$. To conclude the proof of (8.1.1) it is enough to show that the polynomials $v_1, \ldots, v_n$ are algebraically independent. This is however clear: after substituting $\phi_{i,j} = 0$ for $i \neq j$ and $\phi_{i,i} = x_i$, we see that the polynomial v_i specializes to the elementary symmetric function $e_i(x_1, \ldots, x_n)$. ∎

The embedding $A^{\operatorname{GL}(E)} \to A$ induces the orbit map

$$\chi : X \to X/\operatorname{GL}(E) = \mathbf{K}^n$$

sending the matrix ϕ to its characteristic polynomial $\chi(\phi, s)$.

This map has been analyzed in many contexts. We will take the point of view of geometric invariant theory and analyze the nullcone of X – the fiber $\chi^{-1}(0)$. This is the set of matrices with all eigenvalues equal to 0, i.e. the set of nilpotent matrices. By the Jordan canonical form we know that the set $\chi^{-1}(0)$ has finitely many GL(E)-orbits. They correspond to the partitions μ of n. For a partition $\mu = (\mu_1, \ldots, \mu_r)$ we denote by $O(\mu)$ the set of nilpotent matrices with Jordan blocks of dimensions $\mu_1, \ldots, \mu_r$. If $\mu' = (\mu'_1, \ldots, \mu'_s)$ is the partition conjugate to μ, we can describe the orbit $O(\mu)$ as a set of endomorphisms ϕ of E for which

$$\dim \operatorname{Ker} \phi^i = \mu'_1 + \ldots + \mu'_i$$

for $i = 1, \ldots, s$. We denote by Y_μ the closure of the orbit $O(\mu)$ in X.

(8.1.2) Examples.
    (a) $\mu = (n)$. *In this case the set Y_μ is the set of all nilpotent matrices.*
    (b) $\mu = (1^n)$. *In this case Y_μ is just a point 0.*
    (c) *More generally, let $\mu = (p, 1^{n-p})$. The variety Y_μ is a set of nilpotent matrices of rank $< p$.*
    (d) *Let $\mu = (2^i, 1^j)$ where $2i + j = n$. In this case Y_μ is a set of matrices ϕ such that $\phi^2 = 0$ and rank $\phi \le i$.*

(8.1.3) Proposition. *The closure Y_μ is defined set-theoretically by the conditions*

$$\dim \operatorname{Ker} \phi^i \ge \mu'_1 + \ldots + \mu'_i$$

for $i = 1, \ldots, s$.

Proof. The conditions of the Proposition are algebraic and satisfied on $O(\mu)$, so they have to be satisfied on Y_μ. Let us denote by Y'_μ the closed subset of X of endomorphisms ϕ for which all the above inequalities are equalities. In order to show that $Y'_\mu = Y_\mu$ it will be enough to show that Y'_μ is irreducible, of the same dimension as $O(\mu)$. In order to do that we will construct a desingularization of Y'_μ.

Let us consider the flag variety

$$V_\mu = \operatorname{Flag}(\mu'_1, \mu'_1 + \mu'_2, \ldots, \mu'_1 + \ldots + \mu'_{s-1}; E).$$

We denote the typical point in V_μ by $(R_{\mu'_1}, \ldots, R_{\mu'_1 + \ldots + \mu'_{s-1}})$, and we also write in that setup that $R_{\mu'_1 + \ldots + \mu'_s} = E$.

Consider the incidence variety

$$Z_\mu = \{(\phi, (R_{\mu'_1}, \ldots, R_{\mu'_1+\ldots+\mu'_{s-1}})) \in X \times V_\mu \mid \phi(R_{\mu'_1})$$
$$= 0, \forall_{2 \le i \le s} \phi(R_{\mu'_1+\ldots+\mu'_i}) \subset R_{\mu'_1+\ldots+\mu'_{i-1}}\}.$$

Notice that the last conditions imply $R_{\mu'_1+\ldots+\mu'_i} \subset \mathrm{Ker}\,(\phi^i)$.

Now we can consider the diagram

$$
\begin{array}{ccc}
Z_\mu & \subset & X \times V_\mu \\
\downarrow q'_\mu & & \downarrow q_\mu \\
Y'_\mu & \subset & X
\end{array}
$$

We denote by p_μ the projection of $X \times V_\mu$ onto V_μ and its restriction to Z_μ. It is clear that this makes Z_μ a vector bundle on V_μ. Thus we have the exact sequence of vector bundles on V_μ

$$0 \longrightarrow \mathcal{S}_\mu \longrightarrow \mathcal{E} \otimes \mathcal{E}^* \longrightarrow \mathcal{T}_\mu \longrightarrow 0,$$

where $\mathcal{E} \otimes \mathcal{E}^*$ denotes a trivial bundle on V_μ with the fiber $E \otimes E^*$, and Z_μ is a total space of $\mathcal{S}_\mu$.

The fiber of q'_μ over a point from $O(\mu)$ consists of one point. Indeed, if the pair $(\phi, (R_{\mu'_1}, \ldots, R_{\mu'_1+\ldots+\mu'_{s-1}})) \in Z_\mu$ and $\phi \in O(\mu)$, we are forced to have $R_{\mu'_1+\ldots+\mu'_i} = \mathrm{Ker}\,\phi^i$. Since the variety Z_μ is irreducible, we see that Y'_μ is irreducible and of the same dimension that $O(\mu)$. This completes the proof of Proposition (8.1.3). ∎

The construction of the desingularization Z_μ places us in the situation of section 5.1. Thus we get a Koszul complex $\mathcal{K}(\xi_\mu)_\bullet$ of $\mathcal{O}_{X \times V_\mu}$-modules which is a Koszul complex resolving the structure sheaf $\mathcal{O}_{Z_\mu}$. The terms of $\mathcal{K}(\xi_\mu)_\bullet$ are given by the formula

$$\mathcal{K}(\xi_\mu)_\bullet : 0 \to \overset{t}{\bigwedge}(p^*\xi_\mu) \to \ldots \to \overset{2}{\bigwedge}(p^*\xi_\mu) \to p^*(\xi_\mu) \to \mathcal{O}_{X \times V_\mu}$$

where $\xi_\mu = \mathcal{T}_\mu^*$. We also denote $\eta_\mu = \mathcal{S}_\mu^*$. For a vector bundle $\mathcal{V}$ on V_μ we denote by $M^\mu(\mathcal{V})$ the sheaf $\mathcal{O}_{Z_\mu} \otimes p^*(\mathcal{V})$ of $\mathcal{O}_{X \times V_\mu}$-modules. We also recall that by A we denote the polynomial ring $\mathrm{Sym}(E \otimes E^*)$ of regular functions on X.

Applying Theorem (5.1.2) to our situation, we get

(8.1.4) Basic Theorem for Nilpotent Orbits. *For a vector bundle $\mathcal{V}$ on V_μ we define free graded A-modules*

$$F^\mu(\mathcal{V})_i = \bigoplus_{j \ge 0} H^j\left(V_\mu, \overset{i+j}{\bigwedge}\xi_\mu \otimes \mathcal{V}\right) \otimes_k A(-i-j).$$

(a) There exist minimal differentials

$$d_i^{\mu}(V) : F^{\mu}(V)_i \to F^{\mu}(V)_{i-1}$$

of degree 0 such that $F_{\bullet}^{\mu}(V)$ is a complex of graded free A-modules with

$$H_{-i}(F^{\mu}(V)_{\bullet}) = \mathcal{R}^i q_* M(V).$$

In particular the complex $F^{\mu}(V)_{\bullet}$ is exact in positive degrees.

(b) The sheaf $\mathcal{R}^i q_* M^{\mu}(V)$ is equal to $H^i(Z, M^{\mu}(V))$, and it can be also identified with the graded A-module $H^i(V, \mathrm{Sym}(\eta_{\mu}) \otimes V)$.

(c) If $\phi : M^{\mu}(V) \to M^{\mu}(V')(n)$ is a morphism of graded sheaves, then there exists a morphism of complexes

$$f_{\bullet}(\phi) : F^{\mu}(V)_{\bullet} \to F^{\mu}(V')_{\bullet}(n).$$

Its induced map $H_{-i}(f_{\bullet}(\phi))$ can be identified with the induced map

$$H^i(Z, M^{\mu}(V)) \to H^i(Z, M^{\mu}(V'))(n).$$

The calculation of cohomology groups $H^j(V_{\mu}, \bigwedge^{i+j} \xi_{\mu} \otimes V)$ and $H^i(V, \mathrm{Sym}(\eta_{\mu}) \otimes V)$ is much more difficult than in the case of determinantal varieties, or even in the case of hyperdeterminants, considered in chapter 9. The reason is that the bundle ξ_{μ} cannot be expressed conveniently as a tensor product of tautological bundles. In fact this calculation has not been done explicitly even for the case where V is trivial, i.e. for the case of syzygies of the coordinate ring of Y_{μ}. This is a very interesting problem that should be a subject of further research. I believe the full solution is possible and should lead to very interesting combinatorics.

In the remainder of this section we will describe the inductive procedure which allows to give the estimates on the terms of complexes $F_{\bullet}^{\mu}$ in the case of nilpotent orbits. This will be the main tool in the next section when we describe the generators of their defining ideals.

Let us first mention that the constructions of Z_{μ}, ξ_{μ}, and η_{μ} can be done in a relative setting, i.e. when we replace the vector space E by the vector bundle $\mathcal{E}$ over some scheme. Therefore we can talk about the bundles ξ_{μ} and η_{μ} associated to the bundle $\mathcal{E}$ of dimension n. They are respectively a subbundle and a factorbundle of $\mathcal{E} \otimes \mathcal{E}^*$. Since the bundles ξ_{μ} and η_{μ} do not change when we replace $\mathcal{E}$ by $\mathcal{E}^*$, we will denote them by $\xi_{\mu}(\mathcal{E}, \mathcal{E}^*)$ and $\eta_{\mu}(\mathcal{E}, \mathcal{E}^*)$ respectively.

Now we fix n and a partition $\mu = (\mu_1, \ldots, \mu_r)$. We denote by $\hat{\mu}$ the partition $(\mu_1 - 1, \ldots, \mu_r - 1)$. The conjugate partition $\hat{\mu}'$ is the partition $(\mu_2', \ldots, \mu_s')$.

On the flag variety V_μ we have the tautological subbundles $\mathcal{R}_{\mu_1' + \ldots + \mu_i'}$ and the corresponding factorbundles $\mathcal{Q}_{\mu_{i+1}' + \ldots + \mu_s'}$. The indices here denote the dimensions. We will denote $\mathcal{R}_{\mu_1'}$ by $\mathcal{R}$ and $\mathcal{Q}_{\mu_2' + \ldots + \mu_s'}$ by $\mathcal{Q}$. In this setting we have an exact sequence of vector bundles on V_μ:

$$0 \longrightarrow \mathcal{R} \otimes E^* \longrightarrow \xi_\mu \longrightarrow \xi_{\hat\mu}(\mathcal{Q}, \mathcal{Q}^*) \longrightarrow 0,$$

where $\xi_{\hat\mu}(\mathcal{Q}, \mathcal{Q}^*)$ denotes the bundle $\xi_{\hat\mu}$ in the relative setting described above. For each $t \geq 0$ this sequence induces a filtration $\mathcal{F}_i^t$, $0 \leq i \leq t$, on $\bigwedge^t(\xi_\mu)$ such that for each $i \leq t-1$ $\mathcal{F}_i^t \subset \mathcal{F}_{i+1}^t$ and $\mathcal{F}_{i+1}^t/\mathcal{F}_i^t = \bigwedge^{t-i}(\mathcal{R} \otimes E^*) \otimes \bigwedge^i \xi_{\hat\mu}(\mathcal{Q}, \mathcal{Q}^*)$. Indeed, we can define $\mathcal{F}_0 = 0$ and, for each $i \leq t-1$, $\mathcal{F}_{i+1}$ to be the image of the map

$$\bigwedge^{t-i}(\mathcal{R} \otimes E^*) \otimes \bigwedge^i \xi_\mu \to \bigwedge^s \xi_\mu$$

induced by exterior multiplication. This allows us to consider for $0 \leq i \leq t-1$ the exact sequences

$$0 \to \mathcal{F}_i^t \to \mathcal{F}_{i+1}^t \to \bigoplus_{0 \leq i \leq t} \bigwedge^{t-i}(\mathcal{R} \otimes E^*) \otimes \bigwedge^i \xi_{\hat\mu}(\mathcal{Q}, \mathcal{Q}^*) \to 0 \qquad (*)$$

and the associated long sequences of cohomology groups. We will use these sequences to estimate the terms in the complex $F_\bullet^\mu$.

We start with the introduction of the bundle

$$\xi_\mu' = \mathcal{R}_{\mu_1'} \otimes E^* \oplus (\mathcal{R}_{\mu_1' + \mu_2'}/\mathcal{R}_{\mu_1'}) \otimes \mathcal{Q}_{n-\mu_1'}^* \oplus \ldots \oplus$$
$$(E/\mathcal{R}_{\mu_1' + \ldots + \mu_{s-1}'}) \otimes \mathcal{Q}_{\mu_s'}^*$$

The consecutive sequences of type $(*)$ define a filtration on the bundle ξ_μ whose associated graded is the bundle ξ_μ'. We will need a recursive form of $(*)$, which is

$$\xi_\mu' = \mathcal{R} \otimes E^* \oplus \xi_{\hat\mu}'(\mathcal{Q}, \mathcal{Q}^*), \qquad (**)$$

where $\xi_{\hat\mu}'(\mathcal{Q}, \mathcal{Q}^*)$ is a bundle of type ξ' constructed in a relative situation.

We introduce $GL(E)$-modules

$$G_i^\mu := \bigoplus_{j \geq 0} H^j \left(V_\mu, \bigwedge^{i+j} \xi_\mu' \right).$$

It is clear from the long exact sequences of cohomology associated to sequences $(*)$ that the groups $H^j(V_\mu, \bigwedge^{i+j} \xi_\mu)$ are smaller than the groups

$H^j(V_\mu, \bigwedge^{i+j} \xi'_\mu)$. Therefore our first step will be to give a procedure to calculate the terms G_i^μ. It is based on the formula $(**)$, which implies

$$\dot{\bigwedge} \xi'_\mu = \dot{\bigwedge}(\mathcal{R} \otimes E^*) \otimes \dot{\bigwedge} \xi'_{\hat{\mu}}(\mathcal{Q}, \mathcal{Q}^*).$$

The terms of G_i^μ can be calculated in the following way. Let us choose the term of $G_\bullet^{\hat{\mu}}(\mathcal{Q}, \mathcal{Q}^*)$ which is a representation of the type $K_\alpha \mathcal{Q}$ where α is an integral dominant weight for the group $GL(n - \mu'_1)$. Let us assume that this term occurs in homogeneous degree t and in homological degree u. For this term we calculate

$$G(K_\alpha \mathcal{Q}) = \bigoplus_{i,j \geq 0} H^j(\mathrm{Grass}(\mu'_1, E), K_\alpha \mathcal{Q}) \otimes \overset{i+j}{\bigwedge}(\mathcal{R} \otimes E^*), \qquad (?)$$

where the terms corresponding to a given i, j will appear in homogeneous degree $t + i + j$ and in homological degree $u + i$. Notice that the terms of the collection $G(K_\alpha \mathcal{Q})$ will appear as tensor products $K_\beta E \otimes K_\gamma E^*$, but we can decompose them into irreducible representations to make the next step. Now all the collections $G(K_\alpha \mathcal{Q})$ give us the terms of $G_\bullet^\mu$. Notice that the collection $(?)$ comprises the terms of the complex $F(\mathcal{V})_\bullet$ associated to the bundle $\xi = \mathcal{R} \otimes E^*$ with a *twist* $\mathcal{V} = K_\alpha \mathcal{Q}$. These are the twisted complexes with the support in the determinantal variety of $n \times n$ matrices of rank $\leq n - \mu'_1$. Such complexes were considered in section 6.5. These complexes are $GL(E) \times GL(E^*)$-equivariant.

Now we will use our procedure to estimate the weights of the terms of $G_\bullet^\mu$. First we notice that all terms in $G_\bullet^\mu$ are the representations $K_\beta E$ where the sum of all entries in the weight β is equal to 0. To separate positive, zero, and negative entries of β, let us write $\beta = (\sigma, 0^c, \backslash\tau)$ where σ and τ are two partitions. Here for $\tau = (\tau_1, \ldots, \tau_t)$ we let $\backslash\tau = (-\tau_t, \ldots, -\tau_1)$.

(8.1.5) Lemma. *Let n and μ be as above. Let us consider the term $K_\alpha \mathcal{Q}$ of $G_\bullet^{\hat{\mu}}(\mathcal{Q}, \mathcal{Q}^*)$ where α is such that the sum of its entries equals 0 and its last entry is $\geq -\mu'_1$.*

(a) *We consider the twisted complex $F(\mathcal{V})_\bullet$ associated to the bundle $\xi = \mathcal{R} \otimes E^*$, with the twist $\mathcal{V} = K_\alpha \mathcal{Q}$. Then $F(\mathcal{V})_i = 0$ for $i < 0$. Moreover,*

$$F(\mathcal{V})_0 = K_\sigma E \otimes K_\tau E^* \otimes A(-|\sigma|)$$

and

$$F(\mathcal{V})_1 = K_{\sigma, 1^{c+1}} E \otimes K_{\tau, 1^{c+1}} E^* \otimes A(-|\sigma| - c - 1).$$

All terms of $F(V)_\bullet$ are of the form $K_\beta E \otimes K_\gamma E^*$ where β and γ are partitions of the same number, and $\gamma_1 \leq \mu_1'$.

(b) Let us assume that the term $K_\alpha Q$ occurs in $G_i^{\hat\mu}(Q, Q^*)$. Then all terms from $G(K_\alpha Q)$ occur in G_j^μ with $j \geq i$.

Proof. First of all, let us notice that part (b) follows instantly from part (a).

We will prove part (a). Let us write in this proof $t = \mu_1'$ for short. By the Cauchy decomposition (3.2.5) we have

$$\overset{\bullet}{\bigwedge}(\mathcal{R} \otimes E^*) = \bigoplus_\beta K_\beta \mathcal{R} \otimes K_{\beta'} E^*.$$

This means that for each partition β we have to calculate the cohomology groups of

$$K_\alpha Q \otimes K_\beta \mathcal{R} \otimes_{\beta'} E^*,$$

which are the cohomology groups of $K_\alpha Q \otimes K_\beta \mathcal{R}$ tensored with $K_{\beta'} E^*$. To calculate this cohomology for each $\beta = (\beta_1, \ldots, \beta_t)$ we have to apply Bott's theorem (4.1.4) to the sequence

$$z(\beta) = (\alpha, \beta) = (\sigma, 0^c, \backslash\tau, \beta).$$

Let us consider the weight $z(\beta) + \rho$. This is a sequence

$$(\sigma_1 + n - 1, \ldots, \sigma_u + n - u, t + v + c - 1, \ldots,$$
$$t + v, t + v - 1 - \tau_v, \ldots, t - \tau_1, \beta_1 + t - 1, \ldots \beta_t),$$

where we write $\sigma = (\sigma_1, \ldots, \sigma_u)$, $\tau = (\tau_1 \ldots, \tau_v)$ and keep in mind that $n = u + c + v + t$. By (4.1.9) the partitions β giving nonzero contributions to cohomology are these for which the sequence $z(\beta) + \rho$ has no repetitions. For such β we have to reorder our sequence to make it decreasing, and subtract ρ from it. We get a weight $\gamma(\beta)$. The resulting cohomology group will then be $K_{\gamma(\beta)} E \otimes K_{\beta'} E^*$ occurring in the complex $F(V)_\bullet$ in the place $p(\beta) = |\beta| - l(w)$ where w is a permutation needed to reorder $z(\beta) + \rho$.

First we notice that since $t \geq \tau_1$ then all the entries in $z(\beta) + \rho$ are positive which means that all the entries of $\gamma(\beta)$ are nonnegative.

Let us consider two partitions: β and $\gamma = (\beta_1, \ldots, \beta_s + j, \beta_{s+1}, \ldots, \beta_t)$, both giving nonzero contributions to the terms of $F_\bullet(V)$. We will show that $p(\gamma) > p(\beta)$. Indeed, the sequences $z(\beta) + \rho$ and $z(\gamma) + \rho$ differ only in one place, and the term $\beta_s + j + t - s$ in $z(\gamma) + \rho$ can be exchanged with at most $j - 1$ additional numbers compared to the corresponding term $\beta_s + t - s$ in $z(\beta) + \rho$. This will account for an increase of at most $j - 1$ in $l(w)$.

Starting with an arbitrary β we can now use the steps described above to produce terms with smaller β that lie in smaller homological degree. Continuing like this we can make β satisfy the condition $c + v \geq \beta_1$.

On the other hand if $c + v \geq \beta_1$, then

$$\{t + v + c + 1, \ldots, t + v, t + v - 1 - \tau_v, \ldots, t - \tau_1, \beta_1 + t - 1, \ldots, \beta_t\}$$

are $t + v + c$ numbers belonging to $\{0, 1, \ldots, t + v + c - 1\}$. They can be distinct for a unique partition β and it is easy to see that $\beta = \tau'$. It is also clear that $\beta = \tau'$ is the smallest partition giving a nonzero contribution the terms of $F(\mathcal{V})_{\bullet}$. By the previous argument $p(\beta)$ is the smallest for $\beta = \tau'$. It is very easy to check that in fact $p(\tau') = 0$. Similarly we can identify $\gamma = (\tau, 1^c)$ as the only partition with $p(\gamma) = 1$.

It remains to prove the last statement in (a). But this follows if we take into account that the term of $F_{\bullet}(\mathcal{V})$ corresponding to β is $K_{\gamma(\beta)}E \otimes K_{\beta'}E^*$ and that $\beta_1' \leq t$. ∎

(8.1.6) Theorem. *Let μ be a partition of n. Let $F_{\bullet}^{\mu}$ be a complex of A-modules described in (8.1.4).*

(a) The terms F_i^{μ} are zero for $i < 0$.
(b) The term F_0^{μ} equals $A(0)$.
(c) For any representation $K_{\alpha}E$ occurring in $F_{\bullet}^{\mu}$ we have $\alpha_n \geq -\mu_1'$.
(d) The varieties Y_{μ} are normal and they have rational singularities.
(e) The coordinate rings of varieties Y_{μ} are Gorenstein.

Proof. First we notice Theorem (5.1.3) (c) implies that (d) follows from (a) and (b). Also we notice that (e) follows from Theorem (5.1.4). The point is that the maximal exterior power of ξ_{μ} is isomorphic to the canonical bundle on V_{μ}, by Exercise 13, chapter 3.

Therefore it is enough to prove the first three statements of the theorem. We will actually deal with the spaces $G_{\bullet}^{\mu}$.

We will prove the following statements:

(a') The terms G_i^{μ} are zero for $i < 0$.
(b') The term G_0^{μ} equals k.
(c') For any representation $K_{\alpha}E$ occurring in $G_{\bullet}^{\mu}$ we have $\alpha_n \geq -\mu_1'$.

The statements for $F_{\bullet}^{\mu}$ follow because the cohomology groups of exterior powers of ξ_{μ} are smaller than those of exterior powers of ξ_{μ}'.

We argue by induction on the number of parts in μ'.

If μ' has only one part, i.e. $\mu' = (n)$, then Y_μ is the origin and $\xi_\mu = \xi'_\mu$, so $G_i^\mu = \bigwedge^i (E \otimes E^*)$. Then obviously (a') and (b') are satisfied, and (c') follows from the Cauchy formula (2.3.3) (b) and the Littlewood–Richardson rule (2.3.4).

Let us therefore assume that $\mu' = (\mu'_1, \ldots, \mu'_s)$ and that all three statements are true for partitions with at most $s - 1$ parts. Let $\hat{\mu}' = (\mu'_2, \ldots, \mu'_s)$. We consider the terms of the complex $G_\bullet^{\hat\mu}(Q, Q^*)$. By the inductive assumption, all such terms $K_\alpha Q$ occur in nonnegative homological degrees and they satisfy the assumption $\alpha_{n-\mu'_1} \geq -\mu'_2$. This means each term satisfies the assumptions of Lemma (8.1.5). Now statement (a') for μ follows from (8.1.5) (b). We also see that the contribution to G_0^μ can come only from the 0th term of $G_\bullet^{\hat\mu}(Q, Q^*)$. But this consists of one copy of the trivial representation. Therefore by (8.1.5) (a) statement (b') is satisfied for μ. Finally (c') also follows from the last part of (8.1.4) (a) and from the Littlewood–Richardson rule (2.3.4). ∎

In the sequel we will denote by A_μ the coordinate ring of Y_μ. Theorem (8.1.6) implies that the complex $F_\bullet^\mu$ is a minimal free resolution of A_μ treated as an A-module. It also says that the rings A_μ are normal, are Gorenstein, and have rational singularities.

We will denote by J_μ the defining ideal of Y_μ. This means that $A_\mu = A/J_\mu$. Thus the first term of $F_\bullet^\mu$ gives us the information about the minimal generators of the ideal J_μ.

Let us note the following corollary from our calculation of $G_\bullet^\mu$.

(8.1.7) Corollary. *Let μ be a partition of n. The term G_1^μ contains only the representations $K_{(1^j, 0^{n-2j}, (-1)^j)} E$ for $0 \leq j \leq \frac{n}{2}$.*

Proof. Again we use the induction on the number of parts of μ'. For μ' having one part, again the result is true, because $G_i^{(n)} = \bigwedge^i (E \otimes E^*)$. To make an inductive step from $\hat\mu$ to μ, we again use Lemma (8.1.5). ∎

We finish this section with some examples where the complete calculation of the complex $F_\bullet^\mu$ is possible.

(8.1.8) Examples.

(a) Let $\mu' = (n)$. Then $Y_\mu = \{0\}$. The bundle $\xi_\mu = E \otimes E^*$, and therefore

$$F_i^{(n)} = \overset{i}{\bigwedge}(E \otimes E^*) \otimes A(-i)$$

and $F_\bullet^{(n)}$ is the Koszul complex associated to the ideal generated by all variables $\phi_{i,j}$.

(b) Let $\mu' = (1^n)$. The variety $Y_{(1^n)}$ is the set of all nilpotent matrices. We will show that the complex $F_\bullet^{(1^n)}$ is the Koszul complex associated to the ideal generated by the basic invariants $v_1, \ldots, v_n$. This means that

$$F_i^{(1^n)} = \bigoplus_{\{(\epsilon_1,\ldots,\epsilon_n)\ |\ \epsilon_j\in\{0,1\},\ \sum_{j=1}^n \epsilon_j=i\}} A\left(-\sum_{j=1}^n j\epsilon_j\right)$$

We show first that the cohomology groups $G_\bullet^{(1^n)}$ consist of trivial representations only. We proceed by induction on n. For $n = 1$, the statement is obviously true. Let us assume the statement is true for $\mu' = (1^{n-1})$. Using the formula for $G(K_\alpha Q)$, we see that each trivial representation which is a term of $G_\bullet^{(1^{n-1})}(Q_1, Q_1^*)$ contributes to $G_\bullet^{(1^n)}$ the terms

$$\bigoplus_{i,j\geq 0} H^j\left(\text{Grass}(1, E), \bigwedge^{i+j}(\mathcal{R} \otimes E^*)\right),$$

where $\mathcal{R}$ is the tautological subbundle on $\text{Grass}(1, E)$, i.e. the bundle $\mathcal{O}(-1)$. Now it follows from Serre's theorem ([H1, chapter III, Theorem 5.1]) that only two cohomology groups in the above formula are nonzero. These are

$$H^0\left(\text{Grass}(1, E), \bigwedge^0(\mathcal{R} \otimes E^*)\right) = K_{(0)}E$$

and

$$H^{n-1}\left(\text{Grass}(1, E), \bigwedge^n(\mathcal{R} \otimes E^*)\right) = K_{(0)}E.$$

Therefore we get two copies of the trivial representation. Our statement is proved. It follows that the terms of the complex $F_\bullet^{(1^n)}$ consist entirely of trivial representations. This implies that the ideal of functions vanishing on $Y_{(1^n)}$ is generated by $\text{GL}(E)$-invariants, i.e. by $v_1, \ldots, v_n$ (in view of the fact that $B = A^{\text{GL}(E)}$ is a polynomial ring in $v_1, \ldots, v_n$). It remains to show that $v_1, \ldots, v_n$ form a regular sequence in A. This follows from dimensional considerations because $\dim Y_{(1^n)} = \dim Z_{(1^n)} = n^2 - n$.

(c) Let $\mu' = (n - p, 1^p)$ be a hook. The variety Y_μ is the set of nilpotent matrices of rank $\leq p$. We again use the formulas for $G(K_\alpha Q)$. Each term of $G_\bullet^{(1^p)}(Q, Q^*)$ (which has to be a trivial representation) *contributes*

$$\bigoplus_{i,j\geq 0} H^j\left(\text{Grass}(n-p,E), \bigwedge^{i+j}(\mathcal{R}\otimes E^*)\right),$$

where $\mathcal{R}$ and $\mathcal{Q}$ denote respectively the tautological subbundle and factorbundle on $\text{Grass}(n-p,E)$. This gives us the terms of the Lascoux complex resolving $p+1$ order minors of ϕ, described in (6.1.3). Lemma (8.1.5) now shows that the generators of the defining ideal of Y_μ are a subrepresentation of a direct sum of p copies of trivial representations (in homogeneous degrees $1, \ldots, p$) and of the copy of $\bigwedge^{p+1} E \otimes \bigwedge^{p+1} E^*$ in homogeneous degree $p+1$. On the other hand, we have some obvious polynomials vanishing on Y_μ: the minors of degree $p+1$ of the matrix ϕ, and the invariants $v_1, \ldots, v_p$. Moreover, these equations obviously define Y_μ set-theoretically. If we consider the determinantal variety Y_p of matrices of rank $\leq p$, the dimension count shows that $\dim Y_{(n-p,1^p)} = \dim Y_p - p$. This means that the invariants $v_1, \ldots, v_p$ form a regular sequence in the coordinate ring of the determinantal variety Y_p. Now we identify the generators of the defining ideal of Y_μ by induction on the homogeneous degree. In degrees $1, \ldots, p$ the only generators are $v_1, \ldots, v_p$, because they have to come from trivial representations. In degree $p+1$ the minors of degree $p+1$ of ϕ have to be among the minimal generators (because no linear combinations of these minors with coefficients in $\mathbf{K}$ can be in the ideal generated by $v_1, \ldots, v_p$). Our estimate from above of the first term of $F_\bullet^{(n-p,1^p)}$ shows now that the defining ideal of $Y_{(n-p,1^p)}$ is generated by $v_1, \ldots, v_p$ and by $p+1$ order minors of ϕ. The complex $F_\bullet^{(n-p,1^p)}$ being a minimal resolution of this ideal, has to be the tensor product of the Lascoux's resolution (6.1.3) of the ideal of $p+1$ order minors, and of the Koszul complex on $v_1, \ldots, v_p$.

(d) *Let $\mu' = (n-p, p)$ be the partition with two columns. The variety Y_μ consists of matrices ϕ such that rank $\phi \leq p$ and $\phi^2 = 0$. These varieties were first considered by Strickland in [S], where they were called the projector varieties, because of the property $\phi^2 = 0$. Here we just identify the homogeneous components of the coordinate ring of Y_μ. We notice that the flag variety V_μ we are using is in this case just $\text{Grass}(n-p, E)$. It is easy to identify the bundle η_μ with $\mathcal{Q} \otimes \mathcal{R}^*$, where as usual $\mathcal{R}$ and $\mathcal{Q}$ denote respectively the tautological subbundle and factorbundle on $\text{Grass}(n-p, E)$. Applying Theorem (8.1.4)(b) and Theorem (8.1.6), we see that $F_\bullet^\mu$ is a minimal resolution of the coordinate ring of Y_μ which can be identified with*

$H^0(\mathrm{Grass}(n - p, E), \mathrm{Sym}_\bullet(\mathcal{Q} \otimes \mathcal{R}^*))$. *Thus Cauchy's formula (3.2.5) gives*

$$\mathrm{Sym}_d(\mathcal{Q} \otimes \mathcal{R}^*) = \bigoplus_{|\alpha|=d} K_\alpha \mathcal{Q} \otimes K_\alpha \mathcal{R}^*.$$

Using Corollary (4.1.9), we get that for $\alpha = (\alpha_1, \ldots, \alpha_p)$

$$H^0(\mathrm{Grass}(n - p, E), K_\alpha \mathcal{Q} \otimes K_\alpha \mathcal{R}^*) = K_{(\alpha_1,\ldots,\alpha_p,0^{n-2p},-\alpha_p,\ldots,-\alpha_1)} E.$$

The result is that

$$(A_{(n-p,p)})_d = \bigoplus_{|\alpha|=d} K_{(\alpha_1,\ldots,\alpha_p,0^{n-2p},-\alpha_p,\ldots,-\alpha_1)} E.$$

This allows us to see that $A_{(n-p,p)}$ is a factor of A. Cauchy's formula for A gives

$$A_d = \bigoplus_{|\alpha|=d} K_\alpha E \otimes K_\alpha E^*.$$

We see now that if $\alpha = (\alpha_1, \ldots, \alpha_s)$ with $s \geq p + 1$, then the summand $K_\alpha E \otimes K_\alpha E^*$ is contained in $J_{(n-p,p)}$. If $s \leq p$, the only surviving part of $K_\alpha E \otimes K_\alpha E^*$ in $A_{(n-p,p)}$ is the Cartan representation $K_{(\alpha_1,\ldots,\alpha_p,0^{n-2p},-\alpha_p,\ldots,-\alpha_1)} E$ from $K_\alpha E \otimes K_\alpha E^*$. The kernel of the epimorphism

$$K_\alpha E \otimes K_\alpha E^* \to K_{(\alpha_1,\ldots,\alpha_p,0^{n-2p},-\alpha_p,\ldots,-\alpha_1)} E$$

consists of all polynomials having a trace *component, i.e., combinations of polynomials divisible by $w_{s,t} = \sum_{i=1}^n \phi_{i,s}\phi_{t,i}$ or by $v_1 = \sum_{i=1}^n \phi_{i,i}$. Therefore the ideal $J_{(n-p,p)}$ is spanned by the quadratic polynomials of that kind, by the invariant v_1, and by the $p + 1$ order minors of ϕ.*

8.2. The Equations of the Conjugacy Classes of Nilpotent Matrices

We use the inductive procedure from the previous section to get the information about the generators of the defining ideals J_μ of the varieties Y_μ. We preserve the notation from the previous section. In particular, $\mu = (\mu_1, \ldots, \mu_r)$ is a partition of n, and $\hat{\mu} = (\mu_1 - 1, \ldots, \mu_r - 1)$. For such a partition μ we denote the coordinate ring of Y_μ by A_μ. Thus $A_\mu = A/J_\mu$.

We start by exhibiting some explicit polynomials vanishing on Y_μ. By Corollary (8.1.7) we know that the generators of the ideals J_μ consist of the representations of type $K_{(1^j 0^{n-2j},(-1)^j)} E$ for $0 \leq j \leq \frac{n}{2}$. The representations of this type are exactly the ones occurring as composition factors of representations $\bigwedge^p E \otimes \bigwedge^p E^*$ which are linear spans of $p \times p$ minors of the matrix ϕ. The straightening law (3.2.5) tells us that there is a basis of the polynomial ring $A = \text{Sym}(E \otimes E^*)$ consisting of the products of minors of the generic $n \times n$ matrix $\phi = (\phi_{i,j})_{1 \leq i,j \leq n}$. This gives the idea of looking for the polynomials vanishing on Y_μ which are linear combinations of the minors of various sizes of ϕ.

Let p be a number such that $1 \leq p \leq n$, and let $I = (i_1, \ldots, i_p)$, $J = (j_1, \ldots, j_p)$ be two multiindices with entries in the set $[1, n]$. We denote by $\phi_{I,J}$ the $p \times p$ minor of ϕ corresponding to rows $i_1, \ldots, i_p$ and columns $j_1, \ldots, j_p$. The polynomial $\phi_{(I|J)}$ is obviously antisymmetric in the entries $i_1, \ldots, i_p$ and in the entries $j_1, \ldots, j_p$.

We consider the linear span of $p \times p$ minors of ϕ. This is a linear supspace of A_p, which can be identified with $\bigwedge^p E \otimes \bigwedge^p E^*$. If $\{e_1, \ldots, e_n\}$ is a basis of E, and $\{e_1^*, \ldots, e_n^*\}$ is a dual basis of E^*, the tensor $e_{i_1} \wedge \ldots \wedge e_{i_p} \otimes e_{j_1}^* \wedge \ldots \wedge e_{j_p}^*$ is identified with $\phi_{I,J}$.

Using the Littlewood–Richardson rule (2.3.4), we see the following decomposition:

$$\bigwedge^p E \otimes \bigwedge^p E^* = \bigoplus_{0 \leq i \leq \min(p, n-p)} K_{(1^i, 0^{n-2i}, (-1)^i)} E.$$

We will denote the copy of the representation $K_{(1^i, 0^{n-2i}, (-1)^i)} E$ inside the span $\bigwedge^p E \otimes \bigwedge^p E^*$ of $p \times p$ minors by $U_{i,p}$. If $i > \min(p, n-p)$, we set $U_{i,p} = 0$. This means we can rewrite the previous decomposition as

$$\bigwedge^p E \otimes \bigwedge^p E^* = \bigoplus_{0 \leq i \leq \min(p, n-p)} U_{i,p}.$$

Next, we denote by $V_{i,p}$ the subspace of the span $\bigwedge^p E \otimes \bigwedge^p E^*$ of $p \times p$ minors defined as

$$V_{i,p} = U_{0,p} \oplus U_{1,p} \oplus \ldots \oplus U_{i,p}.$$

The point of introducing the spaces $V_{i,p}$ is that they have a simple description in terms of minors of ϕ. The space $V_{i,p}$ is isomorphic as a $\text{GL}(E)$-module to $\bigwedge^i E \otimes \bigwedge^i E^*$, and it can be identified with the image of the map

$$h_{i,p} : \bigwedge^i E \otimes \bigwedge^i E^* \xrightarrow{1 \otimes 1 \otimes t_{p-i}} \bigwedge^i E \otimes \bigwedge^i E^* \otimes \bigwedge^{p-i} E \otimes \bigwedge^{p-i}$$
$$E^* \xrightarrow{m \otimes m} \bigwedge^p E \otimes \bigwedge^p E^*,$$

where $t_{p-i} : \mathbf{K} \to \bigwedge^{p-i} E \otimes \bigwedge^{p-i} E^*$ is the map sending 1 to the $\text{GL}(E)$-invariant t_{p-i}, and $m \otimes m$ denotes exterior multiplication of the first component with 3rd, and of the second with 4-th.

Let us fix two multiindices $P = (p_1, \ldots, p_i)$ and $Q = (q_1, \ldots, q_i)$. We have

$$h_{i,p}(e_{p_1} \wedge \ldots \wedge e_{p_i} \otimes e_{q_1}^* \wedge \ldots \wedge e_{q_i}^*) = \sum_{|J|=p-i} \phi_{(P,J|Q,J)},$$

and so $V_{i,p}$ is the span of such elements for different choices of P and Q.

(8.2.1) Lemma. *Let μ be a partition of n. The elements of the space $V_{i,p}$ vanish identically on the variety Y_μ if and only if*

$$p > \mu_1 + \ldots + \mu_i - i.$$

Proof. Let us first assume that the condition of Lemma (8.2.1) is satisfied. We will show that $V_{i,p}$ vanishes on Y_μ. Let us consider the typical generator

$$\sum_{|j|=p-i} \phi_{(P,J|Q,J)}$$

for fixed subsets P and Q of cardinality i. Since these elements span a $\mathrm{GL}(E)$-stable subspace in A_p, it is enough to show that they vanish on a single matrix from $O(\mu)$. Choosing a matrix from $O(\mu)$ in a canonical Jordan form, it is easy to see that in fact all summands $\phi_{(P,J|Q,J)}$ are zero when evaluated on that matrix.

To prove the other implication, let us assume that the condition of (8.2.1) is not satisfied. This means that

$$p \le \mu_1 + \ldots + \mu_i - i.$$

Let us choose j for which $\mu_j > 1$ and $\mu_{j+1} = 1$ (if $\mu_i > 1$, we choose $j = i$). Now we set

$$P = (1, \mu_1 + 1, \ldots, \mu_1 + \ldots + \mu_j + 1).$$

We can choose the numbers $w_1, \ldots, w_j$ in such way that $1 \le w_m \le \mu_m$ for $m = 1, \ldots, j$, and $w_1 + \ldots + w_j = p + i$. We set

$$Q = (w_1, \mu_1 + w_2, \ldots, \mu_1 + \ldots + \mu_{j-1} + w_j).$$

We consider the polynomial $\sum_{|J|=p-j} \phi_{(P,J|Q,J)}$.

Clearly its value on a matrix from $O(\mu)$ in Jordan canonical form is not zero. Moreover, our element is in $V_{j,p}$, which is contained in $V_{i,p}$ by definition. This completes the proof of the lemma. ∎

Our goal in this section is to prove that the representations $V_{i,p}$ satisfying the condition (8.2.1) are the generators of the ideal J_μ. It is worthwhile to point out right away that they do not define a minimal set of generators of the ideals J_μ. For example, the following result is a simple consequence of Laplace expansion.

(8.2.2) Lemma. *For $i \geq 1$ the representation $V_{i,p+1}$ is contained in the ideal generated by the representation $V_{i,p}$.*

Let us denote by J'_μ the ideal generated by the representations $V_{i,p}$ satisfying the condition in (8.2.1).

Lemma (8.2.2) immediately implies

(8.2.3) Proposition. *The ideal J'_μ is generated by the spaces $U_{i,\mu(i)}$ ($1 \leq i \leq n$) where $\mu(i) = \mu_1 + \ldots + \mu_i - i + 1$ (which are zero if $i > \min(\mu(i), n - \mu(i))$), and by the spaces $U_{0,p}$ (for $1 \leq p \leq n$) which correspond to the invariants t_p.*

We will use the following graphical representation of the representations $U_{i,p}$. We will represent them by an $(n+1) \times ([\frac{n}{2}] + 1)$ matrix whose rows are indexed by $0, 1, \ldots, n$ and columns are indexed by $0, 1, \ldots, [\frac{n}{2}]$. The (p, i)th entry corresponds to $U_{i,p}$. We will treat the entries corresponding to the spaces $U_{i,p}$ which are zero as empty. For a given partition μ we will represent the generators of J'_μ by a matrix $M(\mu)$ whose entry equals 1 if the corresponding $U_{i,p}$ vanishes on Y_μ, equals 0 if the corresponding $U_{i,p}$ does not vanish on Y_μ, and is empty if the corresponding $U_{i,p} = 0$.

(8.2.4) Example. *Let us take $n = 12$, $\mu = (3, 3, 2, 2, 1, 1)$. We have $\mu(1) = 3$, $\mu(2) = 5$, $\mu(3) = 6$, $\mu(4) = \mu(5) = \mu(6) = 7$. This means we have*

$$
M(\mu) = \begin{matrix}
0 \\
1 & 0 \\
1 & 0 & 0 \\
1 & 1 & 0 & 0 \\
1 & 1 & 0 & 0 & 0 \\
1 & 1 & 1 & 0 & 0 & 0 \\
1 & 1 & 1 & 1 & 0 & 0 & 0 \\
1 & 1 & 1 & 1 & 1 & 1 \\
1 & 1 & 1 & 1 & 1 \\
1 & 1 & 1 & 1 \\
1 & 1 & 1 \\
1 & 1 \\
1
\end{matrix}
\;.
$$

The fourth row of the matrix tells us that $U_{0,3}$ and $U_{1,3}$ are in J'_μ and that $U_{2,3}$ and $U_{3,3}$ are not. Lemma (8.2.2) tells us that J'_μ is generated by the $U_{i,p}$'s corresponding to the 1's in the first column of $M(\mu)$ and by the ones corresponding to the highest 1's in each other column. In our example we would conclude that $J'_{(6,4,2)}$ is generated by $U_{0,p}$ ($1 \le p \le 12$) and by $U_{1,3}, U_{2,5}, U_{3,6}, U_{4,7}, U_{5,7}$.

Now we state the main result of this section.

(8.2.5) Theorem. *For each partition μ the ideals J_μ and J'_μ are equal, i.e., J_μ is generated by the spaces $U_{i,\mu(i)}$ (for $1 \le i \le n$) and by the spaces $U_{0,p}$ (for $1 \le p \le n$).*

(8.2.6) Remark. *The set of generators given in (8.2.5) is not claimed to be minimal. We will discuss the minimal sets of generators at the end of this section.*

Proof of Theorem (8.2.5). We proceed by induction on the number s of parts in μ'. If $s = 1$, we have $\mu' = (n)$ and $Y_{(1^n)} = 0$. Therefore the ideal J_μ is generated by the entries $\phi_{i,j}$, i.e., by the representations $U_{0,1}$ and $U_{1,1}$. The combinatorial condition in (8.2.1) tells us that $U_{0,1}$ and $U_{1,1}$ are in $J'_{(n)}$, so we are done.

Let us consider the partition $\mu = (\mu_1, \dots, \mu_r)$. As before we denote $\hat\mu = (\mu_1 - 1, \dots, \mu_r - 1)$. By induction we know that the generators of $J'_{\hat\mu}$ are given by the spaces $U_{i,\hat\mu(i)}$ ($1 \le i \le n - \mu'_1$) and by the spaces $U_{0,p}$ ($1 \le p \le n - \mu'_1$).

Let us consider the complex $F^{\hat\mu}_\bullet$. We can construct this complex in a relative situation, taking the bundle Q on the Grassmannian $\mathrm{Grass}(\mu'_1, E)$ instead of a vector space E. We get a complex $F^{\hat\mu}_\bullet(Q, Q^*)$ of locally free sheaves over the sheaf of algebras $\mathcal{A} := \mathrm{Sym}(Q \otimes Q^*)$ defined over the Grassmannian $\mathrm{Grass}(\mu'_1, E)$. The terms of this complex are the sheaves of type $K_\alpha Q \otimes \mathcal{A}$ where α is a dominant integral weight for the group $\mathrm{GL}(n - \mu'_1)$. Moreover, by (8.1.6)(c) we see that only the weights satisfying $\alpha_1 \le n - \mu'_1$ occur.

Let us fix two terms of the complex $F^{\hat\mu}_\bullet(Q, Q^*)$: the term $K_\alpha Q \otimes \mathcal{A}$ occurring in homological dimension t and the term $K_\beta Q \otimes \mathcal{A}$ occurring in homological degree $t - 1$. The component $d_{\alpha,\beta}$ of the differential from the term $K_\alpha Q \otimes \mathcal{A}$ to the term $K_\beta Q \otimes \mathcal{A}$ comes from the natural map $K_\alpha Q \to K_\beta Q \otimes S_j(Q \otimes Q^*)$.

Let $\mathcal{B}$ denote the sheaf of algebras $\mathrm{Sym}(Q \otimes E^*)$ on $\mathrm{Grass}(\mu'_1, E)$. We can define the complex $\tilde F^{\hat\mu}(Q, Q^*)_\bullet$ of $\mathcal{B}$-modules to be the complex with the same terms as $F_\bullet^{\hat\mu}(Q, Q^*)$ but replacing the differential $d_{\alpha,\beta}$ with its composition with the natural embedding of $K_\beta Q \otimes S_j(Q \otimes Q^*)$ into $K_\beta Q \otimes S_j(Q \otimes E^*)$.

The complex $\tilde F_\bullet^{\hat\mu}(Q, Q^*)_\bullet$ is acyclic. Indeed, the complex $F_\bullet^{\hat\mu}(Q, Q^*)$ is, and locally (on $\mathrm{Grass}(\mu_1', E)$) the complex $\tilde F^{\hat\mu}(Q, Q^*)_\bullet$ has the same differentials as $F_\bullet^{\hat\mu}(Q, Q^*)$, though it is a complex over a polynomial ring with some additional irrelevant variables.

Let us identify the only homology group of $\tilde F^{\hat\mu}(Q, Q^*)_\bullet$. We define the subvariety W_μ of $X \times \mathrm{Grass}(\mu_1', E)$ as follows.

First of all

$$W_\mu \subset \{(\phi, R) \in X \times \mathrm{Grass}(\mu_1', E) \mid \phi|_R = 0\}.$$

An endomorphism ϕ such that $\phi|_R = 0$ induces the morphism of bundles $\phi' : Q \to E$ over $X \times \mathrm{Grass}(\mu_1', E)$. We define $\hat\phi : Q \to Q$ to be the composition of ϕ' with the natural epimorphism $E \to Q$. Now we set

$$W_\mu = \{(\phi, R) \in X \times \mathrm{Grass}(\mu_1', E) \mid \phi|_R = 0, \; \hat\phi \in Y_{\hat\mu}(Q, Q^*)\},$$

where $Y_{\hat\mu}(Q, Q^*)$ is the variety $Y_{\hat\mu}$ constructed in relative situation.

For the remainder of this section we will denote by p (by q) the projection from $X \times \mathrm{Grass}(\mu_1', E)$ onto $\mathrm{Grass}(\mu_1', E)$ (onto X). ∎

(8.2.7) Proposition.

(a) *The homology sheaf* $H_0(\tilde F^{\hat\mu}(Q, Q^*)_\bullet)$ *is equal to the direct image* $p_* \mathcal{O}_{W_\mu}$ *of the structure sheaf* $\mathcal{O}_{W_\mu}$.

(b) *The higher direct images* $R^i q_* \mathcal{O}_{W_\mu}$ *are* 0 *for* $i > 0$ *and* $q_* \mathcal{O}_{W_\mu} = \mathcal{O}_{Y_\mu}$.

Proof. It is clear that W_μ is a reduced variety in $X \times \mathrm{Grass}(\mu_1', E)$. The calculation can be done locally on $\mathrm{Grass}(\mu_1', E)$. The term $\tilde F^{\hat\mu}(Q, Q^*)_0$ equals $\mathcal{B}$, which accounts for the condition $\phi_R = 0$. The reduced equations giving the condition $\phi' \in Y^{\hat\mu}(Q, Q^*)$ are given by images of the elements in the term $\tilde F^{\hat\mu}(Q, Q^*)_1$. This proves part (a).

The map p is affine, so to prove part (b) it is enough to show that the cohomology groups $H^i(\mathrm{Grass}(\mu_1', E), p_* \mathcal{O}_{W_\mu})$ are 0 for $i > 0$ and $H^0(\mathrm{Grass}(\mu_1', E), p_* \mathcal{O}_{W_\mu}) = \mathcal{O}_{Y_\mu}$.

The complex $\tilde F^{\hat\mu}(Q, Q^*)_\bullet$ is an acyclic complex of locally free $\mathcal{B}$-modules with the terms being direct sums of sheaves of the form $K_\alpha \otimes \mathcal{B}Q$.

Let us consider the subvariety $Z = Z_{\mu_1'} \subset X \times \mathrm{Grass}(\mu_1', E)$, which is a total space of the vector bundle $Q^* \otimes E$. In other words,

$$Z = \{(\phi, R) \in X \times \mathrm{Grass}(\mu_1', E) \mid \phi|_R = 0\}.$$

This is a desingularization of the determinantal variety of matrices ϕ of rank $\leq n - \mu'_1$. We considered such varieties in section 6.1. Notice that by Proposition (5.1.1)(b) the sheaf of algebras B is just the direct image $p_*\mathcal{O}_Z$. Therefore, each term $K_\alpha \mathcal{Q} \otimes B$ is the direct image $p_* M(K_\alpha \mathcal{Q})$ of the corresponding twisted module $M(K_\alpha \mathcal{Q})$ associated to the variety Z. Such modules were considered in section 6.5. We also notice that Lemma (8.1.5) says that for all modules $M(K_\alpha \mathcal{Q})$ occurring in the complex $\tilde{F}^{\hat{\mu}}(\mathcal{Q}, \mathcal{Q}^*)_\bullet$ we have $\mathcal{R}^i q_*(M(K_\alpha \mathcal{Q})) = 0$ for $i > 0$. This proves the first part of (b).

To prove the second part of (b), let us calculate the graded Euler characteristic of $\tilde{F}^{\hat{\mu}}(\mathcal{Q}, \mathcal{Q}^*)_\bullet$. By part (a) it is equal to the graded Hilbert function of $q_*\mathcal{O}_{W_\mu}$.

On the other hand, the graded Euler characteristic of $\tilde{F}^{\hat{\mu}}(\mathcal{Q}, \mathcal{Q}^*)_\bullet$ can be calculated in another way. By (8.1.6) we know that the graded Euler characteristic of $\mathcal{O}_{Y_{\hat{\mu}}}$ can be calculated as the graded Euler characteristic of the complex $F^{\hat{\mu}}(\mathcal{Q}, \mathcal{Q}^*)$, i.e. as the Euler characteristic of the Koszul complex on the bundle $p^*(\xi_{\hat{\mu}})$ on $X \times V_\mu$. Thus the Euler characteristic of $\mathcal{O}_{W_\mu}$ can be calculated as the Euler characteristic of the Koszul complex on the bundle $p^*(\xi_{\hat{\mu}} \oplus (\mathcal{R} \otimes E^*))$ on $X \times V_\mu$. But by the exact sequence

$$0 \longrightarrow \mathcal{R} \otimes E^* \longrightarrow \xi_\mu \longrightarrow \xi_{\hat{\mu}'}(\mathcal{Q}, \mathcal{Q}^*) \longrightarrow 0,$$

this is the same as the graded Euler characteristic of $F^\mu_\bullet$, i.e., by (8.1.6), the graded Hilbert function of $\mathcal{O}_{Y_\mu}$. This concludes the proof of part (b) of the lemma. ∎

Let us concentrate on two first terms of the complex $\tilde{F}^{\hat{\mu}}(\mathcal{Q}, \mathcal{Q}^*)_\bullet$.

The term $\tilde{F}^{\hat{\mu}}(\mathcal{Q}, \mathcal{Q}^*)_0$ is just B.

To identify the term $\tilde{F}^{\hat{\mu}}(\mathcal{Q}, \mathcal{Q}^*)_1$, we notice that by the inductive assumption we can assume that the ideal $J_{\hat{\mu}}$ is generated by the representations $U_{0,p}$ $(1 \leq p \leq n')$ and by the representations $U_{i,\hat{\mu}(i)}$ $(1 \leq i \leq n')$, where we denote $n' := n - \mu'_1$. This means that $J_{\hat{\mu}}$ is minimally generated by a subset of these representations. Let us denote this subset by C. This means that

$$\tilde{F}^{\hat{\mu}}(\mathcal{Q}, \mathcal{Q}^*)_1 = \bigoplus_{(i,p) \in C} K_{(1^i, 0^{n'-2i}, (-1)^i)} \mathcal{Q} \otimes B(-p).$$

The map $\partial_{(i,p)}$, from the summand corresponding to $U_{i,p}$ to $\tilde{F}^{\hat{\mu}}(\mathcal{Q}, \mathcal{Q}^*)_0$, is induced by $U_{i,p}$. We consider the map

$$q_*(\partial_{(i,p)}) : q_*(M(K_{(1^i, 0^{n'-2i}, (-1)^i)} \mathcal{Q})) \rightarrow q_*(M(K_{(0^{n'})} \mathcal{Q})).$$

To make our notation more transparent we will denote $N_0 = q_*(M(K_{(0^{n'})}\mathcal{Q}))$ and $N_{(i,p)} = q_*(M(K_{(1^i, 0^{n'-2i}, (-1)^i)}\mathcal{Q}))$.

By Lemma (8.1.5) we know that the higher direct images $\mathcal{R}^i q_*$ applied to the modules N_0 and $N_{(i,p)}$ are 0 for $i > 0$. Moreover, we know that the module N_0 is a coordinate ring of determinantal variety of matrices ϕ of rank $\leq n - \mu'_1$. The module $N_{(i,p)}$ has the presentation

$$
\bigwedge^{i+c+1} E \otimes \bigwedge^{i+c+1} E^* \otimes A(-i-c-1) \to \bigwedge^i E \otimes \bigwedge^i E^* \otimes A(-i)
$$
$$
\to N_{(i,p)} \to 0,
$$

where $c = n' - 2i$.

Since N_0 is the coordinate ring of the determinantal variety, and its defining ideal (generated by the $(n'+1) \times (n'+1)$ minors of ϕ) is clearly contained in J'_μ. We denote the image of J'_μ in N_0 by J''_μ. Theorem (8.2.5) clearly follows from the following lemma.

(8.2.8) Lemma. *Let $(i, p) \in C$. The image of the map $q_*(\partial_{(i,p)})$ is contained in the ideal J''_μ.*

Proof. Let us consider two cases. Either $(i, p) = (0, p)$ or $(i, p) = (i, \hat\mu(i))$. In the first case the map $\partial_{(0,p)}$ is a GL(E)-equivariant map from N_0 to itself. This means it has to be a multiplication by a GL(E)-invariant. Therefore its image is contained in J''_μ.

Let us assume that $(i, p) = (i, \hat\mu(i))$. According to the Theorem (5.1.2)(b) the modules N_0 and $N_{(i,p)}$ can be identified as follows:

$$
N_0 = \bigoplus_{d \geq 0} H^0(\mathrm{Grass}(\mu'_1, E), S_d(\mathcal{Q} \otimes E^*)),
$$
$$
N_{(i,p)} = \bigoplus_{d \geq 0} H^0(\mathrm{Grass}(\mu'_1, E), K_{(1^i, 0^{n'-2i}, (-1)^i)} \mathcal{Q} \otimes S_d(\mathcal{Q} \otimes E^*)).
$$

We know that the module $N_{(i,p)}$ is generated by its component in homogeneous degree i. We calculate the action of $\partial_{(i,p)}$ on the generators of $N_{(i,p)}$. The key statement is

(8.2.9) Lemma. *The map $\partial_{(i,p)}$ factors as follows:*

$$
\bigwedge^i E \otimes \bigwedge^i E^*
$$
$$
\downarrow {\scriptstyle t_p \otimes 1}
$$
$$
\bigwedge^p E \otimes \bigwedge^p E^* \otimes \bigwedge^i E \otimes \bigwedge^i E^*
$$
$$
\downarrow {\scriptstyle w}
$$
$$
\bigwedge^p E \otimes \bigwedge^p E^* \otimes \bigwedge^i E \otimes \bigwedge^i E^*
$$
$$
\downarrow {\scriptstyle \zeta_p \otimes \zeta_i}
$$
$$
S_{i+p}(E \otimes E^*)/(I_{\mu'_1+1})_{i+p},
$$

where w is the identity on the components involving E^* tensored with a $GL(E)$-equivariant map on the components involving E, and $\zeta_p \otimes \zeta_i$ is the product of maps coming from straightening formula.

The composition of the last two maps in the composition (8.2.9) is a $GL(E) \times GL(E^*)$-equivariant map. The rest of the proof of Lemma (8.2.8) (and thus of Theorem (8.2.5)) is based on the following idea. We exhibit an explicit set of generators of the group

$$\text{Hom}_{GL(E) \times GL(E^*)}\left(\bigwedge^p E \otimes \bigwedge^p E^* \otimes \bigwedge^i E \otimes \bigwedge^i E^*,\right.$$

$$\left. S_{i+p}(E \otimes E^*)/(I_{\mu_1'+1})_{i+p}\right),$$

and we show that the image of each of them composed with the map $t_p \otimes 1$ from (8.2.9) is in J_μ'. The precise statements we need are

(8.2.10) Lemma. *The vector space*

$$\text{Hom}_{GL(E) \times GL(E^*)}\left(\bigwedge^p E \otimes \bigwedge^p E^* \otimes \bigwedge^i E \otimes \bigwedge^i E^*,\right.$$

$$\left. S_{i+p}(E \otimes E^*)/(I_{\mu_1'+1})_{i+p}\right)$$

is generated by the elements w_j defined as compositions

$$\bigwedge^p E \otimes \bigwedge^p E^* \otimes \bigwedge^i E \otimes \bigwedge^i E^*$$
$$\downarrow \, 1 \otimes 1 \otimes \hat{\Delta}$$
$$\bigwedge^p E \otimes \bigwedge^p E^* \otimes \bigwedge^j E \otimes \bigwedge^j E^* \otimes \bigwedge^{i-j} E \otimes \bigwedge^{i-j} E^*$$
$$\downarrow \, \hat{m} \otimes 1 \otimes 1$$
$$\bigwedge^{p+j} E \otimes \bigwedge^{p+j} E^* \otimes \bigwedge^{i-j} E \otimes \bigwedge^{i-j} E^*$$
$$\downarrow \, \zeta_{p+j} \otimes \zeta_{i-j}$$
$$S_{i+p}(E \otimes E^*)/(I_{\mu_1'+1})_{i+p}$$

for j satisfying $0 \le j \le i$ and $p + j \le n$. Here $\hat{\Delta}$ denotes the product of two diagonal maps composed with the appropriate permutation of the factors. Similarly $\hat{m}$ is the permutation of the factors composed with the product of two exterior multiplications.

(8.2.11) Lemma. *Let $t_p \otimes 1$ be the first map in the composition (8.2.9). Let w_j be the maps defined in the statement of (8.2.10). Then for each j satisfying $0 \le j \le i$, $p + j \le n$, the image of the composition $w_j(t_p \otimes 1)$ is contained in the ideal J_μ''.*

To conclude the proof of Theorem (8.2.5) it remains to show the statements (8.2.9), (8.2.10), (8.2.11).

Proof of (8.2.10). We will show that the morphisms w_j are a basis of

$$\mathrm{Hom}_{\mathrm{GL}(E)\times\mathrm{GL}(E^*)}\left(\bigwedge^p E \otimes \bigwedge^p E^* \otimes \bigwedge^i E \otimes \bigwedge^i E^*, S_{i+p}(E \otimes E^*)\right).$$

Indeed, decomposing two sides into irreducibles by Pieri's formula and by Cauchy's formula, we see that the common irreducibles are $L_{(p+j,i-j)}E \otimes L_{(p+j,i-j)}E^*$ for each j satisfying $0 \le j \le i$, $p + j \le n$. Each of these irreducibles occurs with multiplicity 1 in both modules. Therefore the basis of

$$\mathrm{Hom}_{\mathrm{GL}(E)\times\mathrm{GL}(E^*)}\left(\bigwedge^p E \otimes \bigwedge^p E^* \otimes \bigwedge^i E \otimes \bigwedge^i E^*, S_{i+p}(E \otimes E^*)\right)$$

is given by the following compositions v_j:

$$\bigwedge^p E \otimes \bigwedge^p E^* \otimes \bigwedge^i E \otimes \bigwedge^i E^* \overset{\mathrm{pr}}{\to} L_{(p+j,i-j)}E \otimes L_{(p+j,i-j)}$$
$$E^* \overset{\mathrm{incl}}{\to} S_{i+p}(E \otimes E^*),$$

where pr and incl denote respectively the $\mathrm{GL}(E) \times \mathrm{GL}(E^*)$-equivariant projection and inclusion. Now it is a direct calculation that the matrix expressing the elements w_j as combinations of v_j's is triangular with nonzero diagonal entries. The lemma follows. ∎

Proof of (8.2.11). Let us consider the composition $w_j(t_p \otimes 1)$. The image is a combination of products of $(p + j) \times (p + j)$ and $(i - j) \times (i - j)$ minors of ϕ, with $(p + j) \times (p + j)$ minors containing the traces on p components. It is enough to show that each such combination of $(p + j) \times (p + j)$ minors is in J'_μ. In order to establish this we need to check the condition of (8.2.1), which in this case reads

$$p + j > \mu_1 + \ldots + \mu_j - j.$$

However, we know that our pair (i, p) corresponds to the generators of $J_{\hat\mu}$, so $p = \hat\mu(i) = \hat\mu_1 + \ldots + \hat\mu_i - i + 1$ and by definition of $U_{(i,p)}$ we have $p \ge i$. We know also that $\hat\mu_k = \mu_k - 1$ as long as $k \le \mu'_1$. This has to happen for all i in question, because otherwise $p + j \ge i + j > \mu'_1$ and we are in J'_μ anyway. Therefore we have

$$p + j = \hat\mu_1 + \ldots + \hat\mu_i - i + 1 + j > \hat\mu_1 + \ldots + \hat\mu_i - i + j$$
$$= \mu_1 + \ldots + \mu_i + j \ge \mu_1 + \ldots + \mu_j + j,$$

and the condition above is proved. ∎

Proof of (8.2.9). It follows from the definition that the map $\partial_{(i,p)}$ is the map induced on the sections by the following composition of maps:

$$K_{(1^i,0^{n'-2i},(-1)^i)}Q$$
$$\downarrow$$
$$\textstyle\bigwedge^i Q \otimes \bigwedge^i Q^*$$
$$\downarrow$$
$$\textstyle\bigwedge^i Q \otimes \bigwedge^i Q^* \otimes \bigwedge^{p-i} Q \otimes \bigwedge^{p-i} Q^*$$
$$\downarrow$$
$$\textstyle\bigwedge^p Q \otimes \bigwedge^p Q^*$$
$$\downarrow$$
$$\textstyle\bigwedge^p Q \otimes \bigwedge^p E^*$$
$$\downarrow$$
$$\mathcal{B}_p = S_p(Q \otimes E^*).$$

Therefore the action of $\partial_{(i,p)}$ on the generators of $N_{(i,p)}$ is given by a composition

$$\textstyle\bigwedge^i E \otimes \bigwedge^i E^* = H^0(K_{(1^i,0^{n'-2i},(-1)^i)}Q \otimes \bigwedge^i Q \otimes \bigwedge^i E^*)$$
$$\downarrow\ {\scriptstyle H^0(U_{i,p}\otimes 1)}$$
$$H^0(\textstyle\bigwedge^p Q \otimes \bigwedge^p Q^* \otimes \bigwedge^i Q \otimes \bigwedge^i E^*)$$
$$\downarrow\ {\scriptstyle H^0(1\otimes i\otimes 1\otimes 1)}$$
$$H^0(\textstyle\bigwedge^p Q \otimes \bigwedge^p E^* \otimes \bigwedge^i Q \otimes \bigwedge^i E^*)\qquad(*)$$
$$\downarrow\ {\scriptstyle H^0(\zeta_p\otimes\zeta_i)}$$
$$H^0(S_{i+p}(Q \otimes E^*))$$
$$\downarrow$$
$$S_{i+p}(E \otimes E^*)/(I_{\mu'_1+1})_{i+p}.$$

Here we denote by i the canonical inclusion of Q^* into E^*, and by ζ_p the embedding $\bigwedge^p F \otimes \bigwedge^p G \to S_p(F \otimes G)$ (with $F := E, G := E^*$) defined in section 3.2.

Let us notice that the last two maps in $(*)$ are $\mathrm{GL}(E) \times \mathrm{GL}(E^*)$-equivariant. Let us describe more precisely the composition of the first two maps in $(*)$. It comes from applying the functor H^0 to the composition

$$\textstyle\bigwedge^i Q \otimes \bigwedge^i E^*$$
$$\downarrow\ {\scriptstyle j_1\otimes 1}$$
$$K_{(1^i,0^{n'-2i},(-1)^i)}Q \otimes \textstyle\bigwedge^i Q \otimes \bigwedge^i E^*$$
$$\downarrow\ {\scriptstyle j_2\otimes 1\otimes 1}$$
$$\textstyle\bigwedge^i Q \otimes \bigwedge^i Q^* \otimes \bigwedge^i Q \otimes \bigwedge^i E^*\qquad(**)$$
$$\downarrow\ {\scriptstyle \wedge t_{p-i}\otimes 1\otimes 1}$$
$$\textstyle\bigwedge^p Q \otimes \bigwedge^p Q^* \otimes \bigwedge^i Q \otimes \bigwedge^i E^*$$
$$\downarrow\ {\scriptstyle 1\otimes i\otimes 1\otimes 1}$$
$$\textstyle\bigwedge^p Q \otimes \bigwedge^p E^* \otimes \bigwedge^i Q \otimes \bigwedge^i E^*,$$

where j_1 denotes and j_2 denote the canonical inclusions.

Let us look at the composition of all maps in (∗∗). We notice that all of them are identities on the factor $\bigwedge^i E^*$ appearing on the right hand side. This means our composition is a tensor product of a composition

$$
\begin{array}{c}
\bigwedge^i \mathcal{Q} \\
\downarrow {\scriptstyle j_1} \\
K_{(1^i, 0^{n'-2i}, (-1)^i)} \mathcal{Q} \otimes \bigwedge^i \mathcal{Q} \\
\downarrow {\scriptstyle j_2 \otimes 1} \\
\bigwedge^i \mathcal{Q} \otimes \bigwedge^i \mathcal{Q}^* \otimes \bigwedge^i \mathcal{Q} \\
\downarrow {\scriptstyle \wedge t_{p-i} \otimes 1} \\
\bigwedge^p \mathcal{Q} \otimes \bigwedge^p \mathcal{Q}^* \otimes \bigwedge^i \mathcal{Q} \\
\downarrow {\scriptstyle 1 \otimes i \otimes 1} \\
\bigwedge^p \mathcal{Q} \otimes \bigwedge^p E^* \otimes \bigwedge^i \mathcal{Q}
\end{array}
$$

tensored with the identity on $\bigwedge^i E^*$.

The map induced by the last composition by applying a functor H^0 is a GL(E)-equivariant map from $\bigwedge^i E = H^0(\bigwedge^i \mathcal{Q})$ to $H^0(\bigwedge^p \mathcal{Q} \otimes \bigwedge^p E^* \otimes \bigwedge^i \mathcal{Q})$. If $p + i \leq n'$, the latter group is $\bigwedge^p E \otimes \bigwedge^p E^* \otimes \bigwedge^i E$. If $p + i > n'$, the latter group is a factor of $\bigwedge^p E \otimes \bigwedge^p E^* \otimes \bigwedge^i E$ by the image of the map

$$
\overset{n'+1}{\bigwedge} E \otimes \overset{p}{\bigwedge} E^* \otimes \overset{p+i-n'-1}{\bigwedge} E \to \overset{p}{\bigwedge} E \otimes \overset{p}{\bigwedge} E^* \otimes \overset{i}{\bigwedge} E,
$$

which is the identity on the second factor, and the composition

$$
\overset{n'+1}{\bigwedge} E \otimes \overset{p+i-n'-1}{\bigwedge} E \overset{\Delta \otimes 1}{\to} \overset{p}{\bigwedge} E \otimes \overset{n'+1-p}{\bigwedge} E \otimes \overset{p+i-n'-1}{\bigwedge}
$$
$$
E \overset{1 \otimes m}{\to} \overset{p}{\bigwedge} E \otimes \overset{i}{\bigwedge} E.
$$

Let us consider the vector space $\mathrm{Hom}_{\mathrm{GL}(E)}(\bigwedge^i E, \bigwedge^p E \otimes \bigwedge^p E^* \otimes \bigwedge^i E)$. We will exhibit an explicit basis of this vector space.

(8.2.12) Lemma. *For all j satisfying $0 \leq j \leq i$, $p + j \leq n$ we define the morphisms $h_j : \bigwedge^i E \to \bigwedge^p E \otimes \bigwedge^p E^* \otimes \bigwedge^i E$ as the compositions*

$$
h_j : \overset{i}{\bigwedge} E \overset{t_p \otimes 1}{\to} \overset{p}{\bigwedge} E \otimes \overset{p}{\bigwedge} E^* \otimes \overset{i}{\bigwedge} E \overset{g_j}{\to} \overset{p}{\bigwedge} E \otimes \overset{p}{\bigwedge} E^* \otimes \overset{p}{\bigwedge} E,
$$

where g_j is the identity on the second factor tensored with the map g'_j defined as a composition

$$\bigwedge^p E \otimes \bigwedge^i E$$
$$\downarrow {\scriptstyle 1 \otimes \Delta}$$
$$\bigwedge^p E \otimes \bigwedge^j E \otimes \bigwedge^{i-j} E$$
$$\downarrow {\scriptstyle m \otimes 1}$$
$$\bigwedge^{p+j} E \otimes \bigwedge^{i-j} E$$
$$\downarrow {\scriptstyle \Delta \otimes 1}$$
$$\bigwedge^p E \otimes \bigwedge^j E \otimes \bigwedge^{i-j} E$$
$$\downarrow {\scriptstyle 1 \otimes m}$$
$$\bigwedge^p E \otimes \bigwedge^i E.$$

Then the maps h_j form a basis of $\mathrm{Hom}_{\mathrm{GL}(E)}(\bigwedge^i E, \bigwedge^p E \otimes \bigwedge^p E^* \otimes \bigwedge^i E)$.

Proof. The vector space $\mathrm{Hom}_{\mathrm{GL}(E)}(\bigwedge^i E, \bigwedge^p E \otimes \bigwedge^p E^* \otimes \bigwedge^i E)$ is canonically isomorphic with $\mathrm{Hom}_{\mathrm{GL}(E)}(\bigwedge^p E \otimes \bigwedge^i E, \bigwedge^p E \otimes \bigwedge^i E)$. The identification is done by associating to a morphism $f \in \mathrm{Hom}_{\mathrm{GL}(E)}(\bigwedge^p E \otimes \bigwedge^i E, \bigwedge^p E \otimes \bigwedge^i E)$ the composition

$$h : \bigwedge^i E \overset{t_p \otimes 1}{\to} \bigwedge^p E \otimes \bigwedge^p E^* \otimes \bigwedge^i E \overset{g}{\to} \bigwedge^p E \otimes \bigwedge^p E^* \otimes \bigwedge^i E,$$

where g is the identity on the second factor tensored with the map f.

It is therefore enough to show that the maps g'_j given in (8.2.12) form a basis of $\mathrm{Hom}_{\mathrm{GL}(E)}(\bigwedge^p E \otimes \bigwedge^i E, \bigwedge^p E \otimes \bigwedge^i E)$. This is however clear. Indeed, decomposing $\bigwedge^p E \otimes \bigwedge^i E$ into irreducibles $L_{(p+j,i-j)}E$, we see that the natural basis of

$$\mathrm{Hom}_{\mathrm{GL}(E)}(\bigwedge^p E \otimes \bigwedge^i E, \bigwedge^p E \otimes \bigwedge^i E)$$

consists of the morphisms

$$u_j : \bigwedge^p E \otimes \bigwedge^i E \overset{\mathrm{pr}}{\to} L_{(p+j,i-j)}E \overset{\mathrm{incl}}{\to} \bigwedge^p E \otimes \bigwedge^i E$$

where pr and incl denote respectively the $\mathrm{GL}(E)$-equivariant projection and inclusion, and j satisfies conditions $0 \le j \le i$, $p + j \le n$. Now it is clear that the transition matrix expressing g'_j's as combinations of u_j's is triangular with the nonzero entries on the diagonal. This concludes the proof of (8.2.12). ∎

Now we come back to the map h we get when applying the functor H^0 to the composition of the maps in (∗∗). Lemma (8.2.12) shows that the map h

can be written as a linear combination of maps h_j. This means that it can be written as a composition

$$h : \bigwedge^i E \overset{{}^t p \otimes 1}{\to} \bigwedge^p E \otimes \bigwedge^p E^* \otimes \bigwedge^i E \overset{h'}{\to} \bigwedge^p E \otimes \bigwedge^p E^* \otimes \bigwedge^i E,$$

where h' is the identity on the second factor tensored with the GL(E)-equivariant map on the remaining two factors.

Since the composition of the first two maps in (∗) is the map h tensored with the identity on the remaining factor $\bigwedge^i E^*$, we can write it in the form (8.2.9) as required. This concludes the proof of (8.2.9) and therefore the proof of Theorem (8.2.5).●

Let us illustrate the inductive step in the proof of Theorem (8.2.5) with the following example.

(8.2.13) Example. *Let us consider* $\mu' = (6, 4, 2)$. *Then* $\hat{\mu}' = (4, 2)$. *We obviously have* dim $E = 12$. *The inductive step consists of pushing down from the Grassmannian* Grass(6, E) *the complex of sheaves* $\tilde{F}^{\hat{\mu}}(Q, Q^*)_\bullet$ *where* Q *is a tautological factorbundle on* Grass(6, E). *The generators of* $J_{(4,2)}$ *can be described as follows:*

$$M((4, 2)) = \begin{matrix} 0 & & & \\ 1 & 0 & & \\ 1 & 1 & 0 & \\ 1 & 1 & 1 & 1 \\ 1 & 1 & 1 & \\ 1 & 1 & & \\ 1 & & & \end{matrix}.$$

This means that $J_{(4,2)}$ is generated by the invariants $U_{0,p}$ ($1 \leq p \leq 3$) and by the representations $U_{1,2}$, $U_{2,3}$, and $U_{3,3}$. The first two terms of $\tilde{F}^{\hat{\mu}}(Q, Q^*)_\bullet$ are easy to describe. The term $\tilde{F}^{\hat{\mu}}(Q, Q^*)_0$ equals $B = \text{Sym}(Q \otimes E^*)$, and $\tilde{F}^{\hat{\mu}}(Q, Q^*)_1$ is a direct sum of the six terms $B \otimes K_{(1^i, 0^{6-2i}, (-1)^i)} Q$ corresponding to the six representations $U_{i,p}$ listed.

We see first of all that the direct image $q_*(B) = A/I_7$. This tells us that 7×7 minors of the generic matrix ϕ are in the ideal $J_{(6,4,2)}$. The invariants $U_{0,p}$ lead to the generators that are the invariants of degrees 1, 2, 3 in A/I_7. The remaining three $U_{i,p}$'s give the following representations. The term $U_{1,2}$ gives a representation $E \otimes E^*$ in degree 3. The term $U_{2,3}$ gives a representation $\bigwedge^2 E \otimes \bigwedge^2 E^*$ in degree 5. The term $U_{3,3}$ gives a representation $\bigwedge^3 E \otimes \bigwedge^3 E^*$ in degree 6. The proof above shows that these three representations have to be contained in the ideal generated by I_7, the invariants $U_{0,p}$

$(1 \le p \le 6)$, and the representations $U_{1,3}$, $U_{2,5}$, and $U_{3,6}$. In fact one would expect them to be respectively $V_{1,3}$, $V_{2,5}$, and $V_{3,6}$. The conclusion in any case is that all generators have to be in $J'_{(6,4,2)}$. This follows from Example (8.2.4).

(8.2.14) Remark. *In the above proof of Theorem (8.2.5) we did not identify precisely the generators coming from the terms $U_{i,p}$ occurring in $\tilde{F}^{\hat{\mu}}(Q, Q^*)_1$. The reader might worry that we did not show that some of them are dependent on others. However, this is not a concern. We are assured by the fact that the only homology of $\tilde{F}^{\hat{\mu}}(Q, Q^*)_\bullet$ is $\mathcal{O}_{W_\mu}$ and by Proposition (8.2.7) that the factor of $A/I_{\mu'_1+1}$ by the images of the terms coming from $\tilde{F}^{\hat{\mu}}(Q, Q^*)_1$ is A/J_μ. Therefore all elements of J'_μ have to be contained in that image. The only concern is whether we get some generators not contained in J'_μ.*

We finish this section with an example related to rectangular partitions.

(8.2.15) Proposition. *Let $n = re$ and let us consider the rectangular partition $\mu = (e^r)$. Then the ideal $J_{(e^r)}$ is generated by the invariants $v_1, \ldots, v_{e-1}$ and by the entries of the matrix ϕ^e. These polynomials form a minimal set of generators of $J_{(e^r)}$.*

Proof. Let us use the inductive procedure used in the proof of (8.2.5) for the family of partitions $\mu = (e^r)$. The partition $\hat{\mu}$ is just $((e-1)^r)$. Thus we can assume by induction that (8.2.15) is true for $\hat{\mu}$.

The inductive assumption means that the first term of $F^{\hat{\mu}}_\bullet$ consists of the trivial representations in homogeneous degrees $1, \ldots, e-1$ and the representation $K_{1,0,\ldots,0,-1}Q$ in homogeneous degree $e-1$. This means that the possible terms in F^μ_1 are the trivial representations in homogeneous degrees $1, \ldots, e-1$, the representation $E^* \otimes E$ in the degree e and the terms coming from $n - \mu_1 + 1 = n - r + 1$ size minors of the matrix ϕ. The generators in degrees $\le e$ have to consist of invariants $v_1, \ldots, v_{e-1}$ and the vanishing of entries of ϕ^e, because these terms have to occur in F^μ_1 and they match the terms we described. Thus the induction implies that the minimal generators of $J(\mu)$ are those listed in (8.2.15) plus possibly some linear combinations of $(n - r + 1) \times (n - r + 1)$ minors of ϕ.

However recall that by (8.2.5) the ideal $J(\mu)$ is generated (nonminimally) by the invariants $U_{0,p}(1 \le p \le n)$ and by the representations $U_{i,ie-i+1}$(for $1 \le i \le r$).

The only possible generator in degree $n - r + 1$ on the above list is the representation $U_{r,n-r+1}$ which does not occur in the span of $n - r + 1$ minors of Φ. This concludes the proof. ∎

A slightly more general result is proved in [W7]. Let $e < n$ and let us consider the division of n by e with the remainder, $n = re + f$ with $0 \le f \le e - 1$. We define the partition $\mu(n, e) = (e^r, f)$. Then the ideal $J_{\mu(n,e)}$ is generated by the invariants $v_1, \ldots, v_{e-1}$ and by the entries of the matrix ϕ^e. These polynomials form a minimal set of generators of $J_{\mu(n,e)}$.

8.3. The Nilpotent Orbits for Other Simple Groups

In this section we investigate the closures of conjugacy classes for other simple groups. We give their explicit desingularization by the collapsing of a homogeneous bundle. We sketch the proof that the normalizations of the closures of conjugacy classes have rational singularities. We use the approach of Broer [Br5].

Let **G** be a simple algebraic group with Lie algebra $\underline{g}$. The group **G** acts on $\underline{g}$ by conjugation. We will still denote this action as a left action.

Let e be a nilpotent element in $\underline{g}$. We denote by **G**e its orbit under the conjugation action, and by $\overline{\mathbf{G}e}$ its closure in $\underline{g}$.

By the Jacobson–Morozov lemma ([Bou, chapter VIII, section 11]) there exist elements $h, f \in \underline{g}$ such that $\{e, h, f\}$ forms an $\underline{sl}_2$-triple in $\underline{g}$, i.e.

$$[h, e] = 2e, \qquad [h, f] = -2f, \qquad [e, f] = h.$$

If $\{e, h', f'\}$ forms another $\underline{sl}_2$-triple, then there exists $g \in \mathbf{G}$ (and even in the connected component of the identity, $C_{\mathbf{G}}(e)^0$, of the centralizer of e), centralizing e, such that $gf = f'$, $gh = h'$.

Let us fix e and the $\underline{sl}_2$-triple $\{e, h, f\}$.

Let V be a finite dimensional $\underline{g}$-module. The action of h induces a grading

$$V = \bigoplus_{i \in Z} V_i, \qquad V_i = \{v \in V \mid h \cdot v = iv\}.$$

The associated filtration

$$\ldots \subset V_{\ge i+1} \subset V_{\ge i} \subset \ldots$$

does not depend on the choice of h, f.

Let us consider $V = \underline{g}$. Then $e \in \underline{g}_2$, $h \in \underline{g}_0$, $f \in \underline{g}_{-2}$. We also clearly have $[\underline{g}_i, \underline{g}_j] \subset \underline{g}_{i+j}$. This means that $\underline{g}_{\ge 0}$ is a Lie algebra of a parabolic group $\mathbf{P} \subset \mathbf{G}$ with a nilpotent radical $\underline{n} = \underline{g}_{>0}$ and the Levi factor $\underline{g}_0$. The subgroup **P** depends only on the filtration and therefore does not depend on the choice of f, h.

We consider the subspace $\underline{g}_{\ge 2}$ of $\underline{g}$. This is clearly a **P**-submodule of $\underline{g}$. We can consider an induced homogeneous vector bundle $Z = \mathbf{G} \times^{\mathbf{P}} \underline{g}_{\ge 2}$ over

G/P. Let us take $X = \underline{g}$, $V = \mathbf{G}/\mathbf{P}$. Then $Z \subset X \times V$. We can consider the diagram

$$
\begin{array}{ccc}
Z & \subset & X \times V \\
\downarrow q' & & \downarrow q \\
Y & \subset & X
\end{array}
$$

where $Y = \overline{\mathbf{G}e}$ and q' is the restriction of the first projection q. The map q' sends the coset of the pair (g, x) to gx. In this situation we have

(8.3.1) Proposition ([KP2], section 7.4). *The map q' makes Z a resolution of singularities of Y.*

Proof. Let us recall that the irreducible $\underline{sl}_2$-modules are just the symmetric powers $S_d F$ where $F = \mathbf{K}^2$. The Lie algebra $\underline{sl}_2 = \underline{sl}(F)$ acts by the derivative action of the conjugation action, i.e. the commutator action. The tangent space at e to $\mathbf{P}e$ is therefore the commutator space $[\underline{g}_{\geq 0}, e] = \underline{g}_{\geq 2}$. It follows that $\mathbf{P}e$ is an open set in $\underline{g}_{\geq 2}$. Since the group $\mathbf{P}$ is uniqely determined by e, we have $\mathbf{P} = g\mathbf{P}g^{-1}$ for each g in the stabilizer $\mathbf{G}_e$. Since each parabolic subgroup equals its normalizer ([Hu2], Corollary B, section 23), we have $\mathbf{G}_e = \mathbf{P}_e$. This means that q' restricted to the open orbit $\mathbf{G}(1, e)$ is an isomorphism, and therefore q' is a birational map. The variety Z is nonsingular, because it is a vector bundle over $\mathbf{G}/\mathbf{P}$. ∎

(8.3.2) Example. *Consider the special linear group $\mathbf{G} = SL(n) = SL(E)$, for $E = \mathbf{K}^n$. The Lie algebra $\underline{g}$ of $\mathbf{G}$ is the set of $n \times n$ matrices of trace 0. We identify the set of $n \times n$ matrices with $\mathrm{Hom}_{\mathbf{K}}(E, E) = E^* \otimes E$. Let us start with the nilpotent element e, which in canonical Jordan form has one Jordan block of size n. Then we can choose a basis $e_1, \ldots, e_n$ of E such that $e(e_1) = 0$ and $e(e_i) = e_{i-1}$ for $i = 2, \ldots, n$. Then the element f can be chosen as follows: $f(e_i) = e_{i+1}$ for $i = 1, \ldots, n - 1$, $f(e_n) = 0$. The element h can be chosen to be the diagonal matrix $h(e_i) = (n + 1 - 2i)e_i$. We introduce the grading on E by letting $\deg(e_i) = n + 1 - 2i$. This induces a grading on $\mathrm{Hom}_{\mathbf{K}}(E, E) = E^* \otimes E$ defined by letting $\deg(e_i^* \otimes e_j) = \deg(e_j) - \deg(e_i)$. The module $\underline{g}_{\geq 2}$ consists then of all upper triangular matrices, and therefore the desingularization constructed above is the same as one constructed in section 8.1.*

Let $\mu = (\mu_1, \ldots, \mu_r)$ be a partition of n. Let us consider the nilpotent element e from the nilpotent orbit $O(\mu)$. We choose a basis $e_1, \ldots, e_n$ of E in such way that

$$
e(e_i) = 0 \quad \text{if } i \in \{1, \mu_1 + 1, \mu_1 + \mu_2 + 1, \ldots\},
$$
$$
e(e_i) = e_{i-1} \quad \text{otherwise.}
$$

This means that for each $j = 1, \ldots, r$ the span of $e_{\mu_1+\ldots+\mu_{j-1}+1}, \ldots, e_{\mu_1+\ldots+\mu_j}$ is the Jordan block of size μ_j. We can choose f to act as follows:

$$f(e_i) = 0 \quad \text{if } i = \mu_1 + \ldots + \mu_j, \qquad f(e_i) = e_{i+1} \quad \text{otherwise.}$$

The element h is a diagonal matrix acting in the j-th block by

$$h(e_{\mu_1+\ldots+\mu_{j-1}+i}) = (\mu_j + 1 - 2i)e_i \qquad \text{for} \quad i = 1, \ldots, \mu_j.$$

Let us introduce the grading on E by letting $\deg(e_{\mu_1+\ldots+\mu_{j-1}+i}) = \mu_j + 1 - 2i$. The grading induced on $\mathrm{Hom}_K(E, E)$ is then given by identifying $\mathrm{Hom}_K(E, E)$ with $E^* \otimes E$ and letting $\deg(e_i^* \otimes e_j) = \deg(e_j) - \deg(e_i)$. This induces a grading on $\underline{g}$. This allows us to define the module $\underline{g}_{\geq 2}$.

Notice that the desingularization given by $\underline{g}_{\geq 2}$ is almost never the same as the one considered in the section 8.1. To see this we consider $n = 3$ and $\mu = (2, 1)$. Then the degrees of e_i are as follows: $\deg(e_1) = 1, \deg(e_2) = -1$, $\deg(e_3) = 0$. Reordering our basis, we can assume that $\deg(e_1) = 1, \deg(e_2) = 0, \deg(e_3) = -1$. We see that the module $\underline{g}_{\geq 2}$ is spanned by $e_3^* \otimes e_1$.

This is a **B**-submodule for the group **B** of upper triangiular matrices. The parabolic subgroup **P** defined above is obviously equal to **B**. The desingularization $\mathbf{G} \times_\mathbf{P} \underline{g}_{\geq 2}$ we defined above is a line bundle over $\mathbf{G/B}$. Let us denote this desingularization by Z_1. Identifying $\mathbf{G/B}$ with the set of flags $R_1 \subset R_2 \subset E$, we can identify our desingularization Z_1 with the set of pairs

$$Z_1 = \{(\phi, R_1, R_2) \in \mathrm{Hom}_K(E, E) \times \mathbf{G/B} \mid \phi(R_2) = 0, \ \phi(E) \subset R_1\}.$$

The desingularization Z_2 constructed in section 8.1 was a two dimensional bundle over the Grassmannian $\mathrm{Grass}(2, E)$. It was defined as the set of pairs

$$Z_2 = \{(\phi, R_2) \in \mathrm{Hom}_K(E, E) \times \mathrm{Grass}(2, E) \mid \phi(R_2) = 0, \ \phi(E) \subset R_2\}.$$

To see that these two desingularizations are different, let us observe that the map that forgets R_1 defines a regular map $Z_1 \to Z_2$. This map is not an isomorphism, because its fiber over a point (ϕ, R_2) such that $\phi = 0$ is clearly isomorphic to $\mathbf{P}^1$. Therefore the desingularizations are different.•

The main result of this section is the following theorem:

(8.3.3) Theorem (Hinich [Hi], Panyushev [Pa2]). *Let* **G** *be a simple group with the Lie algebra* $\underline{g}$. *Let* e *be a nilpotent element in* $\underline{g}$. *We consider the desingularization* Z *of the closure* Y *of the orbit* $\mathbf{G}e$ *constructed above.*

   (a) For all $i > 0$ *we have* $H^i(Z, \mathcal{O}_Z) = 0$.
   (b) The normalization of Y *is a Gorenstein variety with rational singularities.*

Before we prove (8.3.3) we need some preparatory statements.

(8.3.4) Proposition. *Let $\underline{g}$ be a simple Lie algebra with the bracket $[\ ,\ ]$. Then all adjoint orbits in $\underline{g}$ have even dimensions.*

Proof. We denote by $\langle\ ,\ \rangle$ the Killing form on $\underline{g}$. Every element $z \in \underline{g}$ defines an antisymmetric form $(\ ,\ )_z$ on $\underline{g}$ given by $(x, y)_z := \langle z, [x, y]\rangle$. If x is in the radical of $(\ ,\ )_z$, then for all $y \in \underline{g}$ we have

$$\langle z, [y, x]\rangle = \langle [z, x], y\rangle = 0.$$

Since the Killing form is nondegenerate, we have $[z, x] = 0$. This means that the radical of $(\ ,\ )_z$ equals the centralizer $\underline{g}_z$. The induced antisymmetric form on $\underline{g}/\underline{g}_z$ is nondegenerate, and therefore dim $\underline{g}/\underline{g}_z$ is even. However, dim $\mathbf{G}z = \dim \underline{g}/\underline{g}_z$. ∎

(8.3.5) Lemma. *Let $\underline{g}$ and e be as above. Let us consider the grading $\underline{g} = \bigoplus_{i \in \mathbb{Z}} \underline{g}_i$ induced by e, and the associated parabolic subgroup $\mathbf{P}$. We denote $b = \dim \underline{g}_{\geq 2}$, $c := \dim \underline{g}_1$, $d = \dim \mathbf{G}/\mathbf{P}$. Then:*

(a) *The number c is even.*
(b) *There is a nonzero $\mathbf{P}$-equivariant map*

$$s : \underline{g}_{\geq 2} \to \overset{d}{\bigwedge}(\underline{g}/\underline{p})^* \otimes \overset{b}{\bigwedge}\underline{g}_{\geq 2}^*.$$

(c) *$s(ux) = u^{c/2}s(x)$ for all $u \in \mathbf{K}$, $x \in \underline{g}_{\geq 2}$.*

Proof. Let $z \in \underline{g}_2$. The form $(\ ,\ )_z$ defined in the proof of (8.3.4) restricts to a form on $\underline{g}_{-1}$, which can be identified with an element $w_z \in \bigwedge^2 \underline{g}_{-1}^*$. Consider the Levi decomposition $\mathbf{P} = \mathbf{L}\mathbf{P}^u$ of the parabolic group $\mathbf{P}$. The linear map

$$w_\bullet : \underline{g}_2 \to \overset{2}{\bigwedge}\underline{g}_{-1}^*$$

sending z to w_z is $\mathbf{L}$-equivariant. For any $z \in \mathbf{L}e \subset \underline{g}_2$ we have $\underline{g}_z \subset \underline{p} = \underline{g}_{\geq 0}$. From the proof of (8.3.4) it follows that w_z is nondegenerate. This means that c is even and the top exterior power $w_z^{\wedge c/2} \in \bigwedge^c \underline{g}_{-1}^*$ does not vanish. The map

$$s' : \underline{g}_2 \to \overset{c}{\bigwedge}\underline{g}_{-1}^*$$

sending z into $w_z^{\wedge c/2}$ is an $\mathbf{L}$-equivariant polynomial map of homogeneous degree $\frac{c}{2}$. It does not vanish on the open orbit $\mathbf{L}e$. We extend the $\mathbf{L}$-modules

$\underline{g}_2$ and $\bigwedge^c \underline{g}^*_{-1}$ to **P**-modules $\tilde{\underline{g}}_2$ and $\bigwedge^c \widetilde{\underline{g}^*_{-1}}$ by the trivial action of $\mathbf{P}^u$. We have the obvious isomorphisms of **P**-modules

$$\underline{g}_{\geq 2}/\underline{g}_{>2} \simeq \tilde{\underline{g}}_2,$$

$$\overset{c}{\bigwedge} \widetilde{\underline{g}^*_{-1}} \simeq \overset{d}{\bigwedge}(\underline{g}/\underline{p})^* \otimes \overset{b}{\bigwedge} \underline{g}^*_{\geq 2}.$$

We define s as a composition

$$s : \underline{g}_{\geq 2} \to \underline{g}_{\geq 2}/\underline{g}_{>2} \simeq \tilde{\underline{g}}_2 \overset{s'}{\to} \overset{c}{\bigwedge} \widetilde{\underline{g}^*_{-1}} \simeq \overset{d}{\bigwedge}(\underline{g}/\underline{p})^* \otimes \overset{b}{\bigwedge} \underline{g}^*_{\geq 2}.$$

Then s clearly satisfies (b) and (c). ∎

We need one more statement.

(8.3.6) Proposition. *Let* $\mathbf{P} = \mathbf{LP}^u$ *be a Levi decomposition of a parabolic subgroup in* $\mathbf{G}$. *We consider a* **P**-*submodule* $\underline{m}$ *of* $\underline{n} := \mathrm{Lie}(\mathbf{P}^u)$. *We can construct as above a homogeneous bundle* $Z = \mathbf{G} \times^{\mathbf{P}} \underline{m}$ *over* $\mathbf{G/P}$, *which projects onto* $Y = \mathbf{G}\underline{m}$. *Let us assume that* $\dim Z = \dim Y$. *We consider a one dimensional* **P**-*module* $\underline{u}$ *and the associated line bundle* $\mathcal{L} := \mathbf{G} \times^{\mathbf{P}} \underline{u}$ *on* $\mathbf{G/P}$. *Let* p *denote as usual the projection of* Z *onto* $\mathbf{G/P}$. *Assume that the twisted sheaf* $\mathcal{O}_Z \otimes p^*(\mathcal{L})$ *has a global* **G**-*invariant section* s_0. *The section* s_0 *induces the morphism of* $\tilde{s}_0 : \mathcal{O}_Z$-*modules* $\mathcal{O}_Z \to \mathcal{O}_Z \otimes p^*(\mathcal{L})$ *given locally by sending a section* f *to* $f s_0$. *The morphism* $\tilde{s}_0$ *then induces the isomorphism* $\Gamma(\mathcal{O}_Z) \simeq \Gamma(\mathcal{O}_Z \otimes p^*(\mathcal{L}))$.

Proof. Let s be a section of $\mathcal{O}_Z \otimes p^*(\mathcal{L})$. Then $s = f s_0$ where f is a rational function on Z with poles only along the zeros of s_0. We know that $Y' := \mathrm{Spec}\, \mathbf{K}[Z]$ is normal because it has an open orbit whose complement has codimension ≥ 2, and the morphism $Z \to Y'$ is birational. Therefore f can be considered as a rational function on Y' without poles on the open orbit. This means f has no poles on Y', because on the normal variety the set of poles of a function has codimension 1. Thus f is a regular function on Y'. Since by definition of Y' the regular functions on Y' and Z are the same, f can be treated as a regular function on Z. We conclude that the map $\Gamma(\tilde{s}_0) : \Gamma(\mathcal{O}_Z) \to \Gamma(\mathcal{O}_Z \otimes p^*(\mathcal{L}))$ is surjective. This map is obviously injective, which finishes the proof. ∎

Proof of Theorem (8.3.3). Let $\mathbf{G}$, e, $\mathbf{P}$ be as in the statement of the theorem. We denote $d = \dim \mathbf{G/P}$, $c = \dim \underline{g}_{-1}$ and $b = \dim \underline{g}_{\geq 2}$. We take

$\underline{u} := \bigwedge^d (\underline{g}/\underline{p})^* \otimes \bigwedge^b \underline{g}_{\geq 2}^*$. Consider the associated line bundle $\mathcal{L} = \mathbf{G} \times^{\mathbf{P}} \underline{u}$ on $\mathbf{G}/\mathbf{P}$.

There exists $m \in \mathbf{Z}$ such that the canonical sheaf ω_Z is isomorphic to the line bundle $\mathcal{O}_Z \otimes p^*(\mathcal{L})(m)$ with the grading shifted by m. By Lemma (8.3.5) there exists a **P**-equivariant map $s : \underline{g}_{>2} \to \underline{u}$ of degree $\frac{c}{2}$. This means that the sheaf $\mathcal{O}_Z \otimes p^*(\mathcal{L})$ has a nonzero global **G**-invariant section of degree $\frac{c}{2}$ (the image of a constant section of $\underline{g}_{>2}$). This implies that the sheaf ω_Z has a global **G**-invariant section s_0. Now Proposition (8.3.6) gives a morphism of sheaves $\tilde{s}_0 : \mathcal{O}_Z \to \omega_Z(-m + \frac{c}{2})$ which induces an isomorphism on global sections. Applying Proposition (1.2.32), we get that Spec $\mathbf{K}[Z]$ is a Gorenstein variety with rational singularities. We know however that Spec $\mathbf{K}[Z]$ is the normalization of the closure Y of the orbit $\mathbf{G}e$. This concludes the proof of the theorem. ∎

(8.3.7) Remarks. *Broer in a series of beautiful papers [Br1, Br2, Br3] applied the geometric method to deal with the twisted modules supported in nilpotent orbit closures. He also applied the method to decide normality of several nilpotent orbit closures for the exceptional groups.*

Kraft ([Kr2]) classified normal orbit closures for the groups of type G_2. Recently the normal nilpotent orbit closures were classified for groups of type F_4 (Broer, [Br6]) and of type E_6 (Sommers).

8.4. Conjugacy Classes for the Orthogonal Group

In this section F denotes a vector space of dimension n with a nondegenerate symmetric form $\langle \ , \ \rangle$. The special orthogonal group $SO(F)$ is the set of linear automorphisms of F preserving $\langle \ , \ \rangle$, i.e. $\phi \in SO(F)$ if and only if for each $x, y \in F$ we have $\langle \phi(x), \phi(y) \rangle = \langle x, y \rangle$. By definition $SO(F)$ is a subgroup of $GL(F)$.

The corresponding Lie algebra $\underline{so}(F)$ is a subalgebra of the Lie algebra $\underline{gl}(F)$. The morphism $\phi \in \mathrm{Hom}_K(F, F)$ is in $\underline{so}(F)$ if and only if for each $x, y \in F$ we have $\langle \phi(x), y \rangle + \langle x, \phi(y) \rangle = 0$.

Let us choose the hyperbolic basis $e_1, \ldots, e_m, \bar{e}_m, \ldots, \bar{e}_1$ of F (in the case of $n = 2m$), and $e_1, \ldots, e_m, e_0, \bar{e}_m, \ldots, \bar{e}_1$. This means

$$\langle e_i, e_j \rangle = \langle \bar{e}_i, \bar{e}_j \rangle = 0 \qquad \langle e_i, \bar{e}_j \rangle = \delta_{i,j},$$
$$\langle e_0, e_i \rangle = \langle e_0, \bar{e}_i \rangle = 0, \qquad \langle e_0, e_0 \rangle = 1,$$

where $\delta_{i,j}$ denotes the Kronecker delta.

We can write ϕ as a matrix, writing in consecutive columns the images of vectors $-\bar{e}_1, \ldots, -\bar{e}_m, e_0, e_m, \ldots, e_1$ expanded in the basis $e_1, \ldots, e_m, e_0,$ $\bar{e}_m, \ldots, \bar{e}_1$. Then $\phi \in \underline{so}(F)$ if and only if the matrix of ϕ is skew symmetric. This allows us to identify the adjoint representation of $\underline{so}(F)$ with $\bigwedge^2 F$.

Since $\underline{so}(F)$ is a Lie subalgebra of $\underline{gl}(F)$, we might expect that the nilpotent conjugacy classes in $\underline{so}(F)$ will be related to intersections of the nilpotent conjugacy classes in $\underline{gl}(F)$ with $\underline{so}(F)$. The following result is proved in [SS; IV. 2.15].

(8.4.1) Proposition. *Let μ be a partition of n. Let us consider the nilpotent conjugacy class $O(\mu)$ of $\underline{gl}(F)$ corresponding to μ. Then $O(\mu)$ intersects $\underline{so}(F)$ if and only if every even part of μ occurs even number of times. In such case the intersection $O(\mu) \cap \underline{so}(F)$ consists of a single conjugacy class of $SO(F)$.*

For the remainder of this section we denote $P_o(n)$ the set of partitions μ of n in which every even part occurs even number of times. For $\mu \in P_o(n)$ we denote by $C(\mu)$ the corresponding conjugacy class in $\underline{so}(F)$, and by Y_μ its closure.

In order to list the representatives of conjugacy classes for the orthogonal group, we need to exhibit the blocks corresponding to odd rows in our partition and the blocks corresponding to the pairs of even rows. The canonical forms of nilpotent orthogonal endomorphisms corresponding to both kinds of blocks are as follows.

The nilpotent e corresponding to the partition $(n) = (2m + 1)$ has the form

$$e(\bar{e}_1) = 0, \qquad e(\bar{e}_i) = \bar{e}_{i-1} \quad \text{for } i > 0,$$
$$e(e_n) = e_0, \qquad e(e_0) = -\bar{e}_n, e(e_i) = -e_{i+1} \quad \text{for } i < n$$

We can express the action of e by the sequence of arrows

$$\pm e_1 \to \mp e_2 \to \ldots \to e_{n-1} \to -e_n \to e_0 \to \bar{e}_n \to \bar{e}_{n-1}$$
$$\to \ldots \to \bar{e}_2 \to \bar{e}_1 \to 0.$$

If we extend e to an $\underline{sl}_2$ triple $\{e, h, f\}$, the grading induced by e described in section 8.3 is as follows. For the endomorphism e corresponding to the partition $2n$ we have $h(e_i) = (2n + 1 - 2i)e_i$, $h(\bar{e}_i) = (2i - 1 - 2n)\bar{e}_i$. Therefore the degrees of the basis vectors are $\deg e_i = 2n + 1 - 2i$, $\deg \bar{e}_i = 2i - 1 - 2n$.

If $n = 2t$ is even, then the nilpotent e corresponding to the partition (n, n) has the form

$$e(e_i) = e_{i+2} \quad \text{for } i < 2t, \qquad e(e_{2t}) = \bar{e}_{2t+1}, \qquad e(e_{2t+1}) = \bar{e}_{2t},$$
$$e(\bar{e}_i) = -\bar{e}_{i-2} \quad \text{for } i > 2, \qquad e(\bar{e}_2) = e(\bar{e}_1) = 0.$$

We can express the action of e by two sequences of arrows

$$e_1 \to e_3 \to \dots \to e_{2t+1} \to \bar{e}_{2t} \to -\bar{e}_{2t-2} \to \dots \to \pm\bar{e}_2 \to 0,$$
$$e_2 \to e_4 \to \dots \to e_{2t} \to \bar{e}_{2t+1} \to -\bar{e}_{2t-1} \to \dots \to \mp\bar{e}_1 \to 0.$$

Extending e to the $\underline{sl}_2$-triple $\{e, h, f\}$, we see that the element h acts as follows: $h(e_{2i+1}) = 2(t - i)e_{2i+1}$, $h(e_{2i}) = 2(t + 1 - i)e_{2i}$, and $h(\bar{e}_{2i+1}) = -2(t - i)\bar{e}_{2i+1}$, $h(\bar{e}_{2i}) = -2(t + 1 - i)\bar{e}_{2i}$. This gives $\deg e_{2i+1} = 2(t - i)$, $\deg e_{2i} = 2(t + 1 - i)$, and $\deg \bar{e}_{2i+1} = -2(t - i)$, $\deg (\bar{e}_{2i}) = -2(t + 1 - i)$. In this setup the degrees of vectors e_i are always nonnegative and we always have $\deg e_i = -\deg \bar{e}_i$, because $h \in \underline{so}(F)$.

If a partition μ corresponds to the conjugacy class in $\underline{so}(F)$, then we can assign the grading separately in each block, as in Example (8.3.2). We also use the convention that when dealing with several blocks, after assigning the grading, we order e_i's in such a way that $\deg e_i \geq \deg e_{i+1}$.

(8.4.2) Example. *Let us take $n = 8$, $\mu = (3, 2, 2, 1)$. The grading of basis elements is as follows:*

2	0	-2
1	-1	
1	-1	
0		

We order the elements $e_1, \dots e_4, \bar{e}_4, \dots, \bar{e}_1$ so their grading is nonincreasing. We get $\deg e_1 = \deg e_2 = 2$, $\deg e_3 = 1$, $\deg e_4 = \deg e_5 = 0$, with $\deg \bar{e}_i = -\deg e_i$ for $i = 1, \dots, 5$.

This allows us to determine the grading on $\underline{so}(F)$ in all cases.

Identifying the adjoint representation with $\bigwedge^2 F$, we can arrange the weight vectors in it in a triangular grid. In order to describe it, let us introduce the involution $\bar{()}$ of our symplectic basis by requiring that $\bar{\bar{e}}_i = e_i$. The elements of the grid correspond to the entries above the diagonal of our matrix representation of ϕ. If u, v are the elements of our symplectic basis, then the entry corresponding to the row u and the column $\pm v$ will correspond to the weight vector $u\bar{v}$.

For a given conjugacy class $C(\mu)$ we will mark the element of the grid with X if the corresponding weight vector is in $\underline{g}_{\geq 2}$, and with O otherwise. We will denote this grid by $GC(\mu)$.

(8.4.3) Example. *Let $n = 5$, $\mu = (3, 2, 2, 1)$. Then the degrees of the elements e_i are given in Example (8.4.2) and we have*

$$GC(\mu) = \begin{array}{ccccccc} X & X & X & X & O & O & O \\ & X & O & O & O & O & O \\ & & O & O & O & O & O \\ & & & O & O & O & O \\ & & & & O & O & O \\ & & & & & O & O \\ & & & & & & O \end{array}.$$

We continue with several examples of conjugacy classes.

(8.4.4) Example. *Consider the class $C((2^m))$ for even $n = 2m$. This is a class analogous to the previous example. The associated grading gives $\deg e_i = 1$, $\deg \bar{e}_i = -1$ for $i = 1, \ldots, m$. This means that in the grid $GC((2^m))$ the entries marked by X correspond to the vectors $e_i e_j$ for $i \leq j$. For example, if $n = 10$ we get*

$$GC((2^5)) = \begin{array}{ccccccccc} X & X & X & X & O & O & O & O & O \\ & X & X & X & O & O & O & O & O \\ & & X & X & O & O & O & O & O \\ & & & X & O & O & O & O & O \\ & & & & O & O & O & O & O \\ & & & & & O & O & O & O \\ & & & & & & O & O & O \\ & & & & & & & O & O \\ & & & & & & & & O \end{array}.$$

The parabolic subgroup $\mathbf{P}$ is the set of elements fixing a given isotropic subspace of dimension n. The homogeneous space $\mathbf{G}/\mathbf{P}$ is the connected component of the isotropic Grassmannian $\mathrm{IGrass}^+(m, F)$. The desingularization Z constructed in section 8.3 can be identified as

$$Z = \{(\phi, R) \in \underline{so}(F) \times \mathrm{IGrass}^+(m, F) \mid \phi(F) \subset R, \phi(R) = 0\}.$$

Let $\mathcal{R}$ be a tautological subbundle (of dimension m) on $\mathrm{IGrass}^+(m, F)$.
We apply the results of section 5.1 to Z. The vector bundles ξ and η are easily identified: $\eta = \bigwedge^2(F/\mathcal{R})$ and $\xi = \mathrm{Ker}(\bigwedge^2 F \to \bigwedge^2(F/\mathcal{R}))$.

Theorem (5.1.2)(b) implies now that

$$R^i q_* \mathcal{O}_Z = H^i \left(\mathrm{IGrass}^+(m, F), \mathrm{Sym} \left(\bigwedge^2 (F/\mathcal{R}) \right) \right).$$

Using the Cauchy formula (2.3.8)(a), we see that if we denote by $EC(d)$ the set of partitions of $2d$ with every part occurring even number of times,

$$S_d \left(\bigwedge^2 (F/\mathcal{R}) \right) = \bigoplus_{\beta \in EC(d),\ \beta_1' \leq n} K_\beta (F/\mathcal{R}).$$

Theorem (4.3.1) implies that $R^i q_* \mathcal{O}_Z = 0$ for $i > 0$ and that

$$\mathbf{K}[Y_{((2^m))}] = \bigoplus_{d \geq 0} \bigoplus_{\beta \in EC(d),\ \beta_1' \leq m} V_\beta(F),$$

where we identify the partition with at most m parts with the dominant weight for the group $SO(F)$. We have an exact sequence describing the representation $V_\beta(F)$ as a cokernel of a map of Schur functors

$$K_{\beta/(2)} F \to K_\beta F \to V_\beta(F) \to 0$$

(compare Exercise 14 of chapter 6) with the left map being induced by the trace element $\mathrm{tr} = \sum_{1 \leq i \leq m} e_i \bar{e}_i \in S_2 F$.
This means that the defining ideal of $Y_{(2^m)}$ consists of all polynomials which can be expressed as

$$\sum_i (x_1 \wedge e_i)(x_2 \wedge x_3) \ldots (x_{2t-1} \wedge \bar{e}_i) + \sum_i (x_1 \wedge \bar{e}_i)(x_2 \wedge x_3) \ldots (x_{2t-1} \wedge e_i),$$

where $x_1, \ldots, x_{2t-1} \in F$. Since any such polynomial is clearly a product of a polynomial of degree 2 and $t - 2$ polynomials of degree 1, we see that the defining ideal of $Y_{(2^m)}$ is generated by elements of degree 2 of type

$$\sum_i (x_1 \wedge e_i)(x_2 \wedge \bar{e}_i) + \sum_i (x_1 \wedge \bar{e}_i)(x_2 \wedge e_i).$$

(8.4.5) The Nonnormal Orbit. *An important phenomenon, discovered by Kraft and Procesi, is that some of the closures of nilpotent conjugacy classes for symplectic and orthogonal groups are not normal. The smallest example is the orbit $Y_{(3,2,2)}$. Let us consider this case. The grid $GC((3, 2, 2))$ looks as* *follows:*

$$
GC((3,2,2)) =
\begin{matrix}
X & X & X & O & O & O \\
  & X & O & O & O & O \\
  &   & O & O & O & O \\
  &   &   & O & O & O \\
  &   &   &   & O & O \\
  &   &   &   &   & O
\end{matrix}\,.
$$

Let us consider the second symmetric power $S_1(\eta)$. By the Hinich–Panyushev theorem (8.3.3) we know that the only cohomology group that does not vanish is $H^0(G/P, S_1(\eta))$. To calculate this group, it is enough to calculate the Euler characteristic of the bundle $S_1(\eta)$. To do this one can replace the bundle $S_1(\eta)$ with the direct sum of its composition factors of dimension 1. The whole matter becomes an exercise of using Bott's theorem. The result is

$$H^0(G/P, S_2(\eta)) = V_{(1,1,0,0)}(F) \oplus V_{(1,0,0,0)}(F).$$

Applying (5.1.3)(b), we see that the group $H^0(G/P, S_2(\eta))$ is the first graded component of the normalization of the coordinate ring of $Y_{(3,2,2)}$. If this variety were normal, $H^0(G/P, S_1(\eta))$ would be a factor of $\bigwedge^2 F$). Since $\bigwedge^2 F = V_{(1,1,0,0)}(F)$, we see that our orbit closure is not normal.

The normal orbit closures for the orthogonal groups were determined by Kraft and Procesi in [KP2]. They used the method of minimal degenerations. Their result in this case is

(8.4.6) Theorem (Kraft–Procesi). *Let $\mu \in P_o(2n)$. The orbit closure Y_μ is normal if and only if for every $i < j$ such that μ_i and μ_j are even, with $\mu_i > \mu_j$, at least one of the parts $\mu_{i+1}, \ldots, \mu_{j-1}$ is odd.*

All of the above describes the results for the orbits of the orthogonal group. However one can also consider the nilpotent conjugacy classes of the special orthogonal group. Kraft and Procesi prove in [KP2]

(8.4.7) Proposition (Kraft–Procesi). *The conjugacy class $C(\mu)$ for the orthogonal group is also a conjugacy class for the special orthogonal group unless μ is very even, i.e., all parts of μ and μ' are even. For a very even partition μ the conjugacy class $C(\mu)$ for the orthogonal group is a disjoint union of two conjugacy classes $C(\mu)^{(1)}$ and $C(\mu)^{(2)}$ for the special orthogonal group.*

We finish by presenting a conjecture on the generators of the defining ideals of orbit closures Y_μ.

(8.4.8) Conjecture. *Let $\mu \in P_o(n)$. The defining ideal of Y_μ is generated by the representations $V_{(\beta_1,\dots,\beta_n)}(F)$ with $\beta_1 \leq 2$. The generators can be chosen to be subrepresentations of the Schur functors $K_\gamma F$ with $\gamma_1 \leq 2$.*

For the remainder of this section we look at the *very even conjugacy classes*, i.e., we assume that the partitions μ and μ' have only even parts. These orbit closures they have another desingularization which is more convenient for explicit calculations.

Let $\mu = (\mu_1, \dots, \mu_r)$ be a very even partition of $n = 2m$, m even, $\mu_i = 2\nu_i$ for $i = 1, \dots, r$. We denote $\nu = (\nu_1, \dots, \nu_r)$. Note that the parts of ν' are even. Take $V = \mathrm{IFlag}(\nu_1, \nu_1 + \nu_2, \dots, n; F)$. The "desingularization" $\hat{Z}_\nu$ of Y_μ is defined as follows

$$\hat{Z}_\nu = \{(\phi, (R_{\nu_1'}, R_{\nu_1'+\nu_2'}, \dots, R_n)) \in \underline{\mathrm{sp}}(F) \times \mathrm{IFlag}(\nu_1', \nu_1' + \nu_2', \dots, n; F) \mid$$
$$\phi(R_{\nu_1'}) = 0, \; \phi(R_{\nu_1'+\dots+\nu_i'}) \subset R_{\nu_1'+\dots+\nu_{i-1}'} \text{ for } i = 2, \dots, r,$$
$$\phi(R_{\nu_1'+\dots+\nu_{i-1}'}^\vee) \subset R_{\nu_1'+\dots+\nu_i'}^\vee \text{ for } i = 2, \dots, r, \phi(F) \subset R_{\nu_1'}^\vee\}$$

The word desingularization is given above in quotation marks because the variety $\mathrm{IFlag}(\nu_1, \nu_1 + \nu_2, \dots, n; F)$ has two connected components $\mathrm{IFlag}^+(\nu_1, \nu_1 + \nu_2, \dots, n; F)$ and $\mathrm{IFlag}^-(\nu_1, \nu_1 + \nu_2, \dots, n; F)$, depending on whether R_n is in $\mathrm{IGrass}^+(m, F)$ or in $\mathrm{IGrass}^-(m, F)$. The corresponding components $\hat{Z}_\nu^+$ and $\hat{Z}_\mu^-$ give desingularizations of the closures of conjugacy classes $C_\mu^{(1)}$ and $C_\mu^{(2)}$ for the special orthogonal groups. In the sequel we work with $\hat{Z}_\nu^+$ and with $C_\mu^{(1)}$.

(8.4.9) Example. $n = 5$, $\mu = (4, 4, 2, 2)$. *The partition $\nu' = (4, 2)$. The grid corresponding to $\hat{Z}_\nu$ is*

$$\widehat{GC}((4,4,2,2)) = \begin{array}{ccccccccccc} X & X & X & X & X & X & X & X & X & O & O \\ & X & X & X & X & X & X & X & X & O & O \\ & & X & X & X & O & O & O & O & O & O \\ & & & X & X & O & O & O & O & O & O \\ & & & & X & O & O & O & O & O & O \\ & & & & & O & O & O & O & O & O \\ & & & & & & O & O & O & O & O \\ & & & & & & & O & O & O & O \\ & & & & & & & & O & O & O \\ & & & & & & & & & O & O \\ & & & & & & & & & & O \end{array} .$$

Notice that in this grid all entries corresponding to $e_i \wedge e_j$ are marked with X, all entries corresponding to $\bar{e}_i \wedge \bar{e}_j$ are marked with O, and in the region corresponding to $e_i \wedge \bar{e}_j$ we have the same pattern as for the desingularization of the conjugacy class $O(\nu)$ for $GL(n)$. This rule is true for a general very even partition.

The variety $\hat{Z}_\nu^+$ again fits all the assumptions of the setup from chapter 5. Denote the corresponding bundles by $\xi := S_\nu$, $\eta := T_\nu$. Define the partition $\hat{\nu} = (\nu_1 - 1, \ldots, \nu_j - 1)$. Notice that the partition $\hat{\mu} = 2\hat{\nu}$ is again very even.

The idea of the inductive procedure for very even conjugacy classes is to look at the bundle corresponding to the weights in our grid in the columns entirely filled by circles. There are $j = \nu'_1$ such columns. The corresponding bundle S_j lives on $IGrass(j, F)$, and it can be best described by two exact sequences it fits:

$$0 \to S_2 \mathcal{R} \to \mathcal{R} \otimes F \to S_j \to 0, \qquad (*)$$

$$0 \to \bigwedge^2 \mathcal{R} \to S_j \to \mathcal{R} \otimes (F/\mathcal{R}) \to 0. \qquad (**)$$

Notice that the factor $T_j := \bigwedge^2 F/S_j$ can be identified with $\bigwedge^2(F/\mathcal{R})$.

The point now is that we have an exact sequence

$$0 \to S_{\nu'_1} \to S_\nu \to S_{\hat{\nu}}(\mathcal{R}^\vee/\mathcal{R}) \to 0,$$

where $S_{\hat{\nu}}(\mathcal{R}^\vee/\mathcal{R})$ is the bundle $S_{\hat{\nu}}$ for the partition $\hat{\nu}$ in the relative situation where the orthogonal space of dimension $2(n - j)$ is replaced by the bundle $\mathcal{R}^\vee/\mathcal{R}$. Our strategy is again to take a complex $F_\bullet^{\hat{\nu}}$ in a relative situation and for each term of that complex (which is a special orthogonal irreducible $V_{(\alpha_1,\ldots,\alpha_{n-j})}(\mathcal{R}^\vee/\mathcal{R})$) to estimate the terms resulting in the cohomology of $V_{(\alpha_1,\ldots,\alpha_{n-j})}(\mathcal{R}^\vee/\mathcal{R}) \otimes \bigwedge^\bullet S_j$. Notice that these cohomology groups are the terms of a twisted complex $F(V_{(\alpha_1,\ldots,\alpha_{n-j})}(\mathcal{R}^\vee/\mathcal{R}))_\bullet$ supported in the variety Y_j which is the image of the incidence variety

$$Z_j = \{(\phi, R) \in X \times IGrass(j, F) \mid \phi(R) = 0\}.$$

We start with the results we need about the twisted complexes of that kind.

(8.4.10) Proposition. *Assume that $\alpha_1 \leq j$.*

(a) *The complex $F(V_{(\alpha_1,\ldots,\alpha_{n-j})}(\mathcal{R}^\vee/\mathcal{R}))_\bullet$ has terms in nonnegative degree, i.e.,*

$$H^i\left(IGrass(j, F), V_{(\alpha_1,\ldots,\alpha_{n-j})}(\mathcal{R}^\vee/\mathcal{R}) \otimes \bigwedge^m S_j\right) = 0$$

for $i > m$.

(b) *The terms of the complex $F(V_{(\alpha_1,\ldots,\alpha_{n-j})}(\mathcal{R}^\vee/\mathcal{R}))_\bullet$ contain only the representations $V_{(\beta_1,\ldots,\beta_n)}F$ with $\beta_1 \leq j$.*

Proof. We start with property (a). We use the exact sequence $(*)$. The m-th exterior power of the two term complex $S_2\mathcal{R} \to \mathcal{R} \otimes F$ gives by (2.4.7) an acyclic complex

$$0 \to S_m(S_2\mathcal{R}) \to S_{m-1}(S_2\mathcal{R}) \otimes (\mathcal{R} \otimes F)$$

$$\to \ldots \to S_2\mathcal{R} \otimes \overset{m-1}{\bigwedge}(\mathcal{R} \otimes F) \to \overset{m}{\bigwedge}(\mathcal{R} \otimes F) \qquad (?)$$

resolving $\bigwedge^m S_j$. The complex $(?) \otimes V_{(\alpha_1,\ldots,\alpha_{n-j})}(\mathcal{R}^\vee/\mathcal{R})$ provides a resolution of $\bigwedge^m S_j \otimes V_{(\alpha_1,\ldots,\alpha_{n-j})}(\mathcal{R}^\vee/\mathcal{R})$. We split that complex into short exact sequences and use the induced long sequences of cohomology groups. It follows that in order to show (a) it is enough to prove that

$$H^i(\mathrm{IGrass}(j, F), \overset{m-u}{\bigwedge}(\mathcal{R} \otimes F) \otimes S_u(S_2\mathcal{R}) \otimes V_{(\alpha_1,\ldots,\alpha_{n-j})}(\mathcal{R}^\vee/\mathcal{R})) = 0$$

for $i > m + u$.

Decomposing into representations $K_\gamma\mathcal{R}$ using the formula (2.3.8)(b), the Cauchy fomula (2.3.3), and the Littlewood–Richardson rule (2.3.4), we see that in order to prove part (a) of Proposition (8.4.10) we need to show

(8.4.11) Lemma. *Let* $\alpha = (\alpha_1, \ldots, \alpha_{n-j})$ *be a partition such that* $\alpha_1 \le j$. *Let* γ *be a partition of* m. *Then*

$$H^i(\mathrm{IGrass}(j, F), K_\gamma\mathcal{R} \otimes V_{(\alpha_1,\ldots,\alpha_{n-j})}(\mathcal{R}^\vee/\mathcal{R})) = 0$$

for all $i > m$.

Proof. We fix α and look at the set of all γ such that the cohomology

$$H^{l(\alpha,\gamma)}(\mathrm{IGrass}(j, F), K_\gamma\mathcal{R} \otimes V_{(\alpha_1,\ldots,\alpha_{n-j})}(\mathcal{R}^\vee/\mathcal{R})) \ne 0$$

for a unique number $l(\alpha, \gamma)$. We define $N(\alpha, \gamma) = |\gamma| - l(\alpha, \gamma)$. We want to show that $N(\alpha, \gamma) \ge 0$.

We recall that by Corollaries (4.3.7) and (4.3.9) the cohomology of the vector bundle $K_\gamma\mathcal{R} \otimes V_{(\alpha_1,\ldots,\alpha_{n-j})}(\mathcal{R}^\vee/\mathcal{R})$ is calculated by looking at the sequence

$$(-\gamma_j + n - 1, -\gamma_{j-1} + n - 2, \ldots, -\gamma_1 + n - j, \alpha_1 + n - j - 1, \ldots, \alpha_{n-j})$$

and trying to make it decreasing, with the sum of last two entries positive, using the Weyl group action.

Since we have $-\gamma_j + n - 1 > -\gamma_{j-1} + n - 2 > \ldots > -\gamma_1 + n - j$, there exists the smallest s for which $-\gamma_s + n - j + s - 1 < 0$.

The number $l(\alpha, \gamma)$ is calculated as follows. We move $-\gamma_1 + n - j$ to the end by a sequence of exchanges with numbers following it, change its value to its negative, and do the same with $-\gamma_2 + n - j + 1, \ldots, -\gamma_s + n - j + s - 1$. We get a sequence of positive numbers $P(\alpha, \gamma)$, which we reorder to get a decreasing sequence $R(\alpha, \gamma)$. Each exchange of neighboring elements in this process and change of the sign of the last number contributes 1 to $N(\alpha, \gamma)$. Notice that in the sequence $P(\alpha, \gamma)$ the numbers from the positions of $\gamma_1, \ldots, \gamma_s$ occupy the last s spots. Also notice that $\alpha_1 + n - j, \ldots, \alpha_{n-j} + 1$ as well as $-\gamma_{s+1} + (n - j) + s + 1, \ldots, -\gamma_j + n$ are all numbers $\leq n$. Therefore there is a unique $\gamma_1, \ldots, \gamma_s$ for which the resulting sequence is $(n, n - 1, \ldots, 1)$. We claim that the minimum of $N(\alpha, \gamma)$ is achieved for such γ.

Indeed, assume that some of the numbers in $R(\alpha, \gamma)$ are bigger than n. Let u be the smallest number not occurring in $R(\alpha, \gamma)$. We have $u \leq n$. Let m be the smallest number $\geq u$ that occurs in $R(\alpha, \gamma)$ and comes from among the positions corresponding to $\gamma_1, \ldots, \gamma_s$, say from γ_t. Let $u = m - l$. Then there exists a unique partition δ such that $\delta \subset \gamma, |\delta| = |\gamma| - l, R(\alpha, \delta) = R(\alpha, \gamma) \cup \{u\} \setminus \{m\}$. Now δ is obtained from γ in at most l steps. Each step consists of decreasing a part of γ by 1 (which forces some of the following parts to also decrease so we again get a partition), starting with γ_t. At each stage the new repetition in the reordered sequence appears. This repetition has to involve a position coming from $\gamma_1, \ldots, \gamma_s$, so we have to decrease this part of γ next, and so on. We know that $|\delta| = |\gamma| - l$. Comparing the reordering processes for (α, γ) and (α, δ), we see that $l(\alpha, \delta) \geq l(\alpha, \gamma) - l + 1$ and the decrease in the number of exchanges is equal to the number of entries $\{m - 1, \ldots, u + 1\}$ in $R(\alpha, \gamma)$ that did not come from positions corresponding to $-\gamma_s + n - j + s, \ldots, -\gamma_1 + n - j + 1$. Thus $N(\alpha, \delta) \leq N(\alpha, \gamma) - 1$, as desired.

Assume that (α, γ) is such that the reordered sequence is $(n, n - 1, \ldots, 1)$.

We shall prove that for fixed α and for any γ for which $R(\alpha, \gamma) = (n, n - 1, \ldots, 1)$, the partition γ with minimal $N(\alpha, \gamma)$ is the one with $s = 0$. Indeed, let us assume that $s > 0$. Consider the sequence

$$(-\gamma_j + n, -\gamma_{j-1} + n - 1, \ldots, -\gamma_1 + n - j + 1, \alpha_1 + n - j, \ldots, \alpha_{n-j} + 1).$$

We construct a new partition δ as follows. We modify the above sequence by changing the negative entry $-\gamma_s + n - j + s$ to its negative $x = \gamma_s - n + j - s > 0$. Then we order the positive part of this sequence and see that it can be written as

$$(-\delta_j + n, -\delta_{j-1} + n - 1, \ldots, -\delta_1 + n - j + 1, \alpha_1 + n - j, \ldots, \alpha_{n-j} + 1)$$

for a partition δ with $|\delta| = |\gamma| - 2x$. The partition δ has smaller invariant s.

We count the number $l(\alpha, \gamma) - l(\alpha, \delta)$. The difference is that the number $-\gamma_s + n - j + s$ gets exchanged twice with numbers from the set $\alpha_1 + n - j, \ldots, \alpha_{n-j} + 1$ that are in $[1, x - 1]$, then it is reflected to its negative and then gets exchanged additionally with the numbers from the interval $[1, x - 1]$ that preceed it. Therefore we have $l(\alpha, \gamma) - l(\alpha, \delta) \leq 2x - 1$ and thus

$$N(\alpha, \gamma) > N(\alpha, \delta).$$

It remains to show that for fixed α and for partitions γ such that $R(\alpha, \gamma) = (n, n - 1, \ldots, 1)$ and $s = 0$ we have $N(\alpha, \gamma) \geq 0$. But now we see that the last reflection does not occur in exchanging entries. Therefore we are in the situation of Lemma (8.1.5). We see that the choice of γ minimizing $N(\alpha, \gamma)$ is $\gamma = \alpha'$ and that $N(\alpha, \alpha') = 0$. This proves Lemma (8.4.11) and therefore part (a) of Proposition (8.4.10). ∎

We prove part (b) of Proposition (8.4.10). We use the exact sequence (∗∗). We will prove that the cohomology of

$$\bigwedge^{\bullet}(S_2\mathcal{R}) \otimes \bigwedge^{\bullet}(\mathcal{R} \otimes (F/\mathcal{R})) \otimes V_\alpha(\mathcal{R}^\vee/\mathcal{R})$$

does not contain the representations $V_\beta F$ with $\beta_1 > j$. Using the exact sequence

$$0 \to \mathcal{R}^\vee/\mathcal{R} \to F/\mathcal{R} \to F/\mathcal{R}^\vee \to 0,$$

we see that it is enough to show that the cohomology of

$$\bigwedge^{\bullet}(S_2\mathcal{R}) \otimes \bigwedge^{\bullet}(\mathcal{R} \otimes (\mathcal{R}^\vee/\mathcal{R})) \otimes \bigwedge^{\bullet}(\mathcal{R} \otimes (F/\mathcal{R}^\vee)) \otimes V_\alpha(\mathcal{R}^\vee/\mathcal{R})$$

does not contain the representations $V_\beta F$ with $\beta_1 > j$.

Let us look at a typical term

$$K_\beta\mathcal{R} \otimes K_\gamma\mathcal{R} \otimes K_{\gamma'}(\mathcal{R}^\vee/\mathcal{R}) \otimes K_\delta\mathcal{R} \otimes K_{\delta'}(F/\mathcal{R}^\vee) \otimes V_\alpha(\mathcal{R}^\vee/\mathcal{R}).$$

After decomposing the Weyl functors to the symplectic irreducibles and calculating tensor products, we see that the term above will decompose to the summands of the form

$$K_\lambda(F/\mathcal{R}^\vee) \otimes V_\theta(\mathcal{R}^\vee/\mathcal{R}).$$

Such a bundle has a representation $V_\beta F$ with $\beta_1 > j$ in cohomology if a sequence

$$(\lambda_1 + n, \lambda_2 + n - 1, \ldots, \lambda_j + n - j + 1, \theta_1 + n - j, \ldots, \theta_{n-j} + 1)$$

contains either the number $> n + j$ or a number $< -n - j$. Since θ is an integral dominant weight for $\mathrm{Sp}(2n - 2j)$, i.e. a partition, this can happen if one of the following cases occurs:

(a) $\lambda_1 + n > n + j$, i.e. $\lambda_1 > j$,
(b) $\theta_1 + n - j > n + j$, i.e. $\theta_1 > 2j$,
(c) $\lambda_j + n - j + 1 < -n - j$, i.e. $-\lambda_j > 2n + 1$.

Case (a) cannot occur because $K_\lambda(F/\mathcal{R}^\vee)$ is a summand in a tensor product

$$K_\beta \mathcal{R} \otimes K_\gamma \mathcal{R} \otimes K_\delta \mathcal{R} \otimes K_{\delta'}(F/\mathcal{R}^\vee),$$

and the only possible positive weights come from δ'. However, $\delta'_1 \le j$, since $\dim \mathcal{R} = j$.

Case (b) cannot occur because the weight θ comes from a summand in

$$K_{\gamma'}(\mathcal{R}^\vee/\mathcal{R}) \otimes V_\alpha(\mathcal{R}^\vee/\mathcal{R})$$

and $\alpha_1 \le j$ and $\gamma'_1 \le j$ (because the corresponding functor $K_\gamma \mathcal{R}$ is nonzero). Thus $K_{\gamma'}(\mathcal{R}^\vee/\mathcal{R})$ decomposes to the representations $V_\epsilon(\mathcal{R}^\vee/\mathcal{R})$, and in the tensor product

$$V_\epsilon(\mathcal{R}^\vee/\mathcal{R}) \otimes V_\alpha(\mathcal{R}^\vee/\mathcal{R})$$

all weights have to be $\le \alpha + \epsilon$ by Klimyk's formula (exercise 12 of chapter 4), and therefore the first entry of each occurring weight is $\le 2j$.

Case (c) cannot occur because in the tensor product

$$K_\beta \mathcal{R} \otimes K_\gamma \mathcal{R} \otimes K_\delta \mathcal{R} \otimes K_{\delta'}(F/\mathcal{R}^\vee)$$

the last entries are $-\beta_1 \ge -j - 1$, $-\gamma_1 \ge 2j - 2n$, and $-\delta_1 \ge -j$ (because $K_{\delta'}(F/\mathcal{R}^\vee)$ has to be nonzero).

This concludes the proof of part (b) of Proposition (8.4.10). ●

(8.4.12) Corollary.

(a) The term F_0 consists of one copy of the trivial representation in homogeneous degree 0. In particular the variety Z_j is normal, with rational singularities.

(b) The term F_1 contains only the representations $V_\beta(F)$ with $\beta_1 \le 1$. In particular the defining ideal of $Y_j := q(Z_j)$ is generated by such representations. More precisely, the minimal set of generators of the defining ideal of Y_j consists of the vectors in the representation

$$\bigwedge^{2(n-j+1)} F \subset K_{(2^{2(n-j+1)})}F \subset S_{2(n-j+1)}(S_2 F).$$

Proof. The first claim follows from the proof of Lemma (8.4.11). Indeed, the only term contributing to F_0 corresponds to $\gamma = 0$ which, looking at the sequences (?), can appear only for $m = 0$.

The proof of the second part is based on the refinement of the Lemma (8.4.11).

Claim. *The only partition γ with $|\gamma| = m$ such that $H^{m-1}(\mathrm{IGrass}(j, F),$ $K_\gamma \mathcal{R}) \neq 0$ is $\gamma = (2(n - j + 1))$, and the resulting cohomology group is a trivial representation of* $\mathrm{Sp}(F)$.

Proof. We repeat the proof of Lemma (8.4.11) with $\alpha = (0)$. Our partition γ has to lead in one step to the partition $\gamma = 0$, which gives the only partition with $N(\gamma, (0)) = 0$. The sequence $R(\gamma, (0))$ has to be $(n, n - 1, \ldots, 1)$, because if not, then $s > 0$ and the first step in the procedure does decrease s. We see now that $s = 1$; otherwise we need s steps to reduce to $\gamma = 0$. Now we can do a case by case analysis, because the number of possibilities for γ corresponds to possibilities for the number $\gamma_1 - n + j - 1 \in \{n, n - 1, \ldots, n - j + 1\}$. The only viable possibility turns out to be $\gamma_1 - n + j - 1 = n - j + 1$ which implies $\gamma_t = 0$ for $t \geq 2$. ∎

Now we see, using the exact sequence (?) resolving $\bigwedge^m \mathcal{S}_j$, that the only possible contribution to F_1 is $\bigwedge^{2(n-j+1)} F$ in homogeneous degree $2(n - j + 1)$. We easily identify this set of equations with the ones given in (8.4.12) (b) because they vanish on Y_j.

We can finally state the results of our induction.

(8.4.13) Theorem. *Let μ be an even partition, $\mu = 2\nu$.*

   (a) The orbit closure Y_μ is normal, with rational singularities,
   (b) The irreducible representations occurring in the terms of the complex $F_\bullet^\mu$ have highest weights $(\beta_1, \ldots, \beta_n)$ with $\beta_1 \leq \nu_1'$.
   (c) The term F_1^μ contains only the representations $V_\beta(F)$ with $\beta_1 \leq 1$. Therefore the defining ideal of Y_μ is generated by the representations of this kind.

Proof. We use the induction described above, connecting the relative complex $F_\bullet^{\hat{\mu}}$ to the complex $F_\bullet^\mu$. We use the exact sequence

$$0 \to \mathcal{S}_{\nu_1'} \to \mathcal{S}_\nu \to \mathcal{S}_{\hat{\nu}}(\mathcal{R}^\vee/\mathcal{R}) \to 0$$

to estimate the terms in the cohomology of $\mathcal{S}_\nu$. They come from the terms of the cohomology in $V_\alpha(\mathcal{R}^\vee/\mathcal{R}) \otimes \bigwedge^\bullet(\mathcal{S}_{\nu_1'})$. By induction the terms $V_\alpha(\mathcal{R}^\vee/\mathcal{R})$ satisfy the assumption of Proposition (8.4.10), which shows that the i-th term of the complex $F^{\hat{\mu}}(\mathcal{R}^\vee/\mathcal{R})_\bullet$ can produce only terms in the homological degree $\geq i$. Thus, by induction, only the terms in nonnegative homological degree occur, and by Corollary (8.4.12) the term F_0^μ consists of one copy of the trivial representation in homogeneous degree 0. Parts (a) and (b) of the theorem follow.

To prove part (c) let us analyze the term F_1^μ. The terms there can come from the term $F_\bullet$, which has the required form by Corollary (8.4.12)(b), and from the terms from $F_1^{\hat\mu}(\mathcal{R}^\vee/\mathcal{R})$, which by induction are the terms of homological degree zero in $F(V_{(\alpha_1,\dots,\alpha_{n-j})}(\mathcal{R}^\vee/\mathcal{R}))_\bullet$, with weight α for which $\alpha_1 \leq 1$. By the proof of Lemma (8.4.11) the only such term comes from $\gamma = \alpha'$, so (by the use of the complexes $V_\alpha(\mathcal{R}^\vee/\mathcal{R}) \otimes (?)$) the resulting contribution to the possible terms of $F(V_{(\alpha_1,\dots,\alpha_{n-j})}(\mathcal{R}^\vee/\mathcal{R}))_\bullet$ is $K_\alpha F \otimes V_{(0)} F$, which decomposes to the representations of $\mathrm{Sp}(F)$ of the required form. ∎

8.5. Conjugacy Classes for the Symplectic Group

In this section, F denotes a vector space of dimension $2n$ with a nondegenerate antisymmetric form $\langle\ ,\ \rangle$. The symplectic group $\mathrm{Sp}(F)$ is the set of linear automorphisms of F preserving $\langle\ ,\ \rangle$, i.e., $\phi \in \mathrm{Sp}(F)$ if and only if for each $x, y \in F$ we have $\langle\phi(x), \phi(y)\rangle = \langle x, y\rangle$. By definition $\mathrm{Sp}(F)$ is a subgroup of $\mathrm{GL}(F)$.

The corresponding Lie algebra $\underline{\mathrm{sp}}(F)$ is a subalgebra of the Lie algebra $\underline{\mathrm{gl}}(F)$. The morphism $\phi \in \mathrm{Hom}_K(\overline{F}, F)$ is in $\underline{\mathrm{sp}}(F)$ if and only if for each $x, y \in F$ we have $\langle\phi(x), y\rangle + \langle x, \phi(y)\rangle = 0$.

Let us choose the symplectic basis $e_1, \dots, e_n, \bar{e}_n, \dots, \bar{e}_1$ of F. This means

$$\langle e_i, e_j\rangle = \langle \bar{e}_i, \bar{e}_j\rangle = 0, \qquad \langle e_i, \bar{e}_j\rangle = \delta_{i,j},$$

where $\delta_{i,j}$ denotes the Kronecker delta.

We can write ϕ as a matrix, writing in consecutive columns the images of vectors $-\bar{e}_1, \dots, -\bar{e}_n, e_n, \dots, e_1$ expanded in the basis $e_1, \dots, e_n, \bar{e}_n, \dots, \bar{e}_1$. Then $\phi \in \underline{\mathrm{sp}}(F)$ if and only if the matrix of ϕ is symmetric. This allows us to identify the adjoint representation of $\underline{\mathrm{sp}}(F)$ with $S_2 F$.

Since $\underline{\mathrm{sp}}(F)$ is a Lie subalgebra of $\underline{\mathrm{gl}}(F)$, we might expect that the nilpotent conjugacy classes in $\underline{\mathrm{sp}}(F)$ will be related to intersections of the nilpotent conjugacy classes in $\underline{\mathrm{gl}}(F)$ with $\underline{\mathrm{sp}}(F)$. The following result is proved in [SS; IV.2.15].

(8.5.1) Proposition. *Let μ be a partition of $2n$. Let us consider the nilpotent conjugacy class $O(\mu)$ of $\underline{\mathrm{gl}}(F)$ corresponding to μ. Then $O(\mu)$ intersects $\underline{\mathrm{sp}}(F)$ if and only if every odd part of μ occurs an even number of times. In this case the intersection $O(\mu) \cap \underline{\mathrm{sp}}(F)$ consists of a single conjugacy class of $\mathrm{Sp}(F)$.*

For the remainder of this section we denote by $P_s(n)$ the set of partitions μ of $2n$ in which every odd part occurs even number of times. For $\mu \in P_s(n)$ we denote $C(\mu)$ the corresponding conjugacy class in $\underline{sp}(F)$, and by Y_μ its closure.

One might say that the Jordan blocks for the symplectic group are of two kinds: the blocks corresponding to even rows in our partition and the blocks corresponding to the pairs of odd rows. The canonical forms of nilpotent symplectic endomorphisms corresponding to both kinds of blocks are as follows.

The nilpotent e corresponding to the partition $(2n)$ has the form

$$e(\bar{e}_1) = 0, \qquad e(\bar{e}_i) = \bar{e}_{i-1} \quad \text{for } i > 0,$$

$$e(e_n) = -\bar{e}_n, \qquad e(e_i) = -e_{i+1} \quad \text{for } i < n.$$

We can express the action of e by the sequence of arrows

$$\pm e_1 \to \mp e_2 \to \ldots \to e_{n-1} \to -e_n \to \bar{e}_n \to \bar{e}_{n-1} \to \ldots \to \bar{e}_2 \to \bar{e}_1 \to 0.$$

If we extend e to an $\underline{sl}_2$ triple $\{e, h, f\}$, the grading induced by e described in section 8.3 is as follows. For the endomorphism e corresponding to the partition $2n$ we have $h(e_i) = (2n + 1 - 2i)e_i$, $h(\bar{e}_i) = (2i - 1 - 2n)\bar{e}_i$. Therefore the degrees of the basis vectors are $\deg e_i = 2n + 1 - 2i$, $\deg \bar{e}_i = 2i - 1 - 2n$.

If $n = 2t + 1$ is odd, then the nilpotent e corresponding to the partition (n, n) has the form

$$e(e_i) = e_{i+2} \quad \text{for } i < 2t, \qquad e(e_{2t}) = \bar{e}_{2t+1}, \qquad e(e_{2t+1}) = \bar{e}_{2t},$$

$$e(\bar{e}_i) = -\bar{e}_{i-2} \quad \text{for } i > 2, \qquad e(\bar{e}_2) = e(\bar{e}_1) = 0.$$

We can express the action of e by two sequences of arrows

$$e_1 \to e_3 \to \ldots \to e_{2t+1} \to \bar{e}_{2t} \to -\bar{e}_{2t-2} \to \ldots \to \pm\bar{e}_2 \to 0,$$

$$e_2 \to e_4 \to \ldots \to e_{2t} \to \bar{e}_{2t+1} \to -\bar{e}_{2t-1} \to \ldots \to \mp\bar{e}_1 \to 0.$$

Extending e to the $\underline{sl}_2$-triple $\{e, h, f\}$, we see that the element h acts as follows: $h(e_{2i+1}) = 2(t - i)e_{2i+1}$, $h(e_{2i}) = 2(t + 1 - i)e_{2i}$, and $h(\bar{e}_{2i+1}) = -2(t - i)\bar{e}_{2i+1}$, $h(\bar{e}_{2i}) = -2(t + 1 - i)\bar{e}_{2i}$. This gives $\deg e_{2i+1} = 2(t - i)$, $\deg e_{2i} = 2(t + 1 - i)$, and $\deg \bar{e}_{2i+1} = -2(t - i)$, $\deg \bar{e}_{2i} = -2(t + 1 - i)$.

In this setup the degrees of vectors e_i are always nonnegative and we always have $\deg e_i = -\deg \bar{e}_i$, because $h \in \underline{sp}(F)$.

If a partition μ corresponds to the conjugacy class in $\underline{sp}(F)$, then we can assign the grading separately in each block, as in Example (8.3.2). We also use the convention that when dealing with several blocks, after assigning the grading, we order e_i's in such a way that $\deg e_i \geq \deg e_{i+1}$.

(8.5.2) Example. *Let us take* $n = 5$, $\mu = (3, 3, 2, 1, 1)$. *The grading of basis elements is as follows:*

2	0	-2
2	0	-2
1	-1	
0		
0		

We order the elements $e_1, \ldots e_5, \bar{e}_5, \ldots, \bar{e}_1$ *so their grading is nonincreasing. We get* $\deg e_1 = \deg e_2 = 2$, $\deg e_3 = 1$, $\deg e_4 = \deg e_5 = 0$, *with* $\deg \bar{e}_i = -\deg e_i$ *for* $i = 1, \ldots, 5$. *This allows us to determine the grading on* $\underline{sp}(F)$ *in all cases.*

Identifying the adjoint representation with $S_2 F$, we can arrange the weight vectors in it in a triangular grid. In order to describe it, let us introduce the involution $\bar{(\,)}$ of our symplectic basis by requiring that $\bar{\bar{e}}_i = e_i$. The elements of the grid correspond to the entries on or above the diagonal of our matrix representation of ϕ. If u, v are the elements of our symplectic basis, then the entry corresponding to the row u and the column $\pm v$ will correspond to the weight vector $u\bar{v}$.

For a given conjugacy class $C(\mu)$ we will mark the element of the grid with X if the corresponding weight vector is in $\underline{g}_{\geq 2}$, and with O otherwise. We will denote this grid by $GC(\mu)$.

(8.5.3) Example. *Let* $n = 5$, $\mu = (3, 3, 2, 1, 1)$. *Then the degrees of elements* e_i *are given in Example (8.5.2), and we have*

$$
GC(\mu) =
\begin{array}{cccccccccc}
X & X & X & X & X & X & X & O & O & O \\
  & X & X & X & X & X & X & O & O & O \\
  &   & X & O & O & O & O & O & O & O \\
  &   &   & O & O & O & O & O & O & O \\
  &   &   &   & O & O & O & O & O & O \\
  &   &   &   &   & O & O & O & O & O \\
  &   &   &   &   &   & O & O & O & O \\
  &   &   &   &   &   &   & O & O & O \\
  &   &   &   &   &   &   &   & O & O \\
  &   &   &   &   &   &   &   &   & O \\
\end{array}.
$$

We continue with several examples of conjugacy classes.

(8.5.4) Example. *Consider the class* $C((2^n))$. *This is a very interesting conjugacy class. The associated grading gives* $\deg(e_i) = 1$, $\deg(\bar{e}_i) = -1$ *for*

$i = 1, \ldots, n$. *This means that in the grid* $GC((2^n))$ *the entries marked by X correspond to the vectors* $e_i e_j$ *for* $i \leq j$. *For example, if* $n = 5$ *we get*

$$
GC((2^5)) =
\begin{array}{cccccccccc}
X & X & X & X & X & O & O & O & O & O \\
  & X & X & X & X & O & O & O & O & O \\
  &   & X & X & X & O & O & O & O & O \\
  &   &   & X & X & O & O & O & O & O \\
  &   &   &   & X & O & O & O & O & O \\
  &   &   &   &   & O & O & O & O & O \\
  &   &   &   &   &   & O & O & O & O \\
  &   &   &   &   &   &   & O & O & O \\
  &   &   &   &   &   &   &   & O & O \\
  &   &   &   &   &   &   &   &   & O
\end{array}.
$$

The parabolic subgroup **P** *is the set of elements fixing a given isotropic subspace of dimension n. The homogeneous space* **G/P** *is the isotropic Grassmannian* IGrass(n, F). *The desingularization Z constructed in section 8.3 can be identified as*

$$Z = \{(\phi, R) \in \underline{\mathrm{sp}}(F) \times \mathrm{IGrass}(n, F) \mid \phi(F) \subset R, \ \phi(R) = 0\}.$$

Let $\mathcal{R}$ *be a tautological subbundle (of dimension n) on* IGrass(n, F).

We apply the results of section 5.1 to Z. The vector bundles ξ *and* η *are easily identified:* $\eta = S_2(F/\mathcal{R})$ *and* $\xi = \mathrm{Ker}(S_2 F \to S_2 \mathcal{R})$. *Theorem (5.1.2)(b) implies now that*

$$R^i q_* \mathcal{O}_Z = H^i(\mathrm{IGrass}(n, F), \mathrm{Sym}(S_2(F/\mathcal{R}))).$$

Using the Cauchy formula (2.3.8)(a), we see that if we denote by EP(d) *the set of partitions of 2d with even parts, then*

$$S_d(S_2(F/\mathcal{R})) = \bigoplus_{\beta \in \mathrm{EP}(d), \ \beta_1' \leq n} K_\beta(F/\mathcal{R}).$$

Theorem (4.3.1) implies that $R^i q_* \mathcal{O}_Z = 0$ *for* $i > 0$ *and that*

$$\mathbf{K}[C((\bar{2}^n))] = \bigoplus_{d \geq 0} \bigoplus_{\beta \in \mathrm{EP}(d), \ \beta_1' \leq n} V_\beta(F),$$

where we identify the partition with at most n parts with the dominant weight for the group SP(F). *We have an exact sequence describing the representation* $V_\beta(F)$ *as a cokernel of a map of Schur functors*

$$K_{\beta/(1^2)}F \to K_\beta F \to V_\beta(F) \to 0$$

(compare Exercise 4 of chapter 6) with the left map being induced by the trace element tr $= \sum_{1 \leq i \leq n} e_i \wedge \bar{e}_i \in \bigwedge^2 F$.

This means that the defining ideal of $C((\bar{2}^n))$ consists of all polynomials which can be expressed as

$$\sum_i (x_1 e_i)(x_2 x_3)\dots(x_{2t-1}\bar{e}_i) - \sum_i (x_1\bar{e}_i)(x_2 x_3)\dots(x_{2t-1}e_i)$$

where $x_1,\dots,x_{2t-1} \in F$. Since any such polynomial is clearly a product of a polynomial of degree 2 and $t-2$ polynomials of degree 1, we see that the defining ideal of $C((\bar{2}^n))$ is generated by elements of degree 2 of type

$$\sum_i (x_1 e_i)(x_2\bar{e}_i) - \sum_i (x_1\bar{e}_i)(x_2 e_i).$$

(8.5.5) The Nonnormal Orbit. *We give the example of the smallest nonnormal orbit closure for the symplectic group. It is $Y_{(3,3,1,1)}$. Let us analyze this case. The grid $GR(3,3,1,1)$ looks as follows:*

$$GC((3,3,1,1)) =
\begin{array}{cccccccc}
X & X & X & X & X & X & O & O \\
  & X & X & X & X & X & O & O \\
  &   & O & O & O & O & O \\
  &   &   & O & O & O & O & O \\
  &   &   &   & O & O & O & O \\
  &   &   &   &   & O & O & O \\
  &   &   &   &   &   & O & O \\
  &   &   &   &   &   &   & O
\end{array}.$$

Let us consider the second symmetric power $S_2(\eta)$. By the Hinich–Panyushev theorem (8.3.3) we know that the only cohomology group that does not vanish is $H^0(G/P, S_2(\eta))$. To calculate this group, it is enough to calculate the Euler characteristic of the bundle $S_2(\eta)$. To do this one can replace the bundle $S_2(\eta)$ with the direct sum of its composition factors of dimension 1. The whole matter becomes an exercise of using Bott's theorem 55 times. The result is

$$H^0(G/P, S_2(\eta)) = V_{(4,0,0,0)}(F) \oplus V_{(2,2,0,0)}(F) \oplus V_{(1,1,0,0)}(F) \oplus V_{(1,1,1,1)}(F).$$

Applying (5.1.3)(b), we see that the group $H^0(G/P, S_2(\eta))$ is the second graded component of the normalization of the coordinate ring of $Y_{(3,3,1,1)}$. If this variety were normal, $H^0(G/P, S_2(\eta))$ would be a factor of $S_2(S_2 F)$. However using the sequences of exercise 4 of chapter 6, we can easily see that $S_2(S_2 F) = V_{(4,0,0,0)}(F) \oplus V_{(2,2,0,0)}(F) \oplus V_{(1,1,0,0)}(F) \oplus V_{(0,0,0,0)}(F)$. This proves that our closure is not normal.

The normal orbit closures for the symplectic Lie algebra were determined by Kraft and Procesi in [KP2]. They used the method of minimal degenerations. We state their result.

(8.5.6) Theorem (Kraft–Procesi). *Let $\mu \in P_s(2n)$. The orbit closure Y_μ is normal if and only if for every $i < j$ such that μ_i and μ_j are odd, with $\mu_i > \mu_j$, at least one of the parts $\mu_{i+1}, \ldots, \mu_{j-1}$ has to be even.*

Even though the method of minimal degenerations is more effective in describing normality, calculations such as the one above are still useful, as they allow to describe the decomposition to irreducibles of the normalization of the coordinate ring of an orbit closure.

For the remainder of this section we look at the *even conjugacy classes*, i.e., we assume that the partition μ has only even parts. These classes show certain similarities with the conjugacy classes for the general linear group. In particular, they have another desingularization which is more convenient for explicit calculations.

Let $\mu = (\mu_1, \ldots, \mu_r)$ be a partition of $2n$ with only even parts, $\mu_i = 2\nu_i$ for $i = 1, \ldots, r$. We denote $\nu = (\nu_1, \ldots, \nu_r)$.

Take $V = \mathrm{IFlag}(\nu_1, \nu_1 + \nu_2, \ldots, n; F)$. The desingularization $\hat{Z}_\nu$ of $\overline{C(\mu)}$ is defined as follows:

$$\hat{Z}_\nu = \{(\phi, (R_{\nu'_1}, R_{\nu'_1+\nu'_2}, \ldots, R_n)) \in \underline{\mathrm{sp}}(F) \times \mathrm{IFlag}(\nu'_1, \nu'_1 + \nu'_2, \ldots, n; F) \mid$$
$$\phi(R_{\nu'_1}) = 0, \ \phi(R_{\nu'_1+\ldots+\nu'_i}) \subset R_{\nu'_1+\ldots+\nu'_{i-1}} \ \text{for } i = 2, \ldots, r,$$
$$\phi(R^\vee_{\nu'_1+\ldots+\nu'_{i-1}}) \subset R^\vee_{\nu'_1+\ldots+\nu'_i} \ \text{for } i = 2, \ldots, r, \phi(F) \subset R^\vee_{\nu'_1}\}.$$

(8.5.7) Example. *Let $n = 5$, $\mu = (6, 4)$. The partition $\nu' = (2, 2, 1)$. The grid corresponding to $\hat{Z}_\nu$ is*

$$\widehat{GC}((4, 4, 2)) = \begin{matrix}
X & X & X & X & X & X & X & X & O & O \\
  & X & X & X & X & X & X & X & O & O \\
  &   & X & X & X & X & O & O & O & O \\
  &   &   & X & X & X & O & O & O & O \\
  &   &   &   & X & O & O & O & O & O \\
  &   &   &   &   & O & O & O & O & O \\
  &   &   &   &   &   & O & O & O & O \\
  &   &   &   &   &   &   & O & O & O \\
  &   &   &   &   &   &   &   & O & O \\
  &   &   &   &   &   &   &   &   & O
\end{matrix}.$$

Notice that in this grid all entries corresponding to $e_i e_j$ are marked with X, all entries corresponding to $\bar{e}_i \bar{e}_j$ are marked with O, and in the region corresponding to $e_i \bar{e}_j$ we have the same pattern as for the desingularization of the conjugacy class $O(\nu)$ for $GL(n)$. This rule is true for a general even partition.

The variety $\hat{Z}_\nu$ again fits all the assumptions of the setup from chapter 5. Denote the corresponding bundles $\xi := \mathcal{S}_\nu$, $\eta := \mathcal{T}_\nu$. Define the partition $\hat{\nu} = (\nu_1 - 1, \ldots, \nu_j - 1)$.

The idea of the inductive procedure for even conjugacy classes is to look at the bundle corresponding to the weights in our grid in the columns entirely filled by circles. There are $j = \nu_1'$ such columns. The corresponding bundle $\mathcal{S}_j$ lives on $IGrass(j, F)$, and it can be best described by two exact sequences it fits:

$$0 \to \bigwedge^2 \mathcal{R} \to \mathcal{R} \otimes F \to \mathcal{S}_j \to 0, \qquad (*)$$

$$0 \to S_2 \mathcal{R} \to \mathcal{S}_j \to \mathcal{R} \otimes (F/\mathcal{R}) \to 0. \qquad (**)$$

Notice that the factor $\mathcal{T}_j := S_2 F/\mathcal{S}_j$ can be identified with $S_2(F/\mathcal{R})$.

The point now is that we have an exact sequence

$$0 \to \mathcal{S}_{\nu_1'} \to \mathcal{S}_\nu \to \mathcal{S}_{\hat{\nu}}(\mathcal{R}^\vee/\mathcal{R}) \to 0,$$

where $\mathcal{S}_{\hat{\nu}}(\mathcal{R}^\vee/\mathcal{R})$ is the bundle $\mathcal{S}_{\hat{\nu}}$ for the partition $\hat{\nu}$ in the relative situation where the symplectic space of dimension $2(n - j)$ is replaced by the bundle $\mathcal{R}^\vee/\mathcal{R}$. Our strategy is again to take a complex $F_\bullet^{\hat{\nu}}$ in a relative situation and for each term of that complex (which is a symplectic irreducible $V_{(\alpha_1,\ldots,\alpha_{n-j})}(\mathcal{R}^\vee/\mathcal{R})$) to estimate the terms resulting in the cohomology of $V_{(\alpha_1,\ldots,\alpha_{n-j})}(\mathcal{R}^\vee/\mathcal{R}) \otimes \bigwedge^\bullet \mathcal{S}_j$. Notice that these cohomology groups are the terms of a twisted complex $F(V_{(\alpha_1,\ldots,\alpha_{n-j})}(\mathcal{R}^\vee/\mathcal{R}))_\bullet$ supported in the variety Y_j which is the image of the incidence variety

$$Z_j = \{(\phi, R) \in X \times IGrass(j, F) \mid \phi(R) = 0\}.$$

We start with the results we need about the twisted complexes of that kind.

(8.5.8) Proposition. *Assume that $\alpha_1 \leq j$.*

(a) The complex $F(V_{(\alpha_1,\ldots,\alpha_{n-j})}(\mathcal{R}^\vee/\mathcal{R}))_\bullet$ has terms in nonnegative degree, i.e.,

$$H^i\left(IGrass(j, F), V_{(\alpha_1,\ldots,\alpha_{n-j})}(\mathcal{R}^\vee/\mathcal{R}) \otimes \bigwedge^m \mathcal{S}_j\right) = 0$$

for $i > m$.

(b) The terms of the complex $F(V_{(\alpha_1,...,\alpha_{n-j})}(\mathcal{R}^\vee/\mathcal{R}))_\bullet$ *contain only the representations* $V_{(\beta_1,...,\beta_n)}F$ *with* $\beta_1 \leq j$.

Proof. We start with property (a). We use the exact sequence (∗). The m-th exterior power of the two term complex $\bigwedge^2 \mathcal{R} \to \mathcal{R} \otimes F$ gives by (2.4.7) an acyclic complex

$$0 \to S_m\left(\overset{2}{\bigwedge}\mathcal{R}\right) \to S_{m-1}\left(\overset{2}{\bigwedge}\mathcal{R}\right) \otimes (\mathcal{R} \otimes F) \to \dots$$

$$\to \overset{2}{\bigwedge}\mathcal{R} \otimes \overset{m-1}{\bigwedge}(\mathcal{R} \otimes F) \to \overset{m}{\bigwedge}(\mathcal{R} \otimes F) \tag{?}$$

resolving $\bigwedge^m S_j$. The complex (?) $\otimes V_{(\alpha_1,...,\alpha_{n-j})}(\mathcal{R}^\vee/\mathcal{R})$ provides a resolution of $\bigwedge^m S_j \otimes V_{(\alpha_1,...,\alpha_{n-j})}(\mathcal{R}^\vee/\mathcal{R})$. We split that complex into short exact sequences and use the induced long sequences of cohomology groups. It follows that in order to show (a) it is enough to prove that

$$H^i\left(\mathrm{IGrass}(j, F), \overset{m-u}{\bigwedge}(\mathcal{R} \otimes F) \otimes S_u\left(\overset{2}{\bigwedge}\mathcal{R}\right) \otimes V_{(\alpha_1,...,\alpha_{n-j})}(\mathcal{R}^\vee/\mathcal{R})\right) = 0$$

for $i > m + u$.

Decomposing into representations $K_\gamma \mathcal{R}$ using the formula (2.3.8)(a), the Cauchy formula (2.3.3), and the Littlewood–Richardson rule (2.3.4), we see that in order to prove part (a) of Proposition (8.5.8) we need to show

(8.5.9) Lemma. *Let* $\alpha = (\alpha_1, \dots, \alpha_{n-j})$ *be a partition such that* $\alpha_1 \leq j$. *Let* γ *be a partition of* m. *Then*

$$H^i(\mathrm{IGrass}(j, F), K_\gamma \mathcal{R} \otimes V_{(\alpha_1,...,\alpha_{n-j})}(\mathcal{R}^\vee/\mathcal{R})) = 0$$

for all $i > m$.

Proof. We fix α and look at the set of all γ such that the cohomology

$$H^{l(\alpha,\gamma)}(\mathrm{IGrass}(j, F), K_\gamma \mathcal{R} \otimes V_{(\alpha_1,...,\alpha_{n-j})}(\mathcal{R}^\vee/\mathcal{R})) \neq 0$$

for a unique number $l(\alpha, \gamma)$. We define $N(\alpha, \gamma) = |\gamma| - l(\alpha, \gamma)$. We want to show that $N(\alpha, \gamma) \geq 0$.

We recall that by Corollary (4.3.4) the cohomology of the vector bundle $K_\gamma \mathcal{R} \otimes V_{(\alpha_1,...,\alpha_{n-j})}(\mathcal{R}^\vee/\mathcal{R})$ is calculated by looking at the sequence

$$(-\gamma_j + n, -\gamma_{j-1} + n - 1, \dots, -\gamma_1 + n - j + 1, \alpha_1 + n - j, \dots, \alpha_{n-j} + 1)$$

and trying to make it decreasing using the Weyl group action. Since we have $-\gamma_j + n > -\gamma_{j-1} + n - 1 > \ldots > -\gamma_1 + n - j + 1$, there exists the smallest s for which $-\gamma_s + n - j + s < 0$. The number $l(\alpha, \gamma)$ is calculated as follows. We move $-\gamma_1 + n - j + 1$ to the end by a sequence of exchanges with numbers following it, change its value to its negative, and do the same with $-\gamma_2 + n - j + 2, \ldots, -\gamma_s + n - j + s$. We get a sequence of positive numbers $P(\alpha, \gamma)$, which we reorder to get a decreasing sequence $R(\alpha, \gamma)$. Each exchange of neighboring elements in this process and changing of the sign of the last number contribute 1 to $N(\alpha, \gamma)$. Notice that in the sequence $P(\alpha, \gamma)$ the numbers from the positions of $\gamma_1, \ldots, \gamma_s$ occupy the last s spots. We also notice that $\alpha_1 + n - j, \ldots, \alpha_{n-j} + 1$ as well as $-\gamma_{s+1} + (n - j) + s + 1, \ldots, -\gamma_j + n$ are all numbers $\leq n$. Therefore there is a unique $\gamma_1, \ldots, \gamma_s$ for which the resulting sequence is $(n, n - 1, \ldots, 1)$. We claim that the minimum of $N(\alpha, \gamma)$ is achieved for such γ.

Indeed, assume that some of the numbers in $R(\alpha, \gamma)$ are bigger than n. Let u be the smallest number not occurring in $R(\alpha, \gamma)$. We have $u \leq n$. Let m be the smallest number $\geq u$ that occurs in $R(\alpha, \gamma)$ and comes from the positions corresponding to $\gamma_1, \ldots, \gamma_s$, say from γ_t. Let $u = m - l$. Then there exists a unique partition δ such that $\delta \subset \gamma$, $|\delta| = |\gamma| - l$, $R(\alpha, \delta) = R(\alpha, \gamma) \cup \{u\} \setminus \{m\}$. Here δ is obtained from γ in at most l steps. Each step consists of decreasing a part of γ by 1 (which forces some of the following parts to also decrease so we again get a partition), starting with γ_t. At each stage the new repetition in the reordered sequence appears. This repetition has to involve a position coming from $\gamma_1, \ldots, \gamma_s$, so we have to decrease this part of γ next, and so on. We know that $|\delta| = |\gamma| - l$. Comparing the reordering processes for (α, γ) and (α, δ), we see that $l(\alpha, \delta) \geq l(\alpha, \gamma) - l + 1$, and the decrease in the number of exchanges is equal to the number of entries $\{m - 1, \ldots, u + 1\}$ in $R(\alpha, \gamma)$ that did not come from positions corresponding to $-\gamma_s + n - j + s, \ldots, -\gamma_1 + n - j + 1$. Thus $N(\alpha, \delta) \leq N(\alpha, \gamma) - 1$, as desired.

Assume that (α, γ) is such that the reordered sequence is $(n, n - 1, \ldots, 1)$.

We shall prove that for fixed α and for such γ for which $R(\alpha, \gamma) = (n, n - 1, \ldots, 1)$, the partition γ with minimal $N(\alpha, \gamma)$ is the one with $s = 0$. Indeed, let us assume that $s > 0$. Consider the sequence

$$(-\gamma_j + n, -\gamma_{j-1} + n - 1, \ldots, -\gamma_1 + n - j + 1, \alpha_1 + n - j, \ldots, \alpha_{n-j} + 1).$$

We construct a new partition δ as follows. We modify the above sequence by changing the negative entry $-\gamma_s + n - j + s$ to its negative $x = \gamma_s - n + j - s > 0$. Then we order the positive part of this sequence and see that it can be written as

$$(-\delta_j + n, -\delta_{j-1} + n - 1, \ldots, -\delta_1 + n - j + 1, \alpha_1 + n - j, \ldots, \alpha_{n-j} + 1)$$

for a partition δ with $|\delta| = |\gamma| - 2x$. The partition δ has smaller invariant s than γ has.

We count the number $l(\alpha, \gamma) - l(\alpha, \delta)$. The difference is that the number $-\gamma_s + n - j + s$ gets exchanged twice with numbers from the set $\alpha_1 + n - j, \ldots, \alpha_{n-j} + 1$ that are in $[1, x - 1]$, then it is reflected to its negative, and then it gets exchanged additionally with the numbers from the interval $[1, x - 1]$ that preceed it. Therefore we have $l(\alpha, \gamma) - l(\alpha, \delta) \leq 2x - 1$ and thus

$$N(\alpha, \gamma) > N(\alpha, \delta).$$

It remains to show that for fixed α and for partitions γ such that $R(\alpha, \gamma) = (n, n - 1, \ldots, 1)$ and $s = 0$ we have $N(\alpha, \gamma) \geq 0$. But now we see that the last reflection does not occur in exchanging entries. Therefore we are in the situation of Lemma (8.1.5). We see that the choice of γ minimizing $N(\alpha, \gamma)$ is $\gamma = \alpha'$ and that $N(\alpha, \alpha') = 0$. This proves Lemma (8.5.9) and therefore part (a) of Proposition (8.5.8). ■

We prove part (b) of Proposition (8.5.8). We use the exact sequence (∗∗). We will prove that the cohomology of

$$\bigwedge{}^{\bullet}(S_2 \mathcal{R}) \otimes \bigwedge{}^{\bullet}(\mathcal{R} \otimes (F/\mathcal{R})) \otimes V_\alpha(\mathcal{R}^\vee/\mathcal{R})$$

does not contain the representations $V_\beta F$ with $\beta_1 > j$. Using the exact sequence

$$0 \to \mathcal{R}^\vee/\mathcal{R} \to F/\mathcal{R} \to F/\mathcal{R}^\vee \to 0,$$

we see that it is enough to show that the cohomology of

$$\bigwedge{}^{\bullet}(S_2 \mathcal{R}) \otimes \bigwedge{}^{\bullet}(\mathcal{R} \otimes (\mathcal{R}^\vee/\mathcal{R})) \otimes \bigwedge{}^{\bullet}(\mathcal{R} \otimes (F/\mathcal{R}^\vee)) \otimes V_\alpha(\mathcal{R}^\vee/\mathcal{R})$$

does not contain the representations $V_\beta F$ with $\beta_1 > j$.

Let us look at a typical term

$$K_\beta \mathcal{R} \otimes K_\gamma \mathcal{R} \otimes K_{\gamma'}(\mathcal{R}^\vee/\mathcal{R}) \otimes K_\delta \mathcal{R} \otimes K_{\delta'}(F/\mathcal{R}^\vee) \otimes V_\alpha(\mathcal{R}^\vee/\mathcal{R}).$$

After decomposing the Weyl functors to the symplectic irreducibles and calculating tensor products, we see that the term above will decompose to the summands of the form

$$K_\lambda(F/\mathcal{R}^\vee) \otimes V_\theta(\mathcal{R}^\vee/\mathcal{R}).$$

Such a bundle has a representation $V_\beta F$ with $\beta_1 > j$ in cohomology if a sequence

$$(\lambda_1 + n, \lambda_2 + n - 1, \ldots, \lambda_j + n - j + 1, \theta_1 + n - j, \ldots, \theta_{n-j} + 1)$$

contains either a number $> n + j$ or a number $< -n - j$. Since θ is an integral dominant weight for $Sp(2n - 2j)$, i.e. a partition, this can happen if one of the following cases occurs:

(a) $\lambda_1 + n > n + j$, i.e. $\lambda_1 > j$,
(b) $\theta_1 + n - j > n + j$, i.e. $\theta_1 > 2j$,
(c) $\lambda_j + n - j + 1 < -n - j$, i.e. $-\lambda_j > 2n + 1$.

Case (a) cannot occur because $K_\lambda(F/\mathcal{R}^\vee)$ is a summand in a tensor product

$$K_\beta \mathcal{R} \otimes K_\gamma \mathcal{R} \otimes K_\delta \mathcal{R} \otimes K_{\delta'}(F/\mathcal{R}^\vee)$$

and the only possible positive weights come from δ'. However, $\delta'_1 \leq j$, since $\dim \mathcal{R} = j$.

Case (b) cannot occur because the weight θ comes from a summand in

$$K_{\gamma'}(\mathcal{R}^\vee/\mathcal{R}) \otimes V_\alpha(\mathcal{R}^\vee/\mathcal{R}),$$

and $\alpha_1 \leq j$ and $\gamma'_1 \leq j$ (because the corresponding functor $K_\gamma \mathcal{R}$ is nonzero). Thus $K_{\gamma'}(\mathcal{R}^\vee/\mathcal{R})$ decomposes to the representations $V_\epsilon(\mathcal{R}^\vee/\mathcal{R})$, and in the tensor product

$$V_\epsilon(\mathcal{R}^\vee/\mathcal{R}) \otimes V_\alpha(\mathcal{R}^\vee/\mathcal{R})$$

all weights have to be $\leq \alpha + \epsilon$ by Klimyk's formula (exercise 12 of chapter 4), and therefore the first entry of each occurring weight is $\leq 2j$.

Case (c) cannot occur because in the tensor product

$$K_\beta \mathcal{R} \otimes K_\gamma \mathcal{R} \otimes K_\delta \mathcal{R} \otimes K_{\delta'}(F/\mathcal{R}^\vee)$$

the last entries are $-\beta_1 \geq -j - 1$, $-\gamma_1 \geq 2j - 2n$, and $-\delta_1 \geq -j$ (because $K_{\delta'}(F/\mathcal{R}^\vee)$ has to be nonzero).

This concludes the proof of part (b) of Proposition (8.5.8).•

(8.5.10) Corollary.

(a) *The term $F_\bullet$ consists of one copy of the trivial representation in homogeneous degree 0. In particular the variety Z_j is normal, with rational singularities.*

(b) *The terms F_1 contains only the representations $V_\beta(F)$ with $\beta_1 \leq 1$. In particular the defining ideal of $Y_j := q(Z_j)$ is generated by such representations. More precisely, the minimal set of generators of the defining ideal of Y_j consists of the vectors in the representation*

$$\bigwedge^{2(n-j+1)} F \subset K_{(2^{2(n-j+1)})} F \subset S_{2(n-j+1)}(S_2 F).$$

Proof. The first claim follows from the proof of Lemma (8.5.9). Indeed, the only term contributing to F_0 corresponds to $\gamma = 0$, which, as seen from the sequences (?), can appear only for $m = 0$.

The proof of the second part is based on the refinement of Lemma (8.5.9).

Claim. *The only partition γ with $|\gamma| = m$ such that $H^{m-1}(\mathrm{IGrass}(j, F), K_\gamma \mathcal{R}) \neq 0$ is $\gamma = (2(n - j + 1))$, and the resulting cohomology group is a trivial representation of $\mathrm{Sp}(F)$.*

Proof. We repeat the proof of Lemma (8.5.9) with $\alpha = (0)$. Our partition γ has to lead in one step to the partition $\gamma = 0$, which gives the only partition with $N(\gamma, (0)) = 0$. The sequence $R(\gamma, (0))$ has to be $(n, n - 1, \ldots, 1)$, because if not, then $s > 0$ and the first step in the procedure does decrease s. We see now that $s = 1$; otherwise we need s steps to reduce to $\gamma = 0$. Now we can do a case by case analysis, because the number of possibilities for γ corresponds to the possibilities for the number $\gamma_1 - n + j - 1 \in \{n, n - 1, \ldots, n - j + 1\}$. The only viable possibility turns out to be $\gamma_1 - n + j - 1 = n - j + 1$, which implies $\gamma_t = 0$ for $t \geq 2$. ∎

Now we see using the exact sequence (?) resolving $\bigwedge^m S_j$ that the only possible contribution to F_1 is $\bigwedge^{2(n-j+1)} F$ in homogeneous degree $2(n - j + 1)$. We easily identify this set of equations with the ones given in (8.5.10)(b), because they vanish on Y_j.

We can finally state the results of our induction.

(8.5.11) Theorem. *Let μ be an even partition, $\mu = 2\nu$.*

(a) *The orbit closure Y_μ is normal, with rational singularities,*

(b) *The irreducible representations occurring in the terms of the complex*
$F^\mu_\bullet$ *have highest weights* $(\beta_1, \ldots, \beta_n)$ *with* $\beta_1 \leq \nu'_1$.

(c) *The term* F^μ_1 *contains only the representations* $V_\beta(F)$ *with* $\beta_1 \leq 1$.
Therefore the defining ideal of Y_μ *is generated by the representations
of this kind.*

Proof. We use the induction described above, connecting the relative complex
$F^{\hat\mu}_\bullet$ to the complex $F^\mu_\bullet$. We use the exact sequence

$$0 \to \mathcal{S}_{\nu'_1} \to \mathcal{S}_\nu \to \mathcal{S}_{\hat\nu}(\mathcal{R}^\vee/\mathcal{R}) \to 0$$

to estimate the terms in the cohomology of $\mathcal{S}_\nu$. They come from the terms of
the cohomology in $V_\alpha(\mathcal{R}^\vee/\mathcal{R}) \otimes \bigwedge^\bullet(\mathcal{S}_{\nu'_1})$. By induction the terms $V_\alpha(\mathcal{R}^\vee/\mathcal{R})$
satisfy the assumption of Proposition (8.5.8), which shows that the i-th term
of the complex $F^{\hat\mu}(\mathcal{R}^\vee/\mathcal{R})_\bullet$ can produce only the terms in the homological
degree $\geq i$. Thus, by induction, only the terms in nonnegative homological
degree occur, and by Corollary (8.5.10) the term F^μ_0 consists of one copy of
the trivial representation in homogeneous degree 0. Parts (a) and (b) of the
theorem follow.

To prove part (c) let us analyze the term F^μ_1. The terms in this complex come from the term $F_\bullet$, which has the required form by Corollary
(8.5.10)(b), and from the terms $F^{\hat\mu}_1(\mathcal{R}^\vee/\mathcal{R})$, which by induction are the terms
of homological degree zero in $F(V_{(\alpha_1,\ldots,\alpha_{n-j})}(\mathcal{R}^\vee/\mathcal{R}))_\bullet$, with weight α for
which $\alpha_1 \leq 1$. By the proof of Lemma (8.5.9) the only such term comes
from $\gamma = \alpha'$, so (by the use of the complexes $V_\alpha(\mathcal{R}^\vee/\mathcal{R}) \otimes (?)$) the resulting contribution to the possible terms of $F(V_{(\alpha_1,\ldots,\alpha_{n-j})}(\mathcal{R}^\vee/\mathcal{R}))_\bullet$ is $K_\alpha F \otimes
V_{(0)}F$, which decomposes to the representations of $\mathrm{Sp}(F)$ of the required
form. ∎

Let me finish this section with a general conjecture regarding the equations
of the nilpotent orbit closures for the symplectic group.

(8.5.12) Conjecture. *Let* $\mu \in P_s(2n)$. *The defining ideal of* Y_μ *is generated
by representations* $V_{(\beta_1,\ldots,\beta_n)}(F)$ *with* $\beta_1 \leq 2$. *More precisely, the generators
of the defining ideal of* Y_μ *can be chosen as subrepresentations of Schur
functors* $K_{(2^i)}F$ *inside* $S_i(S_2F)$ *for* $1 \leq i \leq 2n$.

(8.5.13) Remark. *Klimek has proved that Conjecture (8.5.12) is true up to a
radical.*

Exercises for Chapter 8

Orbits Corresponding to Special Partitions for Classical Groups

Type B_n.

We assume that F is a vector space of dimension $2n + 1$ and that $\langle \ , \ \rangle$ is a nondegenerate symmetric form. We identify F with F^* by means of the morphism induced by the form.

1. Consider the adjoint representation $X = \bigwedge^2 F$. We identify A with $\text{Sym}(\bigwedge^2 F)$. Show that the space $K_{(2^{2i})}F$ ($1 \leq i \leq n$) contains a unique (up to a scalar) $\text{SO}(F)$-invariant v_{2i}. Prove that the ring of invariants $A^{\text{SO}(F)}$ is isomorphic to the polynomial ring generated by the invariants $v_2, \ldots, v_{2n}$. This is a special case of Chevalley's theorem.

2. Let F be a vector space of dimension $2n + 1$, and let $\langle \ , \ \rangle$ be a nondegenerate symmetric form. We take $\underline{g} = \underline{\text{so}}(2n + 1) = \bigwedge^2 F$. The subregular orbit in $\underline{g}$ is the only orbit of codimension 2 in the nilpotent cone. It corresponds to the partition $\mu = (2n - 1, 1, 1)$. Let $\{e, h, f\}$ be an $\underline{\text{sl}}_2$-triple where e is an element from the subregular orbit. We denote by $\underline{g}_i$ the component of weight i in $\underline{g}$ considered as a representation of $\underline{\text{sl}}_2$.

   (a) Let ξ_n be the vector bundle occurring in the resolution (5.1.1). We saw in the section 8.4 that we have $p^*(\xi_n) = \underline{g}^*_{<2}$. Prove that we have an exact sequence

   $$0 \to \mathcal{R}_1 \otimes F/\mathcal{R}_1 \to \xi_n \to \xi_{n-1}(\mathcal{R}_1^\vee/\mathcal{R}_1) \to 0$$

   where $\mathcal{R}_1$ is a tautological subbundle on the isotropic Grassmannian $\text{IGrass}(1, F)$ (which is actually just a quadric in the projective space of lines in F), and $\xi_{n-1}(\mathcal{R}_1^\vee/\mathcal{R}_1)$ is the bundle corresponding to the subregular orbit in the orthogonal space $\mathcal{R}_1^\vee/\mathcal{R}_1$, in relative situation.

   (b) Use the above exact sequence and the induction on n (starting with $n = 2$) to prove that the only representations occurring in the minimal free resolution of A/J_μ are the trivial representation and the representation $V_1 F$,

   (c) Prove that the orbit closure Y_μ is normal with rational singularities.

3. Prove that the orbit closure $Y_{(2n-1,1,1)}$ of the subregular orbit $O(2n - 1, 1, 1)$ is a complete intersection in the variety Y^a_{2n-2} of skew-symmetric tensors of rank $\leq 2n - 2$, defined by vanishing of $v_2, \ldots, v_{2n-2}$. Show that the minimal resolution of $A/J_{(2n-1,1,1)}$ as an A-module is the

tensor product of Buchsbaum–Eisenbud resolution (6.4.4) and the Koszul complex on $v_2, v_4, \ldots, v_{2n-2}$.

4. Let $\mu = (2r + 1, 1^{2(n-r)})$. Prove that the closure Y_μ is a complete intersection in the skew-symmetric rank variety Y_{2r}^a given by vanishing of $v_2, \ldots, v_{2r}$.

Type C_n.

We assume that F is a vector space of dimension $2n$ and that $\langle \, , \, \rangle$ is a nondegenerate skew-symmetric form. We identify F with F^* by means of the morphism induced by the form.

5. Consider the adjoint representation $X = S_2 F$. We identify A with Sym $(S_2 F)$. Show that the space $K_{(2^{2i})} F$ $(1 \le i \le n)$ contains a unique (up to a scalar) $SP(F)$-invariant v_{2i}. Prove that the ring of invariants $A^{SP(F)}$ is isomorphic to a polynomial ring generated by the invariants $v_2, \ldots, v_{2n}$. This is a special case of Chevalley's theorem.

6. We take $g = \mathrm{sp}(2n) = S_2 F$. The subregular orbit in g is the only orbit of codimension 2 in the nilpotent cone. It corresponds to the partition $\mu = (2n - 2, 2)$. Let $\{e, h, f\}$ be an sl_2-triple where e is an element from the subregular orbit. We denote by g_i the component of weight i in g considered as a representation of sl_2.

   (a) Let ξ_n be the vector bundle occurring in the resolution (5.1.1). We saw in section 8.5 that we have $p^*(\xi_n) = g^*_{<2}$. Prove that we have an exact sequence

   $$0 \to \mathcal{R}_1 \otimes F \to \xi_n \to \xi_{n-1}(\mathcal{R}_1^\vee / \mathcal{R}_1) \to 0$$

   where $\mathcal{R}_1$ is a tautological subbundle on the isotropic Grassmannian $\mathrm{IGr}(1, F)$ (which is actually just a projective space of lines in F), and $\xi_{n-1}(\mathcal{R}_1^\vee / \mathcal{R}_1)$ is the bundle corresponding to the subregular orbit in the symplectic space $\mathcal{R}_1^\vee / \mathcal{R}_1$, in relative situation.

   (b) Use the above exact sequence and induction on n (starting with $n = 2$) to prove that the only representations occurring in the minimal free resolution of A/J_μ are the trivial representation and $V_{1,1} F$,

   (c) Prove that the orbit closure $Y_{(2n-2,2)}$ is normal with rational singularities.

7. Define $Y \subset X$ to be the subvariety

   $$Y = \{\phi \in X \mid \exists R \in \mathrm{IGrass}(2, F), \ \phi(v, w) = 0 \ \forall v \in R, \ w \in X\}.$$

The variety Y has a natural resolution of singularities

$$Z = \{(\phi, R) \in X \times \text{IGrass}(2, F) \mid \phi(v, w) = 0 \;\forall v \in R, w \in X \}.$$

Calculate the resulting complex $F_\bullet$, using the fact that the bundle ξ can be identified with the bundle S_2 used in section 8.5, i.e., it satisfies the exact sequence

$$0 \to \xi \to R \otimes F \to \overset{2}{\bigwedge} R \to 0$$

where R denotes the tautological subbundle on $\text{IGrass}(2, F)$. Prove that Y is normal, with rational singularities and the codimension of Y equals 4. Show that the subregular orbit closure $Y_{(2n-2,2)}$ is a complete intersection in $K[Y]$ given by vanishing of $v_2, \ldots, v_{2n-4}$. The minimal resolution of $A/J_{(2n-2,2)}$ is therefore a tensor product of the minimal resolution of $K[Y]$ and the Koszul complex on $v_2, \ldots, v_{2n-4}$.

8. Let $\mu = (2r, 1^{2(n-r)})$. Prove that Y_μ is a complete intersection in the symmetric rank variety Y^s_{2r-1} given by the vanishing of $v_2, \ldots, v_{2r-2}$.

Type D_n.

We assume that F is a vector space of dimension $2n$ and that $\langle \ , \ \rangle$ is a nondegenerate symmetric form. We identify F with F^* by means of the morphism induced by the form.

9. Consider the adjoint representation $X = \bigwedge^2 F$. We identify A with $\text{Sym}(\bigwedge^2 F)$. Show that the space $K_{(2^{2i})}F$ ($1 \le i \le n$) contains a unique (up to a scalar) $SO(F)$-invariant v_{2i}. Let w_n be the Pfaffian, i.e. the invariant occurring as the subspace $\bigwedge^{2n} F \subset S_n(\bigwedge^2 F)$. Prove that the ring of invariants $A^{SO(F)}$ is isomorphic to a polynomial ring generated by the invariants $v_2, \ldots, v_{2n-2}$ and the invariant w_n. Again this is a special case of Chevalley's theorem.

10. We take $g = \underline{so}(2n) = \bigwedge^2 F$. The subregular orbit in g is the only orbit of codimension 2 in the nilpotent cone. It corresponds to the partition $\mu = (2n - 3, 3)$. Let $\{e, h, f\}$ be an $\underline{sl}_2$-triple where e is an element from the subregular orbit. We denote by $\underline{g}_i$ the component of weight i in $\underline{g}$ considered as a representation of $\underline{sl}_2$.

   (a) Let ξ_n be the vector bundle occurring in the resolution (5.1.1). We saw in section 8.5 that we have $p^*(\xi_n) = \underline{g}^*_{<2}$. Prove that we have an exact sequence

   $$0 \to R_1 \otimes F/R_1 \to \xi_n \to \xi_{n-1}(R_1^\vee/R_1) \to 0$$

where $\mathcal{R}_1$ is a tautological subbundle on the isotropic Grassmannian IGr$(1, F)$ (which is actually just a quadric in the projective space of lines in F), and $\xi_{n-1}(\mathcal{R}_1^{\vee}/\mathcal{R}_1)$ is the bundle corresponding to the subregular orbit in the orthogonal space $\mathcal{R}_1^{\vee}/\mathcal{R}_1$, in relative situation.

(b) Use the above exact sequence and the induction on n (starting with $n = 2$) to prove that the only representations occurring in the minimal free resolution of A/J_{μ} are the trivial representation and the adjoint representation $V_{1,1} F = \bigwedge^2 F$.

(c) Prove that the orbit closure Y_{μ} is normal with rational singularities.

11. Let $\mu = (2r + 1, 1^{2(n-r)-1})$. Prove that the closure Y_{μ} is a complete intersection in the skew symmetric rank variety Y_{2r}^a given by the vanishing of $v_2, \ldots, v_{2r}$.

9

Resultants and Discriminants

This chapter is devoted to the applications of our methods to generalized discriminants and resultants. Here the nature of the applications is twofold. First, the geometric method is useful for knowing in which cases the discriminant type variety has the expected dimension, i.e., it has codimension in the ambient space.

In such cases one can look at the twisted complexes $F(\mathcal{L})_\bullet$ and try to classify those that have only two nonzero terms. The point is that in such case the matrix giving a differential of $F(\mathcal{L})_\bullet$ gives a determinantal expression of the defining variety, i.e. the resultant or discriminant in question. We call such complexes $F(\mathcal{L})_\bullet$ the *determinantal complexes*.

In section 9.1 we give the basic definitions regarding resultants. In section 9.2 we apply our methods to classify (following [WZ]) the determinantal complexes for the resultants of sets of multihomogeneous polynomials.

In section 9.3 we give the basic definitions regarding discriminants. Finally in section 9.4 we consider the case of discriminants of multilinear forms – so-called hyperdeterminants. We give several examples of determinantal complexes, but we also show that for the multilinear forms of general format the determinantal complexes do not exist.

We conclude that section with the analysis of formats of three dimensional matrices for which the hyperdeterminantal variety has codimension bigger that one. In some cases we can investigate normality, Cohen–Macaulayness, and the rational singularities property.

The vector bundles occurring in the Koszul complexes are quite interesting. They are the extensions of a nice tensor product by a line bundle. This makes the analysis more complicated than in the case of determinantal varieties, but much simpler than in the case of nilpotent orbit closures.

9.1. The Generalized Resultants

Let E be a vector space of dimension n over a field $\mathbf{K}$. We consider the affine space X of linear functions on E. We identify X with the space E^*. Let Y be the closed subvariety of X which is a cone over an irreducible projective variety Y' of dimension l in $X' = P(E^*)$. We denote by Y'_0 the set of smooth points in Y'. We assume that Y' is linearly normal and nondegenerate, i.e., the natural map $E^* \to H^0(Y', \mathcal{O}_{Y'})$ is an isomorphism.

The *resultant variety* $\nabla_{Y'}$ is defined as follows. Let

$$\nabla_{Y'}^0 = \{(g_1, \ldots, g_{l+1}) \in E^{l+1} \mid \exists_{y \in Y'_0} \ g_1(y) = \cdots = g_{l+1}(y) = 0\}.$$

Then $\nabla_{Y'}$ is the closure of $\nabla_{Y'}^0$ in E^{l+1}.

We consider the incidence variety Z. We start with

$$Z_0 = \{((g_1, \ldots, g_{l+1}), y) \in E^{l+1} \times Y'_0 \mid g_1(y) = \cdots = g_{l+1}(y) = 0\}.$$

The variety Z is the closure of Z_0 in $E^{l+1} \times Y'$. We denote by q, p the projections of $E^{l+1} \times Y'$ to E^{l+1}, Y' respectively. The restrictions of q and p to Z will be denoted by q' and p'.

(9.1.1) Proposition.
  (a) The variety $\nabla_{Y'}$ has codimension 1 in E^{l+1}.
  (b) The map q' is a birational isomorphism.

Proof. Let us consider the projection p'. It is clear that over the open set Y'_0 the dimension of the fiber equals $(l+1)(n-1)$. Therefore dim $Z = (l+1)(n-1) + l = (l+1)n - 1$. By definition the image $q'(Z) = \nabla$. Since Z_0 and Z are irreducible, it is enough to show that the dimension of a fiber of q' over a generic point is zero. This is clear because we can choose generic $g_1, \ldots, g_l$ so the number of points $y \in Y'$ such that $g_1(y) = \cdots = g_l(y) = 0$ is finite and contained in Y'_0, and then choose g_{l+1} to be a linear form vanishing at one of such points. Then the fiber of q' over $(g_1, \ldots, g_{l+1})$ is finite, and the first statement is proved.

To prove the second statement we observe that by choosing $g_1, \ldots, g_l$ more carefully we can assume that additionally the set $\{y \in Y' \mid g_1(y) = \cdots = g_l(y) = 0\}$ has exactly deg Y' points, and then we can choose g_{l+1} to vanish at precisely one of these points. Therefore the generic fiber of q' has one point. This means that there exists an open set U in $\nabla_{Y'}$ such that if the point $(g_1, \ldots, g_{l+1})$ is in U, then there is precisely one point $y \in Y'$ such that $g_1(y) = \cdots = g_{l+1}(y) = 0$. Sending the point $(g_1, \ldots, g_{l+1})$ to $((g_1, \ldots, g_{l+1}), y)$ gives the inverse of $q'|_U$. This proves the second statement. ∎

We define the *resultant of Y'* to be the irreducible equation $Res(Y')$ of ∇' in E^{l+1}. The resultant is a multihomogeneous polynomial in the coefficients of the forms $g_1, \ldots, g_{l+1}$.

Let E_0^{l+1} denote the subset of E^{l+1} consisting of the $(l+1)$-tuples $(g_1, \ldots, g_{l+1})$ spanning the subspace of dimension $l+1$ in E. Then we have the natural map

$$h : E_0^{l+1} \longrightarrow \mathrm{Grass}(l+1, E)$$

sending the $(l+1)$-tuple $(g_1, \ldots, g_{l+1})$ to the subspace spanned by the g_i's. If $e_1, \ldots, e_n$ is a basis of E and if

$$l_i = \sum_{j=1}^n a_{ij} e_j \qquad \text{for} \quad i = 1, \ldots, l+1,$$

then the homogeneous coordinates on $\mathrm{Grass}(l+1, E)$ are the maximal minors of the $(l+1) \times n$ matrix $\mathcal{A} = (a_{ij})$. For a subset $I \subset [1, n]$, card $I = l+1$, we will denote by M_I the minor of $\mathcal{A}$ corresponding to columns from I.

The condition $(g_1, \ldots, g_{l+1}) \in \nabla_{Y'}$ depends on the subspace spanned by $g_1, \ldots, g_{l+1}$, not by the forms themselves. This means that there exists an irreducible hypersurface $\mathrm{Ch}(Y')$ in $\mathrm{Grass}(l+1, E)$ such that $\nabla_{Y'} \cap E_0^{l+1} = h^{-1} \mathrm{Ch}(Y')$. Denoting by $\Xi_{Y'}$ the irreducible equation of $\mathrm{Ch}(Y')$ in the homogeneous coordinate ring of $\mathrm{Grass}(l+1, E)$, we see that

$$Res(Y')(a_{ij}) = \Xi'(M_I),$$

where Ξ' is any representative of $\Xi_{Y'}$ in the polynomial ring $\mathrm{Sym}(\bigwedge^{l+1}(E^*))$ Therefore $Res(Y')$ is a polynomial in maximal minors of the matrix $\mathcal{A}$. This means that its degree with respect to coefficients of g_i does not depend on i.

Let us mention that the hypersurface $\mathrm{Ch}(Y')$ in $\mathrm{Grass}(l+1, E)$ is called the Chow *cycle of Y'* and the equation $\Xi_{Y'}$ is the Chow *form of Y'*.

We want to apply our technique to the variety Z. By definition we see that Z is given locally by $l+1$ equations in $E^{l+1} \times Y'$. Let us consider the projection $p' : Z \longrightarrow Y'$. The fiber of p' over a point $y \in Y'$ is the set of $(l+1)$-tuples $(g_1, \ldots, g_{l+1})$ from E^{l+1} such that $g_1(y) = \cdots = g_{l+1}(y) = 0$. Let us consider the bundle $\mathcal{O}_{Y'}(1)$. By definition it is a factorbundle of the trivial bundle $E \times Y'$ by the subbundle $\mathcal{R}_{Y'} = \{(g, y) \in E \times Y' | g(y) = 0\}$. Therefore we see that Z can be identified with the vector bundle $\mathcal{Q}_{Y'}^{l+1}$, where $\mathcal{Q}_{Y'} = \mathcal{R}_{Y'}^*$.

We are in a position to apply the general approach from chapter 5. Let us compare the notation. In our application $V = Y', \mathcal{S} = \mathcal{R}_{Y'}^{l+1}, \mathcal{T} = \mathcal{O}_{Y'}(1)^{l+1}$, and therefore $\xi = \mathcal{O}_{Y'}(-1)^{l+1}$ and $\eta = \mathcal{Q}_{Y'}^{l+1}$.

Before we go further let us introduce some notation. The author is simply tired of writing the $(l + 1)$st powers of all the bundles. In order to avoid this problem we identify the space E^{l+1} with the tensor product $W \otimes E$, where W is the space of dimension $l + 1$ over $\mathbf{K}$. The $(l + 1)$-tuple $g_1, \ldots, g_{l+1}$ corresponds to the tensor $\sum_{i=1}^{l+1} w_i \otimes g_i$ where $w_1, \ldots, w_{l+1}$ is a fixed basis of W. Notice however that none of our constructions depend on the choice of a basis $w_1, \ldots w_{l+1}$, because the condition $(g_1, \ldots, g_{l+1}) \in \nabla_{Y'}$ depends only on the subspace spanned by $g_1, \ldots, g_{l+1}$.

In our new notation we can write $\mathcal{S} = W \otimes \mathcal{R}_{Y'}$, $\mathcal{T} = W \otimes \mathcal{O}_{Y'}(1)$, and therefore $\xi = W \otimes \mathcal{O}_{Y'}(-1)$ and $\eta = W \otimes \mathcal{Q}_{Y'}$. Let us also recall that the coordinate ring of $W \otimes E$ is identified with $A = \mathbf{K}[a_{ij}] = \text{Sym}(W^* \otimes E^*)$, where the variable a_{ij} corresponds to the tensor $e_j^* \otimes w_i^*$.

Let us restate Theorem (5.1.2) in this setting.

(9.1.2) Basic Theorem for Resultants. *For a vector bundle $\mathcal{V}$ on Y' we define free graded A-modules*

$$F(\mathcal{V})_i = \bigoplus_{j \geq 0} H^j \left(Y', \bigwedge^{i+j} W \otimes \mathcal{O}_{Y'}(-i - j) \otimes \mathcal{V} \right) \otimes_{\mathbf{K}} A(-i - j).$$

(a) There exist minimal differentials

$$d_i(\mathcal{V}) : F(\mathcal{V})_i \to F(\mathcal{V})_{i-1}$$

of degree 0 such that $F(\mathcal{V})_{\bullet}$ is a complex of graded free A-modules with

$$H_{-i}(F(\mathcal{V})_{\bullet}) = \mathcal{R}^i q'_* M(\mathcal{V}).$$

In particular the complex $F(\mathcal{V})_{\bullet}$ is exact in positive degrees.

(b) The sheaf $\mathcal{R}^i q_* M(\mathcal{V})$ is equal to $H^i(Z, M(\mathcal{V}))$, and it can be also identified with the graded A-module $H^i(V, \text{Sym}(W^* \otimes \mathcal{O}_{Y'}(1)) \otimes \mathcal{V})$.

(c) If $\phi : M(\mathcal{V}) \to M(\mathcal{V}')(m)$ is a morphism of graded sheaves, then there exists a morphism of complexes

$$f_{\bullet}(\phi) : F(\mathcal{V})_{\bullet} \to F(\mathcal{V}')_{\bullet}(m).$$

Its induced map $H_{-i}(f_{\bullet}(\phi))$ can be identified with the induced map

$$H^i(Z, M(\mathcal{V})) \to H^i(Z, M(\mathcal{V}'))(m).$$

We want to apply the complexes $F(\mathcal{L})_{\bullet}$ to find useful expressions for the resultant $Res(Y')$. Let us recall (cf. section 1.3) that we can associate to the complex $F(\mathcal{V})_{\bullet}$ its determinant. Its relation to the resultant is expressed in the following

(9.1.3) Proposition. *The determinant of the complex $F(\mathcal{V})_{\bullet}$ is equal to Res* $(Y')^{\dim \mathcal{V}}$.

Proof. The determinant of a complex $F(\mathcal{V})_{\bullet}$ is by definition a rational function having zeros and poles only at the points where $F(\mathcal{V})_{\bullet}$ is not exact. In our case this is the irreducible variety $\nabla_{Y'}$. Therefore the determinant of $F(\mathcal{V})_{\bullet}$ is a power of $Res(Y')$.

To compute this power let us consider the graded Hilbert function

$$\mathcal{P}(F(\mathcal{V})_{\bullet}, t) = \sum_{i,j \geq 0} (-1)^i t^{i+j} h^j \left(V', \mathcal{V} \otimes \bigwedge^{i+j} \xi \right) (1-t)^{-l-1}$$

of the complex $F(\mathcal{V})_{\bullet}$. Let us write $\mathcal{P}(F(\mathcal{V})_{\bullet}, t) = \sum_{a \geq 0} \mathcal{P}(a) t^a$. Then for $a \gg 0$, $\mathcal{P}(a)$ is a polynomial in a. By (5.1.6) we see that the degree of the determinant of $F(\mathcal{V})_{\bullet}$ equals the coefficient of $\mathcal{P}(a)$ in degree $n(l+1) - 1$ divided by $((l+1)n - 1)!$. By (9.1.2) we see that $\mathcal{P}(F(\mathcal{V})_{\bullet}, t)$ is the alternating sum of graded Hilbert functions of $\mathcal{R}^i q'_* M(\mathcal{V})$. Since all modules $\mathcal{R}^i q'_* M(\mathcal{V})$ are supported in $\nabla_{Y'}$, the degree of $\mathcal{P}(a)$ is $\leq n(l+1) - 1$.

Let us analyze the support of modules $\mathcal{R}^i q'_* M(\mathcal{V})$. For $i > 0$ this is a variety properly contained in $\nabla_{V'}$, because on the open set in Z the map q' is an isomorphism. Therefore for $i > 0$ the modules $\mathcal{R}^i q'_* M(\mathcal{V})$ have support of codimension ≥ 2 in E^{l+1}. The module $\mathcal{R}^0 q'_* M(\mathcal{V})$ is a module supported on $\nabla_{Y'}$, which has rank $\dim \mathcal{V}$. This means that the degree of the determinant of $F(\mathcal{V})_{\bullet}$ is equal to $\dim \mathcal{V} \ \deg(Res(Y'))$, as claimed. ∎

Thanks to the proposition for each line bundle $\mathcal{L}$ on Y', the determinant of the complex $F(\mathcal{L})_{\bullet}$ is equal to $Res(Y')$. This means that the length of the complex $F(\mathcal{L})_{\bullet}$ measures the complexity of calculating a given resultant. The simplest case occurs when the complex $F(\mathcal{L})_{\bullet}$ has only two nonzero terms. Then $F(\mathcal{L})_{\bullet}$ becomes a matrix whose determinant is the resultant of Y'. This explains the following definition: The complex $F(\mathcal{L})_{\bullet}$ is *determinantal* if $F(\mathcal{L})_i = 0$ for $i \neq 0, 1$. In the next section we will classify the determinantal complexes in the case that Y' is the product of projective spaces.

Remark. *The determinantal expressions for various resultants were known classically. One can recall Sylvester's and Bezout's expressions for the resultants of two homogeneous polynomials. It turns out that the complexes $F(\mathcal{L})_{\bullet}$ provide the natural framework for constructing such expressions. In* *fact, all classical expressions known to us can be recovered as special cases of the construction using our complexes. Many new expressions can also be obtained. In fact it seems possible that this construction gives all possible expressions of this kind. This explains why such expressions are our main concern in the sections on resultants and discriminants.*

9.2. The Resultants of Multihomogeneous Polynomials

In this section we apply the results of section 9.1 to the projective spaces. We classify the complexes of type $F(\mathcal{L})_\bullet$ which give the determinantal expressions for the resultants.

Let $Y' = \mathbf{P}^l$. We identify Y' with the set $\mathbf{P}(F)$ of lines in the vector space F of dimension $l + 1$. Let $X_0, \ldots, X_l$ be the basis of F^*. Then the homogeneous coordinate ring of Y' is the polynomial ring $S = \mathbf{K}[X_0, \ldots, X_l]$. We denote by S_d the d-th homogeneous component of S. Denote by $\mathcal{O}(d)$ the d-th twisting sheaf of Serre on Y'. For $d > 0$ the sheaf $\mathcal{O}(d)$ defines the embedding of $\mathbf{P}^l$ into the projective space $X' = \mathbf{P}^{\binom{l+d-1}{l}-1}$, which can be identified with $\mathbf{P}(D_d F)$. The linear form on $\mathbf{P}(D_d F)$ is a homogeneous polynomial of degree d in the variables $X_0, \ldots, X_l$. We can now describe $\nabla_{\mathbf{P}^l}$ in the following way:

$$\nabla_{\mathbf{P}^l} = \{(f_1, \ldots, f_{l+1}) \in W \otimes S_d \mid \exists y \in \mathbf{P}^l \ f_1(y) = \ldots = f_{l+1}(y) = 0\},$$

where W, as in section 9.1, is the vector space of dimension $l + 1$ over $\mathbf{K}$. This means that $\nabla_{\mathbf{P}^l}$ is the *classical resultant variety* and its equation $Res(\mathbf{P}^l)$ is the *classical resultant of $l + 1$ homogeneous polynomials of degree d in $l + 1$ variables.*

Let $\mathcal{L}$ be a line bundle on $\mathbf{P}^l$. It follows from (4.1.3) that $\mathcal{L} = \mathcal{O}(m)$ for some $m \in \mathbf{Z}$. We denote the complex $F(\mathcal{L})_\bullet$ by $F(m)_\bullet$. Recall that the complex $F(m)_\bullet$ is *determinantal* if $F(m)_i = 0$ for $i \neq 0, 1$.

(9.2.1) Theorem. *The complex $F(m)_\bullet$ is determinantal for the resultants of the following types of forms: (a) linear forms, l arbitrary, $d = 1$, $m = -1$, $m = 0$, $m = 1$, (b) binary forms, $l = 1$, $d \geq 2$ arbitrary, $-1 \leq m \leq 2d - 1$, (c) ternary forms, $l = 2$, $d \geq 2$ arbitrary, $d - 2 \leq m \leq 2d - 1$, (d) quaternary forms, $l = 3$, $d \geq 2$ arbitrary, $2d - 3 \leq m \leq 2d - 1$, (e) quadratic forms in five variables, $l = 4$, $d = 2$, $m = 2, 3$, (f) cubic forms in five variables, $l = 4$, $d = 3$, $m = 5$, (g) quadratic forms in six variables, $l = 5$, $d = 2$, $m = 3$.*

Proof. Let us consider the complex $F(m)_\bullet$. We notice that the bundle $\mathcal{O}_{X'}(-1)|_{\mathbf{P}^l} = \mathcal{O}(-d)$ Therefore by (9.1.2) the terms of $F(m)_\bullet$ are given by the formula

$$F(m)_i = \sum_{j \geq 0} H^j(\mathbf{P}^l, \mathcal{O}(m - (i + j)d)) \otimes \bigwedge^{i+j} W \otimes_k A(-i - j).$$

We apply the Serre's theorem ([H1, chapter III, Theorem 5.1]) to figure out in which cases $F(m)_i = 0$ for $i \neq 0, 1$.

First we prove that the restrictions for m given in (a)–(g) are necessary. When $m \geq 2d$, then $H^0(\mathbf{P}^l, \mathcal{O}(m - 2d)) \neq 0$, which gives a contribution to $F(m)_2$. This gives us the condition $m \leq 2d - 1$. If $m - (l - 1)d \leq -l - 1$ then $H^l(\mathbf{P}^l, \mathcal{O}(m - (l - 1)d)) \neq 0$, which gives a contribution to $F_{-1}(m)$. Therefore $m - (l - 1)d \geq -l$, which means that $m \geq (l - 1)d - l$. We can write our restrictions

$$(l - 1)d - l \leq m \leq 2d - 1.$$

For $d = 1$ this obviously gives the restrictions from (a). Let us assume that $d \geq 2$. Then for $l = 1, 2, 3$ we see that this gives precisely the restrictions from (b)–(d). Let $l = 4$. Then the inequality above gives $3d - 4 \leq m \leq 2d - 1$, which obviously implies $d \leq 3$. Substituting $d = 2$, we get $2 \leq m \leq 3$, and substituting $d = 3$ gives $m = 5$, so we get cases (e) and (f). Next let us consider $l = 5$. Then the inequality above gives $4d - 5 \leq m \leq 2d - 1$, which obviously implies $d \leq 2$. Substituting $d = 2$, we get $m = 3$, which gives case (g). Let us assume that $l \geq 6$. Then we have $(l - 1)d - l > 2d - 1$, because for $d = 2$ the value of the left hand side is bigger than the value of the right hand side, and increasing d by 1 increases the left hand side by $l - 1$ and the right hand side by 2. This proves that the restrictions (a)–(g) are necessary.

To prove that these restrictions are sufficient we just exhibit the terms of $F(m)_\bullet$ in the listed cases. All the calculations are simple applications of Serre's theorem. In the formulas below we write the terms $F(m)_1$ and $F(m)_0$. In each term we skip tensoring $\otimes_K A(-j)$. The shift j is in all cases the same as the exterior power of W. In the cases corresponding to (a) we get

l arbitrary, $d = 1, m = 1$: $\bigwedge^1 W \longrightarrow S_1 F,$ (a1)

l arbitrary, $d = 1, m = 0$: $\bigwedge^{l+1} W \longrightarrow S_0 F,$ (a2)

l arbitrary, $d = 1, m = -1$: $D_1 F^* \otimes \bigwedge^{l+1} W \longrightarrow \bigwedge^l W.$ (a3)

The matrix in (a1) is easily identified with the coefficient matrix of our system of linear forms. In (a3) the matrix is just a transpose matrix to (a1). The matrix (a2) demonstrates the well-known property of the determinant of a linear map in terms of the top exterior power.

In the cases corresponding to (b) we get

$$m = 2d - 1: \qquad S_{d-1}F \otimes \overset{1}{\bigwedge} W \longrightarrow S_{2d-1}F, \tag{b1}$$

$$d \le m \le 2d - 2: \qquad (D_{2d-2-m}F^* \otimes \overset{2}{\bigwedge} W) \oplus (S_{m-d}F$$
$$\otimes \overset{1}{\bigwedge} W) \longrightarrow S_m F, \tag{b2}$$

$$m = d - 1: \qquad D_{d-1}F^* \otimes \overset{2}{\bigwedge} W \longrightarrow S_{d-1}F, \tag{b3}$$

$$0 \le m \le d - 2: \qquad S_{2d-2-m}F^* \otimes \overset{2}{\bigwedge} W \longrightarrow (D_{d-2-m}F^*$$
$$\otimes \overset{1}{\bigwedge} W) \oplus (S_m F), \tag{b4}$$

$$m = -1: \qquad D_{2d-1}F^* \otimes \overset{2}{\bigwedge} W \longrightarrow D_{d-1}F^* \otimes \overset{1}{\bigwedge} W. \tag{b5}$$

In the cases (c) we get

$$m = 2d - 1: (D_{d-2}F^* \otimes \overset{3}{\bigwedge} W) \oplus (S_{d-1}F \otimes \overset{1}{\bigwedge} W) \longrightarrow S_{2d-1}F, \tag{c1}$$

$$m = 2d - 2: (D_{d-1}F^* \otimes \overset{3}{\bigwedge} W) \oplus (S_{d-2}F \otimes \overset{1}{\bigwedge} W) \longrightarrow S_{2d-2}F, \tag{c2}$$

$$(D_{3d-3-m}F^* \otimes \overset{3}{\bigwedge} W) \oplus (S_{m-d}F \otimes \overset{1}{\bigwedge} W) \longrightarrow \tag{c3}$$
$$(D_{2d-3-m}F^* \otimes \overset{2}{\bigwedge} W) \oplus S_m F,$$

$$m = d - 1: \qquad (D_{2d-2}F^* \otimes \overset{3}{\bigwedge} W) \longrightarrow \tag{c4}$$
$$(D_{d-2}F^* \otimes \overset{2}{\bigwedge} W) \oplus S_{d-1}F,$$

$$m = d - 2: \qquad (D_{2d-1}F^* \otimes \overset{3}{\bigwedge} W) \longrightarrow \tag{c5}$$
$$(D_{d-1}F^* \otimes \overset{2}{\bigwedge} W) \oplus S_{d-2}F,$$

with the case (c3) corresponding to $d \le m \le 2d - 3$.

In the cases corresponding to (d) we get

$$m = 2d - 1: \quad (D_{2d-3}F^* \otimes \overset{4}{\bigwedge} W) \oplus (S_{d-1}F \otimes \overset{1}{\bigwedge} W) \quad \text{(d1)}$$

$$\longrightarrow (D_{d-3}F^* \otimes \overset{3}{\bigwedge} W) \oplus (S_{2d-1}F),$$

$$m = 2d - 2: \quad (D_{2d-2}F^* \otimes \overset{4}{\bigwedge} W) \oplus (S_{d-2}F \otimes \overset{1}{\bigwedge} W) \quad \text{(d2)}$$

$$\longrightarrow (D_{d-2}F^* \otimes \overset{3}{\bigwedge} W) \oplus (S_{2d-2}F),$$

$$m = 2d - 3: \quad (D_{2d-1}F^* \otimes \overset{4}{\bigwedge} W) \oplus (S_{d-3}F \otimes \overset{1}{\bigwedge} W) \quad \text{(d3)}$$

$$\longrightarrow (D_{d-1}F^* \otimes \overset{3}{\bigwedge} W) \oplus (S_{2d-3}F).$$

In the cases from (e) we get

$$m = 3: \quad (D_2 F^* \otimes \overset{5}{\bigwedge} W) \oplus (S_1 F \otimes \overset{1}{\bigwedge} W) \quad \text{(e1)}$$

$$\longrightarrow (\overset{4}{\bigwedge} W) \oplus (S_3 F),$$

$$m = 2: \quad (D_3 F^* \otimes \overset{5}{\bigwedge} W) \oplus (\overset{1}{\bigwedge} W) \quad \text{(e2)}$$

$$\longrightarrow (D_1 F^* \otimes \overset{4}{\bigwedge} W) \oplus (S_2 F).$$

In the case from (f) we get

$$m = 5: \quad (D_5 F^* \otimes \overset{5}{\bigwedge} W) \oplus (S_2 F \otimes \overset{1}{\bigwedge} W) \quad \text{(f1)}$$

$$\longrightarrow (D_2 F^* \otimes \overset{4}{\bigwedge} W) \oplus S_5 F.$$

Finally the case g gives

$$m = 3: \quad (D_3 F^* \otimes \overset{6}{\bigwedge} W) \oplus (S_1 F \otimes \overset{1}{\bigwedge} W) \quad \text{(g1)}$$

$$\longrightarrow (D_1 F^* \otimes \overset{5}{\bigwedge} W) \oplus S_3 F.$$

This proves our theorem, because all listed complexes have only two nonzero terms. ∎

Next we generalize our theorem to the product of projective spaces.

Let us fix dimensions $l_1, \ldots, l_r$ and positive degrees $d_1, \ldots, d_r$. Let $l = l_1 + \ldots + l_r$. We define $V' = \mathbf{P}^{l_1} \times \ldots \times \mathbf{P}^{l_r}$. We identify $\mathbf{P}^{l_i}$ with the set of lines in the vector space F_i of dimension $l_i + 1$. Let us consider V' as a projective variety embedded in the projective space X' whose homogeneous coordinates are the sections of the line bundle $\mathcal{O}(d_1, \ldots, d_r)$. This means we consider the space $E = S_{d_1} F_1^* \otimes \ldots \otimes S_{d_r} F_r^*$ and then we define, as in the previous section,

$$\nabla_{V'} = \{(g_1, \ldots, g_{l+1}) \in E^{l+1} \mid \exists y \in Y' \; g_1(y) = \cdots g_{l+1}(y) = 0\}.$$

In this case we call the resultant $Res(V')$ the *resultant of $l + 1$ multigraded homogeneous polynomials* $g_1, \ldots, g_{l+1}$. As in section 9.1, we introduce the vector space W of dimension $l + 1$ and we identify E^{l+1} with $W \otimes E$.

We will apply the results from section 9.1 to investigate the determinantal expressions for this resultant. The bundle $\mathcal{L} = \mathcal{O}(m_1, \ldots, m_r)$, where $m = (m_1, \ldots, m_r) \in Z^r$. We denote by $F(m)_\bullet$ the complex $F(\mathcal{L})_\bullet$ in that case. Let us describe the complex $F(\mathcal{L})_\bullet$ more precisely. The bundle $\mathcal{O}_{X'}(-1)|_{Y'} = \mathcal{O}(-d_1, \ldots, -d_r)$, so the bundle ξ can be written $\xi = W \otimes \mathcal{O}(-d_1, \ldots, -d_r)$. Setting

$$K_{j,p}(m) = \overset{p}{\bigwedge} W \otimes H^j(Y', \mathcal{O}(m_1 - p \, d_1, \ldots, m_r - p \, d_r)) \otimes_{\mathbf{K}} A(-p),$$

the terms of the complex $F(m)_\bullet$ are

$$F(m)_i = \sum_{j \geq 0} K_{j,i+j}(m).$$

Our first goal is to characterize the tuples $(l_1, \ldots, l_r)$ and $(d_1, \ldots d_r)$ for which there exists $(m) \in \mathbf{Z}^r$ such that the complex $F(m)_\bullet$ is determinantal. For each pair of positive integers (l, d) we define its *defect* $\delta(l, d) := l - \lceil \frac{l}{d} \rceil$, where $\lceil x \rceil$ is the smallest integer $\geq x$. Clearly the defect is always nonnegative.

(9.2.2) Theorem. *A determinantal complex $F(m)_\bullet$ exists if and only if $\delta(l_i, d_i) \leq 2$ for each $i = 1, \ldots, r$.*

Before proving (9.2.2), we give the list of pairs (l, d) with $\delta(l, d) \leq 2$, i.e. those which give rise to a determinantal complex. The proof of the following proposition is straightforward.

(9.2.3) Proposition.

    (a) A pair (l, d) of positive integers has defect 0 if and only if $\min(l, d) = 1$.

    (b) For every integer $t \geq 1$ the list of all pairs (l, d) with $\delta(l, d) = t$ consists of all pairs $(t + 1, d)$, $d \geq t + 1$, and the finite number of pairs

satisfying the inequalities $t + 2 \leq l \leq 2t + 1$, $\frac{l}{l-t} \leq d \leq \frac{l-1}{l-t-1}$. *In particular, the pairs with defect* 1 *are* $(2, d)$ *with* $d \geq 2$ *and* $(3, 2)$; *the pairs with defect* 2 *are* $(3, d)$ *with* $d \geq 3$ *and* $(4, 2)$, $(4, 3)$, $(5, 2)$.

Proof. Let us fix the r-tuples $(l_1, \ldots, l_r)$ and $(d_1, \ldots, d_r)$. We consider the twist $m = (m_1, \ldots, m_r)$. First we describe the terms of $F(m)_\bullet$ more explicitly. In doing so we use the following conventions. We write $S_{-n} F$ for $(S_n F)^*$. We also write $S(d_1, \ldots, d_r)$ for $S_{d_1} F_1 \otimes S_{d_2} F_2 \otimes \ldots \otimes S_{d_r} F_r$. For $k = 1, \ldots r$ let P_k denote the set of integers p such that $m_k/d_k < p \leq (m_k + l_k)/d_k$. We write $p < P_k$ if $p < m_k/d_k$; similarly, we write $p > P_k$ if $p > (m_k + l_k)/d_k$. We use this notation even when $P_k = \emptyset$. If $p \notin P_k$, then we put $j(p) = p - \sum_{k:p>P_k} l_k$ and define integers $n_1(p), \ldots, n_r(p)$ as follows:

$$n_k(p) = \begin{cases} m_k - p d_k & \text{if } p < P_k, \\ m_k + l_k + 1 - p d_k & \text{if } p > P_k. \end{cases}$$

In what follows we abbreviate $[a, b] := \{a, a + 1, \ldots, b\}$.

(9.2.4) Proposition. *Let* $p \in [0, l + 1]$.

(a) *If* $p \in \bigcup_{k=1}^r P_k$, *then* $K_{j,p}(m) = 0$ *for all integers* j.
(b) *If* $p \notin \bigcup_{k=1}^r P_k$, *then there is exactly one value of* j *such that* $K_{j,p}(m) \neq 0$, *namely* $j = j(p)$.
(c) *In the situation of* (b),

$$K_{j(p),p}(m) = S(n_1(p), \ldots, n_r(p)) \otimes \bigwedge^p W.$$

Proof. We apply Künneth's formula to the product of the projective spaces. We get

$$H^j(Y', \mathcal{O}(n_1, \ldots n_r)) = \bigoplus_{j_1+\ldots+j_r=j} \bigotimes_{k=1}^r H^{j_k}(\mathbf{P}(F_k^*), \mathcal{O}(n_k)).$$

By Serre's theorem ([H1], chapter III, Theorem 5.1), each group $H^{j_k}(\mathbf{P}(F_k^*), \mathcal{O}(n_k))$ is computed as follows:

$$H^{j_k}(\mathbf{P}(F_k^*), \mathcal{O}(n_k)) = \begin{cases} S_{n_k} F_k & \text{if } n_k \geq 0, j_k = 0, \\ S_{n_k+l_k+1} F_k & \text{if } n_k < -l_k, j_k = l_k, \\ 0 & \text{otherwise,} \end{cases}$$

and the proof of (9.2.4) is complete. ■

It follows from the definitions that

$$\text{card } P_k \le \left\lceil \frac{l_k}{d_k} \right\rceil = l_k - \delta(l_k, d_k). \qquad (*)$$

Let $P = \bigcup_{k=1}^r P_k$ and $Q = [0, l+1] \setminus P$. Consider the function $j : Q \to Z$ defined by $j(p) = p - \sum_{k:p > P-k} l_k$. By Proposition (9.2.4) the condition that $F(m)_\bullet$ is determinantal is equivalent to the condition that $j(Q)$ consists of two consecutive integers. Let $a = \min(Q)$, $b = \max(Q)$.

(9.2.5) Lemma. *If m is determinantal, then:*

(a) $j(Q) = \{0, 1\}$, i.e., $K_i(m) = 0$ for $i \ne 0, 1$.
(b) If $P_k < a$ or $b < P_k$, then $\delta(l_k, d_k) = 0$. Furthermore,

$$\bigcup_{k \,:\, P_k < a} P_k = [0, a-1], \qquad \bigcup_{k \,:\, b < P_k} P_k = [b+1, l+1],$$

both decompositions disjoint.

Proof. Let

$$L_0 = \sum_{k \,:\, P_k < a} l_k, \qquad L_1 = \sum_{k \,:\, a < P_k < b} l_k, \qquad L_2 = \sum_{k \,:\, b < P_k} l_k.$$

By definition, $j(p) = a - L_0$, $j(b) = b - L_0 - L_1 = b - l + L_2$, so $j(b) - j(a) = b - a - l + L_0 + L_2$. By our choice of a and b, the sets $[0, a-1]$ and $[b+1, l+1]$ are contained in P. By $(*)$, this implies

$$a \le \text{card}\left(\bigcup_{k:P_k < a} P_k \right) \le \sum_{k:P_k < a} \text{card}(P_k) \le L_0, \qquad (**)$$

and similarly that $l + 1 - b \le L_2$. It follows that $j(b) - j(a) \ge b - a - l + a + (l + 1 - b) = 1$. Since m is determinantal, the latter inequality must become an equality. This implies $L_0 = a$, $L_2 = l + 1 - b$, and so $j(a) = 0$, $j(b) = 1$, which proves (a). Going through the chain of inequalities in $(**)$ and the similar chain of inequalities for the inequality $l + 1 - b \le L_2$ yields (b). ∎

Lemma (9.2.5) (b) implies in particular that $P \subset [0, l+1]$. We represent Q as the disjoint union of its connected components, so two points $a < b$ lie in the same component if and only if there is no index k such that $a < P_k < b$. Let $[a_0, b_0], [a_1, b_1], \ldots, [a_s, b_s]$ be the components of Q listed from left to right; we adopt the convention that $[a_0, b_0] = \emptyset$ if $0 \in P$, and likewise $[a_s, b_s] = \emptyset$ if $l + 1 \in P$. Let $\epsilon_i = b_i - a_i + 1$ for $i \in [0, s]$ be the sizes of components of Q. All ϵ_i are nonnegative integers; by our convention, ϵ_0, ϵ_s can be equal to 0, while the other ϵ_i are positive.

Each component of $P = [0, l + 1] \setminus Q$ is a union of some sets P_k. It follows that $1 \leq s \leq r$. We define a surjection $\phi : [1, r] \to [1, s]$ by the requirement that $b_{i-1} < P_k < a_i$ for $\phi(k) = i$. For $i = 1, \ldots, s$ let

$$L_i = \sum_{k:\phi(k)=i} l_k, \qquad \delta_i = \sum_{k:\phi(k)=i} \delta(l_k, d_k). \qquad (?)$$

Let $\Delta_i := L_i - (a_i - b_{i-1} - 1)$. The same argument as in $(**)$ shows that

$$\Delta_i \geq \delta_i \qquad (??)$$

with the equality if and only if the sets P_k for $\phi(k) = i$ are mutually disjoint and each of them makes $(**)$ an equality.

For our next statement we need the *dominance partial order*

$$(\alpha_0, \alpha_1, \ldots, \alpha_s) \leq (\beta_0, \beta_1, \ldots, \beta_s)$$

on integer vectors, defined by the conditions

$$\alpha_0 \leq \beta_0, \qquad \alpha_0 + \alpha_1 \leq \beta_0 + \beta_1, \ldots,$$
$$\alpha_0 + \alpha_1 + \ldots + \alpha_{s-1} \leq \beta_0 + \beta_1 + \ldots + \beta_{s-1},$$
$$\alpha_0 + \alpha_1 + \ldots + \alpha_s = \beta_0 + \beta_1 + \ldots + \beta_s.$$

(9.2.6) Lemma. *The statement that $F(m)_\bullet$ is determinantal is equivalent to the following inequalities:*

$$(\Delta_1, \Delta_2, \ldots, \Delta_s, 2) \leq (\epsilon_0, \epsilon_1, \ldots, \epsilon_s) \leq (2, \Delta_1, \ldots, \Delta_s).$$

Proof. First of all we have

$$\sum_{i=0}^{s} \epsilon_i + \sum_{i=1}^{s} (L_i - \Delta_i) = \text{card}([0, l + 1]) = l + 2 = 2 + \sum_{i=1}^{s} L_i,$$

which can be rewritten as

$$\sum_{i=0}^{s} \epsilon_i = 2 + \sum_{i=1}^{s} \Delta_i.$$

This is a necessary condition for the dominance order inequalities (9.2.6). Remembering the definitions, we compute for $i = 1, \ldots, s$

$$j(a_i) = \epsilon_0 + \epsilon_1 + \ldots + \epsilon_{i-1} - \Delta_1 - \Delta_2 - \ldots - \Delta_i,$$
$$j(b_{i-1}) = \epsilon_0 + \epsilon_1 + \ldots + \epsilon_{i-1} - 1 - \Delta_1 - \Delta_2 - \ldots - \Delta_{i-1}.$$

By Lemma (9.2.5) m is determinantal if and only if $j(a_i) \geq 0$, $j(b_{i-1}) \leq 1$ for all i. But these inequalities are equivalent to (9.2.6). ∎

(9.2.7) Lemma. *Suppose two nonnegative integer vectors $(\epsilon_0, \epsilon_1, \ldots, \epsilon_s)$ and $(\Delta_1, \Delta_2, \ldots, \Delta_s)$ satisfy (9.2.6). Then $\epsilon_i, \Delta_i \leq 2$ for all i.*

Proof. Let (A_i) denote the inequality

$$\epsilon_0 + \epsilon_1 + \ldots + \epsilon_{i-1} \geq \Delta_1 + \Delta_2 + \ldots + \Delta_i,$$

and (B_i) denote the inequality

$$\Delta_1 + \Delta_2 + \ldots + \Delta_i + 2 \geq \epsilon_0 + \epsilon_1 + \ldots + \epsilon_i.$$

Adding (A_i) and (B_i) yields $2 \geq \epsilon_i$; adding (A_i) and (B_{i-1}) yields $2 \geq \Delta_i$. ∎

Lemma (9.2.6) together with (?) and (??) proves that the conditions $\delta(l_k, d_k) \leq 2$ in Theorem (9.2.2) are necessary for the existence of the determinantal complex $F(m)_\bullet$. To prove that they are sufficient we give a refinement of Lemma (9.2.7).

(9.2.8) Lemma. *The following conditions on a nonnegative integer vector $(\Delta_1, \ldots, \Delta_s)$ are equivalent:*

   (a) There is a nonnegative integer vector $(\epsilon_0, \epsilon_1, \ldots, \epsilon_s)$ satisfying (9.2.6) such that $\epsilon_i \geq 1$ for $i = 1, \ldots, s - 1$.

   (b) $\Delta_i \leq 2$ for $i = 1, \ldots, s$, and for every two indices i, j such that $1 < i < j < s$ and $\Delta_i = \Delta_j = 0$ there exists an index k such that $i < k < j$ and $\Delta_k = 2$.

Proof. We introduce new variables $g_i = \sum_{j=0}^{i} \epsilon_j$ for $i = 0, \ldots, s - 1$. Condition (a) can be rewritten as follows:

   (a') There is an integer vector $(g_0, g_1, \ldots, g_{s-1})$ satisfying the inequalities

$$\sum_{j=0}^{i+1} \Delta_j \leq g_i \leq 2 + \sum_{j=0}^{i} \Delta_j \qquad \text{for} \quad i = 0, \ldots, s - 1,$$

$$1 \leq g_i - g_{i-1} \leq 2 \qquad \text{for} \quad i = 1, \ldots, s - 1.$$

The consistency conditions for the system of inequalities above (with arbitrary constraints for all g_i and $g_i - g_{i-1}$) take the form (b') $\Delta_i \leq 2$ for $i = 1, \ldots, s$, and

$$\Delta_i + \Delta_{i+1} + \ldots + \Delta_j \geq j - i$$

for all i, j such that $1 < i < j < s$.

(The necessity of (b') follows from the chain of inequalities

$$2 + \sum_{k=0}^{j} \Delta_k \geq g_j = g_{i-2} + \sum_{k-i-2}^{j-1} (g_{k+1} - g_k) \geq \sum_{k=0}^{i-1} \Delta_k + (j - i + 2)$$

resulting from (a').)

It remains to show that (b') is equivalent to (b). We can rewrite (b') as

$$\sum_{k=i}^{j} (\Delta_k - 1) \geq -1. \tag{b''}$$

Assuming that $0 \leq \Delta_i \leq 2$ for all i, we see that the left hand side of (b'') is equal to $\mathrm{card}\{k \in [i, j] : \Delta_k = 2\} - \mathrm{card}\{k \in [i, j] : \Delta_k = 0\}$. The condition that this difference cannot be less than -1 means that there is at least one term equal to 2 between any two zero terms in the sequence $\Delta_2, \Delta_3, \ldots, \Delta_{s-1}$. But this is exactly condition (b). Lemma (9.2.8) is proven. ∎

Now we can complete the proof of Theorem (9.2.2). Suppose that $\delta(l_k, d_k) \leq 2$ for each $k = 1, \ldots, r$. According to Lemmas (9.2.6) and (9.2.8), all possible determinantal complexes $F(m)_\bullet$ can be obtained by the following procedure:

(a) Choose $s \in [1, r]$, and a surjection $\phi : [1, r] \to [1, s]$ such that

$$\delta_i := \sum_{k\,:\,\phi(k)=i} \delta(l_k, d_k) \leq 2 \qquad \text{for} \quad i = 1, \ldots, s.$$

(b) Choose an integer vector $(\Delta_1, \ldots, \Delta_s)$ satisfying (??) and the conditions of Lemma (9.2.8).

(c) Choose an integer vector $(\epsilon_0, \epsilon_1, \ldots, \epsilon_s)$ as in Lemma (9.2.8) (a).

(d) Consider the disjoint decomposition of $[0, l + 1]$ into the intervals of integers of cardinalities (from left to right) $\epsilon_0, L_1 - \Delta_1, \epsilon_1, L_2 - \Delta_2, \ldots, L_s - \Delta_s, \epsilon_s$, where the L_is are given by (?). As above, let $[a_i, b_i]$ denote the interval of cardinality ϵ_i in the decomposition. Finally, choose $m = (m_1, \ldots, m_r)$ so that for $i = 1, \ldots, s$ and $\phi_k = i$ we have $b_{i-1} < P_k < a_i$ and $[b_{i-1} + 1, a_i - 1] = \bigcup_{k\,:\,\phi(k)=i} P_k$.

Here is one possible construction satisfying all the above conditions. We choose s and $\phi : [1, r] \to [1, s]$ in such a way that ϕ sends all the indices k

with $\delta(l_k, d_k) = 0$ to the same index i_0 and ϕ is injective on $\{k \in [1, r] : \delta(l_k, d_k) > 0\}$. Let $\Delta_i = \delta_i$ for all i. The condition of Lemma (9.2.8)(b) holds, because there is at most one index i with $\Delta_i = 0$. Choose an arbitrary vector $(\epsilon_0, \epsilon_1, \ldots, \epsilon_s)$ as in Lemma (9.2.8)(a). It remains to find $m = (m_1, \ldots, m_r)$ as in part (d) above. If $i \neq i_0$, then there is exactly one k such that $\phi(k) = i$, and we find m_k from the condition $(m_k + l_k)/d_k = a_i - 1$. Since $\text{card}([b_{i-1} + 1, a_i - 1]) = L_i - \Delta_i = l_k - \delta(l_k, d_k) = \lceil l_k/d_k \rceil$, it follows that $P_k = [b_{i-1} + 1, a_i - 1]$, as required. As for the remaining index i_0 (if it exists at all), we have $\text{card}([b_{i_0-1} + 1, a_{i_0} - 1]) = \sum_{k : \delta(l_k, d_k)=0} l_k$. Therefore we can choose m_k with $\delta(l_k, d_k) = 0$ in such a way that if I_k is the interval of cardinality l_k with right end $(m_k + l_k)/d_k$, then the I_k form a (disjoint) decomposition of $[b_{i_0-1} + 1, a_{i_0} - 1]$. We have constructed a determinantal complex $F(m)_\bullet$, which completes the proof of Theorem (9.2.2). ∎

9.3. The Generalized Discriminants

Let E be a vector space of dimension n over a field $\mathbf{K}$. We treat the vector space E as the affine space X over $\mathbf{K}$. Let Y be the closed subvariety of X which is a cone over a smooth projective variety Y' in $\mathbf{P}(E)$ of dimension r. We assume that Y' is linearly normal and nondegenerate, i.e., the natural map $E \to H^0(Y', \mathcal{O}_{Y'})$ is an isomorphism.

We define $(Y')^\vee \subset \mathbf{P}(E^*)$ to be the projectively dual variety to Y'. By definition $(Y')^\vee$ consists of hyperplanes in $\mathbf{P}(E)$ tangent to Y' at some point of Y'. We denote by $Y^\vee$ the cone over $(Y')^\vee$. The variety $Y^\vee$ is a closed subvariety in the dual affine space $X^\vee$ which can be identified with E^*. If codim $Y^\vee = 1$, then the homogeneous polynomial Δ_Y generating the ideal of functions vanishing on $Y^\vee$ is called the discriminant of Y. The polynomial Δ_Y is defined up to a nonzero constant. If codim $Y^\vee > 1$, we define Δ_Y to be a nonzero constant.

The geometric approach developed in chapter 5 applies to the variety $Y^\vee$. Indeed, Let us consider the variety $Z \subset X^\vee \times Y'$ defined as follows:

$$Z = \{(H, y) \in X^\vee \times Y' \mid H|_{T_{Y',y}} = 0\}$$

As before, we denote the projection from $X^\vee \times Y'$ to $X^\vee$ by q and its restriction to Z by q'. The projection from $X^\vee \times Y'$ to Y' is denoted by p. Finally p' denotes the restriction of p to Z.

It is clear that the first projection $p' : Z \longrightarrow Y'$ makes Z a vector bundle over Y', and that the image of the second projection $q' : Z \longrightarrow X^\vee$ is equal to $Y^\vee$.

By definition Z is a total space of a bundle ξ of 1-jets on Y'. More precisely, we have the exact sequence

$$0 \longrightarrow \mathcal{O}_{Y'}(-1) \longrightarrow \xi \longrightarrow T(Y')(-1) \longrightarrow 0,$$

where $T(Y')$ denotes the tangent bundle on Y'. There is a simple way to describe $\xi(1)$ geometrically. For any $y \in Y'$ the fiber $\xi(1)_y$ is the tangent space to the cone Y over Y'.

We get a diagram

$$\begin{array}{ccc} Z & \subset & X^{\vee} \times Y' \\ \downarrow q' & & \downarrow q \\ Y^{\vee} & \subset & X^{\vee} \end{array}$$

which is a special case of the basic diagram from section 5.1 with V, Y, X specializing to Y', $Y^{\vee}$, and $X^{\vee}$ respectively. We can now restate the basic theorem (5.1.2) for the special case of discriminants. Let us denote by A the coordinate ring on $X^{\vee}$. This is a polynomial ring that can be identified with the symmetric algebra $\mathrm{Sym}(E)$. For a vector bundle $\mathcal{V}$ on Y' we consider the sheaf $M(\mathcal{V}) := \mathcal{O}_Z \otimes p^*(\mathcal{V})$ on $X^{\vee} \times Y'$.

(9.3.1) Basic Theorem for Discriminants. *For a vector bundle $\mathcal{V}$ on Y' we define free graded A-modules*

$$F(\mathcal{V})_i = \bigoplus_{j \geq 0} H^j \left(Y', \overset{i+j}{\bigwedge} \xi \otimes \mathcal{V} \right) \otimes_k A(-i-j).$$

(a) There exist minimal differentials

$$d_i(\mathcal{V}) : F(\mathcal{V})_i \to F(\mathcal{V})_{i-1}$$

of degree 0 such that $F(\mathcal{V})_\bullet$ is a complex of graded free A-modules with

$$H_{-i}(F(\mathcal{V})_\bullet) = \mathcal{R}^i q_* M(\mathcal{V}).$$

In particular the complex $F(\mathcal{V})_\bullet$ is exact in positive degrees.

(b) The sheaf $\mathcal{R}^i q_* M(\mathcal{V})$ is equal to $H^i(Z, M(\mathcal{V}))$, and it can be also identified with the graded A-module $H^i(V, \mathrm{Sym}(\eta) \otimes \mathcal{V})$.

(c) If $\phi : M(\mathcal{V}) \to M(\mathcal{V}')(m)$ is a morphism of graded sheaves, then there exists a morphism of complexes

$$f_\bullet(\phi) : F(\mathcal{V})_\bullet \to F(\mathcal{V}')_\bullet(m).$$

Its induced map $H_{-i}(f_\bullet(\phi))$ can be identified with the induced map

$$H^i(Z, M(\mathcal{V})) \rightarrow H^i(Z, M(\mathcal{V}'))(m).$$

We want to apply the complexes $F(\mathcal{V})_\bullet$ to find expressions for the discriminant Δ_Y. Let us recall (cf. section 1.3) that we can associate to the complex $F(\mathcal{V})_\bullet$ its determinant. As in section 9.1, the determinant of $F(\mathcal{V})_\bullet$ is a power of the discriminant.

(9.3.2) Proposition. *The determinant of the complex $F(\mathcal{V})_\bullet$ is equal to $(\Delta_Y)^{\dim \mathcal{V}}$.*

Proof. The proof is the same as that of Proposition (9.1.3). The determinant of a complex $F(\mathcal{V})_\bullet$ is by definition a rational function having zeros and poles only at the points where $F(\mathcal{V})_\bullet$ is not exact. In our case this is the irreducible variety $Y^\vee$. Therefore the determinant of $F(\mathcal{V})_\bullet$ is a power of Δ_Y.

To compute this power we consider the graded Hilbert function

$$\mathcal{P}(F(\mathcal{V})_\bullet, t) = \sum_{i,j \geq 0} (-1)^i t^{i+j} (1-t)^{-l-1} h^j(V', \mathcal{V} \otimes \overset{i+j}{\bigwedge} \xi)$$

of the complex $F(\mathcal{V})_\bullet$. Let us write $\mathcal{P}(F(\mathcal{V})_\bullet, t) = \sum_{a \geq 0} \mathcal{P}(a) t^a$. Then for $a \gg 0$, $\mathcal{P}(a)$ is a polynomial in a. Since the dimension of Z equals $n-1$, we can distinguish two cases:

(i) $\dim Y^\vee = \dim X^\vee - 1$,
(ii) $\dim Y^\vee < \dim X^\vee - 1$.

It turns out (cf. [Ka]) that in case (i) the map q' is a birational isomorphism. Let us first assume that case (i) occurs. We consider the coefficient of $\mathcal{P}(a)$ in degree $n-1$, divided by $(n-1)!$. By (5.1.6) we see that this coefficient is equal to the degree of the determinant of $F(\mathcal{V})_\bullet$. By (9.3.1) (a) we see that $\mathcal{P}(F(\mathcal{V})_\bullet, t)$ is the alternating sum of graded Hilbert functions of $\mathcal{R}^i q'_* M(\mathcal{V})$. Since all modules $\mathcal{R}^i q'_* M(\mathcal{V})$ are supported in $Y^\vee$, the degree of $\mathcal{P}(a)$ is $\leq n-1$. Let us analyze the support of modules $\mathcal{R}^i q'_* M(\mathcal{V})$. For $i > 0$ this is a variety properly contained in $Y^\vee$, because on the open set in Z the map q' is an isomorphism. Therefore for $i > 0$ the modules $\mathcal{R}^i q'_* M(\mathcal{V})$ have support of codimension ≥ 2 in $X^\vee$. The module $\mathcal{R}^0 q'_* M(\mathcal{V})$ is a module supported on $Y^\vee$ which has rank $\dim \mathcal{V}$. This means that the degree of the determinant of $F(\mathcal{V})_\bullet$ is equal to $\dim \mathcal{V} \deg(\Delta_Y)$, as claimed. This completes the proof of the proposition in case (i). In case (ii) the complex $F(\mathcal{V})_\bullet$ has the homology with support of codimension ≥ 2, so by (1.3.5) its determinant has degree 0, i.e., it is a nonzero constant. ∎

The proof of this proposition implies instantly the following formula for the degree of the hyperdiscriminant.

(9.3.3) Corollary. *The degree of the hyperdiscriminant is given by the following formula:*

$$\deg \Delta_Y = \sum_{i,j} (-1)^{i+1} h^j \left(Y', \bigwedge^{i+j} \xi \otimes \mathcal{L} \right).$$

Proof. By the previous corollary the degree of Δ_Y equals to the coefficient of t^{n-1} in the graded Hilbert function of the complex $F(\mathcal{L})_\bullet$. ∎

Thanks to the proposition, for each line bundle $\mathcal{L}$ on Y' the determinant of the complex $F(\mathcal{L})_\bullet$ is equal to Δ_Y. In the case when the complex $F(\mathcal{L})_\bullet$ has only two nonzero terms, it becomes a matrix whose determinant is the discriminant of Y'. We call the complex $F(\mathcal{L})_\bullet$ *determinantal* if $F(\mathcal{L})_i = 0$ for $i \neq 0, 1$.

Another application of the complexes $F(\mathcal{L})_\bullet$ is the formula for the codimension of the dual variety $Y^\vee$. Let us take $\mathcal{L} = \mathcal{O}_{Y'}$, and let us denote the complex $F(\mathcal{L})_\bullet$ by $F_\bullet$. Then we have the following

(9.3.4) Theorem. $\operatorname{codim} Y^\vee = \max\{i \mid F_i \neq 0\}$.

Proof. Let us consider the canonical sheaf Ω_Z. By the adjunction formula ([H1, Proposition II.8.20]), $\Omega_Z = \Omega_{X^\vee \times Y'} | Z \otimes \bigwedge^{n+1} \xi^*$. Since $X^\vee$ is just the affine space, $\Omega_{X^\vee \times Y'} = p^* \Omega_{Y'}$. Let us consider the locally free resolution of Ω_Z as an $\mathcal{O}_{X^\vee \times Y'}$-module given by the complex $p^*(\Omega_{Y'} \otimes \bigwedge^{n+1} \xi^*) \otimes \bigwedge^{\cdot} \mathcal{S}$. Pushing down this resolution, we get the complex $D(F)$ of A-modules with the term

$$D(F)_i = \bigoplus_{j \geq 0} H^j \left(Y', \Omega_{Y'} \otimes \bigwedge^{n+1-i-j} \xi^* \right) \otimes A(-i-j),$$

whose cohomology equals $H_{-i}(D(F)) = \mathcal{R}^i q_*(\Omega_Z)$. By the Grauert–Riemenschneider theorem $\mathcal{R}^i q_*(\Omega_Z) = 0$ for $i > \dim Z - \dim Y^\vee$. Since the differentials in $D(F)$ have the entries in the maximal ideal of A, this means that $D(F)_i = 0$ for $i < \dim Y^\vee - \dim Z$. Now, using the Serre duality on Y', we see that F_i is the dual of $D(F)_{1-i}$. This means that $F_i = 0$ for $i > 1 - \dim Y^\vee + \dim Z = \operatorname{codim} Y^\vee$. To prove the theorem it is enough to show that for $i = \dim Z - \dim Y^\vee$ we have $\mathcal{R}^i q_*(\Omega_Z) \neq 0$. However, this is easy. Indeed, after shrinking $Y^\vee$ we may assume that p is smooth projective. Then the result follows from the upper semicontinuity theorem ([H1, III.12.11]) and the adjunction formula ([H1, II.8.20]), since each fiber Z_y is smooth of dimension i, so $H^i(Z_y, \Omega_{Z_y})$ is one dimensional (hence nonzero) by Serre duality. ∎

9.4. The Hyperdeterminants

In this section we consider an important example of discriminants – the discriminants of multidimensional matrices. They are called *hyperdeterminants*.

Let $F_1, \ldots F_r$ be r vector spaces over $\mathbf{K}$ of dimensions $m_1, \ldots, m_r$ respectively.

We start with the vector space $E = F_1 \otimes F_2 \otimes \ldots \otimes F_r$. For each $i = 1, \ldots, r$ let $x_1^{(i)}, \ldots x_{m_i}^{(i)}$ be a basis of F_i^*. An arbitrary element of E^* (which will be identified canonically with $F_1^* \otimes \ldots \otimes F_r^*$) can be written as

$$U = \sum_{1 \leq i_1 \leq m_1} \cdots \sum_{1 \leq i_r \leq m_r} u_{i_1, \ldots, i_r} x_{i_1}^{(1)} \ldots x_{i_r}^{(r)},$$

and we will identify it with the r-dimensional matrix $U = (u_{i_1, \ldots, i_r})$ of the format $m_1 \times \ldots \times m_r$. In the affine space $X = E$ we consider the subvariety Y of totally decomposable tensors, i.e. the variety of tensors of type $y_1 \otimes y_2 \otimes \ldots \otimes y_r$ where $y_i \in F_i$ for $i = 1, 2, \ldots, r$. The variety Y is the cone over the product of projective spaces $Y' = \mathbf{P}(F_1) \times \ldots \times \mathbf{P}(F_r)$ embedded in $\mathbf{P}(E)$ by the multiple Segre embedding. We can apply the construction from the previous section to the variety Y. We get the variety $Y^\vee$ contained in $X^\vee = E^*$. Let us look at the variety $Y^\vee$ more closely. For each $j = 1, \ldots, r$ let $e_1^{(j)}, \ldots, e_{m_j}^{(j)}$ be the basis in F_j dual to the basis $x_1^{(j)}, \ldots, x_{m_j}^{(j)}$ chosen above. We notice that the variety Y' is the homogeneous space for the group $\mathbf{G} = \operatorname{GL}(F_1) \times \ldots \times \operatorname{GL}(F_r)$. Let us choose the distinguished point $y_0 = e_1^{(1)} \otimes \ldots \otimes e_1^{(r)} \in Y$. Therefore the variety $Y^\vee$ has the following description:

$$Y^\vee = \{U \in E^* \mid \exists g \in \mathbf{G} \text{ such that } gU|_{T_{y_0,Y}} = 0\}.$$

Next we describe the tangent space $T_{y_0,Y}$.

(9.4.1) Proposition. *The tangent space $T_{y_0,Y}$ is spanned by the vector $e_1^{(1)} \otimes \ldots \otimes e_1^{(r)}$ and the vectors $e_1^{(1)} \otimes \ldots \otimes e_1^{(j-1)} \otimes e_i^{(j)} \otimes e_1^{(j+1)} \otimes \ldots \otimes e_1^{(r)}$ for $j = 1, \ldots, r$ and $i = 2, \ldots, m_j$.*

Proof. The vector $y_0 = e_1^{(1)} \otimes \ldots \otimes e_1^{(r)}$ has to be in $T_{y_0,Y}$ because Y is a cone. The space $T_{y_0,Y}$ contains all the vectors of the type $e_1^{(1)} \otimes \ldots \otimes e_1^{(j-1)}$ $\otimes e_i^{(j)} \otimes e_1^{(j+1)} \otimes \ldots \otimes e_1^{(r)}$ for $j = 1, \ldots, r$ and $i = 2, \ldots, m_j$. Indeed, such a vector is the derivative of a curve $p(t) = e_1^{(1)} \otimes \ldots \otimes e_1^{(j-1)} \otimes (e_1^{(j)} + te_i^{(j)})$ $\otimes e_1^{(j+1)} \otimes \ldots \otimes e_1^{(r)}$ contained in Y. These vectors span $T_{y_0,Y}$, because Y is nonsingular at y_0. ∎

Let $Y_0^\vee$ be the set of forms on E vanishing on $T_{y_0,Y}$. Identifying E^* with the set of multidimensional matrices, we have

$$Y_0^\vee = \{U \in E^* \mid u_{i_1,\ldots,i_r} = 0 \text{ if at least } r - 1 \text{ numbers } i_j \text{ equal } 1\}.$$

This means that $Y_0^\vee$ is the set of multilinear forms which, after dehomogenizing around y_0 (i.e. setting $x_1^{(j)} = 1$ for $j = 1, \ldots, r$) have neither free nor linear terms. We say that such a multilinear form is singular at y_0.

The homogeneity of Y' also implies

$$Y^\vee = \mathbf{G}Y_0^\vee,$$

which gives a simple description of $Y^\vee$.

We apply the construction from the previous section to the variety $Y^\vee$. This gives us for every line bundle $\mathcal{L}$ on Y' a family of complexes $F_\bullet(\mathcal{L})$. Since $Y' = \mathbf{P}(F_1) \times \ldots \times \mathbf{P}(F_r)$, the line bundle $\mathcal{L} = \mathcal{O}(n_1, \ldots, n_r)$ for some $(n_1, \ldots, n_r) \in \mathbf{Z}^r$. We will denote the complex $F(\mathcal{L})_\bullet$ by $F_\bullet(n_1, \ldots, n_r)$. Let us recall that $F_\bullet(n_1, \ldots, n_r)$ is a complex of graded free A-modules with the support in $Y^\vee$. Here $A = \text{Sym}(E) = \text{Sym}(F_1 \otimes \ldots \otimes F_r)$. In order to be able to compute our complexes we need a more precise description of the bundle ξ.

Let us recall that the bundle ξ is a bundle of 1-jets on Y', so it satisfies the exact sequence

$$0 \longrightarrow \mathcal{O}_{Y'}(-1) \longrightarrow \xi \longrightarrow T(Y')(-1) \longrightarrow 0.$$

For $j = 1, \ldots, r$ we identify $\mathbf{P}(F_j)$ with the Grassmannian of lines in F_j. We write the tautological sequence

$$0 \longrightarrow \mathcal{R}_j \longrightarrow F_j \longrightarrow \mathcal{Q}_j \longrightarrow 0$$

with $\dim \mathcal{R}_j = 1$ and $\dim \mathcal{Q}_j = m_j - 1$. The tangent space to $\mathbf{P}(F_j)$ can be canonically identified with $\mathcal{R}_j^* \otimes \mathcal{Q}_j$ by (3.3.5), and the twist (-1) means tensoring with $\mathcal{R}_1 \otimes \ldots \otimes \mathcal{R}_r$. Therefore we have an exact sequence

$$0 \longrightarrow \mathcal{R}_1 \otimes \ldots \otimes \mathcal{R}_r \longrightarrow \xi \longrightarrow \bigoplus_{j=1}^r \Xi_j \longrightarrow 0,$$

where $\Xi_j = \mathcal{R}_1 \otimes \ldots \otimes \mathcal{R}_{j-1} \otimes \mathcal{Q}_j \otimes \mathcal{R}_{j+1} \otimes \ldots \otimes \mathcal{R}_r$

We notice that the bundle ξ can be expressed as an extension of vector bundles in many different ways. One useful way to do it is to notice that the preimage of the subbundle Ξ_j in ξ is just $\mathcal{R}_1 \otimes \ldots \otimes \mathcal{R}_{j-1} \otimes F_j \otimes \mathcal{R}_{j+1} \otimes \ldots \otimes \mathcal{R}_r$. Therefore for each $j = 1, \ldots, r$ we get an exact sequence

$$0 \longrightarrow \mathcal{R}_1 \otimes \ldots \otimes \mathcal{R}_{j-1} \otimes F_j \otimes \mathcal{R}_{j+1} \otimes \ldots \otimes$$

$$\mathcal{R}_r \longrightarrow \xi \longrightarrow \bigoplus_{k \neq j} \Xi_k \longrightarrow 0. \qquad (*)$$

Our goal is to compute the complexes $F(n_1, \ldots, n_r)_\bullet$. This means we have to calculate the cohomology groups of the exterior powers of ξ twisted by $\mathcal{L}$. Notice that we face here a new problem (compared with the calculations from chapter 6 and section 9.2). Before, the calculation of the cohomology of twisted exterior powers could be done directly from Bott's or Serre's theorem. Here the same would be true if the sequence $(*)$ above split. Since it does not, we will use the following trick.

Let us take $\xi' = (\mathcal{R}_1 \otimes \ldots \otimes \mathcal{R}_r) \oplus \bigoplus_{j=1}^r \Xi_j$, and let us calculate the cohomology of $\bigwedge^\bullet(\xi') \otimes \mathcal{L}$. This can be done by Bott's theorem. The cohomology of $\bigwedge^\bullet(\xi) \otimes \mathcal{L}$ is smaller because it has the filtration whose associated graded object is $\bigwedge^\bullet(\xi') \otimes \mathcal{L}$ and therefore there are some cancellations coming from the connecting homomorphisms. In order to figure out which representations cancel out, we will use the sequences $(*)$ in the following way. We introduce the bundles

$$\xi_j = (\mathcal{R}_1 \otimes \ldots \otimes \mathcal{R}_{j-1} \otimes F_j \otimes \mathcal{R}_{j+1} \otimes \ldots \otimes \mathcal{R}_r) \oplus \bigoplus_{k \neq j} \Xi_k.$$

Again the cohomology of $\bigwedge^\bullet(\xi_j) \otimes \mathcal{L}$ "approximates" the cohomology of $\bigwedge^\bullet(\xi) \otimes \mathcal{L}$, but it is in general too big.

(9.4.2) The Principle of Cancellation. *If we have two copies of an irreducible representation occurring in $H^i(\bigwedge^\bullet(\xi') \otimes \mathcal{L})$ and in $H^{i+1}(\bigwedge^\bullet(\xi') \otimes \mathcal{L})$ which we suspect cancel out in $H^\bullet(\bigwedge^\bullet(\xi) \otimes \mathcal{L})$, we look for them in $H^\bullet(\bigwedge^\bullet(\xi_j) \otimes \mathcal{L})$ (in the same place there). If for some $j = 1, \ldots, r$ we cannot find these representations there, this means they indeed do not occur in the cohomology of $\bigwedge^\bullet(\xi) \otimes \mathcal{L}$.*

Because of this complication, we shall not be able to compute the terms of all complexes $F(\mathcal{L})_\bullet$ in the case of hyperdeterminants. Our goals in this section are as follows. First we will use our complexes to calculate the codimension of the variety $Y^\vee$. In the case when the codimension of $Y^\vee$ equals 1, we will give some interesting examples of determinantal expressions for hyperdeterminants of low formats. In the case of codim $Y^\vee > 1$, we will discuss the problem of the defining equations of $Y^\vee$.

Let us look at the specialization of the formula (9.3.3) to the hyperdeterminants. We notice that since the calculations there are based on the Hilbert polynomial in the Euler characteristic of the complex $F(\mathcal{L})_\bullet$, the result stays the same if we replace the bundle ξ by any of the bundles ξ_k or the bundle ξ'. Therefore the complication described above does not occur in calculations of the degree of $Y^\vee$, and we can do it in any special case using Bott's theorem.

(9.4.3) Proposition. *For every $k = 1, \ldots, r$ and for every $\mathcal{L}$ we have*

$$\deg \Delta_Y = \sum_{i,j}(-1)^{i+1}h^j\left(Y', \bigwedge^{i+j}(\xi_k) \otimes \mathcal{L}\right).$$

(9.4.4) Remarks. *The formula above has the drawback of being the alternating sum of positive terms. This makes it impossible to see quickly whether the degree is positive or not. The combinatorial formula for the degree in the case of hyperdeterminants is given by Gelfand, Kapranov, and Zelevinski in ([GKZ, chapter 14, Theorem 2.5]). The terms of this formula are however not easy to understand.*

We can use our complexes to calculate the codimension of $Y^\vee$.

(9.4.5) Theorem. *Let us order the spaces $F_1, \ldots, F_r$ in such way that $m_1 \geq m_2 \geq \ldots \geq m_r$. Then*

$$\text{codim } Y^\vee = \max(1, m_1 - m_2 - \ldots - m_r + r - 1).$$

Proof. We consider the complex $F_\bullet$. By Theorem (9.3.4) it is enough to prove that if $d = \max(1, m_1 - m_2 - \ldots - m_r + r - 1)$ then $F_d \neq 0$ and $F_e = 0$ for $e > d$. We consider the bundle $\xi_1 = (F_1 \otimes \mathcal{R}_2 \ldots \otimes \mathcal{R}_r) \oplus \bigoplus_{j=2}^{r} \Xi_j$. We write $F_s^{(1)} = \bigoplus_{i \geq 0} H^i(V', \bigwedge^{i+s}(\xi_1))$. We know from previous remarks that F_s is smaller than $F_s^{(1)}$ for all s.

Let us use Bott's theorem to calculate the cohomology of the exterior powers of ξ_1,

$$\bigwedge^a(\xi_1) = \bigoplus_{(a_1,\dots,a_r)\,:\,\sum a_i = a} \bigwedge^{a_1}(F_1) \otimes \bigotimes_{j=2}^{r} \bigwedge^{a_j} \Xi_j.$$

Let us denote $\mathcal{F}(a_1,\dots,a_r) = \bigwedge^{a_1}(F_1) \otimes \bigotimes_{j=2}^{r} \bigwedge^{a_j} \Xi_j$. Then we have

$$\mathcal{F}(a_1,\dots,a_r) = \bigwedge^{a_1}(F_1) \otimes \bigotimes_{j=2}^{r} \bigwedge^{a_j} \mathcal{Q}_j \otimes \bigotimes_{j=1}^{r} \mathcal{R}_j^{a-a_j}.$$

By Künneth's formula, such a cohomology is the tensor product of cohomologies of r factors corresponding to the spaces F_j, each of which can be computed by Bott's theorem.

Let us look at the cohomology of $\mathcal{F}(a_1,\dots,a_r)$. The weight corresponding to the space F_1 of this bundle is $(0^{m_1-1}, a - a_1)$. We apply Bott's algorithm (4.1.5) to this weight. We are looking for the possibilities of getting a nonzero cohomology. There are two such possibilities: either there are no exchanges in applying Bott's algorithm, or the last place goes to the first place. Let us analyze these cases.

In the first case we have $a_2 + \dots + a_r = 0$. Then the weight corresponding to F_j is $(0^{m_j-1}, a_1)$. This in turn gives a nonzero contribution either when $a_1 = 0$ (in which case we are dealing with $H^0(\bigwedge^0(\xi'))$ occurring in $F_0^{(1)}$) or when $a_1 \geq m_j$ (in which case we get a nonzero representation in the term $H^{m_2+\dots+m_r-r+1}(\bigwedge^{a_1}(\xi'))$). This occurs in $F_{a_1-m_2-\dots-m_r}^{(1)}$; therefore the maximal possible index is $m_1 - m_2 - \dots - m_r + r - 1$.

Let us look at the second case. By assumption $a_2 + \dots + a_r \geq m_1$. The weight corresponding to F_j is $(1^{a_j}, 0^{m_j-a_j-1}, a - a_j)$. For such a weight there are three cases: either there are no exchanges, or the last place goes to place $a_j + 1$, or the last place goes to the first place.

The first case is impossible. Indeed, since $a_2 + \dots + a_r \geq m_1$, we see that at least two of these indices are positive, so $a - a_j$ is positive for each j.

Let us assume that for some $j > 1$ the second case occurs. This means that $a = m_j$. Since $a_2 + \dots + a_r \geq m_1$, this is possible only if $m_1 = m_j$ and if $a = m_1$. Let s be such that $m_1 = \dots = m_s > m_{s+1}$. Then we see easily that the second case occurs also for all $j = 2, \dots, s$ and that the third case occurs for all $j = s + 1, \dots, r$. Therefore such a term occurs in the cohomology group H^t with $t = m_1 - 1 + \sum_{k=1}^{s}(m_j - a_j - 1) + \sum_{j=s+1}^{r}(m_j - 1)$. This means that our term occurs in $F_s^{(1)}$ with $s = m_1 - \sum_{j=1}^{r}(m_j - 1) + \sum_{j=2}^{s} a_j$.

We have

$$s \le 1 - \sum_{j=2}^{r} (m_j - 1) + \sum_{j=2}^{r} a_j \le 1 .$$

Therefore all such terms occur in places $F_s^{(1)}$ with $s \le 1$.

Let us finally assume that for all $j = 2, \ldots, r$ the third case occurs. Then our term occurs in the cohomology group H^t for $t = \sum_{k=1}^{r} (m_j - 1)$. Therefore such a term occurs in the place $F_s^{(1)}$ with $s = \sum_{j=1}^{r} a_j - \sum_{j=1}^{r} (m_j - 1)$. Therefore again $s \le 1$. Notice that if we take $a_1 = m_1$, $a_j = m_j - 1$ for $j = 2, \ldots, r$, we get a term in $F_1^{(1)}$ which certainly occurs also in F_1, because it comes from the top exterior power of ξ, which is just a line bundle to which we can apply Bott's theorem directly.

We established that all terms $F_s^{(1)}$ are 0 for $s > \max(1, m_1 - m_2 - \ldots - m_r + r - 1)$, so the same is true for F_s. If $1 \ge m_1 - m_2 - \ldots - m_r + r - 1$, we also know from the last remark that $F_1 \ne 0$. It remains to show that for $1 < m_1 - m_2 - \ldots - m_r + r - 1$ the term we found in $F_s^{(1)}$ with $s = m_1 - m_2 - \ldots - m_r + r - 1$ does occur in F_s. Let us recall that this term came from the cohomology of $\mathcal{F}(a_1, \ldots, a_r)$ with $a_1 = m_1$ and $a_j = 0$ for $j = 2, \ldots, r$. However, the resulting representation occurs only once in $H^\bullet(\bigwedge^{m_1}(\xi_1))$, so it cannot cancel out in the spectral sequence. This completes the proof of (9.4.5). ∎

(9.4.6) Remark. *This result can be proven in many ways. It is a special case of the results of Knop and Mentzel [KM]. It was also deduced by Gelfand, Kapranov, and Zelevinski ([GKZ, chapter 14, Theorem 1.3]) from their expression for the degree of hyperdeterminant.*

The above result is the basis for introducing the following distinction. We assume that $m_1 \ge \ldots \ge m_r$. Following [GKZ], we say that the format $m_1 \times \ldots \times m_r$ is

(1) *exterior*, if $m_1 > m_2 + \ldots + m_r - r + 2$;
(2) *boundary*, if $m_1 = m_2 + \ldots + m_r - r + 2$;
(3) *interior*, if $m_1 < m_2 + \ldots + m_r - r + 2$.

We notice that codimension of $Y^\vee$ equals 1 precisely for the interior and boundary formats. This makes the distinction of the exterior case very natural. The reason for distinguishing the boundary case is that in this case the variety $Y^\vee$ is also a resultant variety.

To make this statement precise, let us assume that $m_1 = m_2 + \ldots + m_r - r + 2$ (i.e., we are dealing with the boundary case), and let us consider the space $E' = F_2 \otimes \ldots \otimes F_r$. Let $l_j = m_j - 1$ and $l = l_2 + \ldots + l_r$. We consider the cone $\hat{Y} \subset E'$ over $Y' = \mathbf{P}(F_2) \times \ldots \times \mathbf{P}(F_r)$. In section 9.2 we introduced the resultant variety $\nabla_{\hat{Y}'} \subset W \otimes E^*$, where W is the vector space of dimension $l + 1$. Then dim $W = \dim F_1$, and we can identify W with F_1^*.

(9.4.7) Proposition. *Under the above identifications the dual variety $Y^\vee$ is equal to the resultant variety $\nabla_{\hat{Y}'}$.*

Proof. By definition the variety $\nabla_{\hat{Y}'}$ can be defined as follows:

$$\nabla_{\hat{Y}'} = \left\{ g = \sum_{i=1}^{l+1} x_i \otimes g_i \in F_1^* \otimes E'^* \, | \right.$$

$$\left. \exists y \in Y \text{ such that } g_1(y) = \cdots = g_{l+1}(y) = 0 \right\}.$$

Since the variety $\hat{Y}'$ is homogeneous under the action of the group $\mathcal{H} = \mathrm{GL}(F_2) \times \ldots \times \mathrm{GL}(F_r)$, we can write

$$\nabla_{\hat{Y}'} = \mathcal{H} \nabla_{\hat{Y}'0},$$

where

$$\nabla_{\hat{Y}'0} = \left\{ g = \sum_{i=1}^{l+1} x_i \otimes g_i \in F_1^* \otimes E'^* \, | \, g_i(e_1^{(2)} \otimes \ldots \otimes e_1^{(r)}) = 0 \right.$$

$$\left. \text{for } i = 1, \ldots, r \right\}.$$

Now it is clear by definition of $Y^\vee$ that $Y^\vee \subset \nabla_{\hat{Y}'}$. But both are irreducible varieties of codimension 1 in $W \otimes E'^*$, so they have to be equal. ∎

This makes the results of the section 9.2 on multilinear forms applicable to the hyperdeterminants of boundary formats.

We proceed with some examples of determinantal expressions for the hyperdeterminants of interior formats. We will deal here mainly with three dimensional matrices.

(9.4.8) Example. $m \times m \times 2$ *matrix.*
   Let us take $r = 3$ and dim $F_1 = \dim F_2 = m$, dim $F_3 = 2$. The matrix of this format is

$$U = \sum_{i=1}^{m} \sum_{j=1}^{m} \sum_{k=1}^{2} u_{ijk} x_i^{(1)} x_j^{(2)} x_k^{(3)}.$$

We can treat the matrix U as a pair of two $m \times m$ matrices $U_1 = (u_{ij1})$ and $U_2 = (u_{ij2})$. Then the hyperdeterminant can be described as follows. Let us form an $m \times m$ matrix

$$\hat{U} = U_1 x_1^{(3)} + U_2 x_2^{(3)}.$$

The entries of $\hat{U}$ are linear forms in $x_1^{(3)}, x_2^{(3)}$, so the determinant $\det \hat{U}$ is a binary form in $x_1^{(3)}, x_2^{(3)}$ of degree m. We claim that

$$\Delta_Y = \mathrm{disc}(\det \hat{U}).$$

To prove the claim let us start with the matrix $U \in Y^\vee$. After the change of basis we can assume that $u_{i11} = u_{1j1} = u_{11k} = 0$ for all i, j, k. This means that in the matrix $\hat{U}$ the term $\hat{U}_{1,1} = 0$ and all other terms of $\hat{U}$ in the first row and the first column are divisible by $x_2^{(3)}$. This means that $\det \hat{U}$ is divisible by the square of $X_2^{(3)}$, so its discriminant vanishes. Therefore Δ_Y is a factor of $\mathrm{disc}(\det \hat{U})$. Since one can easily see that $\mathrm{disc}(\det \hat{U})$ is not identically zero, it is enough to show that the degrees of Δ_Y and of $\mathrm{disc}(\det \hat{U})$ are the same. We will do this using the formula (9.3.3). Let us look at the complex $F_\bullet$. We use the bundle ξ_1 to estimate the terms of $F_\bullet$:

$$\bigwedge^a (\xi_1) = \bigoplus_{a_1+a_2+a_3=a} \bigwedge^{a_1} F_1 \otimes \bigwedge^{a_2} \mathcal{Q}_2 \otimes \bigwedge^{a_3} \mathcal{Q}_3 \otimes$$
$$\mathcal{R}_1^{a-a_1} \otimes \mathcal{R}_2^{a-a_2} \otimes \mathcal{R}_3^{a-a_3}.$$

Easy calculation shows that the following four terms have nonzero cohomology:

(1) $a_1 = a_2 = a_3 = 0$ gives the trivial representation in $H^0(Y', \bigwedge^0 \xi_1)$;

(2) $a_1 = m, a_2 = a_3 = 0$ gives the representation $\bigwedge^m F_1 \otimes \bigwedge^m F_2 \otimes K_{(m-1,1)} F_3$ in $H^m(Y', \bigwedge^m \xi_1)$;

(3) $a_1 = 0, a_2 = m - 1, a_3 = 1$ gives the representation $\bigwedge^m F_1 \otimes \bigwedge^m F_2 \otimes K_{(m-2,2)} F_3$ in $H^m(Y', \bigwedge^m \xi_1)$;

(4) $a_1 = m, a_2 = m - 1, a_3 = 1$ contributes $(\bigwedge^m F_1)^{\otimes 2} \otimes (\bigwedge^m F_2)^{\otimes 2} \otimes K_{(2m-2,2)} F_3$ in $H^{2m-1}(Y', \bigwedge^{2m} \xi_1)$.

All four representations are irreducible and different, so there can be no further cancellations when computing the cohomology of exterior powers of ξ. The first three representations give the contribution to F_0, and the fourth *one occurs in F_1. This means that the complex $F_\bullet$ looks as follows:*

$$K_{(2m)}F_1 \otimes K_{(2m)}F_2 \otimes K_{(2m-2,2)}F_3 \otimes A(-2m)$$
$$\downarrow$$
$$\textstyle\bigwedge^m F_1 \otimes \bigwedge^m F_2 \otimes K_{(m-1,1)}F_3 \otimes A(-m)\oplus$$
$$\textstyle\bigwedge^m F_1 \otimes \bigwedge^m F_2 \otimes K_{(m-2,2)}F_3 \otimes A(-m) \oplus A.$$

The formula (9.3.3) gives $\deg \Delta = 2m(2m-3) - m(m-1) - m(m-3) = 2m(m-1)$, *which obviously equals to the degree of* $\mathrm{disc}(\det \hat{U})$. *This completes the proof of our claim.*

Let us notice that the complex $F_\bullet$ is determinantal. Therefore it is interesting to see what kind of determinantal expression of Δ_Y we get. This is easy to guess. The matrix $\hat{U}$ corresponds to the representation $\bigwedge^m F_1 \otimes \bigwedge^m F_2 \otimes S_m F_3$ in $\mathrm{Sym}_m(F_1 \otimes F_2 \otimes F_3)$. The discriminant of $\det \hat{U}$ is the resultant of its two partial derivatives, which are the binary forms of degree $m-1$. Taking the expression (b4) from section 9.2 for $d = m-1$ and for the shift $m = 0$, we get the same terms as in our complex $F_\bullet$ (up to powers of determinantal representation $\bigwedge^2 F_3$). Indeed, we can identify the space W from section 9.2 with F_3, and then we have $F_3 \otimes S_{m-1}F_3 = K_{(m-1,1)}F_3 \oplus K_{(m-2,2)}F_3$ and $\bigwedge^2 F_3 \otimes K_{(2m-3,1)}F_3 = K_{(2m-2,2)}F_3$.

(9.4.9) Example. $(m+1) \times m \times 3$ matrix. *Let us consider the bundle $\mathcal{L} = \mathcal{O}(0, -1, -m+2)$. We will identify $\mathcal{L}$ with $\bigwedge^{m-1} Q_2 \otimes (\bigwedge^3 Q_3)^{\otimes n-2}$. We have*

$$\overset{a}{\bigwedge}(\xi_1) \otimes \mathcal{L} = \bigoplus_{a_1+a_2+a_3=a} \overset{a_1}{\bigwedge} F_1 \otimes \overset{a_2}{\bigwedge} Q_2 \otimes \overset{a_3}{\bigwedge}$$
$$Q_3 \otimes \mathcal{R}_1^{a-a_1} \otimes \mathcal{R}_2^{a-a_2} \otimes \mathcal{R}_3^{a-a_3} \otimes \mathcal{L}.$$

The cohomology of the piece corresponding to (a_1, a_2, a_3) can be calculated by applying Bott's algorithm (4.1.5) to the triple weight $(0^m, a - a_1)$, $(2^{a_2}, 1^{m-1-a_2}, a - a_2)$, $(m-2, m-2, a - a_3)$ and tensoring the result by $\bigwedge^{a_1} F_1$. A simple calculation gives us the following pieces with nonzero cohomology:

(1) $a_1 = a_2 = a_3 = 0$ *gives* $\bigwedge^{m-1} F_2 \otimes K_{(m-2,m-2,0)}F_3$ *in* $H^0(\bigwedge^0(\xi_1) \otimes \mathcal{L})$;

(2) $a_1 = 1, a_2 = a_3 = 0$ *gives* $F_1 \otimes \bigwedge^m F_2 \otimes K_{(m-2,m-2,1)}F_3$ *in* $H^0(\bigwedge^1(\xi_1) \otimes \mathcal{L})$;

(3) $a_1 = m+1, a_2 = a_3 = 0$ *gives* $\bigwedge^{m+1} F_1 \otimes K_{(2m)}F_2 \otimes K_{(m-1,m-1,m-1)}F_3$ *in* $H^{m+1}(\bigwedge^{m+1}(\xi_1) \otimes \mathcal{L})$;

(4) $a_1 = 0, a_2 = m - 1, a_3 = 2$ *gives* $\bigwedge^{m+1} F_1 \otimes K_{(2^m)} F_2 \otimes$
$K_{(m-1,m-1,m-1)} F_3$ *in* $H^m(\bigwedge^{m+1}(\xi_1) \otimes \mathcal{L})$;

(5) $a_1 = m, a_2 = m - 1, a_3 = 2$ *gives* $K_{(2^m,1)} F_1 \otimes K_{(3^m)} F_2 \otimes$
$K_{(2m-3,m,m)} F_3$ *in* $H^{2m+1}(\bigwedge^{2m+1}(\xi_1) \otimes \mathcal{L})$;

(6) $a_1 = m + 1, a_2 = m - 1, a_3 = 2$ *gives* $K_{(2^{m+1})} F_1 \otimes K_{(4,3^{m-1})} F_2 \otimes$
$K_{(2m-2,m,m)} F_3$ *in* $H^{2m+1}(\bigwedge^{2m+2}(\xi_1) \otimes \mathcal{L})$.

Notice that representations (1), (2), (5), (6) have to survive in the cohomology of $\bigwedge^\bullet(\xi) \otimes \mathcal{L}$, but representations (3) and (4) could cancel out in the spectral sequence. We will show that indeed they do. Let us assume that (3) and (4) do not cancel out. Then the complex $F(\mathcal{L})_\bullet$ has the six representations listed above in its terms with (1), (3), (5) occurring in $F(\mathcal{L})_0$ and (2), (4), (6) occurring in $F(\mathcal{L})_1$. At that point we know that the complex $F(\mathcal{L})_\bullet$ gives a matrix whose determinant equals its hyperdeterminant. Let us investigate the differential in our complex. We look at representation (4). It occurs in the homogeneous degree $m + 1$. Therefore only the nonzero part of the differential can go from piece (4) to piece (1). This differential has to be a $\mathbf{G}$-equivariant map from $\bigwedge^{m+1} F_1 \otimes K_{(2^m)} F_2 \otimes K_{(m-1,m-1,m-1)} F_3$ to $\bigwedge^{m-1} F_2 \otimes K_{(m-2,m-2,0)} F_3 \otimes A_{m+1}$. However, $A_{m+1} = \mathrm{Sym}_{m+1}(F_1 \otimes F_2 \otimes F_3)$. Using the Littlewood–Richardson rule, we see that such differential has to come from the occurrence in $\mathrm{Sym}_{m+1}(F_1 \otimes F_2 \otimes F_3)$ of the representation $\bigwedge^{m+1} F_1 \otimes K_{(2,1^{m-1})} F_2 \otimes K_{(m-1,1,1)} F_3$. However, by applying Cauchy's formula twice we see that this representation does not occur in $\mathrm{Sym}_{m+1}(F_1 \otimes F_2 \otimes F_3)$, because the partitions $(2, 1^{m-1})$ and $(m - 1, 1, 1)$ are not conjugate to each other. Therefore the restriction of the differential of $F(\mathcal{L})_\bullet$ to representation (4) is 0, so our matrix has to have a zero column. This gives a contradiction. We conclude that the pieces (3) and (4) cancel out in the spectral sequence, so the only nonzero cohomology groups of $\bigwedge^\bullet(\xi) \otimes \mathcal{L}$ are:

(1) $H^0(\bigwedge^0(\xi) \otimes \mathcal{L}) = \bigwedge^{m-1} F_2 \otimes K_{(m-2,m-2,0)} F_3$,

(2) $H^0(\bigwedge^1(\xi) \otimes \mathcal{L}) = F_1 \otimes \bigwedge^m F_2 \otimes K_{(m-2,m-2,1)} F_3$,

(3) $H^{2m+1}(\bigwedge^{2m+1}(\xi) \otimes \mathcal{L}) = K_{(2^m,1)} F_1 \otimes K_{(3^m)} F_2 \otimes K_{(2m-3,m,m)} F_3$,

(4) $H^{2m+1}(\bigwedge^{2m+2}(\xi) \otimes \mathcal{L}) = K_{(2^{m+1})} F_1 \otimes K_{(4,3^{m-1})} F_2 \otimes K_{(2m-2,m,m)} F_3$.

Therefore the complex $F_\bullet(\mathcal{L})$ looks as follows:

$$K_{(2^{m+1})} F_1 \otimes K_{(4,3^{m-1})} F_2 \otimes K_{(2m-2,m,m)} F_3 \otimes A(-2m - 2)$$
$$\oplus F_1 \otimes \wedge^m F_2 \otimes K_{(m-2,m-2,1)} F_3 \otimes A(-1)$$
$$\downarrow$$
$$K_{(2^m,1)} F_1 \otimes K_{(3^m)} F_2 \otimes K_{(2m-3,m,m)} F_3 \otimes A(-2m - 1)$$
$$\oplus \wedge^{m-1} F_2 \otimes K_{(m-2,m-2,0)} F_3 \otimes A.$$

We can easily calculate the degree of hyperdeterminant in this case. It equals

$$(2m + 2)(m)\binom{m}{2} + (m + 1)\binom{m-1}{2}$$
$$- (2m + 1)(m + 1)\binom{m-1}{2} = 2m(m^2 - 1).$$

Let us discuss the differential in $F(\mathcal{L})_\bullet$. It has one zero block, two linear blocks, and one block of degree $2m + 2$. The two linear blocks are easy to describe. The first one is a $\mathbf{G}$-equivariant map

$$K_{(2^{m+1})}F_1 \otimes K_{(4, 3^{m-1})}F_2 \otimes K_{(2m-2, m, m)}F_3$$
$$\longrightarrow K_{(2^m, 1)}F_1 \otimes K_{(3^m)}F_2 \otimes K_{(2m-3, m, m)}F_3 \otimes A_1,$$

and by Pieri's formulas there is (up to a nonzero scalar) only one such map, coming from the obvious diagonalizations. Similarly, there is (up to a scalar) only one $\mathbf{G}$-equivariant map

$$F_1 \otimes \bigwedge^m F_2 \otimes K_{(m-2, m-2, 1)}F_3 \longrightarrow \bigwedge^{m-1} F_2 \otimes K_{(m-2, m-2, 0)}F_3 \otimes A_1,$$

again induced by obvious diagonalizations. The block of degree $2m + 2$ does not have a clear description. It would be very interesting to find a general recipe to find the differentials like this one explicitly.

(9.4.10) Example. $m \times m \times 3$ matrix. *Here the hyperdeterminant has a similar interpretation to that in the case of $m \times m \times 2$ matrices. Let us take $r = 3$ and $\dim F_1 = \dim F_2 = m$, $\dim F_3 = 3$. The matrix of this format is*

$$U = \sum_{i=1}^{m} \sum_{j=1}^{m} \sum_{k=1}^{3} u_{ijk} x_i^{(1)} x_j^{(2)} x_k^{(3)}.$$

We can treat the matrix U as a triple of $m \times m$ matrices $U_1 = (u_{i,j,1})$, $U_2 = (u_{i,j,2})$, and $U_1 = (u_{i,j,3})$. Then the hyperdeterminant can be described as follows.

Let us form an $m \times m$ matrix

$$\hat{U} = U_1 x_1^{(3)} + U_2 x_2^{(3)} + U_3 x_3^{(3)}.$$

The entries of $\hat{U}$ are linear forms in $x_1^{(3)}, x_2^{(3)}, x_3^{(3)}$, so the determinant $\det \hat{U}$ is a ternary form in $x_1^{(3)}, x_2^{(3)}$ of degree m. Again we have

$$\Delta_Y = \mathrm{disc}(\det \hat{U}).$$

The proof is the same as the proof given for $m \times m \times 2$ matrices. First we notice that for a singular matrix U we change the basis in F_1, F_2, F_3

so $u_{1,1,k} = u_{1,j,1} = u_{i,1,1} = 0$ *for all* i, j, k. *Then we notice that for such* U *the* $(1, 1)$ *entry of the matrix* $\hat{U}$ *is* 0 *and all entries in the first row and column of* $\hat{U}$ *are the combinations of* $x_2^{(3)}$ *and* $x_3^{(3)}$. *This means that the point* $(1 : 0 : 0)$ *is the singular point of the hypersurface* $\det \hat{U} = 0$ *in* $\mathbf{P}^2$. *Therefore* $\operatorname{disc}(\det \hat{U}) = 0$.

It remains to show that $\deg \Delta_Y = \deg \operatorname{disc}(\det \hat{U})$. *This can be done by calculating any of the complexes* $F(\mathcal{L})_\bullet$. *Since for the calculation of the degree the cancellations in the spectral sequence do not matter, this is not difficult, and we leave it to the reader.*

At this point one could suspect that the same method of calculating the hyperdeterminant works for $m \times m \times t$ matrices. Indeed, the first part of our reasoning works in general exactly as above. However, the two degrees are not the same. The polynomial $\operatorname{disc}(\det \hat{U})$ turns out to be reducible. The extra factors are related to the components of the singular locus of $Y^\vee$ (compare [GKZ, chapter 14, section 4]).

This relation between hyperdeterminant and discriminant was discovered a long time ago by Schläfli and was called by Gelfand, Kapranov, and Zelevinski the Schläfli method. They conjecture that this method gives the hyperdeterminant only for $m \times m \times 2$, $m \times m \times 3$, and $2 \times 2 \times 2 \times 2$ matrices. This is easy to check in the cases of small formats using the complexes $F(\mathcal{L})_\bullet$ as in the previous examples.

We continue with the description of the hyperdeterminants in the boundary cases. Here we have several choices of complexes $F(\mathcal{L})_\bullet$, giving different determinantal expressions for hyperdeterminant.

(9.4.11) Example. *The determinantal expression for the three dimensional matrices in the boundary in the subboundary case.*

Let us consider three dimensional matrices of the format $m_1 \times m_2 \times m_3$, *with* $m_1 = m_2 + m_3 - 1$. *Let us take* $\mathcal{L} = \mathcal{O}(-1, 2m_3 - 4, m_2 - 3)$. *Let us use the notation from the proof of Theorem* (9.4.5). *Let us look at the cohomology of* $\mathcal{F}(a_1, a_2, a_3) \otimes \mathcal{L}$. *The weight corresponding to the first coordinate here is* $((-1)^{m_1-1}, a - a_1)$. *This means the we get nonzero terms only if in Bott's algorithm the last place goes to the first one. This in turn means that* $a - a_1 \geq m_1 - 1$. *But* $a - a_1 = a_2 + a_3 \leq (m_2 - 1) + (m_3 - 1) \leq m_1 - 1$. *Thus for all nonzero terms we have* $a_2 = m_2 - 1$, $a_3 = m_3 - 1$. *Let us consider the contribution of terms* $\mathcal{F}(a_1, m_2 - 1, m_3 - 1) \otimes \mathcal{L}$. *The weights corresponding to that term on the second and third coordinate are* $((2m_3 - 3)^{m_2-1}, a_1 + m_3 - 1)$ *and* $((m_2 - 2)^{m_3-1}, a_1 + m_2 - 1)$ *respectively. From this we see that the terms with* $0 \leq a_1 \leq m_3 - 2$ *vanish because of the third coordinate weight,* *and the terms with $m_3 - 1 \le a_1 \le m_2 + m_3 - 3$ vanish because of the second coordinate. The only remaining terms are those for $a_1 = m_2 + m_3 - 2$ and $a_1 = m_2 + m_3 - 1$. The complex $F(\mathcal{L})_\bullet$ consists therefore of two terms*

$$\bigwedge^{m_1-1} F_1 \otimes K_{((2m_3-2)^{m_2})} F_2 \otimes K_{(2m_2-2,(m_2-1)^{m_3-1})} F_3 \quad = \quad F_1^* \otimes S_{m_2-1} F_3,$$

$$\uparrow$$

$$\bigwedge^{m_1} F_1 \otimes K_{(2m_3-1,(2m_3-2)^{m_2-1})} F_2 \otimes K_{(2m_2-1,(m_2-1)^{m_3-1})} F_3 \quad = \quad F_2 \otimes S_{m_2} F_3,$$

with both identifications coming from disregarding the powers of determinantal representations. The differential

$$F_2 \otimes S_{m_2} F_3 \to F_1^* \otimes S_{m_2-1} F_3 \otimes A_1$$

comes from identifying the trilinear form U with the generic map $\Psi(U)$: $F_2 \otimes F_3 \otimes A \to F_1^* \otimes A$. The differential in question is just the composition

$$F_2 \otimes S_{m_2} F_3 \otimes A \overset{1 \otimes \Delta}{\to} F_2 \otimes F_3 \otimes S_{m_2-1} F_3 \otimes A \overset{\Psi(U)}{\to} F_1^* \otimes S_{m_2-1} F_3 \otimes A.$$

This construction generalizes to r-dimensional matrices in boundary format.

(9.4.12) Example. *The hyperdeterminant in boundary format.*

Let us consider the format $m_1 \times m_2 \times \ldots \times m_r$ with $m_1 = m_2 + \ldots + m_r - r + 2$. Choose $\mathcal{L} = \mathcal{O}(u_1, \ldots, u_r)$ with $u_1 = -1, u_j = m_2 + \ldots + m_{j-1} + 2m_{j+1} + \ldots + 2m_r - 2r + j$ for $j = 2, \ldots, r$. Then the same reasoning as above implies that the complex $F(\mathcal{L})_\bullet$ can be identified with a map

$$S_{n_2+1} F_2 \otimes S_{n_3+} F_3 \otimes \ldots \otimes S_{n_r+1} F_r \otimes A \to F_1^* \otimes S_{n_2} F_2 \otimes \ldots \otimes S_{n_r} F_r \otimes A$$

for $n_2 = 1, n_j = m_2 + \ldots + m_{j-1} - j + 2$ for $j = 3, \ldots, r$. The map here is a composition of the tensor products of symmetric diagonals with the map $\Psi(U)$, which is a generic map

$$\Psi(U) : F_2 \otimes \ldots \otimes F_r \otimes A \to F_1 \otimes A$$

corresponding to the form U. These determinantal expressions were first described in ([GKZ, chapter 14, Proposition 3.2]).

We have another choice of twists leading to quite different determinantal expressions.

(9.4.13) Example. *Bondal's expression for the hyperdeterminant of a three dimensional matrix in boundary format.*

Consider three dimensional matrices in boundary format $m_1 \times m_2 \times m_3$. Let us take $\mathcal{L} = \mathcal{O}(-1, m_3 - 2, 2m_2 - 3)$. Let us analyze the terms of the complex $F(\mathcal{L})_\bullet$. Again we use the notation from the proof of Theorem (9.4.5). Let us look at the cohomology of $\mathcal{F}(a_1, a_2, a_3) \otimes \mathcal{L}$. The weight corresponding to the first coordinate here is $((-1)^{m_1-1}, a - a_1)$. This means the we get nonzero terms only if in Bott's algorithm the last place goes to the first one. This in turn means that $a - a_1 \geq m_1 - 1$. But $a - a_1 = a_2 + a_3 \leq (m_2 - 1) + (m_3 - 1) \leq m_1 - 1$. Thus for all nonzero terms we have $a_2 = m_2 - 1$, $a_3 = m_3 - 1$.

Let us consider the contribution of terms $\mathcal{F}(a_1, m_2 - 1, m_3 - 1) \otimes \mathcal{L}$. The weights corresponding to that term on the second and third coordinate are $((m_3 - 1)^{m_2-1}, a_1 + m_3 - 1)$ and $((2m_2 - 2)^{m_3-1}, a_1 + m_2 - 1)$ respectively. From this we see that the terms with $1 \leq a_1 \leq m_2 - 1$ vanish because of the second coordinate weight, and the terms with $m_2 \leq a_1 \leq m_2 + m_3 - 2$ vanish because of the third coordinate. The only remaining terms are those for $a_1 = 0$ and $a_1 = m_2 + m_3 - 1$. The complex $F(\mathcal{L})_\bullet$ consists therefore of two terms

$$K_{((m_3-1)^{m_2})} F_2 \otimes K_{((2m_2-2)^{m_3-1}, m_2-1)} F_3 \quad = \quad S_{m_2-1} F_3^*$$
$$\uparrow$$
$$\bigwedge^{m_1} F_1 \otimes K_{(2m_3-1, (m_3)^{m_2-1})} F_2 \otimes K_{(2m_2-1)^{m_3}} F_3 \quad = \quad \bigwedge^{m_1} F_1 \otimes S_{m_3-1} F_2$$

with both identifications coming from disregarding the powers of determinantal representations.

The differential

$$\bigwedge^{m_1} F_1 \otimes S_{m_3-1} F_2 \otimes A \to S_{m_2-1} F_3^* \otimes A$$

is homogeneous of degree m_1. It is in fact easy to describe. There is only one (up to a nonzero scalar) $GL(F_1) \times GL(F_2) \times GL(F_3)$-equivariant map of the desired form. The point is that, using the Cauchy formulas (2.3.3)(a) and (b), we find out that the representation $\bigwedge^{m_1} F_1 \otimes K_{(m_3, 1^{m_2-1})} F_2 \otimes K_{(m_2, 1^{m_3-1})} F_3$ occurs in $\mathrm{Sym}_{m_1}(F_1 \otimes F_2 \otimes F_3)$ with multiplicity one. The embedding

$$\bigwedge^{m_1} F_1 \otimes K_{(m_3, 1^{m_2-1})} F_2 \otimes K_{(m_2, 1^{m_3-1})} F_3 \to \mathrm{Sym}_{m_1}(F_1 \otimes F_2 \otimes F_3)$$

induces the equivariant map

$$\bigwedge^{m_1} F_1 \otimes K_{(m_3, 1^{m_2-1})} F_2 \otimes A \to K_{(m_2, 1^{m_3-1})} F_3^* \otimes A,$$

which, up to tensoring with determinant representations, is our differential.

(9.4.14) Example. *Generalized Bondal's expression for the hyperdeterminant of a matrix in boundary format.*

Example (9.4.13) generalizes naturally to r-dimensional matrices of boundary format. Let us consider the format $m_1 \times \ldots \times m_r$ with $m_1 = m_2 + \ldots + m_r - r + 2$. The line bundle we choose is $\mathcal{L} = \mathcal{O}(u_1, \ldots, u_r)$ with $u_1 = -1, u_j = 2m_2 + \ldots + 2m_{j-1} + m_{j+1} + \ldots + m_r - j - 1$. The same reasoning as in the three dimensional case shows that the complex $F(\mathcal{L})_\bullet$ is just a map

$$S_{m_2-1}F_3^* \otimes S_{m_2+m_3-2}F_4^* \otimes \ldots \otimes S_{m_2+\ldots+m_{r-1}-r+2}F_r^* \otimes A$$

$$\uparrow$$

$$\bigwedge^{m_1} F_1 \otimes \bigwedge^{m_2} F_2 \otimes S_{m_3+\ldots+m_r-r+2}F_2 \otimes \ldots \otimes \bigwedge^{m_{r-1}} F_{r-1} \otimes$$
$$S_{m_r-1}F_{r-1} \otimes \bigwedge^{m_r} F_r \otimes A$$

with the map of degree m_1 in the entries of U.

We will say that a format $m_1 \times \ldots \times m_r$ is *subboundary* if $m_1 = m_2 + \ldots + m_r - r + 1$, i.e. if the format $(m_1 + 1) \times m_2 \times \ldots \times m_r$ is a boundary format.

(9.4.15) Example. *The hyperdeterminant of a three dimensional matrix of a subboundary case has a determinantal expression.*

Let us consider three dimensional matrices of the subboundary format $m_1 \times m_2 \times m_3$, i.e. $m_1 = m_2 + m_3 - 2$. Let us take $\mathcal{L} = \mathcal{O}(-1, m_3 - 2, 2m_2 - 3)$.

As before, we use the notation from the proof of Theorem (9.4.5). Let us look at the cohomology of $\mathcal{F}(a_1, a_2, a_3) \otimes \mathcal{L}$. The weight corresponding to the first coordinate here is $((-1)^{m_1-1}, a - a_1)$. This means the we get nonzero terms only if in Bott's algorithm the last place goes to the first one. This in turn means that $a - a_1 \geq m_1 - 1$. But $a - a_1 = a_2 + a_3 \leq (m_2 - 1) + (m_3 - 1) \leq m_1$. Thus the pair (a_1, a_2) is $(m_2 - 2, m_3 - 1)$, $(m_2 - 1, m_3 - 2)$, or $(m_2 - 1, m_3 - 1)$.

Let us consider the contribution of terms $\mathcal{F}(a_1, m_2 - 2, m_3 - 1) \otimes \mathcal{L}$. The weights corresponding to that term on the second and third coordinates are $((m_3 - 1)^{m_2-2}, m_3 - 2, a_1 + m_3 - 1)$ and $(2m_2 - 2)^{m_3-1}, a_1 + m_2 - 2)$ respectively. From this we see that the surviving terms are:

(1) The term with $a_1 = 1$, giving the term with the partitions

$$(1, 0^{m_1-1}), \qquad ((m_3 - 1)^{m_2}), \qquad ((2m_2 - 2)^{m_3-1}, m_2 - 1)$$

on three coordinates.

(2) The term with $a_1 = m_2$, giving the term with the partitions

$$(1^{m_2}, 0^{m_1 - m_2}), \qquad (m_3^{m_2 - 1}, m_3 - 1), \qquad ((2m_2 - 2)^{m_3})$$

on three coordinates.
Next we analyze the contributions of the terms $\mathcal{F}(a_1, m_2 - 1, m_3 - 2)$ $\otimes \mathcal{L}$. The partitions on the second and third coordinate are $((m_3 - 1)^{m_2 - 1}, a_1 + m_3 - 2)$ and $((2m_2 - 2)^{m_3 - 2}, 2m_2 - 3, a_1 + m_2 - 1)$ respectively. The surviving terms are:
(3) The term with $a_1 = 0$, with the partitions

$$(0^{m_1}), \qquad ((m_3 - 1)^{m_2 - 1}, m_3 - 2),$$
$$((2m_2 - 2)^{m_3 - 2}, 2m_2 - 3, m_2 - 1)$$

on three coordinates.
(4) The term with $a_1 = 1$, with the partitions

$$(1, 0^{m_1 - 1}), \qquad ((m_3 - 1)^{m_2}), \qquad ((2m_2 - 2)^{m_3 - 2}, 2m_2 - 3, m_2)$$

on three coordinates.
Finally we deal with the contributions of the terms $\mathcal{F}(a_1, m_2 - 1, m_3 - 1) \otimes \mathcal{L}$. The partitions on the second and third coordinate are $((m_3 - 1)^{m_2 - 1}, a_1 + m_3 - 1)$ and $(2m_2 - 2)^{m_3 - 1}, a_1 + m_2 - 1)$ respectively. The only surviving term is
(5) The term with $a_1 = 0$, with the partitions

$$(1, 0^{m_1 - 1}), \qquad ((m_3 - 1)^{m_2}), \qquad ((2m_2 - 2)^{m_3 - 1}, m_2 - 1)$$

on three coordinates.

Now we show that terms (1) and (5) cancel in the spectral sequence. This follows from considering the exact sequence

$$0 \to \mathcal{R}_1 \otimes \mathcal{F}_2 \otimes \mathcal{R}_3 \to \xi \to \mathcal{Q}_1 \otimes \mathcal{R}_2 \otimes \mathcal{R}_3 \oplus \mathcal{R}_1 \otimes \mathcal{R}_2 \otimes \mathcal{Q}_3 \to 0.$$

One checks readily that the representation occuring in terms (1) and (5) does not occur in the cohomology of

$$\mathcal{L} \otimes \bigwedge^{\bullet} (\mathcal{R}_1 \otimes \mathcal{F}_2 \otimes \mathcal{R}_3 \oplus \mathcal{Q}_1 \otimes \mathcal{R}_2 \otimes \mathcal{R}_3 \oplus \mathcal{R}_1 \otimes \mathcal{R}_2 \otimes \mathcal{Q}_3).$$

The conclusion is that the complex $F(\mathcal{L})_{\bullet}$ is the following map:

$$K_{((m_3 - 1)^{m_2 - 1}, m_3 - 2)} F_2 \otimes K_{((2m_2 - 2)^{m_3 - 2}, 2m_2 - 3, m_2 - 1)} F_3 \otimes A$$

$$\uparrow$$

$$F_1 \otimes K_{((m_3 - 1)^{m_2})} F_2 \otimes K_{((2m_2 - 2)^{m_3 - 2}, 2m_2 - 3, m_2)} F_3 \otimes A$$
$$\oplus \bigwedge^{m_2} F_1 \otimes K_{((m_3)^{m_2 - 1}, m_3 - 1)} F_2 \otimes K_{((2m_2 - 2)^{m_3})} F_3 \otimes A.$$

The differential is easy to describe. The part from the upper term is linear and therefore describable using Pieri's formula. It is given by the product of the maps described in exercise 17 of chapter 2 (in the case where one of the partitions has one box).

The part of the differential from the lower term has degree m_2. Let us try to exhibit the equivariant map

$$\bigwedge\nolimits^{m_2} F_1 \otimes K_{((m_3)^{m_2-1}, m_3-1)} F_2 \otimes K_{((2m_2-2)^{m_3})} F_3$$

$$\uparrow \theta$$

$$K_{((m_3-1)^{m_2-1}, m_3-2)} F_2 \otimes K_{((2m_2-2)^{m_3-2}, 2m_2-3, m_2-1)} F_3 \otimes \mathrm{Sym}_{m_2}(F_1 \otimes F_2 \otimes F_3)$$

that gives the generators. By exercise 12 of chapter 2 we see that the only representation $K_\lambda F_3$ such that $K_{((2m_2-2)^{m_3})} F_3$ occurs in $K_{((2m_2-2)^{m_3-2}, 2m_2-3, m_2-1)} F_3 \otimes K_\lambda F_3$ is $K_{(m_2-1,1)} F_3$. Also, the multiplicity of $K_{((2m_2-2)^{m_3})} F_3$ in this tensor product is equal to one. This means that the image of the map θ has to be contained in $K_{((m_3-1)^{m_2-1}, m_3-2)} F_2 \otimes K_{((2m_2-2)^{m_3-2}, 2m_2-3, m_2-1)} F_3$ tensored with the subrepresentation of $\mathrm{Sym}_{m_2}(F_1 \otimes F_2 \otimes F_3)$ of the form $\bigwedge\nolimits^{m_2} F_1 \otimes V \otimes K_{(m_2-1,1)} F_3$, where V is some homogeneous polynomial representation of $\mathrm{GL}(F_2)$ of degree m_2. However, using the Cauchy formula (2.3.3) (a) and (b), we see that there is only one irreducible subrepresentation in $\mathrm{Sym}_{m_2}(F_1 \otimes F_2 \otimes F_3)$ of this form, namely $\bigwedge\nolimits^{m_2} F_1 \otimes K_{(2,1^{m_2-2})} F_2 \otimes K_{(m_2-1,1)} F_3$, which occurs in $\mathrm{Sym}_{m_2}(F_1 \otimes F_2 \otimes F_3)$ with multiplicity one. This determines the differential in our complex up to a nonzero scalar.

The conclusion from the examples above is that there seems to be no direct connection between the hyperdeterminant of the matrix in the boundary format and its submatrices in the subboundary format that we get by omitting a slice in the direction of F_1. This means that if the analogue of Laplace expansion exists, it is not a straightforward generalization of the two dimensional case.

(9.4.16) Remark. *The example (9.4.15) generalizes to the r-dimensional matrices of subboundary format. Let us look at the format $m_1 \times \ldots \times m_r$ with $m_1 = m_2 + \ldots + m_r - r + 1$. Let us take $\mathcal{L} = \mathcal{O}(u_1, \ldots, u_r)$ where $u_1 = -1$, $u_j = 2m_2 + \ldots + 2m_{j-1} + m_{j+1} + \ldots + m_r - j$ for $2 \le j \le r$. Then, as was observed in [W3, Example 4.4], the complex $F(\mathcal{L})_\bullet$ gives a determinantal expansion of the hyperdeterminant. The complex $F(\mathcal{L})_\bullet$ has $r - 2$ irreducible representations in homogeneous degree m_1 and in homological degree 0. In homological degree 1 there are $r - 2$ irreducible representations* *in homogeneous degree $m_1 + 1$, and $r - 2$ irreducible representations in homogeneous degrees $m_1 + \ldots + m_j - j + 1$ ($2 \leq j \leq r - 1$).*

(9.4.17) Remarks.
  (a) *It is a very interesting problem to obtain the analogue of the results of section 9.2. The first question is for which formats there exists a bundle $\mathcal{L}$ such that the complex $F(\mathcal{L})_\bullet$ is determinantal. It seems that the determinantal complexes occur for the formats that are close to the boundary case. We have seen above that they exist for the boundary and subboundary cases. It is also proven in [W3] that for the format $m \times m \times m$ for $m \geq 8$ there is no line bundle $\mathcal{L}$ such that the complex $F(\mathcal{L})_\bullet$ is determinantal. More generally, Galindo [Ga] has shown that for three dimensional matrices the determinantal expressions for the hyperdeterminant occur only for formats close to boundary formats.*
  (b) *If the above problem is solved, the next interesting question is the classification of determinantal complexes. The combinatorics of this seems to be quite tricky.*

We close this section with a discussion of the case of $m \times 2 \times 2$ matrices, i.e. the smallest case where $Y^\vee$ has codimension bigger than 1. We just investigate the complex $F_\bullet$. It turns out that it is almost a free resolution of the coordinate ring of $Y^\vee$. We use that fact to get the information about the defining ideal of $Y^\vee$.

Let us take $r = 3$, $\dim F_1 = m$, $\dim F_2 = \dim F_3 = 2$. We compute the complex $F_\bullet$. Let us start with the bundle ξ_1. As before, we have

$$\bigwedge(\xi_1) = \bigoplus_{a_1+a_2+a_3=a} \overset{a_1}{\bigwedge} F_1 \otimes \overset{a_2}{\bigwedge} \mathcal{Q}_2 \otimes \overset{a_3}{\bigwedge} \mathcal{Q}_3 \otimes \mathcal{R}_1^{a-a_1} \otimes \mathcal{R}_2^{a-a_2} \otimes \mathcal{R}_3^{a-a_3}.$$

Let us look at the weight corresponding to F_1. It is $(0^{m-1}, a_2 + a_3)$. However, now $m \geq 4$ and $a_2 \leq 1$, $a_3 \leq 1$. This means that the only possibility for nonzero cohomology is $a_2 = a_3 = 0$. In such a case the two other weights are $(0, a_1)$. Therefore the summand with $a_1 = a_2 = a_3$ contributes a trivial representation to $H^0(\bigwedge^0 \xi_1)$, and the summands with $a_1 \geq 2$, $a_2 = a_3 = 0$ contribute to $H^2(\bigwedge^{a_1} \xi_1)$. Since all odd cohomology groups are zero, there are no cancellations in the spectral sequence, and all the terms appear in $F_\bullet$. We have proved

(9.4.18) Proposition. *The complex* $F_{\bullet}$ *has the following form:*

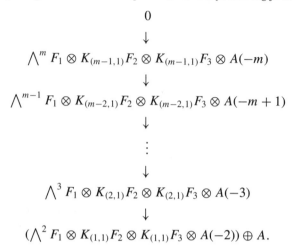

$$0$$
$$\downarrow$$
$$\bigwedge^{m} F_1 \otimes K_{(m-1,1)} F_2 \otimes K_{(m-1,1)} F_3 \otimes A(-m)$$
$$\downarrow$$
$$\bigwedge^{m-1} F_1 \otimes K_{(m-2,1)} F_2 \otimes K_{(m-2,1)} F_3 \otimes A(-m+1)$$
$$\downarrow$$
$$\vdots$$
$$\downarrow$$
$$\bigwedge^{3} F_1 \otimes K_{(2,1)} F_2 \otimes K_{(2,1)} F_3 \otimes A(-3)$$
$$\downarrow$$
$$(\bigwedge^{2} F_1 \otimes K_{(1,1)} F_2 \otimes K_{(1,1)} F_3 \otimes A(-2)) \oplus A.$$

The complex $F_{\bullet}$ *is the minimal resolution of the normalization of* $\mathbf{K}[Y^{\vee}]$*, and that normalization has rational singularities.*

Proof. The terms of the complex are calculated above. The last statement follows from (5.1.3). ∎

The term F_0 consists of two representations, which means that $\mathbf{K}[Y^{\vee}]$ is not normal. Let us look at this normalization a little bit more closely. We want to see how it can be seen as an algebra. We will look at the algorithm due to Grauert and Remmert and to de Jong, described in section 1.2.4.

Notice that the cokernel $C = \bar{\mathcal{O}}_{Y^{\vee}}/\mathcal{O}_{Y^{\vee}}$ has the presentation

$$\bigwedge^{3} F_1 \otimes K_{(2,1)} F_2 \otimes K_{(2,1)} F_3 \otimes A(-3)$$
$$\rightarrow \bigwedge^{2} F_1 \otimes K_{(1,1)} F_2 \otimes K_{(1,1)} F_3 \otimes A(-2) \rightarrow C \rightarrow 0.$$

Disregarding the twist by $\bigwedge^2 F_2 \otimes \bigwedge^2 F_3$, we see that the presenting map is $\mathrm{GL}(F_1) \times \mathrm{GL}(F_2 \otimes F_3)$-equivariant. Let $\hat{U}$ be a matrix of the map $F_1 \otimes A(-1) \rightarrow F_2^* \otimes F_3^* \otimes A$ associated to the generic form U. The support of the cokernel $C = \bar{\mathcal{O}}_{Y^{\vee}}/\mathcal{O}_{Y^{\vee}}$ is given by 3×3 minors of the matrix $\hat{U}$. Let $R = A/I(Y^{\vee})$, and let $J \subset R$ denote the ideal in R generated by 3×3 minors of $\hat{U}$.

(9.4.19) Proposition. *The normalization* $\mathbf{K}[\bar{Y}^{\vee}]$ *is isomorphic to* $\mathrm{Hom}_R(J, J)$*. This means that the algorithm of Grauert and Remmert and de Jong produces the normalization in one step.*

Proof. First we treat the case $m = 3$. The complex $F_\bullet$ is just a map of free modules

$$d_1 = \begin{array}{c} \bigwedge^3 F_1 \otimes K_{(2,1)}F_2 \otimes K_{(2,1)}F_3 \otimes A(-3) \\ \downarrow \\ (\bigwedge^2 F_1 \otimes K_{(1,1)}F_2 \otimes K_{(1,1)}F_3 \otimes A(-2)) \oplus A. \end{array}$$

Explicitly the matrix of this map can be written as

$$\begin{pmatrix} u_{3,1,1} & u_{3,1,2} & u_{3,2,1} & u_{3,2,2} \\ -u_{2,1,1} & -u_{2,1,2} & -u_{2,2,1} & -u_{2,2,2} \\ u_{1,1,1} & u_{1,1,2} & u_{1,2,1} & u_{1,2,2} \\ M_{2,2} & M_{2,1} & -M_{1,2} & -M_{1,1} \end{pmatrix},$$

where $M_{i,j}$ is the 3×3 minor of the matrix

$$\hat{U} = \begin{pmatrix} u_{3,1,1} & u_{3,1,2} & u_{3,2,1} & u_{3,2,2} \\ -u_{2,1,1} & -u_{2,1,2} & -u_{2,2,1} & -u_{2,2,2} \\ u_{1,1,1} & u_{1,1,2} & u_{1,2,1} & u_{1,2,2} \end{pmatrix}$$

obtained by skipping a column containing the entries with indices (i, j) in the second and third places. The linear part of the map is identified directly from Pieri's formula. The last row can be identified if we observe that $K_{(2,1)}F_2 \otimes K_{(2,1)}F_3 = \bigwedge^3(F_2 \otimes F_3)$ and by the Cauchy formula the map has to be given by 3×3 minors of the 3×4 matrix obtained from U by flattening two directions into one.

In fact the hyperdeterminant Δ can be calculated in terms of the matrix d_1 as its determinant, and it equals

$$\Delta = 2(M_{1,1}M_{2,2} - M_{1,2}M_{2,1}).$$

Let $f_{1,2}$, $f_{1,3}$, $f_{2,3}$ be the generators of $\bigwedge^2 F_1 \otimes K_{(1,1)}F_2 \otimes K_{(1,1)}F_3 \otimes A(-2)$. We want to interpret them as endomorphisms of J. But the matrix d_1 gives us the syzygies between $f_{i,j}$ and the identity endomorphism of J. Using Cramer's rule, we get

$$f_{k,l}(M_{i,j}) = \Delta_{k,l}^{i,j},$$

where $\Delta_{k,l}^{i,j}$ denotes the 3×3 minor of d_1 which we get by skipping the row corresponding to (k, l) and the column corresponding to (i, j). This means that $f_{k,l}$ is represented in the field of fractions of R by the fraction $f_{k,l} = \Delta_{k,l}^{i,j}/M_{i,j}$. It is easy to see this fraction does not depend on the choice of (i, j), or, equivalently, that $f_{k,l}$ is well defined by the above formulas.

We would also like to get the presentation of the normalization $\mathbf{K}[\bar{Y}^\vee]$. Let us consider the polynomial ring $R[T_0, T_{1,2}, T_{1,3}, T_{2,3}]$ and the epimorphism

$$\theta : R[T_0, T_{1,2}, T_{1,3}, T_{2,3}] \to \mathbf{K}[\bar{Y}^\vee]$$

given by $\theta(T_0) = \mathrm{id}$, $\theta(T_{k,l}) = f_{k,l}$. By (1.2.16) we know that the kernel of the homomorphism θ is generated by linear and quadratic relations. The linear relations come from the syzygies between the elements and therefore are given by the columns of the matrix d_1. The quadratic relations can be identified with the help of representation theory. According to (1.2.16) these relations have to have the form

$$T_{k,l} T_{s,t} - \sum_{(x,y)}^{s} a_{(x,y)}^{(k,l)(s,t)} T_{x,y} + c_{(k,l)(s,t)} T_0$$

with $(k, l)(s, t) \in \{(1, 2), (1, 3), (2, 3)\}$.

We claim that the coefficients $a_{(x,y)}^{(k,l)(s,t)}$ are all zero. Indeed, looking at the situation from equivariant point of view, these coefficients, if nonzero, would generate a representation W inside $\mathrm{Sym}_2(F_1 \otimes F_2 \otimes F_3)$ such that $W \otimes \bigwedge^2 F_1 \otimes \bigwedge^2 F_2 \otimes \bigwedge^2 F_3$ would have a nonzero map into $S_2(\bigwedge^2 F_1) \otimes K_{(2,2)}F_2 \otimes K_{(2,2)}F_3$. It is easily seen from Pieri's and Cauchy's formulas that such a representation W does not exist.

It remains to identify the coefficients $c_{(k,l)(s,t)}$. These coefficients have to span the unique representation $K_{(2,2)}F_1 \otimes K_{(2,2)}F_2 \otimes K_{(2,2)}F_3$ inside $S_4(F_1 \otimes F_2 \otimes F_3)$ and therefore are determined up to a nonzero scalar.

This concludes the proof of (9.4.14) in the case $m = 3$.

Let us move to the general case. We denote by $f_{k,l}$ ($1 \le k < l \le m$) the generator of F_0 corresponding to $e_k \wedge e_l$ (where $\{e_1, \ldots, e_m\}$ is a basis of F_1). We define the homomorphism

$$\theta : R[T_0, \{T_{k,l}\}] \to \mathbf{K}[\bar{Y}^\vee]$$

by $\theta(T_0) = \mathrm{id}$, $\theta(T_{k,l}) = f_{k,l}$.

Again we need to identify the elements $f_{k,l}$ as endomorphisms of J and as fractions, and we need to work out the linear and quadratic relations generating the kernel of θ. We denote by $M_{i,j}^{p,q,r}$ ($1 \le p < q < r \le m$, $1 \le i, j \le 2$) a 3×3 minor of $\hat{U}$ corresponding to columns p, q, r with the row corresponding to (i, j) omitted.

We notice that for $m \ge 4$ the ring R has among its relations the 4×4 minors of the matrix $\hat{U}$. We still have the relations expressing $f_{k,l}$ as fractions for $m = 3$. Let us describe the representation generated by the elements $\Delta_{k,l}^{i,j}$ (with i, j, k, l allowed to come from the set $\{1, 2, 3\}$). We know that all such elements are linear combinations of products of $f_{k,l}$ and of $M_{i,j}^{p,q,r}$. Such a representation is contained in R and is an image of the tensor product $\bigwedge^2 F_1 \otimes L_{(2)} F_2 \otimes L_{(2)} F_3 \otimes \bigwedge^3 F_1 \otimes K_{(2,1)} F_2 \otimes K_{(2,1)} F_3$. However, such an image must consist of the single irreducible representation $K_{(2,2,1)} F_1 \otimes K_{(3,2)} F_2 \otimes K_{(3,2)} F_3$. Indeed, all the other irreducibles in this tensor product involve a row of length ≥ 4 in the partition corresponding to F_1 and therefore are contained in the ideal of 4×4 minors of $\hat{U}$, so they are identically zero in R. Now it is clear that the products of $f_{k,l}$ and the minors $M_{i,j}^{p,q,r}$ span an irreducible representation which is contained in R. Furthermore, for every k, l and i, j, p, q, r the element $f_{k,l}$ can be expressed as an element $\Delta_{(i,j)(k,l)}^{p,q,r}$ of R divided by $M_{i,j}^{p,q,r}$. Now we can define

$$f_{k,l}(M_{i,j}^{p,q,r}) = \Delta_{(i,j)(k,l)}^{p,q,r},$$

and this defines an endomorphism of J.

The relations defining the kernel of θ can be described as in the case $m = 3$. ∎

In order to get information about the defining ideal of $Y^\vee$ for $m \times 2 \times 2$ matrices we have to investigate the cokernel $C = \overline{\mathcal{O}_{Y^\vee}} / \mathcal{O}_{Y^\vee}$ more closely. From our computation of $F_\bullet$ it follows that the cokernel C has the presentation

$$\bigwedge^3 F_1 \otimes K_{(2,1)} F_2 \otimes K_{(2,1)} F_3 \otimes A(-3)$$

$$\rightarrow \bigwedge^2 F_1 \otimes K_{(1,1)} F_2 \otimes K_{(1,1)} F_3 \otimes A(-2) \rightarrow C \rightarrow 0.$$

Here we can use the knowledge about the twisted modules supported in determinantal varieties. Since dim $F_2 = $ dim $F_3 = 2$, we see that, writing $H = F_2 \otimes F_3$ and disregarding the twist by the one dimensional representation $\bigwedge^2 F_2^* \otimes \bigwedge^2 F_3^*$, we get the following presentation:

$$\bigwedge^3 F_1 \otimes H \otimes A(-3) \rightarrow \bigwedge^2 F_1 \otimes A(-2) \rightarrow C \rightarrow 0.$$

This module can be resolved using the techniques of section 6.5. Let us consider the Grassmannian Grass$(m - 2, F_1)$ with the tautological sequence

$$0 \longrightarrow \mathcal{R} \longrightarrow F_1 \longrightarrow \mathcal{Q} \longrightarrow 0,$$

where dim $\mathcal{R} = m - 2$, dim $\mathcal{Q} = 2$. We introduce the affine space $X = \text{Hom}_k (H, F_1^*)$, which we will identify with $F_1^* \otimes H^*$. Notice that Grass$(m - 2, F_1)$ can be canonically identified with Grass$(2, F_1^*)$ by associating to the subspace R the dual Q^* of the quotient $Q = F_1/R$.

We consider the subvariety of $Z \subset X \times \mathrm{Grass}(m - 2, F_1)$ given by

$$Z = \{(\phi, Q^*) \mid \mathrm{Im}\,\phi \subset Q^*\}.$$

This is precisely the construction from section 6.5 after specializing F and G to H and F_1^* respectively (and changing $F \otimes G^*$ into $G^* \otimes F$ everywhere). The variety Z is the desingularization of the subvariety $Y \subset X$ of maps ϕ of rank ≤ 2. We repeat the construction of the Koszul complex, noting that the bundle ξ becomes $\mathcal{R} \otimes H$.

We consider now the twisted complex $F(\mathcal{L})_\bullet$ where $\mathcal{L} = \bigwedge^2 Q$. In order to distiguish the new $F_\bullet$ from the old one we will denote it by $\hat{F}(\mathcal{L})_\bullet$. Then the terms of the complex $\hat{F}(\mathcal{L})_\bullet$ become

$$\hat{F}(\mathcal{L})_i = \bigoplus_{j \geq 0} H^j\left(\mathrm{Grass}(m - 2, F_1), \bigwedge^{i+j}(\mathcal{R} \otimes H) \otimes \bigwedge^2 Q\right) \otimes A(-i - j).$$

Using Cauchy's formula, we see that the nonzero cohomology groups of the bundles $\bigwedge^\bullet(\mathcal{R} \otimes H) \otimes \bigwedge^2 Q$ are

(1) $\bigwedge^{j+2} F_1 \otimes \bigwedge^j H$ in $H^0(\bigwedge^j(\mathcal{R} \otimes H) \otimes \bigwedge^2 Q)$;
(2) for each partition α of n, $\alpha_1 \leq 2$, $K_{(2^3, \alpha)} F_1 \otimes L_{(4, \alpha)} H$ in $H^2(\bigwedge^{n+4} (\mathcal{R} \otimes H) \otimes \bigwedge^2 Q)$.

We see that $\hat{F}(\mathcal{L})_i = 0$ for $i < 0$. Moreover, $\hat{F}(\mathcal{L})_0 = \bigwedge^2 F_1 \otimes A(-2)$ and $\hat{F}(\mathcal{L})_1 = \bigwedge^3 F_1 \otimes H \otimes A(-3)$. Therefore the complex $\hat{F}(\mathcal{L})_\bullet$ gives the resoluton of the module C.

The map $\overline{\mathcal{O}_{Y^\vee}} \to C$ induces a map of free resolutions $\Gamma : F_\bullet \to \hat{F}(\mathcal{L})$ which is unique up to a homotopy. Of course the map Γ can be choosen to be G-equivariant.

Now the cone of Γ is a nonminimal resolution of $\mathcal{O}_{Y^\vee}$. By definition the map Γ sends the representation $\bigwedge^2 F_1 \otimes S_{(1,1)} F_2 \otimes S_{(1,1)} F_3 \otimes A(-2)$ isomorphically to $\hat{F}(\mathcal{L})_0$. Then it follows easily by induction that the map Γ sends the term F_j isomorphically into the term $\bigwedge^{j+2} F_1 \otimes K_{(j+1,1)} F_2 \otimes K_{(j+1,1)} F_3 \otimes A(-j - 2)$ sitting inside $\bigwedge^{j+2} F_1 \otimes \bigwedge^j H \otimes A(-j)$ twisted by $\bigwedge^2 F_2 \otimes \bigwedge^2 F_3$.

This means that in the minimal resolution of $\mathcal{O}_{Y^\vee}$ these terms cancel out. The conclusion is that the j-th term of the minimal resolution of $\mathcal{O}_{Y^\vee}$ equals

$$\bigoplus_{|\gamma| = j+1, \gamma \neq (j+1)} \bigwedge^{j+3} F_1 \otimes \bigwedge^2 F_2 \otimes \bigwedge^2 F_3 \otimes K_\gamma F_2 \otimes K_\gamma F_3$$
$$\otimes A(-j - 3)$$
$$\bigoplus_{|\beta| = j-1, \beta_1 \leq 2} K_{(2,2,2,\beta)} F_1 \otimes K_{(3,3)} F_2 \otimes K_{(3,3)} F_3 \otimes L_\beta(F_2 \otimes F_3)$$
$$\otimes A(-|\beta| - 6).$$

Let us collect the results we can deduce from this formula.

(9.4.20) Proposition. *Let U be a generic $m \times 2 \times 2$ matrix. Let $A =$ Sym$(F_1 \otimes F_2 \otimes F_3)$ be a coordinate ring on the entries of U.*

(a) *The homological dimension of $\mathbf{K}[Y^\vee]$ as an A-module is $2m - 5$. The codimension of this module equals $m - 2$, so $\mathbf{K}[Y^\vee]$ is never perfect.*

(b) *The defining ideal $I_{Y^\vee}$ is generated by two irreducible representations. The first one, isomorphic to $\bigwedge^4 F_1 \otimes K_{(2,2)} F_2 \otimes K_{(2,2)} F_3$ in degree 4, corresponds to 4×4 minors of the $m \times 4$ matrix we get from our $m \times 2 \times 2$ matrix by writing one $m \times 2$ slice under another. The second one, isomorphic to $K_{(2^3)} F_1 \otimes K_{(3,3)} F_2 \otimes K_{(3,3)} F_3$ in degree 6, is generated as a $\mathbf{G}$-module by the hyperdeterminants of $3 \times 2 \times 2$ submatrices of our matrix.*

Proof. The first statement is clear. To prove the second we notice that the formula above shows that the first term of the minimal resolution of $\mathbf{K}[Y^\vee]$ corresponding to the generators of the defining ideal $I_{Y^\vee}$ of $Y^\vee$ equals

$$K_{(2^3)} F_1 \otimes K_{(3,3)} F_2 \otimes K_{(3,3)} F_3 \otimes A(-6)$$
$$\oplus \bigwedge^4 F_1 \otimes K_{(2,2)} F_2 \otimes K_{(2,2)} F_3 \otimes A(-4).$$

To identify these generators we have to find the corresponding sets of polynomials vanishing on $Y^\vee$. Both representations identified in the proposition vanish on $Y^\vee$ and occur in the right degrees, so they have to generate the ideal $I_{Y^\vee}$. ∎

(9.4.21) Remark. *This description of the defining ideal seems to generalize to other formats. In [W3] the analogous statement is proven for $m \times 3 \times 2$ matrices. It would be very interesting to find the generators of the ideal $I_{Y^\vee}$ in general.*

This concludes our collection of examples involving hyperdeterminants. I think it illustrates well the state of affairs here. There are still many open problems and possibilities for interesting research.

Exercises for Chapter 9

1. Let G be a simple group, and let $\underline{g}$ be its Lie algebra. Let $X = V_\lambda$ be the highest weight representation of highest weight λ. We define $Y_\lambda^\vee$ to be the dual variety of the closure Y of the highest weight vector in X. We identify Y with the cone over a homogeneous space G/P for the suitable

parabolic subgroup in G. Prove that $Y_\lambda^\vee$ has a desingularization of type described in section 9.3. The bundle $\xi = \underline{n} \times^P G$, where $\underline{n}$ is a rational P-module whose weights are $-\lambda$ and $-\lambda + \alpha$, where α runs through all positive roots not in P.

2. Knop and Menzel classified in [KM] the highest weights λ for simple groups for which the discriminant $Y_\lambda^\vee$ is not a hypersurface. The most interesting exceptional case for simple groups is the half-spinor representation for the group of type D_5. This is a representation of dimension 16 whose weights are $(\pm\frac{1}{2}, \pm\frac{1}{2}, \pm\frac{1}{2}, \pm\frac{1}{2}, \pm\frac{1}{2})$ with an even number of minus signs. The highest weight is $(\frac{1}{2}, \frac{1}{2}, \frac{1}{2}, \frac{1}{2}, \frac{1}{2})$. The orbit of the highest weight vector is the cone over a variety of pure spinors. Calculate the complex $F_\bullet$ to see that the dual variety is again the variety of pure spinors and has codimension 5. In fact, the terms of the complex are

$$F_0 = A, \qquad F_1 = V(\omega_1) \otimes A(-2), \qquad F_2 = V(\omega_5) \otimes A(-3),$$
$$F_3 = V(\omega_4) \otimes A(-5), \qquad F_4 = V(\omega_1) \otimes A(-6), \qquad F_5 = A(-8),$$

where we use Bourbaki's notation for the fundamental weights, i.e. $\omega_i = (1^i, 0^{5-i})$ for $i = 1, 2, 3$, $\omega_4 = (\frac{1}{2}, \frac{1}{2}, \frac{1}{2}, \frac{1}{2}, \frac{1}{2})$, $\omega_5 = (\frac{1}{2}, \frac{1}{2}, \frac{1}{2}, \frac{1}{2}, \frac{-1}{2})$.

Discriminants of Adjoint Representations

3. Let $\underline{g} = \underline{sl}(n)$. Consider the dual variety $Y^\vee$ of the orbit Y of the highest weight vector in $\underline{g}$.
    (a) Show that the bundle ξ_n occurring in the Koszul complex related to the desingularization of $Y^\vee$ contained as an incidence variety in $X \times V$, where $V = \mathrm{Fl}(1, n-1; E)$, can be described by the exact sequences

    $$0 \to \mathcal{R}_1 \otimes \mathcal{Q}_1^* \to \mathcal{R}_1 \otimes E^* \oplus E \otimes \mathcal{Q}_1^* \to \xi_n' \to 0,$$
    $$0 \to \xi_n \to \xi_n' \to \mathcal{O}_V \to 0.$$

    (b) Show that the complex $F_\bullet$ has only two terms

    $$F_0 = \bigoplus_{0 \le i \le n-2} A(-i),$$
    $$F_1 = \bigoplus_{n \le i \le 2n-2} A(-i).$$

    Deduce that the degree of the discriminant equals n^2.
    (c) Prove that the equation of $X^\vee$ is the discriminant of the characteristic polynomial of an $n \times n$ matrix.

4. Let $g = \underline{so}(2n + 1)$. Consider the cone $Y^\vee$ over the dual variety of the orbit Y' of the highest weight vector in $\underline{g}$.

   (a) Identify the bundle ξ_n occurring in the Koszul complex related to the desingularization of $Y^\vee$ contained as an incidence variety in $X \times V$, where $V = \text{IGrass}(2, F)$. We identify V with the homogeneous space $\mathbf{G}/\mathbf{P}$. The bundle $\xi_n = \underline{n} \times^{\mathbf{P}} \mathbf{G}$, where $\underline{n}$ is a rational $\mathbf{P}$-module with the weights $-\epsilon_1 \pm \epsilon_j$ $(3 \leq j \leq n)$, $-\epsilon_2 \pm \epsilon_j$ $(3 \leq j \leq n)$, and $-\epsilon_1 - \epsilon_2$, $-\epsilon_1$, $-\epsilon_2$,

   (b) Prove that the Euler characteristic of $\bigwedge^i \xi_n$ equals to 0 for i odd or $i = 2n - 2$, 1 for i even, $0 \leq i \leq 2n - 4$, and -1 for i even, $2n \leq i \leq 4n - 6$. All representations in the cohomology of exterior powers of ξ_n are trivial, and the only nonzero terms of the complex $F_\bullet$ are F_0 and F_1.

   (c) Deduce that the degree of the discriminant equals $(2n - 2)n$. Identify the discriminant.

5. Let $\underline{g} = \text{sp}(2n)$. Consider the cone $Y^\vee$ over the dual variety of the orbit Y' of the highest weight vector in $\underline{g}$.

   (a) Identify the bundle ξ_n occurring in the Koszul complex related to the desingularization of $Y^\vee$ contained as an incidence variety in $X \times V$, where $V = \text{IGrass}(1, F) = \mathbf{P}(F)$. We identify V with the homogeneous space $\mathbf{G}/\mathbf{P}$. Since Y is $\text{GL}(F)$-stable, so is $Y^\vee$.

   (b) Identify $Y^\vee$ with the set of symmetric matrices not of maximal rank,

   (c) Deduce that the discriminant can be identified with the determinant.

6. Let $g = \underline{so}(2n)$. Consider the cone $Y^\vee$ over the dual variety of the orbit Y' of the highest weight vector in $\underline{g}$.

   (a) Identify the bundle ξ_n occurring in the Koszul complex related to the desingularization of $Y^\vee$ contained as an incidence variety in $X \times V$, where $V = \text{IGrass}(2, F)$. We identify V with the homogeneous space $\mathbf{G}/\mathbf{P}$. The bundle $\xi_n = \underline{n} \times^{\mathbf{P}} \mathbf{G}$, where $\underline{n}$ is a rational $\mathbf{P}$-module with the weights $-\epsilon_1 \pm \epsilon_j$ $(3 \leq j \leq n)$, $-\epsilon_2 \pm \epsilon_j$ $(3 \leq j \leq n)$, and $-\epsilon_1 - \epsilon_2$,

   (b) Prove that the Euler characteristic of $\bigwedge^i \xi_n$ equals $\delta_{i,n-2} - \delta_{i,3n-4}$ for i odd, $1 + \delta_{i,n-2} - \delta_{i,3n-4}$ for i even $(0 \leq i \leq 2n - 3)$, and $-1 + \delta_{i,n-2} - \delta_{i,3n-4}$ for i even $(2n - 2 \leq i \leq 4n - 6)$ ($\delta_{i,j}$ denotes the Kronecker delta). Prove that all representations in the cohomology of exterior powers of ξ_n are trivial and that the only nonzero terms of the complex $F_\bullet$ are F_0 and F_1.

   (c) Deduce that the degree of the discriminant equals $(2n - 2)n$. Identify the discriminant.

Determinantal Expressions for Powers of the Resultant

7. Let F be a vector space of dimension $l + 1$. Consider the resultant variety $\nabla_{\mathbf{P}^l}$ in $X = S_d(F^*) \otimes W$ (as defined in section 9.2). We have a complex $F_{\bullet}$ associated to the desingularization Z of $\nabla_{\mathbf{P}^l}$ contained in $X \times \mathbf{P}^l$. The vector bundle ξ is just $\mathcal{O}_{\mathbf{P}^l}(-d) \otimes W$, where W is a vector space of dimension $l + 1$. Denote by

    $$0 \to \mathcal{R} = \mathcal{O}_{\mathbf{P}^l}(-1) \to F \times \mathcal{O}_{\mathbf{P}^l} \to \mathcal{Q} \to 0$$

    the tautological sequence on $\mathbf{P}^l$. Let us fix two numbers $0 \le b_1 < b_2 \le l$. Define the vector bundle $\mathcal{V}$ on $\mathbf{P}^l$ to be $\mathcal{V} = K_{\lambda(b_1,b_2)}\mathcal{Q}$, where $\lambda(b_1, b_2)$ is the partition

    $$((l + 1)(d - 1) + 1, l(d - 1) + 1, \ldots, (b_2 + 1)(d - 1) + 1, (b_2 - 1)$$
    $$(d - 1), \ldots, (b_1 + 1)(d - 1), (b_1 - 1)(d - 1) - 1, \ldots, (d - 1) - 1, -1).$$

    (a) Prove that the complex $F_{\bullet}(\mathcal{V})$ has only two terms. One term has homological degree 0 and homogeneous degree b_1; the other term has homological degree 1 and homogeneous degree b_2.

    (b) Use the Weyl character formula to show that the dimensions of both terms are the same, so the complex $F_{\bullet}$ can be interpreted as a square matrix over the polynomial ring $A = \mathrm{Sym}(D_d F \otimes W^*)$ with entries of degree $b_2 - b_1$,

    (c) Prove that the determinant of $F_{\bullet}$ is equal to $Res(\mathbf{P}^l)^{\mathrm{rank}\ \mathcal{V}}$.

8. Let $b_2 = l + 1$, $b_1 = 0$. Prove that $\lambda(0, l + 1) = (d - 1)\rho$. The complex $F_{\bullet}$ is a square matrix of size $d^{l(l+1)/2}$. Its entries have homogeneous degree $l + 1$. They can be written as polynomials in maximal minors of the $(l + 1) \times \binom{d+l}{l}$ matrix formed by the coefficients of $l + 1$ polynomials in $l + 1$ variables of degree d.

9. Let $b_1 = 0$, $b_2 = 1$. We have

    $$\lambda(0, 1) = ((l + 1)(d - 1) + 1, \ldots, 2(d - 1) + 1).$$

    The complex $F(K_{\lambda(0,1)}\mathcal{Q})_{\bullet}$ has just two nonzero terms,

    $$F_0 = K_{\lambda(0,1)}F \otimes A,$$
    $$F_1 = K_{((l+1)(d-1)+1,\ldots,2(d-1)+1,d)}F \otimes W^* \otimes A(-1).$$

    Calculate the ranks of both modules and the power of the discriminant that equals $\det(F(\mathcal{V})_{\bullet})$.

References

[AB1] Akin, Kaan; Buchsbaum, David A. Characteristic free representation theory of general linear group. Adv. Math. 58 (1985), no. 2, 149–200.

[AB2] Akin, Kaan; Buchsbaum, David A. Characteristic free representation theory of general linear group II. Homological considerations. Adv. Math. 72 (1988), no. 2, 171–210.

[ABW1] Akin, Kaan; Buchsbaum, David A.; Weyman, Jerzy. Resolutions of determinantal ideals: the submaximal minors. Adv. in Math. 39 (1981), no. 1, 1–30.

[ABW2] Akin, Kaan; Buchsbaum, David A.; Weyman, Jerzy. Schur functors and Schur complexes. Adv. in Math. 44 (1982), no. 3, 207–278.

[An] Andersen, Janet Determinantal rings associated with symmetric matrices: a counterexample, Thesis, Univ. of Minnesota, 1992.

[Ar1] Artale, M. Syzygies of a certain family of generically imperfect modules. J. Algebra 167 (1994), 233–257.

[Ar2] Artale, M. Perfection and representation of GL(N). J. Algebra 168 (1994), no. 3, 695–727.

[B1] Birkenhake, Christina, Linear systems on projective spaces, Manuscripta Math. 88 (1995), 177–184.

[Bo1] Boffi, Giandomenico. The case of the submaximal minors: a finishing touch. Adv. in Math. 68 (1988), no. 1, 64–84.

[Bo2] Boffi, Giandomenico. The universal form of the Littlewood–Richardson rule. Adv. in Math. 68 (1988), no. 1, 40–63.

[Bo3] Boffi, Giandomenico. Bilinear forms and (hyper-) determinants. Adv. in Math. 123 (1996), no. 1, 91–103.

[BS] Boffi, Giandomenico; Sánchez, Rafael. On the resolutions of the powers of the Pfaffian ideal. J. Algebra 152 (1992), no. 2, 463–491.

[BBG] Boffi, G.; Bruns, W.; Guerrieri, A. On the Jacobian ideal of a trilinear form. J. Algebra 197 (1997), no. 2, 521–534.

[B] Borel, A. Linear Algebraic Groups, 2nd ed., Grad. Texts in Math. 129, Springer-Verlag, 1991.

[Bou] Bourbaki, N. Elements of Mathematics. Lie Groups and Lie Algebras, chapters I–IX, Springer-Verlag, Berlin, 1989.

[Br1] Broer, Bram. Line bundles on the cotangent bundle of the flag variety. Invent. Math. 113 (1993), no. 1, 1–20.

[Br2] Broer, Bram. Normality of some nilpotent varieties and cohomology of line bundles on the cotangent bundle of the flag variety. Lie Theory and Geometry, 1–19, Progr. Math., 123, Birkhäuser Boston, Boston, MA, 1994.

[Br3] Broer, Abraham. A vanishing theorem for Dolbeault cohomology of homogeneous vector bundles. J. Reine Angew. Math. 493 (1997), 153–169.

[Br4] Broer, Abraham. Lectures on decomposition classes. Representation Theories and algebraic Geometry (Montreal, PQ, 1997), NATO Adv. Sci. Inst. Ser. C Math. Phys. Sci. 514, 39–83, Kluwer Academic, Dordrecht, 1998.

[Br5] Broer Abraham, Hilbert Series in Invariant Theory, Thesis, Rijksuniversiteit, Utrecht, 1991.

[Br6] Broer, Abraham Normal nilpotent varieties in F_4. J. Algebra 207 (1998), no. 2, 427–448.

[BV] Bruns, Winfried; Vetter, Udo. Determinantal Rings, Monografas de Matematica [Mathematical Monographs] 45, Instituto de Matematica Pura e Aplicada, Rio de Janeiro, 1988.

[BE1] Buchsbaum, David A.; Eisenbud, David. What makes a complex exact? J. Algebra 25 (1973), 259–268.

[BE2] Buchsbaum, David A.; Eisenbud, David. Generic free resolutions and a family of generically perfect ideals. Adv. in Math. 18 (1975), no. 3, 245–301.

[BE3] Buchsbaum, David A.; Eisenbud, David. Algebra structures for finite free resolutions, and some structure theorems for ideals of codimension 3. Amer. J. Math. 99 (1977), no. 3, 447–485.

[BD] Bucur I., Deleanu A. Introduction to the Theory of Categories and Functors, Pure and Appl. Math. XIX, Wiley, 1968.

[C] Conrad, B. Grothendieck Duality and Base Change, Springer Lecture Notes in Math. 1750, 2000.

[D] Demazure, Michel une démonstration algébrique d'un théorème de Bott, Invent. Math. 5 1968, 349–356.

[Da] Daszkiewicz, A. On the invariant ideals of the symmetric algebra $S.(V \oplus \Lambda^2 V)$. J. Algebra 125 (1989), no. 2, 444–473.

[DGJP] Decker, W.; de Jong, T.; Greuel, G.-M.; Pfister, G. The normalization: a new algorithm, implementation and comparisons. Computational methods for representations of groups and algebras (Essen, 1997). Progress in Mathematics 173, 177–185, Birkhaüser, Basel.

[DEP1] De Concini, C.; Eisenbud, David; Procesi, C. Young diagrams and determinantal varieties. Invent. Math. 56 (1980), no. 2, 129–165.

[DEP2] De Concini, Corrado; Eisenbud, David; Procesi, Claudio. Hodge Algebras (with a French summary), Astérisque 91. Soc. Math. de France, Paris, 1982.

[DP] De Concini, Corrado; Procesi, Claudio. Symmetric functions, conjugacy classes and the flag variety. Invent. Math. 64 (1981), no. 2, 203–219.

[DS] De Concini, Corrado; Strickland, Elisabetta. On the variety of complexes. Adv. in Math. 41 (1981), no. 1, 57–77.

[DC] Dieudonné J.; Carrell, J. Invariant Theory, Old and New. Academic Press, New York, 1971.

[DRS] Doubilet, P.; Rota, G.-C.; Stein, J. Foundations of combinatorics IX. Combinatorial methods in invariant theory, Stud. Appl. Math. 53 (1974), 185–216.

[EN1] Eagon, J. A.; Northcott, D. G. Ideals defined by matrices and a certain complex associated with them. Proc. Roy. Soc. Ser. A 269 (1962), 188–204.

[EN2] Eagon, J. A.; Northcott, D. G. Generically acyclic complexes and generically perfect ideals. Proc. Roy. Soc. Ser. A 299 (1967), 147–172.

[ES] Eisenbud, David; Schreyer, Frank. Sheaf cohomology and free resolutions over exterior algebra. Preprint, 2000.

[FW1] Fukui, Toshizumi; Weyman, Jerzy. Cohen–Macaulay properties of Thom–Boardman strata. I. Morin's ideal. Proc. London Math. Soc. (3) 80 (2000), no. 2, 257–303.

[FW2] Fukui, Toshizumi; Weyman, Jerzy. Modified Morin ideals. Preprint, to appear in Proc. London Math. Soc. 2000.

[Ga] Galindo, Laura. Existence of determinantal expressions for hyperdeterminants of three dimensional matrices. Thesis, University Central, Caracas, 2001.

[GKZ] Gelfand, I. M.; Kapranov, M. M.; Zelevinsky, A. V. Discriminants, Resultants, and Multidimensional Determinants, Math. Theory and Appl., Birkhäuser, Boston, 1994.

[GM] Gelfand, S.; Manin, Y. Methods of Homological Algebra (transl. from the Russian), Springer-Verlag, Berlin, 1996.

[Go] Godement, Roger. Topologie Algébrique et Théorie des Faisceaux, 3rd ed., Actualités Sci. et Ind. 1252, Publ. Inst. Math. Univ. Strasbourg XIII, Hermann, Paris, 1973.

[GT] Goto, Shiro; Tachibana, Sadao. A complex associated with a symmetric matrix. J. Math. Kyoto Univ. 17 (1977), no. 1, 51–54.

[GR] Grauert, H.; Riemenschneider, O. Verschwindungssätze fur analytische Kohomologiegruppen auf komplexen Räumen, Inv. Math. 11 (1970), 263–292.

[GRe] Grauert, H.; Remmert, R.; Analytische Stellenalgebren, Springer, 1971.

[GL] Green, M.; Lazarsfeld, R. Some results on the syzygies of finite sets and algebraic curves. Compos. Math. 67 (1988), no. 3, 301–314.

[Gr] Grothendieck, A., Elements de Géometrie Algébrique II, III, Publ. Math. I.H.E.S., 8, 11, 1961.

[GN] Gulliksen, T. H.; Negard, O. G. Un complexe résolvant pour certains ideaux dete minantiels. C. R. Acad. Sci. Paris Ser. A 274, 16–18 (1972).

[H1] Hartshorne, Robin. Algebraic Geometry, Graduate Texts in Mathematics 52, Springer-Verlag, New York, 1977.

[H2] Hartshorne, Robin. Residues and Duality (lecture notes of a seminar on the work of A. Grothendieck, given at Harvard 1963/64, with an appendix by P. Deligne), Lecture Notes in Math. 20, Springer-Verlag, New York, 1966.

[HK] Hashimoto, Mitsuyasu; Kurano, Kazuhiko. Resolutions of determinantal ideals: n-minors of $(n + 2)$-square matrices. Adv. Math. 94 (1992), no. 1, 1–66.

[Ha1] Hashimoto, Mitsuyasu. Determinantal ideals without minimal free resolutions. Nagoya Math. J. 118 (1990), 203–216.

[Ha2] Hashimoto, Mitsuyasu. Resolutions of determinantal ideals: t-minors of $(t + 2) \times n$ matrices. J. Algebra 142 (1991), no. 2, 456–491.

[Ha3] Hashimoto, Mitsuyasu. Resolutions of determinantal ideals – a counterexample on symmetric matrices. Proceedings of the 26th Symposium on Ring Theory (Tokyo, 1993), 43–51, Okayama Univ., Okayama, 1993.

[Ha4] Hashimoto, Mitsuyasu. Determinantal ideals and their Betti numbers – a survey (in Japanese). Modern Aspects of Combinatorial Structure on Convex

Polytopes (Kyoto, 1993), Sūrikaisekikenkyūsho Kōkyūroku No. 857, 40–50, 1994.

[Ha5] Hashimoto, Mitsuyasu. Relations on Pfaffians: number of generators. J. Math. Kyoto Univ. 35 (1995), no. 3, 495–533.

[Ha6] Hashimoto, Mitsuyasu. Second syzygy of determinantal ideals generated by minors of generic symmetric matrices. J. Pure Appl. Algebra 115 (1997), no. 1, 27–47.

[Ha7] Hashimoto, Mitsuyasu. Homological aspects of equivariant modules: Matijevic–Roberts and Buchsbaum–Rim. Commutative Algebra, Algebraic Geometry, and Computational Methods (Hanoi, 1996), 259–302, Springer-Verlag, Singapore, 1999.

[HKu] Herzog, J.; Kunz, E. Der Kanonische Modul Eines Cohen–Macaulay Rings, Lect. Notes in Math. 238, Springer-Verlag, 1971.

[He] Hesselink, W. Desingularizations of varieties of nullforms. Invent. Math. 55, 141–163 (1979).

[Hi] Hinich, V. On the singularities of nilpotent orbits. Israel J. Math. 73 (1991), no. 3, 297–308.

[HE] Hochster, M.; Eagon, John A. Cohen–Macaulay rings, invariant theory, and the generic perfection of determinantal loci. Amer. J. Math. 93 (1971), 1020–1058.

[HR] Hochster, Melvin; Roberts, Joel L. Rings of invariants of reductive groups acting on regular rings are Cohen–Macaulay. Adv. in Math. 13 (1974), 115–175.

[Hu1] Humphreys, J. Introduction to Lie Algebras and Representation Theory, Springer-Verlag, New York, 1972.

[Hu2] Humphreys, J. Linear algebraic groups, corr. 2nd printing, Graduate Texts in Math. 21, Springer-Verlag, New York, 1981.

[Jm] James, Gordon. The Representation Theory of the Symmetric Groups, Lecture Notes in Math. 682, Springer-Verlag, Berlin, 1978.

[Ja] Jantzen, Jens. Carsten Representations of Algebraic Groups, Pure and Appl. Math. 131, Academic Press, Harcourt Brace Jovanovich, Boston, 1987.

[JPW] Józefiak, T.; Pragacz, P.; Weyman, J. Resolutions of determinantal varieties and tensor complexes associated with symmetric and antisymmetric matrices. Young Tableaux and Schur Functors in Algebra and Geometry (Toruń, 1980), 109–189, Astérisque 87–88, Soc. Math. France, Paris, 1981.

[dJ] de Jong, T. An algorithm for computing the integral closure, J. Symbolic Comp. 26 (1998), 273–277.

[J] Józefiak, Tadeusz. Ideals generated by minors of a symmetric matrix. Comment. Math. Helv. 53 (1978), no. 4, 595–607.

[JP1] Józefiak, Tadeusz; Pragacz, Piotr. Syzygies de Pfaffiens (in French). C. R. Acad. Sci. Paris Sr. A–B 287 (1978), no. 2, A89–A91.

[JP2] Józefiak, Tadeusz; Pragacz, Piotr. Ideals generated by Pfaffians. J. Algebra 61 (1979), no. 1, 189–198.

[K] Kac, Victor. Some remarks on nilpotent orbits. J. Alg. 64 (1980), no. 1, 190–213.

[Ka] Katz, N. Pinceaux de Lefschetz: théorème d'existence, SGA 7 11, Exposé XVII. Lecture Notes in Math. 340, 212–253, Springer-Verlag, 1973.

[Ke0] Kempf, George. The singularities of certain varieties in the Jacobian of a curve, Thesis, Columbia University, 1971.

[Ke1] Kempf, George R. Images of homogeneous vector bundles and varieties of complexes. Bull. Amer. Math. Soc. 81 (1975), no. 5, 900–901.

[Ke2] Kempf, George R. Linear systems on homogeneous spaces. Ann. of Math. (2) 103 (1976), no. 3, 557–591.

[Ke3] Kempf, George R. Vanishing theorems for flag manifolds. Amer. J. Math. 98 (1976), no. 2, 325–331.

[Ke4] Kempf, George R. On the collapsing of homogeneous bundles. Invent. Math. 37 (1976), no. 3, 229–239.

[Ke5] Kempf, George R. Instability in invariant theory. Ann. of Math. (2) 108 (1978), no. 2, 299–316.

[Ke6] Kempf, George R. Varieties with rational singularities. The Lefschetz centennial conference, Part I (Mexico City, 1984), Contemp. Math. 58, 179–182, Amer. Math. Soc., Providence, RI, 1986.

[Ke7] Kempf, George R. Equations of isotropy. Group actions and invariant theory (Montreal, 1988), CMS Conf. Proc. 10, 85–91, Amer. Math. Soc., Providence, RI, 1989.

[KKMSD] Kempf, G.; Knudsen, Finn Faye; Mumford, D.; Saint-Donat, B. Toroidal Embeddings. I. Lecture Notes in Mathematics 339, Springer-Verlag, New York, 1973.

[Kl] Klimek, J. A resolution of the square of a determinantal ideal associated to a symmetric matrix. Fund. Math. 130 (1988), no. 2, 101–111.

[KKW] Klimek, J.; Kraśkiewicz, W.; Weyman, J. A free resolution of a symplectic rank variety. J. Algebra 196 (1997), no. 2, 475–489.

[KM] Knop, Friedrich; Menzel, Gisela. Duale Varietten von Fahnenvarietten (in German) [Dual varieties of flag varieties]. Comment. Math. Helv. 62 (1987), no. 1, 38–61.

[Kr1] Kraft, Hanspeter. Geometrische Methoden in der Invarianthentheorie (in German). Aspekte der Mathematik D1. Vieweg, Braunschweig/Wiesbaden, 1985.

[Kr2] Kraft, Hanspeter. Closures of conjugacy classes in G_2. J. Algebra 126 (1989), no. 2, 454–465.

[KP1] Kraft, Hanspeter; Procesi, Claudio. Closures of conjugacy classes are normal. Inv. Math. 53 (1979), no. 3, 227–247.

[KP2] Kraft, Hanspeter; Procesi, Claudio. On the geometry of conjugacy classes in classical groups. Comm. Math. Helv. 57 (1982), no. 4, 539–602.

[K1] Kurano, Kazuhiko. The first syzygies of determinantal ideals. J. Algebra 124 (1989), no. 2, 414–436.

[K2] Kurano, Kazuhiko. On relations on minors of generic symmetric matrices. J. Algebra 124 (1989), no. 2, 388–413.

[K3] Kurano, Kazuhiko. Relations on Pfaffians. I. Plethysm formulas. J. Math. Kyoto Univ. 31 (1991), no. 3, 713–731.

[K4] Kurano, Kazuhiko Relations on Pfaffians. II. A counterexample. J. Math. Kyoto Univ. 31 (1991), no. 3, 733–742.

[Ku] Kutz, Ronald Cohen–Macaulay rings and ideal theory in rings of invariants of algebraic groups, Trans. Amer. Math. Soc. 194 (1974), 115–129.

[Ku1] Kustin, Andrew R. Ideals associated to two sequences and a matrix. Comm. Algebra 23 (1995), no. 3, 1047–1083.

[Ku2] Kustin, Andrew R. Complexes associated to two vectors and a rectangular matrix. Mem. Amer. Math. Soc., to appear.

[KU] Kustin, Andrew R.; Ulrich, Bernd. A family of complexes associated to an almost alternating map, with applications to residual intersections. Mem. Amer. Math. Soc. 95 (1992), no. 461.

[L] Lang, Serge. Algebra, 3rd ed. Addison–Wesley, Reading, MA, 1993.

[LM] Landsberg, Joseph; Manivel, Laurent. Series of Lie Groups, Math. AG/0203241.

[L1] Lascoux, Alain. Thése, Paris, 1977.

[L2] Lascoux, Alain. Syzygies des varits déterminantales (in French). Adv. in Math. 30 (1978), no. 3, 202–237. (Reviewer: S. L. Kleiman.) 14M12 (13D25 14M10).

[Li1] Lichtenstein, Woody. A system of quadrics describing the orbit of the highest weight vector. Proc. Amer. Math. Soc. 84 (1982), no. 4, 605–608.

[Mc] McGovern, W. Rings of regular functions on nilpotent orbits and their covers. Invent. Math. 97 (1989), 209–217.

[M] Maeda, Takashi. Minimal free resolution of the third Veronese subring of three variables. Ryukyu Math. J. 12 (1999), 9–30.

[Ma] Manivel, Laurent. On the syzygies of flag manifolds. Proc. Amer. Math. Soc. 124 (1996), no. 8, 2293–2299.

[MD] Macdonald, I. G. Symmetric Functions and Hall Polynomials, Clarendon Press, Oxford, 1979.

[MR] Mehta, Vickram; Ramanathan Frobenius. Splitting and cohomology vanishing for Schubert varieties. Ann. Math. (2) 122 (1985), no. 1, 27–40.

[MvK] Mehta, Vickram; van der Kallen, Wilberd. A simultaneous Frobenius splitting for closures of conjugacy classes of nilpotent matrices. Comp. Math. 84 (1992), no. 2, 211–221.

[Mo] Morin, B. Calcul jacobien. Ann. Sci. Ec. Norm. Sup. 8 (1975), 1–98.

[N] Nagata M. A generalization of the embedding problem of an abstract variety in a complete variety. J. Math. Kyoto Univ. 3 (1963), 89–102.

[Ne] Neeman A. The Grothendieck duality theorem via Bousfield's techniques and Brown representability. J. Amer. Math. Soc. 9 (1996), 205–236.

[Ni] Nielsen H.A., Tensor functors of complexes. Aarhus Univ. Preprint Series, no. 15, 1977–1978.

[No] Northcott, D. G. Finite free resolutions. Cambridge Tracts in Mathematics, No. 71. Cambridge University Press, Cambridge-New York-Melbourne, 1976.

[OP] Ottaviani G.; Paoletti R. Syzygies of Veronese embeddings, Compos. Math. 125 (2001), no. 1, 31–37.

[Pa1] Panyushev, D. I. The structure of the canonical module and the Gorenstein property for some quasihomogeneous varieties (in Russian). Mat. Sb. (N.S.) 137(179) (1988), no. 1, 76–89, 143.

[Pa2] Panyushev, D. I. Rationality of singularities and the Gorenstein property of nilpotent orbits (in Russian). Funktsional. Anal. i Prilozhen. 25 (1991), no. 3, 76–78; translation, Functional Anal. Appl. 25 (1991), no. 3, 225–226 (1992).

[Pn] Poon, K. Y. A resolution of certain perfect ideals defined by some matrices, Thesis, University of Minnesota, 1973.

[Po] Porras, Olga. Rank varieties and their resolutions. J. Algebra 186 (1996), no. 3, 677–723.

[P] Pragacz, Piotr. Characteristic free resolution of $n - 2$ order Pfaffians of $n \times n$ antisymmetric matrix. J. Algebra 78 (1982), no. 2, 386–396.

[PW1] Pragacz, Piotr; Weyman, Jerzy. On the construction of resolutions of determinantal ideals: a survey. Séminaire d'algèbre Paul Dubreil et Marie-Paule Malliavin, 37ème Année (Paris, 1985), Lecture Notes in Math. 1220, 73–92, Springer-Verlag, New York, 1986.

[PW] Pragacz, Piotr; Weyman, Jerzy. Complexes associated with trace and evaluation. Another approach to Lascoux's resolution. Adv. in Math. 57 (1985), no. 2, 163–207.

[R1] Roberts, P. A minimal free complex associated to the minors of a matrix, preprint, 1978.

[RW] Roberts, Joel; Weyman, Jerzy. A short proof of a theorem of M. Hashimoto. J. Algebra 134 (1990), no. 1, 144–156.

[R2] Roberts, Paul. On the construction of generic resolutions of determinantal ideals. Young Tableaux and Schur Functors in Algebra and Geometry (Toruń, 1980), Astérisque 87–88, 353–378, Soc. Math. France, Paris, 1981.

[Ro] Ronga, Felice. Le calcul des classes duales aux singularités da Boardmann d'ordre deux. Comment. Math. Helv. 47 (1972), 15–35.

[Rot] Rotman, J. An Introduction to Homological Algebra (in English). Pure and Applied Math. 85, XI, 376 Academic Press, New York, 1979.

[SK] Sato, M.; Kimura, T. A classification of irreducible prehomogeneous vector spaces and their relative invariants. Nagoya Math. J. 65 (1977), 1–155.

[SW] Shimozono, Mark; Weyman, Jerzy. Bases for coordinate rings of conjugacy classes of nilpotent matrices. J. Algebra 220 (1999), no. 1, 1–55.

[SkW] Skowroński A.; Weyman, J. The algebras of semi-invariants of quivers. Transform. Groups 5 (2000), no. 4, 361–402.

[SS] Springer, T.; Steinberg R. Conjugacy classes. Seminar on Algebraic Groups and related Finite Groups, Lecture Notes in Math. 131, Springer-Verlag, Berlin, 1970.

[S] Strickland, Elisabetta. On the variety of projectors. J. Algebra 106 (1987), no. 1, 135–147.

[To] Towber, Jacob Two new functors from modules to algebras, J. Algebra 47 (1977), no. 1, 80–104.

[TW] Tchernev, A. Weyman, J. Free resolutions for Polynomial Functors, preprint, 2000.

[TT] Thomason, R. W.; Trobaugh, T. Higher algebraic K-theory of schemes and of derived categories. The Grothendieck Festschrift, Vol. III, Progr. Math. 88, 369–453, Birkhäuser, 1990.

[T] Turnbull, I. The Theory of Determinants, Matrices and Invariants. Blackie, Glasgow, 1929.

[W1] Weyman, Jerzy. The equations of strata for binary forms. J. Algebra 122 (1989), no. 1, 244–249.

[W2] Weyman, Jerzy. The equations of conjugacy classes of nilpotent matrices. Invent. Math. 98 (1989), no. 2, 229–245.

[W3] Weyman, Jerzy. Calculating discriminants by higher direct images. Trans. Amer. Math. Soc. 343 (1994), no. 1, 367–389.

[W4] Weyman, Jerzy. On the Hilbert functions of multiplicity ideals. J. Algebra 161 (1993), no. 2, 358–369.

[W5] Weyman, Jerzy. Gordan ideals in the theory of binary forms. J. Algebra 161 (1993), no. 2, 370–391.

[W6] Weyman, J. The Grothendieck group of GL(F) × GL(G)-equivariant modules over the coordinate ring of determinantal varieties. Colloq. Math. 76 (1998), no. 2, 243–263.

[W7] Weyman, Jerzy. Two results on equations of nilpotent orbits. J. Algebraic Geom. 11 (2002), no. 4, 791–800.

[WZ] Weyman, Jerzy; Zelevinsky, Andrei. Determinantal formulas for multigraded resultants. J. Algebraic Geom. 3 (1994), no. 4, 569–597.

Notation Index

367

Subject Index

algebra, exterior, 3
algebra, symmetric, 5
algebra, divided power, 7
Auslander-Buchsbaum formula, 14, 16
Auslander-Buchsbaum-Serre theorem, 15
arm length, 8
arm, of a box, 8

Basic theorem, 138
Basic Theorem for Discriminants, 329
Basic Theorem for nilpotent orbits, 254
Basic Theorem for Resultants, 316
Bezout's expression for resultant, 317
binary forms, 157–58
binary forms with p-tuple roots, 157–58
Bondal's expression for hyperdeterminant, 344
Bondal's expression for hyperdeterminant, generalized, 346
Bott's algorithm, 113
Bott's theorem, 113, 123, 133
Bott's theorem for partial flag variety, 114
Brauer-Klimyk formula, 134–35
Buchsbaum-Eisenbud acyclicity criterion, 18
Buchsbaum-Eisenbud complex, 192
Buchsbaum-Eisenbud structure theorem, 31

canonical bundle, on a flag variety, 109
canonical module, 21
canonical sheaf, 21
Cartan piece of a tensor product, 82
Cauchy formulas, 59–60, 95–6, 106
character, of a representation, 55
Chow cycle, 315
Chow form, 315
codimension, of a module, 13
Cohen-Macaulay complex, 140
Cohen-Macaulay module, 14
Cohen-Macaulay module, maximal, 14, 223–24
Cohen-Macaulay ring, 14, 17
Cohen-Macaulay sheaf, 24

compactification, of a morphism, 22
complete intersection, 13
complex, minimal, 17
condition LP, 61
conjugacy classes, for $GL(E)$, 253
conjugacy classes, for $\underline{so}(F)$, 284
conjugacy classes, for $\underline{sp}(F)$, 297

d-tuple embedding of projective space, 236
degeneration sequence, 154–56
degree, of line bundle, 117
deJong-Grauert-Remmert algorithm, 19, 350–53
Demazure pairing, 118
depth, of a module, 12
determinant, of a complex, 27
determinantal complex, 140, 313
determinantal ideal, 160
determinantal ideal, skew symmetric, 187
determinantal ideal, symmetric, 175
determinantal variety, 160
determinantal variety, skew symmetric, 187
determinantal variety, symmetric, 175
diagonal, exterior, 3
diagonal, symmetric, 5
diagonal, divided power, 7
differentials in complexes $F(V)_\bullet$, 152–54
differentials in resolutions of determinantal ideals, 221
differentials in resolutions of skew symmetric determinantal ideals, 222–223
differentials in resolutions of symmetric determinantal ideals, 222
dimension, of a module, 12
discriminant, 328, 356–57
dual variety, 328
duality for proper morphisms, 23
dualizing complex, 21
Durfee square, 8

Eagon-Northcott complex, 164
Eagon-Northcott complex, relative, 205

369